Technological Change and Skill Development in Sudan

Samia Mohamed Nour

Technological Change and Skill Development in Sudan

Springer

Samia Mohamed Nour
Department of Economics
University of Khartoum
Khartoum
Sudan

UNU-MERIT - School of Business
and Economics
Maastricht University
Maastricht
The Netherlands

ISBN 978-3-642-32810-7 ISBN 978-3-642-32811-4 (eBook)
DOI 10.1007/978-3-642-32811-4
Springer Heidelberg New York Dordrecht London

Library of Congress Control Number: 2012954245

Printed on acid-free paper

Springer is part of Springer Science+Business Media (www.springer.com)

Acknowledgments

I sincerely thank Allah, my God, the Most Gracious, Most Merciful for enabling me to complete this post-doctoral research project and book successfully. I would like to extend my deep thanks, gratitude and appreciation to all people contributed to the successful completion of my research that presented in this book.

First and foremost, I am delighted to gratefully acknowledge and thank my supervisor, Professor Joan Muysken (the head of the Macroeconomic Section, Department of Economics, School of Business and Economics (SBE), Maastricht University), for his very kind support and cooperation in facilitating my research visit, without which it would have been impossible to arrange. I am extremely grateful and extremely indebted to Professor Muysken for his excellent supervision, intellectual guidance, kind interest, sincere efforts and for his time he has kindly spent to guide and review my research with great interest, care and attention. I sincerely thank him for his invaluable comments and suggestions that contributed significantly to the improvement and completion of my research. I am also pleased to thank Dr. Thomas Ziesemer for his very kind cooperation, continuous support and excellent initial comments that helped me to improve and complete my research.

I am delighted to gratefully acknowledge and thank the Arab Fund for Economic and Social Development (AFESD), Kuwait, for its 1-year full research grant and fellowship and continued generous support through the duration of my postdoctoral research project. I sincerely thank its Distinguished Scholar Awards Programme (Zamalat) for the great confidence they put in my abilities and my research and for

---

Disclaimer: The Study presented in this book was conducted at Maastricht Economic and Social Research Institute on Innovation and Technology (UNU-MERIT), School of Business and Economics, Maastricht University, Maastricht, the Netherlands, and fully sponsored by the Arab Fund for Economic and Social Development, Distinguished Scholar Awards and Post-Doctoral Fellowships Programme (Zamalat), the Arab Fund for Economic and Social Development (AFESD), Kuwait. The views, analysis and policy recommendations of this Report are those of the author and do not necessarily reflect the views and policies of the United Nations University, Maastricht Economic and Social Research Institute on Innovation and Technology (UNU-MERIT), or UNU Affiliated Institutes, or the views and policies of the Arab Fund for Economic and Social Development.

the pride, privilege and honour they have given me by selecting me as a grantee recipient among the ten Arab researchers recipients in its eleventh session in 2010, and one of the 66 Arab researcher recipients of the Arab Fund for Economic and Social Development Distinguished Scholar Awards since its inception in 1997. I gratefully thank H.E. Dr. Abdulatif Yousef Al-Hamad, Director General of AFESD, Dr. Imad Limam, Senior Economist AFESD, Mr. Ahmed Osman, Director of the Technical Department, and all colleagues in the AFESD Distinguished Scholar Awards Programme (Zamalat) for their generous support and kind cooperation during my fellowship. I would like to thank AFESD for its leading role and outstanding contribution to continuously supporting Arab social and economic development and also for introducing this innovative and unique initiative to build bridges and achieve mutual benefits between Arab and foreign universities, facilitate the transfer of knowledge and technology to the Arab countries and support capacity building in the Arab region by honouring and supporting Arab Ph.D. holders in different fields of specialisation who are currently working in Arab universities with scholarship opportunities to conduct advanced research and/or lecture in some of the world's best universities.

I am delighted also to gratefully thank the former and current directors of the School of Business and Economics for their kind cooperation and facilitation of my research visit. I gratefully thank Professor Luc Soete (Director of UNU-MERIT, School of Business and Economics, University of Maastricht) for both giving me the opportunity to be part of his team of high quality international research staff and for his continuous support and kind cooperation. I acknowledge with thanks the host institute at UNU-MERIT, the School of Business and Economics at the University of Maastricht, for the cooperation and hospitality I received during my visiting research fellowship. I am extremely grateful to the host institute for enabling my research visit to benefit greatly from practical implementation of the approach of internationalisation of research and knowledge, sharing success and learning by doing. I am very grateful to the host institute for giving me the opportunity of being a visiting research fellow affiliated and involved with the truly global, internationally orientated and prestigious University of Maastricht in the Netherlands.

I would like to acknowledge that the successful completion of my research project would not have been possible without the extraordinary cooperation and full support of the Arab Fund for Economic and Social Development (AFESD), Professor Joan Muysken, UNU-MERIT of the School of Business and Economics, Maastricht University, the Netherlands and Khartoum University, Sudan. I am again grateful to Professor Muysken, UNU-MERIT, AFESD and Khartoum University for helping me to learn more, improve my understanding, enhance my capacity especially in the area of academic production, and for having greatly facilitated and fully supported my research, making it a truly productive, fruitful and enjoyable experience personally and professionally.

I wish to thank all the academic and administrative staff at Maastricht University, SBE and UNU-MERIT. I also thank Ms. Anja Brouwers and all her colleagues at Knowledge Center of International Staff for their kind cooperation and excellent support to facilitate administrative arrangements for my visiting research fellowship, employment and research work at Maastricht University. I would like to thank all UNU-MERIT administrative staff for their kind cooperation, support and offering me with all administrative support and facilities that benefited my research stay at UNU-MERIT. I would like to extend my gratitude and special thanks to Wilma Coenegrachts, Eveline in de Braek, Monique Raedts, Marc Vleugels, Herman Pijpers, Howard Hudson, Eric Engelen, Ad Notten, Mourik Jan Heupink, and Silvana de Sanctis and Fleur Keune – secretaries to Prof. Muysken at the Department of Economics, School of Business and Economics – for their assistance and kind cooperation. I would like to thank all academic staff, researchers and colleagues at UNU-MERIT for their kind cooperation during my research visit and for helping me to improve my research and share knowledge with outstanding and highly qualified professional academic colleagues from different countries. I would like to especially thank my good colleagues Dr. Mamata Parhi (India), Dr. Michiko Iizuka (Japan), Ms. Kirsten Haaland (Norway), Ms. Kristine Farla (the Netherlands) and my office mates Ms. Paula Nagler (Germany), Ms. Ozge Biligili (Trukey) and Ms. Sepideh Yousefzadeh (Iran) for their friendship, good support and kind cooperation. I wish to thank the participants at the UNU-MERIT seminar (September 27, 2011) and the participants at the UNU-MERIT workshop (October 21, 2011), for their useful comments to improve my research. I would like to thank Ms. Emma Brewin for her kind cooperation and help in editing my research report. I am delighted to gratefully thank my brother Sami Satti (Engineer, Executive and Art Director of Celeste Media, Advertising and Publishers Limited Company, Khartoum, Sudan) for his very kind cooperation to design the cover page of the report.

During the fieldwork and data collection I received considerable help from many people. I would like to thank Ms. Amna Omer, Sudan Central Bureau of Statistics, for facilitating access to the population census database and Ms. Mahasin Yagoub, Ministry of Industry, for facilitating access to the industrial database. I thank Mr. Imad Khatmi, Ministry of Industry, and part time researchers Mr. Yasir Hassan and Dr. Khidir Siddig for their helpful cooperation in the arrangement and collection of the survey questionnaire from firms. I thank Mr. Salah Saeed, Ministry of Higher Education, Mr. Salah Arabi, Sudan Academy for Administrative Sciences, Dr. Khalid Osman, Youth Training Center, and Mr. Ismail Alazhari, Department of Vocational Training, Ministry of Labour, for their kind cooperation with the collection of macro survey. I also thank all the respondents to the three surveys.

I am delighted to thank the former and current vice chancellors of Khartoum University. I thank Dr. Ahmed Elshiehk and Dr. Hassn Hag Ali, the former and current deans of the Faculty of Economic and Social Studies, and Dr. Fadia Khalil and Dr. Layla Simsa, the former and current heads of the Economics Department, for their excellent support and excellent cooperation to facilitate my sabbatical leave. I would like to thank Dr. Hisham Mohamed Hassan for his excellent cooperation and excellent research assistance and useful comments. I would like to thank my colleagues Dr. Fatma Ibnouf and Ms. Ahlam Hamza for their good comments and kind support, my colleagues Ms. Duria Ahmed, Mr. Mahmoud Abudallah, Prof. Mohamed Hashim Awad, Prof. Ali Abdel Gadir Ali, Dr. Ibrahim

Sobahi, Dr. Elwasila Saeed, and all my colleagues at the Faculty of Economic and Social Studies, Khartoum University, for their good support. I wish to extend my thank to all my Sudanese friends in Maastricht, Ms. Hanan Mohamed, Ms. Gootalgolob Mohamed, Dr. Aziza Mukhayer, Ms. Hind Mohamed, Ms. Manal Mohamed, Ms. Awatif Mohamed, and their families for their help and warm support during my stay in Maastricht.

I thank my grandparents, my aunts, uncles and their families. Last but not least, my everlasting gratitude, appreciation and sincere deep thanks to my kind, beloved, wonderful mother Alawia and father Satti, my beloved sisters, Rashida, Khalda, Eiman and Lana and my kind brothers, Rashid, Osman, Khalid, Sami and their families, for their sincere prayers, warm, generous, steady support and faithful encouragement of my long learning journey.

Abstract (English Abstract)

This research discusses the need for skill development and interaction with technological change in Sudan. Different from the Sudanese literature, a novel element in our analysis is that we use new primary data from macro and firm surveys undertaken in Sudan in 2010 and we provide a new contribution and fill a significant gap in the Sudanese literature by examining several hypotheses on the causes and consequences of low skill and technology indicators and the transfer of knowledge at the macro and micro levels in Sudan. We recognise upskilling as a necessary condition for the fulfilment of economic stabilisation, balanced development, economic diversification, technological development, poverty alleviation, reduction of unemployment and restructuring of the labour market. We find that the interaction between the deficient educational system caused by the low quality of education, at the macro level, and the high incidence of unskilled workers, at the micro level, leads to low skill and technology levels, poor provision of training, severe skill mismatch, low transfer of knowledge, high dependence on foreign technologies at the macro and micro levels and poor industrial performance at the micro level. We explain that the poor technological capability can be attributed to low Research and Development (R&D) efforts; lack of R&D culture; low skill levels; lack of resources, transfer of knowledge and cooperation between universities and firms; and lack of entrepreneurial perspective. We provide a new contribution and fill a gap in Sudanese literature by examining the industrial performance indicators defined by three different sets of economic and productivity indicators, activity indicators and profitability indicators in Sudan. We find that the performance of the industrial firms is most probably immensely undermined by the shortage of capital, shortage of skilled workers and also by the lack of entrepreneurial perspective and lack of entrepreneurial culture. We show that the provision of education and training are seriously undermined by the low commitment to the standardised international adequacy, equity, and efficiency criterion related to the supply and demand sides of education and training policies in Sudan. We present supply–demand analysis of the causes and consequences of low skills and technology levels and the link between them at both the macro and micro levels. We fill the gap in the Sudanese literature by highlighting the importance of external effects of schooling and the transfer of knowledge and examining the factors hindering, and those contributing toward, enhancing the transfer of knowledge. Our results show a surprisingly contradicting view and inconsistency concerning the incidence and transfer of knowledge at the macro–micro levels. Finally, we recommend further efforts to be made to improve provision of education, training, skills and technology indicators and transfer of knowledge at macro–micro levels that are all essential for economic growth and development in Sudan.

June, 2012 Samia Satti Osman Mohamed Nour[1]

[1] Corresponding Author: Dr. Samia Satti Osman Mohamed Nour, Visiting Research Fellow – University of Maastricht, School of Business and Economics, UNU-MERIT, Maastricht, the Netherlands; and Assistant Professor of Economics, Economics Department, Faculty of Economic and Social Studies, Khartoum University, Khartoum, Sudan. E-mail: samiasatti@yahoo.com; samia_satti@hotmail.com. The first draft of this paper was presented at the UNU-MERIT seminar (September 27, 2011) and at the UNU-MERIT workshop (October 21, 2011), and was originally prepared within the author's research project "Technological Change and Skill Development: the case of Sudan" during the author's time as a visiting research fellow at the University of Maastricht, School of Business and Economics, UNU-MERIT, Maastricht, the Netherlands. The author gratefully acknowledges the Arab Fund for Economic and Social Development, Kuwait, for research grant and fellowship and University of Maastricht, School of Business and Economics, UNU-MERIT, for the good hospitality during her visiting research fellowship. The author would like to gratefully thank Professor Joan Muysken, Dr. Hisham Mohamed Hassan, Dr. Fatma Ibnouf and the participants at the UNU-MERIT seminar (September 27, 2011) and the UNU-MERIT workshop (October 21, 2011), for their useful comments on the first draft of this book. All the usual disclaimers apply.

Abstract (Arabic Abstract)

بسم الله الرحمن الرحيم

خلاصة الدراسة (باللغة العربية)

تطوير المهارات والتغيرات التكنولوجية فى السودان

يناقش هذا البحث أهمية تنمية المهارات والتفاعل مع التغيرات التكنولوجية في السودان. وقد أكدت نتائج البحث على أهمية (نهج أو إتباع) سبل وسياسات تطوير المهارات وتنمية الموارد البشرية على كافة المستويات فى السودان مع التركيز على التعليم والتدريب لأهميتهما فى تطوير المهارات وتنمية الموارد البشرية ومن ثم التنمية الاقتصادية. أكدت نتائج البحث كذلك على أهمية (إتباع) سبل تطوير المهارات وإستعمال التكنولوجيا والتاثير الإيجابي للتغييرات التكنولوجية على المستوى الجزئي خاصة على القطاع الصناعى فى السودان. تكمن أهمية هذه الدراسة في مساهمتها فى زيادة الوعى والادراك لاهمية تطوير المهارات والتنمية البشرية وأهمية التغيرات التكنولوجية وبناء المقدرات التكنولوجية ودورها فى خلق وتحويل المعرفة وتنمية القطاع الصناعى ومن ثم المساهمة فى التنمية الاقتصادية المستدامة فى السودان. اختلاف هذه الدراسة عن الدراسات المشابهة فى الأدبيات السودانية تأتي من خلال تميزها بتقديم مساهمة اصيلة وجديدة لملء فجوة كبيرة في الأدبيات السودانية حيث أنها قدمت تحليلاً كمياً ونوعياً لجانبى العرض والطلب وكذلك تحليلاً اكثر شمولاً للأسباب والنتائج المترتبة على تدنى مستوى المهارات ومؤشرات التكنولوجيا ونقل المعرفة والعلاقة بينهما على المستويين الكلي والجزئي في السودان. وقد استخدمت الدراسة تعريفاً ومنهجاً اكثر شمولاً باعتماد العديد من المؤشرات لقياس مستوى المهارات ومؤشرات التكنولوجيا، وإعتمدت على استخدام البيانات الأولية من المسوحات على المستويين الكلي والجزئي والتي أجريت في السودان في العام 2010. أكدت الدراسة على أهمية تحسين المهارات كشرط ضروري لتحقيق الاستقرار الاقتصادي، والتنمية المتوازنة، والتنويع الاقتصادي، والتطور التكنولوجي، وتخفيف حدة الفقر، والحد من البطالة وإعادة هيكلة سوق العمل فى السودان. كما بينت الدراسة أن التفاعل بين قصور النظام التعليمي بسبب تدنى جودة التعليم – على المستوى الكلي – ووجود نسبة عالية من العمالة غير الماهرة – على المستوى الجزئي – قد أدى الى تدنى مستويات المهارات ومؤشرات التكنولوجيا، وقلة فرص التدريب، وعدم تطابق المهارات، وقلة المقدرة التكنولوجية، والاعتماد الكبير على التكنولوجيا الأجنبية على المستويين الكلي والجزئي وضعف الأداء الصناعي على المستوى الجزئي. والمساهمة الاصيلة والجديدة لهذه الدراسة لملء فجوة لمثل هذه الدراسات في الادبيات السودانية أنها أوضحت أن ضعف المقدرة التكنولوجية يُعزى إلى قلة جهود البحث والتطوير (R&D) لقلة الانفاق وقلة الوعى والثقافة باهمية البحث والتطوير (R&D)، وتدنى مستوى المهارات، ونقص الموارد، ومحدودية نقل المعرفة والتعاون بين الجامعات والمنشات الصناعية، وقلة الوعى باهمية دور المنظم الصناعي. من خلال دراسة مؤشرات الأداء الصناعي وذلك باستخدام العديد من المؤشرات الشائعة الإستعمال فى الأدبيات الحديثة لقياس الأداء الصناعي، ساهمت هذه الدراسة في ملء فراغ كبير في هذا الإطار في الدراسات المشابهة في الأدبيات السودانية، حيث بحثت الدراسة المؤشرات الاقتصادية ومؤشرات الإنتاجية ومؤشرات الربحية ومؤشرات النشاط في السودان. وتوصلت الدراسة إلى أن وجود العديد من المعوقات على سبيل المثال نقص رأس المال ونقص العمالة الماهرة، وقلة وجود ثقافة المبادرة الفردية وقلة الوعى باهمية دور المنظم الصناعى لادارة المشاريع فى القطاع الصناعى جميعها على الارجح قد أدت الى ضعف أداء المنشات الصناعية بشكل كبير. من نتائج هذه الدراسة أن العديد من المعوقات ذات الصلة بقلة الالتزام بالمعايير الدولية الموحدة من حيث الكفاية والعدالة والكفاءة المتصلة بجانبي العرض والطلب في

سياسات التعليم والتدريب قد قوضت وبشكل خطير توفير فرص التعليم والتدريب في السودان. وكما إن المساهمة الاصيلة والجديدة لهذه الدراسة أنها سلطت الضوء على أهمية الآثار الخارجية للتعليم ونقل المعرفة ودراسة العوامل التي تعوق، والاخرى المساهمة في تعزيز الآثار الخارجية للتعليم ونقل المعرفة، هذا وقد أكدت نتائج الدراسة على أهمية تعزيز الآثار الخارجية للتعليم ونقل المعرفة. كما أظهرت نتائج الدراسة التناقض المثير للدهشة فى وجهات النظر بشأن حدوث ونقل المعرفة على المستويين الكلي والجزئي. بناءً على نتائج هذه الدراسة توصي الدراسة ببذل مزيد من الجهود لتحسين توفير فرص وجودة التعليم والتدريب، وتطوير المهارات، وتحسين اداء مؤشرات التكنولوجيا وتشجيع الآثار الخارجية للتعليم وتعزيز نقل المعرفة وتحسين التوافق بين مؤسسات القطاعيين العام والخاص على المستويين الكلي والجزئي من اجل تحسين النمو الاقتصادي، والتنمية البشرية، والتنمية المستدامة في السودان.

Contents

Part I Introduction

List of Abbreviations

AHDR	Arab Human Development Report
API	Arab Planning Institute
CIS	Community Innovation Survey
DAL	DAL Group of Companies
EPO	European Patents Office
ERF	Economic Research Forum for the Arab Countries, Iran and Turkey
ESCWA	Economic and Social Commission for Western Asia
FDI	Foreign Direct Investment
FTER	Full–time equivalent researcher
GCC	Gulf Cooperation Council
GDP	Gross Domestic Products
GNP	Gross National Product
ICT	Information and Communication Technology
ILO	International Labour Organization
IMF	International Monetary Fund
ISCO	International Standards Classification of Occupations
ISIC	International Standard Industrial Classification
IT	Information Technology
ITU	International Telecommunication Union
KSC	Kenana Sugar Company
MENA	Middle East and North Africa
MNC	Multinational Corporations
OECD	Organisation for Economic Co-operation and Development
R&D	Research and Development
S&T	Science and Technology
SITC 7	Standard International Trade Classification (SITC)
TAI	Technology Achievement Index
TFP	Total Factor Productivity
TNCs	Transnational Corporations
UAE	The United Arab Emirates
UIS	UNESCO Institute of Statistics

UN	The United Nations
UNCTAD	The United Nations Conference on Trade and Development
UNDP	United Nations Development Programme
UNESCO	United Nations Educational, Scientific and Cultural Organization
USPTO	US Patent and Trademark office web site
WB	World Bank
WDI	World Development Indicators – World Bank
WEO	World Economic Outlook –International Monetary Fund
WITSA	World Information Technology and Services Alliance

List of Figures

List of Tables

Part I
Introduction

Chapter 1
General Introduction

Abstract The aim of this Chapter is to give a short general overview of the research problem and to briefly show the importance, relevance, objectives, questions, hypotheses and the general structure of the research.

1.1 Introduction

The aim of this Chapter is to give a short general overview of the research problem and to briefly show the importance, relevance, objectives, questions, hypotheses and the general structure of the research.

1.2 Research Problem, Importance, Relevance and Method

Economists have long recognised the essential role of technical progress in the creation, acceleration and sustainability of economic growth and improvement of the quality of life in any society. Both the endogenous growth theories and empirical literature acknowledge the importance of human capital accumulation for economic growth. Endogenous growth literature also elaborates on the interaction and complementary relationship between technological progress and human capital to create, reinforce, accelerate and sustain economic growth (cf. Lucas 1988; Romer 1990; Freeman and Soete 1997). Moreover, economists also highlight the role of high levels and quality of skills as critical factors for competitiveness associated with the rapid progress of globalisation and fast technological progress in developed and developing countries. Considerable debate in the literature is on the effects of human capital and the diffusion and transfer of technology to developing countries to accelerate the catching-up process. For instance, the Nelson and Phelps (1966) model allows human capital levels to affect the speed of technological catch-up and diffusion. Romer

The study presented in this book refers to Sudan before separation.

S. Mohamed Nour, *Technological Change and Skill Development in Sudan*,
DOI 10.1007/978-3-642-32811-4_1, © Springer-Verlag Berlin Heidelberg 2013

(1990) has also argued that the level of human capital may have an influence on growth of technological innovation both directly and through its effect on the speed of the catch-up process. Benhabib and Spiegel (1994) adapted the Nelson and Phelps (1966) framework to allow for the catching-up of technology with that of the leading countries. In their view, the level of education not only enhances the ability of a country to develop its own technological innovations, but also its ability to adapt and implement technologies developed elsewhere.

Endogenous growth theory predicts that in the long run economic growth at the aggregate level is determined by endogenous sources of technological change: human capital, learning by doing, spillovers of knowledge and the external effect of human capital. Endogenous growth literature revealed several robust facts and interesting implications that paved the way for growth and it also provided some insights for a possible role for government policy. The literature on endogenous growth and public policy presents an argument for government intervention to promote the accumulation of technology, human capital and hence growth rate. The most popular view in the literature is that the rationale for government intervention is basically related to the idea that knowledge (in the form of technical progress or accumulation of human capital) is a public good, which is non-rival and partially excludable. As in Romer (1990) and Barro and Sala-i-Martin (1995) these two features imply unbounded growth and incomplete appropriability of knowledge, and raise the possibility of knowledge spillovers across firms and hence the whole economy. While the feature of spillovers of knowledge supports endogenous growth, it also creates a form of externality and implies that private investments generate a positive external effect and the private returns from investment tend to be lower than the social returns. The outcomes tend not to be Pareto optimal but suboptimal that require government intervention to correct the distortion; the social optimum can be achieved by many instruments, such as providing subsidies to improve the accumulation of technology and human capital. In the endogenous growth literature some studies explicitly model the importance of technology and human capital for endogenous growth but only implicitly indicate a role for public policy. For instance, while the Lucas (1988) model emphasises investment in human capital, it only implicitly allows for a role for public policy by subsidies (Haslinger and Ziesemer 1996: 230). Moreover, the Arrow (1962) learning-by-doing and Romer (1986) models imply an indirect intervention: an investment tax credit that increased the accumulation of capital necessarily also increased the accumulation of technology (Romer 1990: S94).

According to Ziesemer's (1987) interpretation, T. W. Schultz (1964) presents a pioneering theoretical justification for central role for government interference to promote public investment and emphasis their long run effects on growth and development. The theory of Schultz (1964) reveals that the provision of public factors, such as basic education and basic scientific research, is necessary for human capital formation and this would drive technical progress. Therefore, technical progress depends on human capital and the production of human capital requires public factors, such as basic education and basic scientific research financed through a linear income tax rate. Several studies emphasise a role for government intervention and a positive impact of public provision of education and training (cf. Azariadis and Drazen 1990; Barro and Sala-i-Martin 1992, 1995; Jones 1998; Chatterji 1995; Haslinger and Ziesemer 1996; Otani and Villanueva 1990; Ziesemer 1990, 1991, 1995).

Few studies examine the practical relevance of the models of growth enhancing policies, particularly for developing countries. For instance, Haslinger and Ziesemer (1996) indicate that most of the models of publicly financed investment in human capital are basically intended for industrialised not developing countries. In their view, in developing countries, raising the publicly financed investment is hampered by the lack of well-developed institutional setup to use the instruments of taxation, mainly because of substantial engagement in non-monetised activities, large informal sectors, extreme poverty and different effects from the prevalent regressive trade tax (Haslinger and Ziesemer 1996: 240–241). Given these practical limitations for developing countries, in the recent years there has been a growing body of literature on the role of public policies and government intervention to promote human capital and technological capabilities in developing countries. For instance, Lall (1999) discusses strategies to develop skills, technology and capabilities in developing countries and identifies autonomous (Korea and Taiwan) and foreign direct investment (FDI)-led targeted strategy (Singapore and Malaysia) and the FDI-led market-led strategy (Mexico and Thailand) (Lall 1999: 9, 10).

In this research we use the framework of and perspectives from the new growth literature to investigate the relevance and importance of skill upgrading and technological development, and the interaction between these for economic development in Sudan. The importance of the country increased after the recent exploitation of its oil. In recent years the increasing dependence on oil has led to high fluctuation in economic growth, for instance, the gross domestic product (GDP) growth rates increased from 6 % in 2003 to 10.5 % in 2007; however, the global financial crisis and related shock in 2008 and 2009 resulted in low global oil prices, stagnating domestic oil production and caused a reduction in GDP growth rate that dropped from 10.5 % in 2007 to 7.8 % and 6.1 % in 2008 and 2009 respectively. While oil has recently contributed to the improvement of the country's economic performance, the recent heavy dependence on it may lead to serious challenges for Sudan since oil is an exhaustible resource and – because of the instability of oil prices – revenue from oil is uncertain and volatile. Moreover, the increased wealth from oil encourages migration to the country, consequently, migrant workers may start to replace domestic workers and so contribute to unemployment in the labour market in the country.[1]

The GDP and GDP per capita of Sudan are still growing, but their annual growth rates are either stagnant and/or have even declined. So far, the country suffers from serious political instability and serious conflicts, in addition to several serious

[1] For the empirical investigation in this research we focus on Sudan as a case study of the Arab countries, due to easy accessibility to data, information and facilities for the fulfillment of the fieldwork. According to the World Bank classification, Sudan is amongst the lower medium income, least developing and heavily indebted countries.

economic problems such as high rates of inflation and unemployment and a high incidence of poverty. Over the past years, the increasing contribution of the oil sector has been used to improve the country's economic performance, but, in our view, the big question is whether or not this will work in the future. We feel this may be somewhat doubtful in light of the recent current stagnation in education, particularly tertiary education and also because of the negative impact and loss of most of the oil reserves (70 %) and oil revenues (50 %) following the official secession of Southern Sudan. In addition, the high share of foreign capital indicates that Sudan is currently an attracting economy. But it may lose this position and become even poorer if both the oil industry and the complementary ones decline and/or if foreign capital moves out and there is less work in the government sector. Notwithstanding these structural shifts, and whether the agriculture sector will remain the country's main driver of the economy or the services and industrial sectors can fill the gap, it is evident high levels of education will be needed. Therefore, it is quite essential for the government to improve investment and enrolment in education and skill upgrading.

The economic growth and sustainable development strategy in Sudan depends on both a shift in the focus from an oil resources-based economy to a technology and skill-based economy and to economic diversification. A key part of this strategy is also to achieve peace and political and economic stabilisation, to reduce unemployment and poverty, and to manage the economy away from dependency on the import of technology and high skilled workers in favour of domestic skilled workers. Overcoming these strategic problems and challenges confronting economic development and thereby achieving long run economic growth and sustainable development in Sudan, depends on five main strategies: alleviation of poverty; achieving economic diversification; reducing unemployment and restructuring the labour market; building local technological development and enhancing self-reliance on domestic capital and workers and achieving stabilised, sustained and balanced economic growth and development strategies. In our opinion the success/fulfilment of these strategies is contingent upon the development of adequate and appropriate skills, skill upgrading and efficient educational and training policies and building of local technological capabilities in Sudan.

In our view, although the development of local technologies is a costly process and likely somewhat problematic to be funded from the oil revenues in the short term, notably in light of the loss of 50 % of Sudanese economy's oil reserves after the official secession of Southern Sudan, but the development of local technologies can be encouraged by offering further incentives to motivate private investment in skills, technology and technical education. Moreover, to some extent, Sudan has an option to remain importing technologies and to specialise in fields other than producing machinery and transport equipment (SITC 7). Sudan shows poor technology achievement index, deficient skill and technology indicators and a substantial gap when compared to rapidly advanced countries (cf. Lall 1999; UNDP 2001). Therefore, the need for the development of local skills and building local technologies are important not only for fulfilling the above strategies, but also for shortening the gap, building local capacity, improving productivity and competitiveness in the international market.

In our view, the implementation of these strategies is eminently impeded by the deficient educational system, the serious skills mismatch and the lack of incentives in Sudan. One should note that many previous studies in the Sudanese literature have thoroughly investigated the causes and implications of the serious problems and improper political and economic performance (cf. Awad 1991a, b; Ali 1990). However, the impacts of the deficiencies in the educational system and serious skills mismatch have not received adequate investigation in these studies, despite the well-known stylised facts in the endogenous growth literature that highlight the essential endogenous effect of education for accumulation of human capital and economic growth.

For that reason, our analysis in this research provides many new and interesting results. Different from other studies in the Sudanese literature (Al-Sanousi 1999; Beshir 1969; Suleiman 2007; Jalal al-Din 2002), we identify upskilling as a key element for the fulfilment of the five current strategies in Sudan. Furthermore, we provide a more comprehensive investigation since we discuss the skills problem from two different perspectives: combining the impacts of the deficient educational system and the high incidence of unskilled workers. In addition, our analysis goes beyond the available Sudanese literature and presents a more elaborate and in-depth analysis to assess skill and technological performance in Sudan since we use a very comprehensive set of indicators that is often used in the new growth literature. We use these indicators to analyse the causes and consequences of low skill and technology levels, the link between them and the implications for skills mismatch, and the lack of local efforts for technological development and the consequent dependence on foreign technologies at both the macro and micro levels. In addition, in light of the recent literature that highlights the role of diffused technologies, our analysis uses a broader definition of technological change that considers the role of diffusion in fostering economic growth and we assess the role of imported technologies in promoting local technologies and local skills. This definition is particularly relevant for our analysis since Sudan is highly dependent on imported technologies and recently relatively on imported skilled workers to manage them. Our analysis also addresses the policy issues, stresses the role of both public and private educational and training policies and the need for incentives, harmony and collaboration between public-private institutions in upgrading skills and fostering human capital accumulation in Sudan. Finally, our analysis fills the gap in the Sudanese literature since we highlight the importance of knowledge and external effects of schooling/the transfer of knowledge, and we explore the factors hindering and those contributing toward enhancing the transfer of knowledge at both the macro and micro levels.

Moreover, our research contributes to the few recent studies in the Arab region that call for upskilling and the interaction between skills and technology (Muysken and Nour 2006; Nour 2005a, b). We explain the deficiencies in the educational and training systems, their impacts on declining industrial performance and on skills mismatch (Al-Sulayti 2002; Gray 1999; Abdelkarim and Haan 2002; Suleiman 2007;

Jalal al-Din 2002). We show the impacts on poor technological level that dependence on foreign technologies and the impacts of technologies transfer (El Sabaa 1997; Haan 1999), macro–micro duality concerning knowledge transfer, upskilling and training policies.

To investigate the research problem we focus on Sudan as a case study of the Arab countries and we use the descriptive, comparative and statistical methods of analysis, and we use a combination of primary and secondary data covering both the macro and micro levels and the results from the macro and firm surveys (2010) that were held in Sudan in 2010. In addition, the surveys data was supported by ten face-to-face interviews with firm managers and five interviews with the officials. The firm survey (2010) on 'Technological Change and Skill Development in Sudan's Manufacturing Sector' aims to assess skill and technology indicators and the impacts of unskilled workers amongst the food, textile, chemical and metal small, medium and large size establishments in Sudan. The macro survey (2010) on 'Skill Creation, Human Resources Development and Policy Intervention' was sent to policy makers and experts in public and university institutions to examine the causes and consequences of low skills and the deficient educational system in Sudan. In addition, we conducted the R&D Survey (2010), which is small survey on research and development (R&D) based on 25 face-to-face interviews with officials policy makers and experts in the government and academic staff in the public and private universities, to examine the causes and consequences of poor R&D activities at the macro level. All primary and secondary data were collected personally.

1.3 Objectives, Hypotheses and Questions of the Research

Based on the above background, the central themes of this research are: the required skill formation and upskilling of the workers, together with their interaction with technological change; and the deficient educational system and their implications. First, our analysis aims to provide an empirical investigation of the causes and consequences of low skill and technology indicators, in particular, the causes and consequences of the deficient educational system and the implications of the serious skills mismatch at both the macro and micro levels. Second, we examine the interaction between the low skill and technology indicators, the relationships between skill, upskilling and technology indicators, skills mismatch, the uses and impacts of ICT and differences, defined by firm size (small, medium, large) and industry (food, textile, chemical, metal), at the micro/firm level. Third, we examine the factors hindering and those contributing toward enhancing the transfer of knowledge/external schooling effects at the macro and micro levels. Finally, we highlight the need for implementation of consistent policies, increasing incentives and collaboration between public and private educational and training institutions to enhance skill upgrading, local technological development, economic development and transfer of knowledge.

Grounded on these objectives, our research attempts to answer three sets of questions:

1. What are the major causes and consequences of low skill and technological levels in Sudan? What are the major implications of the deficient educational system and the high incidence of unskilled workers at the macro–micro levels in Sudan?
2. Does the external effect of schooling/transfer of knowledge occur in Sudan? If not, why does it not yet occur? What are the major factors hindering and those contributing toward enhancing the transfer of knowledge at both the macro and micro levels in Sudan?
3. What are the major policies for upgrading skill, reforming the educational system, enhancing the provision of training and the development of local technologies at both the macro and micro levels in Sudan?

Based on the research questions and objectives, the major hypotheses to be tested in this research are:

1. Sudan needs to promote local skills and local technologies in order to implement the five strategies of reducing poverty; achieving economic diversification; reducing unemployment and restructuring the labour market; building local technological capacity; and achieving long-term stabilised, sustainable and balanced economic growth and development.
2. In the short- and medium-term, Sudan is unable to rely on local technologies and will remain heavily dependent on foreign technologies.
3. (a) The interaction between the deficient educational system and the high incidence of unskilled workers leads to low skill and technology levels and many other serious implications.

   (b) The deficient educational system is caused by low quality of education and leads to: (1) poor provision of training; (2) low skill levels; (3) skills mismatch; (4) low transfer of knowledge/external schooling effect; (5) weak technology indicators; (6) dependence on foreign technologies at the macro and micro levels and (7) poor industrial performance at the micro level.

   (c) The major causes of low levels of local technology are low/lack of R&D activities due to a lack of funding, skills, transfer of knowledge, networks and collaboration between universities and industry/firms.
4. (a) The observed differences in actual and required skill levels, i.e. the high skill requirements and the prevalent low skill levels (due to high share of unskilled workers) lead to skills mismatch and contribute to industrial performance and productivity decline at the micro level/across firms.

   (b) An increase in skill level (share of highly skilled in total employment and in firm size) leads to improved relationships between actual and required education and experience and wages.

(c) An increase in skill level (share of highly skilled in total employment and in firm size) leads to improvements in the complementary relationships between skill, upskilling and technology (ICT).

5. The use of ICT has positive but insignificant/inconclusive effect at the micro level/across firms.
6. The transfer of knowledge/external schooling effects is unsuccessful at the micro and macro levels. The major reasons behind the low transfer of knowledge/external schooling effects are low educational qualifications, and deficient educational and training systems. The major consequences are the lack of networks and collaboration between universities and firms, low R&D efforts and low technology indicators.
7. Knowledge has positive impacts at the macro–micro levels; it can be enhanced by many factors.

   (a) At the macro level codified knowledge and tacit knowledge are positively correlated with economic growth (GDP growth rate) and are positively correlated with schooling.
   (b) At the macro level codified knowledge (the total spending on R&D) and tacit knowledge (the total number of full time equivalent researchers (FTER)) are positively correlated with each other and also with technology (patents).
   (c) At the micro (firm) level tacit knowledge is positively correlated with technology (Information and Communication Technology (ICT)), upskilling (training), profit, productivity, output and output diversification.
   (d) At the micro (firm) level tacit knowledge is positively correlated with market size (firm size; capital; and investment) and firm age.

8. Sudan needs to enhance both the public and private educational and training policies:

   (a) Skill development depends on: (1) reforming the educational system; (2) enhancing the provision of training; (3) planning skill needs and matching educational output with market needs; (4) enhancing the transfer of knowledge/schooling effect; and (5) incentives and collaboration between public and private institutions.
   (b) The promotion of local technology and adoption of appropriate foreign technologies and the interaction between both to foster economic growth in Sudan depends on skill development. In particular, an enhancement of: (1) skill upgrading: educational and training systems; (2) R&D activities; (3) the transfer of knowledge/schooling effects; (4) network systems; and (5) incentives to motivate collaboration between universities and firms and between public and private institutions.
   (c) Both upskilling policies (educational and training policies) and transfer of knowledge are consistent at both the macro–micro levels and across public-private sectors.

1.4 Structure of the Research

Considering the research problem, aims, questions and hypotheses presented above, it is convenient in this Chapter to set out the structure of the research. This research is composed of four parts and ten chapters structured in the following way. Part I presents the introduction and motivation of the research and includes both Chaps. 1 and 2. Chapter 1 presents an introduction and briefly shows the aims, importance, relevance, hypotheses, questions and the general structure of the research. Chapter 2 explains some stylised facts about Sudan that help to investigate more extensively the research problem along with other strategic problems and challenges confronting economic development, the impacts of oil and the Dutch Disease phenomenon in the structure of the labour markets and economy. In addition, this chapter aims to examine the structural problems related to the labour market and unemployment problem and to assess and elaborate the low skill indicators and to show some stylised facts that justify and highlight the need for skill upgrading and development in Sudan.

Part II presents the conceptual and theoretical framework and includes Chap. 3, which defines the concepts and describes the measures of technological change and human capital (education) and briefly explains the theoretical and empirical literature on the relationship between human capital, technological changes and economic growth. The purpose of this survey is to provide a background for our study, mainly to highlight the endogenous effects of technical change and human capital as confirmed in the endogenous growth literature to motivate the empirical analysis in the next chapters.

Part III presents the empirical application and includes Chaps. 4, 5, 6, 7 and 8. Chapter 4 defines the methods of data collection; identifies the motives for performing the macro and firm surveys and selection of a case study; specifies the selection of the sample and the composition, operation, coverage, advantages and limitations of the survey data; and shows the structure and design of the questionnaire. Chapter 5 uses the data and results from the firm and macro surveys to examine the serious implications of the interaction between the deficient educational system and skills mismatch. We use the results from the macro survey to show the causes of the deficient educational system and consequences on low skill levels, poor provision of training, skills mismatch and lack of/low transfer of knowledge at the macro level. In addition, we use the results from the firm survey to illustrate the implications of the high incidence of unskilled workers on causing low skill levels, poor provision of training, skills mismatch, poor technology indicators and a heavy dependence on foreign technologies. Furthermore, we investigate from the micro–macro perspectives the transfer of knowledge, upgrading of skill and technology and their potential implications, and we also present a more comprehensive assessment of technology and skill indicators at the micro level. Chapter 6 extends our analysis in Chap. 5 by assessing and elaborating the low skill and technology indicators at the macro level; explaining the gap in Sudan compared to developed and developing countries and showing some stylised facts that justify and highlight the need for skill upgrading and technological development in Sudan. Chapter 7 aims to broaden our analysis in Chaps. 5 and 6 by providing an in-depth analysis of skill and technology indicators, the relationship between them, and the implications of the prevalence of low-skilled workers on skills mismatch and poor industrial performance at the micro level. We use the data from the firm survey (2010) to examine the relationships between skills (actual and required education and experience) and wages; between skill, upskilling (ICT training) and technology (ICT); and between technology (ICT) and input–output indicators across firms. We also compare the relevance of our results to the theoretical framework in Chap. 3 and the findings concerning these relationships in the new growth literature. Chapter 8 extends our analysis in Chap. 5 on the transfer of knowledge. We use the data from the firm survey (2010) at the micro level and some secondary data at the macro level to discuss the importance/impacts of knowledge in Sudan. In particular, we check the relevance of some stylised facts about the importance/impact of knowledge at the micro–macro levels to the findings in knowledge literature.

Finally, Part IV presents the policies, recommendations and conclusions and includes Chaps. 9 and 10. Chapter 9 concludes our analysis by discussing educational and training policies, since our earlier analysis in Chap. 5 investigates the causes and consequences of deficient educational and training systems, the lack of knowledge transfer and upskilling, and the results set in Chaps. 7 and 8 imply the importance of a good education. From that perspective, therefore, Chap. 9 discusses the supply–demand sides and the implications of educational and training policies in Sudan. In addition, we use the results from the macro and firm surveys (2010) to examine the macro–micro views and suggestions for relevant mechanisms and policies for skill upgrading via an enhancement of the educational system, provision of training and transfer of knowledge/external schooling effects at the macro–micro levels in Sudan. Finally, Chap. 10 summarises and compares the main findings with the results in the general literature and contributions to the Sudanese literature and concludes with policy recommendations.

References

Abdelkarim, A., & Haan, H. (2002, January). *Skills and training in the UAE: The need for the dimensions of institutional intervention* (Policy Research Series No. 5). Dubai: Centre for Labour Market Research and Information.

Ali, A. G. A. (1990). *From dependency to dependency*. Cairo: Centre for Future Studies. in Arabic.

Al-Sanousi, T. A. (1999). Literacy and adult education in Sudan: Problems and future prospects. *Mahawir, Journal for Sudanese Studies*, *2*, 26–35, Omdurman: M. O. Beshir Centre for Sudanese Studies, Omdurman Ahalia University.

Al-Sulayti, H. A. (2002). *Education and human development in the GCC countries: An analytical study* (Strategic Studies No. 71). Abu Dhabi: The Emirates Center for Strategic Studies and Research (ECSSR). in Arabic.

Arrow, K. J. (1962). The economic implication of learning by doing. *The Review of Economic Studies, 29*, 155–173.

Awad, M. H. (1991a). *The Sudan and the IMF* (Book Series No. 6). Khartoum: Development Studies and Research Centre (DSRC), Khartoum University, in collaboration with Fredrich Eberit Foundation.

Awad, M. H. (1991b, November). *The politics of economic adjustment: The Sudanese case* (Seminars Series No. 94). Khartoum: Development Studies and Research Centre (DSRC), Khartoum University, in collaboration with Fredrich Eberit Foundation.

Azariadis, C., & Drazen, A. (1990). Threshold externalities in economic development. *Quarterly Journal of Economics, 105*(2), 501–526.

Barro, R. J., & Sala-i-Martin, X. (1992). Public finance in models of economic growth. *The Review of Economic Studies, 59*(4), 645–661.

Barro, R. J., & Sala-i-Martin, X. (1995). *Economic growth.* New York: McGraw Hill.

Benhabib, J., & Spiegel, M. M. (1994). The role of human capital in economic development: Evidence from aggregate cross-country data. *Journal of Monetary Economics, 34,* 143–173. New York 1994.

Beshir, M. O. (1969). *Education development in Sudan 1989–1956.* London: Oxford University Press.

Chatterji, M. (1995, September). Training subsidies, technical progress and economic growth. *The Manchester School, LXIII*(3). Blackwell Publishers Ltd./Victoria University of Manchester.

El Sabaa. S. (1997, November) Linking technology to industrial development in the United Arab Emirates. *Trade and Industry Magazine* – Dubai, *22*(263), 16–29.

Firm Survey. (2010, June). *Technological change and skill development: A comparative study of chemical, food, metal and textile small, medium and large scale enterprises in the Sudan.* Khartoum.

Freeman, C., & Soete, L. (1997). *The economic of industrial innovation* (3rd ed.). London: Cassell. pp 2, 24, 404–405.

Gray, L. (1999, April). *Private sector training and employment: Evaluation and intervention strategies* (Specialist Rep. No. 4). Dubai: The Ministry of Labour and Social Affairs National Strategy for Labour Force Development and Employment Project.

Haan, H. (1999, January/March). *The UAE: Assessment of technology use in some selected sectors – Towards introduction and explaining capital and skill intensive technologies* (Specialist Rep. No. 6). Dubai/Amsterdam: National Human Resources Development and Employment Strategy Project.

Haslinger, F., & Ziesemer, T. (1996). Endogenous growth and distributional conflict. In C. Gore, G. Köhler, U. Peter Reich, & T. Ziesemer (Eds.), *Questioning development* (pp. 223–240). Marburg: Metropolis Verlag.

Jalal Al-Din, M. A. (2002). *The relationship between higher education and the world of work and production.* Omdurman: Omdurman Ahlia University, Mohamed Bashier Omer Centre for Sudanese studies and Friedrich Ebert Stiftung.

Jones, C. (1998). *Introduction to economic growth.* New York/London: W. W. Norton and Company.

Lall, S. (1999). *Competing with labour: Skills and competitiveness in developing countries* (Issues in Development Discussion Paper No. 31, p. 17). Geneva: Development Policies Department, ILO.

Lucas, R. E. (1988). On the mechanics of economic development. *Journal of Monetary Economics, 22*, 3–42. North Holland.

Macro Survey. (2010, June). *Skill creation, human resources development and policy intervention: Interviews with policy makers and experts in the Sudan.* Khartoum.

Muysken, J., & Nour, S. (2006). Deficiencies in education and poor prospects for economic growth in the Gulf countries: The case of the UAE. *Journal of Development Studies*, *42*(6), 957–980. UK: Taylor and Francis Ltd.

Nelson, R., & Phelps, E. (1966). Investment in humans, technological diffusion and economic growth. *American Economic Review: Papers and Proceedings, LVI, 61*, 69–75.

Nour, S. (2005a, August). *Science and technology (S&T) development indicators in the Arab Region: A comparative study of Arab Gulf and Mediterranean countries* (Published in UNU-INTECH Discussion Paper Series UNU-INTECHDPS2005-3). Maastricht.

Nour, S. (2005b, August). Science and technology (S&T) development indicators in the Arab Region: A comparative study of Arab Gulf and Mediterranean countries. *The Journal of Science, Technology and Society, 10*(2), 249–274. SAGE Publications. Maastricht: Published in UNU-INTECH Discussion Paper Series UNU-INTECHDPS2005-3.

Otani, I., & Villanueva, D. (1990). Long-term growth in developing countries and its determinants: An empirical analysis. *World Development, 18*(6), 769–783.

R&D Survey. (2010, June). *Sudan R&D Survey: Efforts for enhancing R&D in the Sudan: Interview with policy makers in the government and academic staff in public and private universities in Sudan*. Khartoum.

Romer, P. M. (1986). Increasing returns and long run growth. *Journal of Political Economy, 94*(5), 1002–1037.

Romer, P. M. (1990). Endogenous technological change. *Journal of Political Economy, 98*(5, Pt. 2), 71–102.

Schultz, T. W. (1964). *Transforming traditional agriculture*. New Haven: Yale University Press.

Suleiman, A. A. (2007). Employment of training and education to serve economic and social development. Chapter 5. In *Future of economic and social development in Sudan* (pp. 121–123). Khartoum: Sudan Currency Printing Press Company Ltd, (in Arabic).

UNDP. (2001). *Summary Human Development Report (2001): Making new technologies work for human development*. New York/Oxford: UNDP/Oxford University Press.

Ziesemer, T. (1987). *Economic theory of underdevelopment*. Regensburg: Transfer Verlag.

Ziesemer, T. (1990). Public factors and democracy in poverty analysis. *Oxford Economic Papers, 42*(Special issue on public economics), 268–280. (Reprinted in *Taxation, private information and capital*, by P. J. N. Sinclair & M. D. E. Slater, Eds., Oxford: Clarendon Press).

Ziesemer, T. (1991, Winter). Human capital, market structure and taxation in a growth model with endogenous technical progress. *Journal of Macroeconomics, 13*(1), 47–68.

Ziesemer, T. (1995). Endogenous growth with public factors and heterogeneous human capital producers. *Finanzarchiv,* Neue Folge, *52*(1), 1–20.

Chapter 2
Economic Development Challenges in Sudan and the Need for Skill Upgrading and Technological Development

Abstract This chapter explains the general socioeconomic characteristics of Sudan and strategic problems for development in the country, and discusses the impact of oil and the opportunities and challenges for enhancing economic development in Sudan, the strategic problems facing the labour market in Sudan and highlighting the need for skill development. This chapter uses new data on population, employment and unemployment based on Sudan Central Bureau of Statistics (Fifth Sudan Population and Housing Census 2008, Sudan Central Bureau of Statistics, Khartoum, 2010) to examine four stylised facts related to the high unemployment rate; interpretate unemployment crisis from two different endogenous and exogenous perspectives due to endogenous and exogenous causes; the high incidence of unemployment among youth population and the large mismatch between educational qualifications – supply – and labour market requirements-demand in Sudan. The findings in this Chapter support the first hypothesis in Chap. 1 that Sudan needs to promote local skills and local technologies in order to facilitate implementation of the five strategies of reducing poverty; achieving economic diversification; reducing unemployment and restructuring the labour market; building local technological capacity and achieving long-term stabilised, sustainable and balanced economic growth and development.

2.1 Introduction

In Chap. 1 we introduced the research problem; the aim of this chapter is twofold: first, to present a background to motivate the research by explaining some stylised facts to examine more extensively the research problem along with other strategic

---

Nour, S. (2011, June). Assessment of the impacts of oil: Opportunities and challenges for economic development in Sudan. *The Journal of African Review of Economics and Finance*, 2(2), 122–148. African Center for Economics and Finance, Published by Print Services Unit, Rhodes University, Grahamstown, South Africa.

S. Mohamed Nour, *Technological Change and Skill Development in Sudan*,
DOI 10.1007/978-3-642-32811-4_2, © Springer-Verlag Berlin Heidelberg 2013

problems confronting economic development in Sudan. Second it aims to highlight the need for upskilling and technological development in Sudan. This chapter investigates the impact of oil in enhancing economic development in Sudan. The aim is twofold: first to examine evidence related to the positive impact and opportunities offered by oil for enhancing economic development in Sudan, and then to consider the negative impacts and challenges on economic development. We provide a comprehensive analysis, improve understanding and fill the gaps in the Sudanese literature by examining our hypothesis that oil has created a mixed positive–negative impact on Sudanese economy.

In this research we use the framework of and perspectives from the new growth literature to investigate the relevance and importance of skill upgrading and technological development and the interaction between these for economic development in Sudan. Explaining the case of Sudan is significant because of the recent acceleration in growth and structural change in Sudanese economy after the exploitation of oil in the country. According to the World Bank (2008) Sudan is one of the newest significant oil producing countries in the world, and the third largest oil producer in Sub-Saharan Africa behind Nigeria and Angola. As a result of oil exploitation, the structure of Sudanese economy has shifted over time from being predominantly reliant on agriculture for growth and exports to its current reliance on the oil sector (World Bank 2008). In recent years the increasing dependence on oil has led to stable economic growth. Consequently, Sudan's real economic growth averaged around 9 % during 2005–2006, putting Sudan among the fastest growing economies in Africa (World Bank 2008). But while oil has recently contributed to the improvement of economic performance and FDI in Sudan, the recent heavy dependence on oil may lead to challenges, as oil is an exhaustible resource and because the revenue from oil is uncertain and volatile due to the instability of oil prices. Moreover, the increasing dependence on oil raises questions such as the incidence of the Dutch Disease phenomenon.[1]

Moreover, both the growing inflow of FDI and the increased wealth from oil has encouraged migration to Sudan. Consequently, migrant workers have increased in the labour market, particularly in the private sector, which leads to several problematic features such as low skill levels and contribution to the growing unemployment rate. In addition, Sudan suffers from structural problems related to the lack of political stability, continuous conflict, regional disparities due to imbalanced development strategies, poverty, high unemployment rate, and lack of incentives, which has also affected the structure of labour market. Moreover, skill and technology indicators in Sudan show poor performance and a substantial gap when compared to international standards. Hence, upskilling and technological development become imperative to overcome the strategic problems and challenges confronting economic development in Sudan.

---

[1] "The Dutch Disease is a process in which the discovery of natural resources causes a country to experience a 'change in the group of reference' from one that aim at generating a trade surplus in manufacturing to one that able to generate a trade surplus in primary commodities. The country experiencing this disease also shows differences between employment in manufacturing. The process of de-industrialisation due to the discovery of natural resources, mainly natural gas apparent from the case of Holland". Cf. Palma (2003), p. 21.

The rest of this chapter is organised in the following way: Sect. 2.2 explains the general socio-economic characteristics of Sudan; Sect. 2.3 examines the stylised facts on the positive and negative impact of oil in enhancing economic development in Sudan; Sect. 2.4 discusses the strategic problems facing the labour market in Sudan and highlights the need for skill upgrading and technological development; Sect. 2.5 concludes.

2.2 Economic Characteristics and Strategic Problems for Development in Sudan

Technological change and skill development are often closely related not only to the resources directly devoted to their development but also to the whole economic structure that supports them. Therefore, before assessing the need for technological development and skill development in Sudan it is useful to start by explaining the general political context and socio-economic characteristics of the country. Since the structure of Sudanese economy is now related to oil, it will be useful to examine the impact of oil in the following section. Next, we show the structural problems related to the labour market, skill, technology and productivity, and we attempt to link these to the general socio-economic characteristics of Sudan.

2.2.1 General Political Context and Socio-Economic Characteristics of Sudan

Before assessing the impact of oil in enhancing economic development, it is useful to start by explaining the general socio-economic characteristics of Sudan. Next, we examine the link between the general socio-economic characteristics and the impact of oil in Sudanese economy.

The political context in Sudan is characterised by a long history of political instability, continuing civil wars and complex conflict between the north and the south.[2] Even after the independence of Southern Sudan, Sudan still endures political

[2] As for the political context since independence in 1956 and over the past five decades, Sudan was ruled by three civilian governments (1956–1958, 1964–1969 and 1985–1989) and three military governments (1958–1964, 1969–1985; 1989–2010). Sudan suffered from political instability, as the three short-lived civilian governments were often removed and overthrown by the military governments. For instance, the first civilian government after independence (1956–1958) was overthrown in 1958 by the Abbud Military Government (1958–1964); the second elected civilian government (1964–1969) was overthrown in 1969 by the Nimeiri Military Government (1969–1985); and once again the third elected civilian government (1985–1989) was overthrown in 1989 by the Al Bashir Military Government (1989–2005). Since the signing and implementation of the Comprehensive Peace Agreement (CPA) in January 2005, Sudan has been ruled by the Government of National Unity (GNU), which represents a power-sharing government between the National Congress Party (NCP) of the north and Sudan People's Liberation Movement (SPLM) of the south. The implementation of the CPA implies several important agreed issues, which included

instability, a lack of sound and systematic institutions and a lack of a commitment to implementing long-term sustainable and balanced economic development plans and strategies.[3] This implies that the interaction between these political, economic and institutional factors together have unfortunately continued to contribute to a low standard of economic development in Sudan, as we explain below in the next sections.

The general socio-economic characteristics of Sudan indicate great diversity between Sudan compared to other African, Arab and world countries in terms of population, standard of economic development defined by Gross National Income (GNI) and GDP per capita and human development index. Table 2.1 below explains that on average Sudan has a higher population coupled with a lower standard of economic development. The World Bank classification of economies puts Sudan among the lower-middle income bracket and the United Nations Development Programme (UNDP) puts Sudan among the low human development, and poor and highly indebted economies. Moreover, the United Nations Development Programme-Human Development Index (UNDP-HDI) shows that the average life expectancy, literacy rate and combined enrolment ratios of Sudan are lower than those of other Arab and world countries. Furthermore, Sudan has continued to suffer from macro-economic instability, high rates of poverty, unemployment and debt. Despite the high and increasing inflow of FDI to Sudan (increased from –0.2 % of GDP in 1990 to 8.4 % of GDP in 2005), the country suffered from the high increase in debt services both as percentage of GDP (0.4 %–1.4 %) and as percentage of exports (8.7–6.5 %) over the period 1990–2005. That was probably because like most African countries, Sudan's economy has relied heavily on a large influx of foreign aid from different sources; Sudan is among the top ten recipients of gross Official Development Assistance during 1990–2007 (see UNDP 2007).

the formation of the Government of National Unity (GNU) in 2005, which represents a power-sharing government between the National Congress Party (NCP) of the north and Sudan People's Liberation Movement (SPLM) of the south. Moreover, the implementation of the CPA implies several other important agreed issues, which included the establishment of an interim transitional period of autonomous rule for the South for 6 years (2005–2011), followed by self-determination for Southern Sudan and a referendum that was held in January 2011, in which Southern Sudan decided on secession from the north. In July 2011 Southern Sudan officially gained its independence from Sudan. As for government and politics, the politics of Sudan takes place in the framework of a federal presidential representative democratic republic; the judiciary is independent and obtained by the Constitutional Court and the legislative power is vested in both the government and in the two chambers, the National Assembly (lower) and the Council of States (upper). The bicameral National Legislature is the official Sudanese parliament and consists of 500 appointed members. Before the secession of Southern Sudan, Sudan was divided into 26 states, which in turn were subdivided into 87 districts; the ten states in Southern Sudan were subdivided into 84 counties. The states are: Al Gezira, Al Qadarif, Blue Nile, Central Equatoria, East Equatoria, Jonglei, Kassala, Khartoum, Lakes, North Bahr al Ghazal, North Darfur, North Kurdufan, Northern, Red Sea, River Nile, Sennar, South Darfur, South Kurdufan, Unity, Upper Nile, Warab, West Bahr al Ghazal, West Darfur, West Equatoria and White Nile.

[3] In Sudan the available natural resources include agricultural, water and rivers, in addition mineral resources include petroleum and crude oil, natural gas, gold, silver, asbestos, manganese, gypsum, mica, zinc, iron, lead, uranium, copper, kaolin, cobalt, granite, nickel, tin, chrome, and aluminum.

Table 2.1 General socio-economic characteristics of the Sudan[1]

Country	Population[a,b] (millions) (2010)	Gross national income (GNI) per capita (PPP[c] US$)	Human development index[a] (%)	Life expectancy [a] (years)	Mean years of schooling	Expected years of schooling	Adult literacy rate [a] (% aged 15 and above)	Population with at least secondary education (% ages 25 and older)	Tertiary enrolment ratio (% of tertiary school-age population) gross (%)
	2010 (a)	2010 (b)	2010 (c)	2010 (a)	2010 (d)	2010 (e)	2005–2008a (e)	2010 (d)	2001–2009a (e)
Sudan[b]	43.2	2,051	0.379	58.9	2.9	4.4	69.3	11.5	5.9 b
Latin America and the Caribbean	582.7	10,642	0.704	74.0	7.9	13.7	91.1	32.5	36.7
Arab States	348.2	7,861	0.588	69.1	5.7	10.8	72.1		22.7
Europe and Central Asia	410.3	11,462	0.702	69.5	9.2	13.6	97.5	65.1	54.2
East Asia and the Pacific	1,974.3	6,403	0.643	72.6	7.2	11.5			20.9
South Asia	1,719.1	3,417	0.516	65.1	4.6	10.0	62.4	21.6	12.8
Sub-Saharan Africa	808.8	2,050	0.389	52.7	4.5	9.0	62.4		5.5
OECD	1,026.3	37,077	0.879	80.3	11.4	15.9		73.8	71.4
High human development	1,052.4	12,286	0.717	72.6	8.3	13.8	92.3	41.0	43.2
Medium human development	3,597.3	5,134	0.592	69.3	6.3	11.0	80.7		17.6
Low human development	1,099.0	1,490	0.393	56.0	4.1	8.2	61.2	14.3	6.0
Least developed countries	854.7	1,393	0.386	57.7	3.7	8.0	59.9		5.4
World	6,908.7	10,631	0.624	69.3	7.4	12.3			25.7

[1]The World Bank and United Nations Development Programme (UNDP) Human Development Report classify world countries differently according to income level. We use the World Bank classification of economies that puts Sudan in the lower middle-income category or group

Source: (a) UNDP (2010a). Notes: [a]2007, [b]2008, [c]*PPP* purchasing power parity: pp. 145–146, 186–187, 195–196

(a) UNDESA (2009d), (b) Based on data on GNI per capita and GDP per capita in PPP US dollars (current and constant prices) from World Bank (2010g) and implied growth rates of GDP per capita from IMF (2010a), (c) Calculated based on data from UNDESA (2009d), Barro and Lee (2010), UNESCO Institute for Statistics (2010a), World Bank (2010g) and IMF (2010a), (d) Barro and Lee (2010), (e) UNESCO Institute for Statistics (2010a)

The structure of Sudanese economy has long been characterised by a small share of industry, notably manufacturing, and a high share of agriculture and service sectors in GDP and employment. The share of agriculture in GDP increased from 30.3 % in 1990 to 49.8 % in 1999 and then declined to 31.1 % in 2009; the share of the services in GDP declined from 54.4 % in 1990 to 34.4 % in 1999 and then increased to 45.0 % in 2009; the share of industry in GDP increased from 15.4 % in 1990 to 15.8 % in 1999 and then increased to 23.9 % in 2009 (see Table 2.2 below). In 1999 Sudan began exporting oil and since then has become increasingly dependent on oil exports to the extent that the economy has turned into an oil dependent economy.[4] Since the late 1990s the implementation of macro-economic reforms policies, along with the positive contribution of oil to Sudan economy since 1999, has caused a rapid increase in real economic growth, GDP and GDP per capita incomes (see Table 2.2 and Fig. 2.1 below). Consequently, Sudan has moved from a low income economy into a lower medium income economy according to World Bank classifications. But while the increasing dependence on oil has had some positive effects, it has also sparked a number of negative impacts and raises questions such as the incidence of the Dutch Disease phenomenon as we explain in the next sections.

2.3 Overview on the Importance and Impact of Oil in Sudan

Based on the above background on the socio-economic characteristics of Sudanese economy and since its structure is now closely linked to oil, in this section it is useful to examine our hypothesis on the mixed positive–negative impact of oil on Sudanese economy. Before explaining the positive and negative impact of oil on Sudanese economy, it is useful to start with a historical background about the structure of investment in oil and show the role of China in investment in the oil sector in Sudan.

2.3.1 Overview on the Importance and Historical Background About Oil in Sudan

According to the World Bank (2008) Sudan is one of the newest significant oil producing countries in the world, and the third largest oil producer in Sub-Saharan Africa behind Nigeria and Angola (see Fig. 2.2). The major oil production fields are located in Southern Sudan but the major oil refineries, ports and pipelines are located in Sudan. Due to this conflict, oil exploration has mostly been limited to the central and south-central regions of Sudan. The institutional structure of the oil sector in

[4] Sudan oil output is estimated at 500,000 barrels per day (2007) and oil reserves at five billion barrels (2005). See WB-DTIS (2008), p. 2. Moreover in 2005, the Sudanese Energy Ministry estimated total oil reserves at five billion barrels.

Table 2.2 The performance, structure and structural change in Sudan economy (1990–2009). Basic indicators of labour force, unemployment and inflation rates in Sudan over the period (1990–2008)

	GDP								Balance of trade			Structure of Sudan economy (share of sectors in GDP)		
Year	Total	Growth rate	Per capita	Per capita growth rate	Inflation rate	Unemployment rate	Exchange rate	Balance of payment	Exports	Imports	Balance	Agriculture	Industry	Services
1990	244.7	5.4	47.7	0	41	14.2	0.45	−76.1	374	618.4	−244.4	30.3	15.4	54.4
1991	276.8	7.5	81	69.8	62.7		0.69	−101.3	308.7	890.3	−581.6	28.7	17.6	53.9
1992	4,327.8	6.5	17.2	−78.7	105.4		9.7	−58.1	319.3	820.9	−501.6	33.7	17.1	49.1
1993	5,862.1	4.5	37.6	118.4	115	10.3	16.1	37.7	417.3	944.9	−527.6	37.9	17.4	44.5
1994	6,351.2	1	72.5	92.7	96.3		29.6	17.9	535.6	1,059.6	−524	40.1	16.4	43.5
1995	9,880.7	5.9	151.7	109.4	177.2		55.9	−58.1	555.7	1,184.8	−629.1	43.1	15.8	41.1
1996	8,259.3	5.9	375.9	147.7	76.3	14.3	125	−63.9	620.3	1,504.5	−884.2	44.9	14.5	40.6
1997	10,684.8	6.3	563.7	50	52.6	18.1	156.9	−36.2	594.2	1,421.9	−827.7	47.6	15.1	37.2
1998	11,513.7	6.4	743.7	31.9	28.2	15.1	198.8	25.1	595.7	1,732.2	−1136.5	48.6	15	36.2
1999	10,325	6.7	892.3	20	6.4	15.7	252	111.5	780.1	1,256.2	−476.1	49.8	15.8	34.4
2000	11,242.2	8	1,083.1	21.4	8.5	15.2	257.2	81.5	1,807	1,553	254	46.4	21.4	32.2
2001	12,596.5	6.7	1,274.0	17.6	4.8	15.0	257.3	−90.04	1,547	1,457	90	45.6	22.8	31.6
2002	3,924	6.5	1,457.4	14.4	8.3	15.8	236	198.72	1,949	2,179.22	−230.11	46	23.2	30.9
2003	4,549	6	1,656.4	13.7	7.7	16.3	261	422.6	2,542.2	2,536.1	6.07	44	24.1	30.3
2004	5,278	7.2	1,991.2	20.2	8.5	16.3	258	730.2	3,777.75	3,586.18	191.57	40	28.0	32.0
2005	6,283	8	2,421.2	21.6	8.5	16.2	245.6	530.5	4,824.3	5,946.0	−1121.7	39.0	28.0	32.0
2006	22,217.	10.0	2,719.0	12.3	7.2	17.3	2.0248	−208.6	5,656.6	7,104.0	−1,448.1	36.8	27.5	35.7
2007	22,21	10.5	3,059.2	12.5	8.1	19.4	2.0308	−282	8,879.2	7,722.4	1,156.8	35.3	30.6	34.1
2008	26.03	7.8	3,262.6	6.6	14.3	20.7	2,09	21.1	11,670.5	8,229.4	3,441.1	29.3	29.2	41.5
2009	27.63	6.1			11.2		2,32	−502.2	7,833.7	8,528.0	−694.3	31.1	23.9	45.0

Sources: (1) Sudan Ministry of Finance and National Economy (1997–2007), (2) The Central Bank of Sudan Annual Report (Various Issues: 1999–2009) (3) Sudan Central Bureau of Statistics: Sudan Ministry of the Cabinet – Central Bureau of Statistics: Sudan statistical year book: Sudan statistics 1990–2008: pp. 39–43. (4) Sudan Ministry of Labour and Administration Reform – Department of Planning and Monitoring and Follow-Up. (5) Central Bureau of Statistics – Migration and Labour Force Survey 1996. (6) Central Bureau of Statistics – Department of Internal Commerce and Pricing. (7) Own calculation based on Sudan Central Bureau of Statistics Population Census Data (2010): The Fifth Sudan Population and Housing Census (2008)

Figures for 1998 from Ministry of Finance and National Economy – Annual Economic Survey 2000, Table 7-2, p. 10.

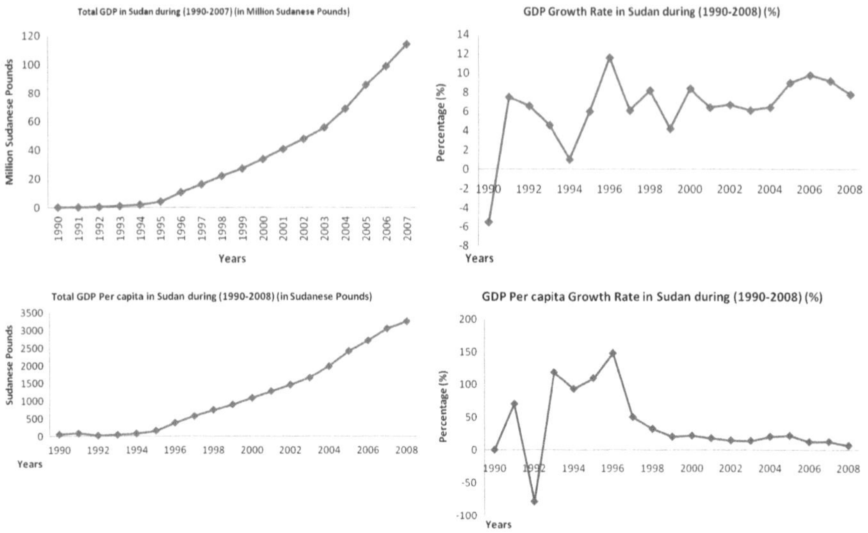

Fig. 2.1 Total and growth rates of GDP and GDP per capita in Sudan during (1990–2008) (Millions Sudanese pounds) and (%) (Source: Adapted from Sudan Central Bureau of Statistics: Sudan Ministry of the Cabinet – Central Bureau of Statistics: Sudan statistical year book: Sudan statistics 1990–2008: pp. 39–43)

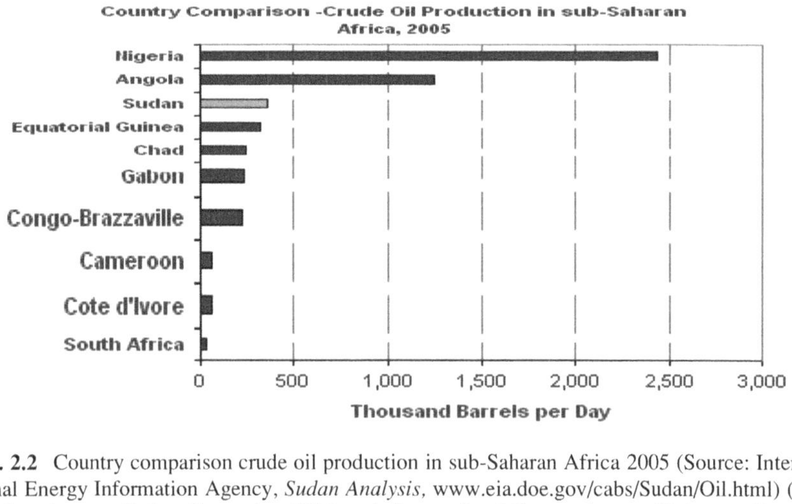

Fig. 2.2 Country comparison crude oil production in sub-Saharan Africa 2005 (Source: International Energy Information Agency, *Sudan Analysis*, www.eia.doe.gov/cabs/Sudan/Oil.html) (See Oil fact sheet on Sudan, September 2006, p. 1, International Energy Agency Estimates; International Energy Annual, IPM)

Sudan indicates that the oil industry is regulated by the Ministry of Energy and Mining, yet the Ministry of Finance and National Economy and National Petroleum Commission are also involved. Sudanese oil sector includes several foreign international oil companies with a long history of investing in oil exploration and production in Sudan. More recently, the sector reflects increasing involvement of national and foreign companies.[5] Foreign oil producing companies involved in Sudan's oil sector are primarily from Asia organised under the consortium of the Greater Nile Petroleum Operating Companies (GNPC), led by the China National Petroleum Corporation (CNPC), which owns the largest single share in the GNPC consortium (40 %); followed by Malaysia's Petronas (30 %); India's Oil and Natural Gas Corporation (25 %), and Sudanese Government's Sudapet company (5 %).[6]

2.3.2 The Role of China in the Exploration, Production and Export of Sudanese Oil

As the major player in Sudanese oil industry, China uses a combination of investment, trade, aid flows and diplomacy to maintain access to oil resources in Sudan.

For the period 2000–2007, we find that China contributed more than one third of total foreign investment across all sectors (38.67 %), and almost half of total foreign investment in the petroleum sector (47.63 %). By contrast, China's FDI contribution is marginal in the industrial (0.56 %), services (0.08 %) and agricultural (0.02 %) sectors. Therefore, Chinese investments are largely concentrated in the petroleum sector (99.90 %), as compared to industrial (0.07 %), services (0.03 %) and agricultural (0.0001 %) sectors.[7] The importance of Chinese investment in the oil sector in Sudan compared to that of other Asian countries over the period 1999–2008 is demonstrated by China's large share in oil concessions (6–95 %), total oil investment (47.3 %), upstream oil investment (43.8 %), downstream oil investment (56.9 %), oil pipe lines (47.6 %), oil refinery (50 %), petrochemicals (95 %), oil refinery and petrochemicals (51 %) and oil marketing, industry and manufacturing (12.5 %) (see Table 2.3 below).[8] Moreover, the significant Chinese

[5] Sudan National Petroleum Corporation (Sudapet) develops joint ventures with foreign companies in downstream projects. However, due to its limited technical and financial resources, the company takes a minor role in large upstream development projects.

[6] India's Oil and Natural Gas Corporation has acquired Talisman's Energy (Canada) interest for 25 %. See 'Oil Fact Sheet on Sudan' (September 2006), produced by C. Pinaud for UnderstandingSudan.org, pp. 1–2; Salih (2004), pp. 21–40.

[7] See Sudan Ministry of Investment unpublished statistics and data from the feasibility studies (2009).

[8] According to the Ministry of Energy and Mining (2008), among the Asian countries China contributes by significant share in investment and concessions in the oil sector that includes many Chinese companies involved in many blocks over the period (1999–2008). For example, we observe a significant share of the Chinese companies, namely, (China National Petroleum Company (CNPC) (40 %) of total concession of the Greater Nile Petroleum Operating Company (GNPOC); CNPC (41 %) of total concession of Petrodar Petroleum Operating Company; Sinopec (6 %) of total concession of Petrodar Petroleum Operating Company; CNPC (95 %) of total concession of China National Petroleum Company International Sudan (GNPCIS); Petroenergy (40 %) of total concession of Group of Companies and Petroenergy (35 %) of total concession of

Table 2.3 The share of China in total Asian countries concession and investment, in oil sector in Sudan (1999–2008) (%)

Items	China oil company	Share of China in total (%)
(1) Oil concession		
Greater Nile Petroleum Operating Company (GNPOC)	China National Petroleum Company (CNPC)	40
Petrodar Petroleum Operating Company (PDOC)	CNPC	41
Petrodar Petroleum Operating Company (PDOC)	SINOPEC	6
China National Petroleum Company International Sudan (GNPCIS)	CNPC	95
Group of Companies	PETROENERGY	40
Red Sea Oil Company	PETROENERGY	35
(2) Oil investment in		
(a) Up-stream oil investment	CNPC and SINOPEC	43.8
(b) Down-stream oil investment	CNPC and SINOPEC	56.9
Average total up-stream and down-stream oil investment	CNPC and SINOPEC	47.3
(c) Investment in oil pipe lines	CNPC and SINOPEC	47.6
(d) Investment in oil refinery	(CNPC)	50
(e) Investment in petrochemicals	(CNPC)	95
(f) Investment in oil refinery and petrochemicals	(CNPC)	51
(g) Investment in marketing, industry and manufacturing of oil	KANDOC PETROCHEMICAL	12.5

Source: Sudan Ministry of Energy and Mining 2008

investment in the oil sector in Sudan has spurred the trade relationship between Sudan and China, which in turn has benefited both the Chinese and Sudanese economies as demonstrated by the large volume of exports (US$ 39.241 million) and imports (US$ 11.576 million) between Sudan and China over the period 1997–2010, and the large average share of China in total Sudanese exports (69.56 %) and imports (15.67 %) over the period 2000–2010.[9] According to data

Red Sea Oil Company. Moreover, the Sudan Ministry of Energy and Mining 'Unpublished Report' (2008), indicates that out of the Asian countries' total investment (84.4 %) in Sudan, the share of China (Chinese CNPC + SINOPEC companies) is largest in total oil investment (47.3 %), upstream oil investment (43.8 %) and downstream oil investment (56.9 %). In addition to the large share of China (47.6 %: CNPC (45.2 %) and SINOPEC (2.4 %)) in total Asian countries' investment in Sudanese oil pipe lines during 1999–2008, China has a large share (CNPC: 50 %) and partnership with the Sudanese government in investment in oil refinery and in petrochemicals (CNPC: 95 %), refinery and petrochemicals (CNPC: 51 %) and in marketing, industry and manufacturing of oil (Kandoc petrochemical: 12.5 %) of the total of Asian countries' investment in the oil sector in Sudan over the period 1999–2008.

[9] See Sudan Ministry of Finance and National Economy 'Unpublished Report' (2008); Central Bank of Sudan '44th Annual Report' (2004), Appendix No. XVI, pp. 188–189; and '48th Annual Report' (2008), Appendix No. XVI-B-XVIIB, pp. 158–164.

from the Central Bank of Sudan for the period 2000–2010, China's share in Sudanese total exports to all foreign countries ranged from 44 % to over 80 %; its share in Sudanese total imports from all foreign countries ranged from 6 % to over 30 %; and its share in Sudanese petroleum exports to all foreign countries ranged from 58.87 % to over 87.7 % (see Table 2.4 below). Over the period 1999–2010, petroleum dominated Sudan's exports to China (99.4 %), while non-oil exports to China represented only a small share (0.6 %). China is therefore the largest importer of Sudan's petroleum (80.07 %), leaving Sudan's petroleum exports to other countries at only (19.93 %). Furthermore, the significant investment of China in the oil sector in Sudan motivated China to increase its aid and development assistance, loans and grants to Sudan. For instance, of the total loans and grants transmitted to Sudan over the period 1999–2009, compared to other countries, the share from China is significant and ranges from 7 % to 76 % during those years. The average shares increased from 33 % to 45 % and 58 % during the periods 2002–2007, 2004–2007 and 2005–2007 respectively (see Table 2.4 below).[10]

2.3.3 *Overview on the Oil Impact, Opportunities and Challenges for Development in Sudan*

Based on the above background, since the structure of Sudan's economy is now related to oil, this section examines the impact of oil on Sudanese economy, by explaining first the positive impact of oil and opportunities for development and then explaining the negative impact of oil and challenges of development in Sudan (see Table 2.5 below).

2.3.3.1 Oil and the Opportunities for Development in Sudan

This section explains how oil created various positive effects and opportunities for development in Sudan. These include the effect of oil in satisfying domestic demand and achievement of self-sufficiency, increasing government resources, revenues and spending, economic growth (GDP growth and composition), foreign trade (volume and structure of exports), balance of trade, balance of payment, FDI and social development in Sudan.

---

[10] Nour (2011) indicates that Chinese aid to Sudan is tied/related to trade, FDI and importance of oil to Chinese economy and that the increase in the inflow of Chinese aid and development assistance in the form of loans has caused mixed positive and negative impacts for Sudanese economy over the period 1997–2007, by providing alternative complementary sources of finance to complement the shortage of domestic capital and financing development projects, but by increasing Sudan's debts to China from 0.9 % in 1999 to 13.45 % in 2007. Despite the global economic crisis China has reaffirmed its commitment to maintain further aid and development assistance to Sudan, to maintain the economic interests of its access to oil in the country.

Table 2.4 The trend and share of China in petroleum and total exports from Sudan and in total loans and grants to Sudan (1999–2010) (%)

Share of China in total (%)	1999	2000	2001	2002	2003	2004	2005	2006	2007	2008	2009	2010[a]	(2000–2010)[b]
Petroleum export (%)		58.87	72.78	85.03	84.99	80.64	80.86	82.30	86.16	78.85	82.63	87.7	80.07
Total export (%)	0.07	44.12	59	65.74	69.31	66.89	71.04	74.87	81.95	75.02	75.77	81.42	69.56
Total loans and grants (%)	17			7	8	7	76	24	73	3.35	27.44		
Average total loans and grants (%) (1999–2007) (2007–2009)[c]	24	24	28	33	38	45	58	49	73	38	35		

Sources: (1) Adapted from Sudan Ministry of Foreign Trade and Central Bank of Sudan Annual Foreign Trade Statistical Digest various issues (1999–2010): 2006: p. 20, p. 38. 2005: p. 38, p.20. 2004: p. 20, p. 39. 2002: p. 9, p. 24. 2000: p. 9, p. 24. (2) Adapted from the Central Bank of Sudan Annual Reports (1999–2007), Ministry of International Cooperation and Ministry of Finance and National Economy

[a]January 2010–March 2010.

[b]Average (2000–2010).

[c]For calculation of the average share of China in total loans and grants (1999–2007) and (2007–2009) we use the year 2007 as a reference year because it witnessed the largest inflow of China aid and development assistance to Sudan over the period (1999–2009).

Table 2.5 The impact of oil in Sudan economy and macroeconomic indicators in Sudan (1999–2010)

	1999	2000	2001	2002	2003	2004	2005	2006	2007	2008	2009
Revenue (% of GDP)	8 %	11.5 %	10.7 %	11.9 %	16.0 %	19.7 %	21.7 %	20.0 %	19.9 %	19.3 %	13.5 %
Expenditure (% of GDP)	8.9 %	12.2 %	11.6 %	12.8 %	15.3 %	18.2 %	23.4 %	24.3 %	23.0 %	17.8 %	16.3 %
Fiscal deficit (% of GDP)	−0.9 %	−0.7 %	−0.9 %	−0.8 %	1 %	1.5 %	−1.8 %	−4.3 %	−3.1 %	1.6 %	−2.8 %
Oil exports (% of GDP)	1 %	9.5 %	10 %	10 %	12 %	14 %	15 %	14 %	18 %	7.3 %	7.2 %
Total oil export	689 %	1,350,757	1,376,666	1,510,857	2,047,705	3,100.5	4,187.4	5,087.2	8,418.5	11,094.1	7,131.20
Total non oil export	1,164	455.9	322	438.3	4,949.5	677.3	636.9	569.4	460.7	576.4	702.5
Total exports	1,853	1,806.7	1,698.7	1,949.1	2,542.2	3,777.8	4,824.3	5,656.6	8,879.2	11,670.5	7,833.70
Total imports	1,256.2	1,553	1,457	2,179.22	2,536.1	3,586.18	5,946.0	7,104.0	7,722.4	8,229.4	8,528.0
Trade deficit	−476.1	254	90	−230.11	6.07	191.57	−1121.7	−1,448.1	1,156.8	3,441.1	−694.3
Balance of payment deficit	111.5	81.5	−90.04	198.72	422.6	730.2	530.5	−208.6	−282	21.1	−502.2
Share of oil exports (%)	37 %	74.8 %	81 %	77.5 %	80.6 %	82.1 %	87 %			95.1 %	91.0 %
Share of non oil export (%)	63 %	25.2 %	19 %	22.5 %	19.4 %	17.9 %	13 %			4.9 %	9.0 %
Total revenue	109,015	334.0	366.3	474.9	715.0	1,029.0	1,218.4	15,075	18,462.4	24,707.9	20,045.6
Total oil revenues	15.7	143.8	149.7	200.6	399.0	502.9	608.6	7557	10,047.6	15,996.7	9,596.2
Total non oil revenues	108,999.3	190.2	216.6	274.3	316.0	526.1	609.8	7,518	8,414.8	8,711.20	10,449.4
Share of oil in total revenues (%)	0.01 %	43 %	41 %	42.3 %	40 %	49 %	50 %	50 %	54.4 %	64.7 %	47.9 %
Share of non-oil in total revenues (%)	99.09	57 %	59 %	57.7 %	60 %	51 %	50 %	50 %	45.6 %	35.3 %	52.1 %
Share of oil in GDP (%)	1 %	6.8 %	7.9 %	9.1 %	9.6 %	14.6 %	15.1 %	15.1 %	20.1 %	18.2 %	
Share of oil revenues in GDP (%)		4.6 %	4.3 %	5.1 %	8.8 %	9.5 %	9.9 %				
Share of current spending in total spending (%)		85 %	81.8 %	73 %	74.9 %	71.9 %	78.5 %	80.61 %	82.99 %	87.45 %	85.29 %
Share of development spending in total spending (%)		15 %	18.2 %	27 %	25.1 %	28.1 %	21.5 %	19.39 %	17.01 %	12.55 %	14.71 %
Share of current spending in GDP (%)		9.70 %	9.90 %	10.60 %	12.00 %	15.10 %	18.10 %				
Share of development spending in GDP (%)		1.7 %	2.2 %	2.6 %	4 %	5.8 %	4.5 %	6.4 %	7.2 %	7.4 %	
Share of total spending in GDP (%)		11.4 %	12.1 %	13.2 %	16 %	20.9 %	22.6 %				
Net FDI		392	574	713.2	1,349.2	1,511.1	2,304	35,341	24,256		

Source: Adapted from the Central Bank of Sudan and Ministry of Finance and National Economy Annual Reports (various issues)

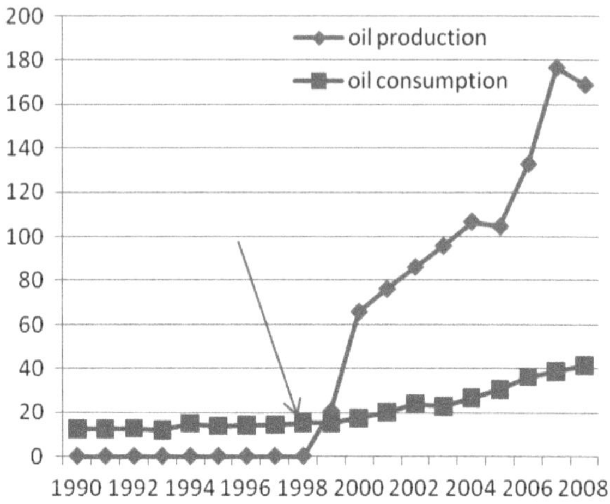

Fig. 2.3 Sudan's oil production and consumption 1990–2008 (Source: Adapted from Sudan Ministry of Energy and Mining Statistics)

Beginning with the impact of oil production, we find that the local production of oil created important positive effects and opportunities by enabling Sudan to gain self-sufficiency in oil to satisfy domestic demand. Moreover, the local production of oil enables the government to stop the high costs previously required for the importation of oil to satisfy local demand and to mobilise the accumulated saving from the surplus amount of capital to be allocated for funding other domestic needs.[11] Furthermore, the local production and exportation of oil implies that Sudan shifted from an oil importing economy into an oil exporting economy (see Fig. 2.3 above). For instance, in 2001 more than half of Sudanese crude oil was exported (51 %) while the rest was used to satisfy local consumption (49 %).[12]

Moreover, the positive impact of oil on government financial resources is observed from the increasing share of Sudanese government in oil revenues from partnerships with foreign oil producing companies in Sudan. For instance, the rise in oil production has led to a rapid continuous increase in the share of Sudanese government in total oil production and revenues from 23 % in 2000 to 75 % in 2005. Moreover, oil revenues have enabled Sudanese government to cover half of the total costs spent in the establishment of the Khartoum refinery.[13] The government's share in total oil revenue is influenced by the interaction between the output effect and the price effect, notably the government's share in total oil revenue increases in line with the increase of oil production and increase of oil prices in the international market and vice-versa. For instance, during the period 1997–2008 the increase in oil production led to an increase in the government's share from 25 % in 1999 to 75 %

[11] See Salih (2004), p. 166.

[12] About 40 % of oil is shipped to China (Salih 2004, p. 94). Sudan's crude oil exports have increased sharply since the completion of a major oil export pipeline in 1999. In 2004, oil imports were reported at 0 bbl/day. Sudanese domestic oil consumption averaged 82,000 bbl/d in 2005, which was a 15 % increase over the 70,000 bbl/d consumed in 2004. Return from oil exports to Sudan was US$ 500 million and US$ 600 million in 2000 and 2001 respectively. Salih (2004), p. 91.

[13] See Ministry of Finance and National Economy (2006) 'The Performance of Sudan Economy 2000–2005' (April, 2006), pp. 16–18. See also Salih (2004), pp. 182, 93.

in 2005, but the share declined to 56.7 % in 2009 due to the negative effect of the great decline in oil prices due to the decline in demand in the international market linked to the global economic crisis in 2009.

Furthermore, oil has created a positive impact on foreign trade as perceived from the volume and structure of exports, balance of trade and balance of payment. We find that oil has a positive impact on the balance of trade and balance of payments, because after the export of oil in 2001 the chronic deficit reported in the balance of trade and balance of payment turned into a surplus for the first time since independence. But this surplus in the balance of payments could not be sustained, and immediately turned into a deficit, most probably due to the increase in imports of capital goods. The same applied for the balance of trade, since oil exports represented about 95 % of total exports, it led to a positive impact in the balance of trade over the period 2000–2009 as the chronic deficit in the balance of trade turned into surplus in 2000, 2003, 2004, 2007 and 2008. While total exports grew dramatically from 7 % of GDP in 1996 to 14 % in 2006, imports remained higher at 16 % of GDP and led to a trade deficit averaging 2 % of GDP since 1999. The oil export boom raised the value of total exports from US$ 620 million in 1996 to US$ 4,522 million (1996 prices) in 2006, representing a more than 700 % increase over the decade. The large import demand of the country, the huge transport costs and other expenses related to oil operation, and the weak performance of the non-oil exports contributed to the current account deficit. The size of the current account and balance of payments deficit during 1999–2006 were however smaller compared to pre-oil exportation levels.[14]

Moreover, oil has led to a significant positive impact on GDP as perceived from the impact of oil in the structure of Sudanese economy and macro-economic indicators as measured by the share of oil in GDP, its growth rate and its composition. For instance, we observe the increasing impact of oil as measured by the rapid and continuous increase in the contribution of oil sector in GDP from 1 % in 1999 to 10 % in 2004. Furthermore, oil has led to positive impacts in real GDP growth, for instance, the average rate of growth of GDP increased from 6.2 % to 6.8 %, 8 %, 10 %, 9 % and 9.6 % over the periods 1997–1999, 2000–2009, 2005, 2006, 2005–2007 and 2006–2008 respectively, putting Sudan among the fastest growing economies in the region; Fig. 2.4 below shows that Sudan is a top growth performer in the region, with oil playing a pivotal role. Oil has led to structural change in the composition of GDP, as the dividends from oil exportation have caused major transformations and structural changes in the economy. The structure of Sudanese economy has shifted over time from being predominantly reliant on agriculture for growth and exports to its current reliance on the oil sector (see Figs. 2.5 and 2.6 below).[15]

[14] See WB-DTIS (2008), p. 7.

[15] Notwithstanding these structural shifts, agriculture remains Sudan's main driver of employment, especially outside the country's top urban areas of Khartoum and Port Sudan. Near to 35 % of Sudan's GDP comes from agricultural production, which employs 80 % of the workforce. See WB-DTIS (2008), pp. 1–3. See also 'Oil Fact Sheet on Sudan' (September 2006), p. 1.

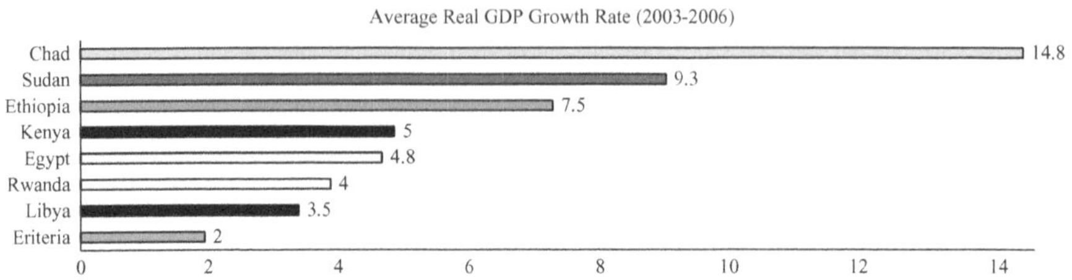

Fig. 2.4 Average real GDP growth rate in Sudan compared to other African countries during the period (2003–2006) (Source: The World Development Indicators (WDI)/IMF/World Bank Staff Estimate 2008)

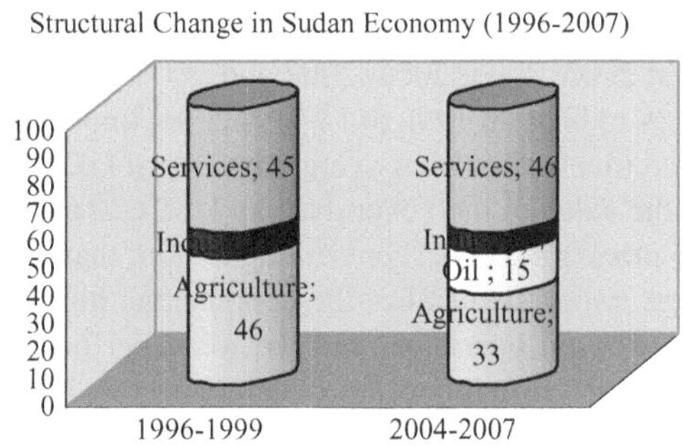

Fig. 2.5 Structural change in Sudan economy (1996–2007) (Source: The World Development Indicators (WDI)/IMF/World Bank Staff Estimate 2008)

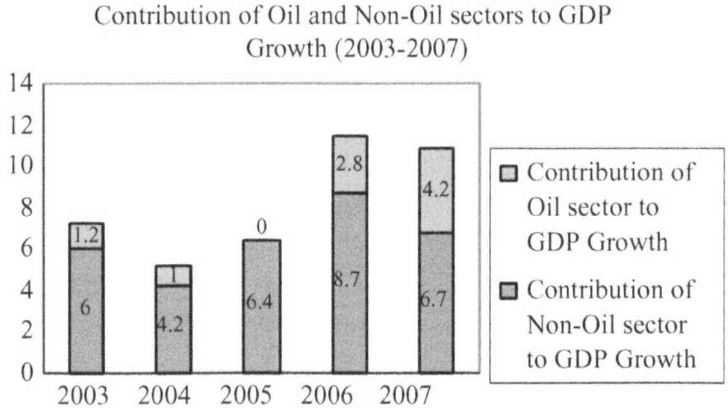

Fig. 2.6 Contribution of oil and non-oil sectors to GDP growth in Sudan (2003–2007) (Source: The World Development Indicators (WDI)/IMF/World Bank Staff Estimate 2008)

Unlike other typical oil economies, in Sudan the impact of oil on non-oil sectors (agriculture, industry and services) remains very limited. This is noticeable from the composition or the sectoral share in GDP, as due to increasing share of oil in GDP, the contribution of industrial sectors (including oil, quarrying and mining, manufacturing, electricity and water and construction) to GDP increased to 32.3 %

in 2008 compared to 21.7 % over the period 1999–2004, while the contribution of both agricultural and service sectors declined, which clearly indicates the limited impact of oil in the agricultural and service sectors over the short run. However, over the long run these sectors may benefit from the impact of oil sector development. Currently, the impact of oil is limited to only three branches of the services sector. For example, the oil sector has led recent growth, both in terms of direct value-added to the economy as well as the associated investment boom and boost to services such as transport and construction. The emergence of the oil sector adds directly to GDP and has induced growth in certain service sectors. The construction sector has grown by about 10 % per annum since 1999 and has been the fastest growing sector in recent years, even surpassing the growth in the oil industry. Trade, restaurants and hotels have also flourished, mainly in the country's capital, and generated about one fifth of non-oil domestic product during 1996–2006.

Furthermore, the positive impact of oil in the government's public budget is perceived from the contribution of oil revenues in public finances and budget as it leads to a significant increase in government revenues and spending over the period 1999–2009. For instance, we observe the large and increasing share of oil revenues in total revenues that grew from 43 % in 2000 to 50 % and 66 % in 2006 and 2008 respectively. Despite continuous government efforts to increase the share of non-oil revenues in total revenues, the share of oil revenues in total revenues remains significant at about 50 % over the period 1999–2004, but this share declined significantly to 34 % in 2009, most probably due to the impact of the global economic crisis. Therefore, this implies the urgent need to avoid the heavy dependence on oil revenues. On the other hand the impact of oil on government expenditure is obvious from the increasing share of oil in public spending. Moreover, development spending also increased as its share in public expenditures increased from 21 % over the period 1996–1999 to 24 % over the period 2000–2004. But despite the increase in development expenditure from public expenditure from 9 % in 1999 to 31 % in 2004, its share declined and flattened out to 24 % within total public spending over the period 2006–2009.[16]

Oil created a positive impact on FDI to Sudan and motivated the increase in FDI inflow to Sudan. The implementation of economic reform policies, liberalisation and privatisation in the late 1990s, together with the exploitation of oil in 1999, and the Investment Encouragement Act of 2003, all encouraged high and increasing inflow of FDI to Sudan (see Table 2.5 above). In particular, the exploitation of oil in 1999 encouraged the inflow of FDI. For instance, according to the Arab Human Development Report (2003) the estimated net FDI flow to Sudan increased from US$ 392 million in 2000 to US$ 574 million in 2001.[17] In addition, we find that the volume of investment increased over the period 1996–2004 from US$ 251.3 to US$ 1,381

[16] See 'Sudan Factsheet Human Rights and Oil Workshop – January 31' (2003), p. 2.

[17] See the UNCTAD 'World Investment Report' (2002). See also 'Arab Human Development Report' (2003), Table 5.1, p. 102.

million, which implies that the rate of growth is near to about 500 %.[18] Moreover, in 2006, the levels of FDI in Sudan were amongst Africa's highest with over US$ 3.5 billion. From annual averages of US$ 100–200 million prior to 2000, in 2006 net FDI and portfolio inflows were US$ 3.5 billion, though tailing off to US$ 3 billion in 2007.[19,20] For 2009 however, FDI inflow decreased due to the global shock resulting in lower global oil prices, stagnating domestic oil production and related reduction in government spending. Due to increasing investment in oil, the sectoral distribution implies that the large share of FDI was concentrated on the energy and mining sector (74.7 %, 73 %), followed by industry (9.1 %, 10 %), agriculture (8.6 %, 2 %) and services sector (7.6 %, 15 %).[21] This implies that oil enables Sudan to emerge as one of the highest recipients of FDI in the African and Arab regions.

Moreover, concerning the impact of oil in enhancing capacity building, we are aware of the fact that it may be useful to depart from the analysis of the general standardised approach of examining only the macro-economic impact of oil, and to use a more indepth analysis to examine the effect of production and export of oil (natural resource-based exports) on capacity-building including education, training, science and technology (S&T) and R&D infrastructure and the growth and development trajectory of Sudanese economy. But our attempt to briefly examine the impact of oil on capacity building is constrained by the lack of reliable data at the macro and micro levels and also by the fact that Sudan is a relatively new exporter. We find that most probably the impact of oil in capacity building including education, training, S&T and R&D infrastructure might still be very limited as the country is a relatively new exporter since 1999. Furthermore, the impact on oil in the development expenditures implies that it is not at all clear and is somewhat problematic to distinguish the share and growth of spending on education, training and R&D that were mainly attributed to production and export of oil. It is clear that at the macro level the share of spending on education and R&D as a percentage of GDP most probably remained almost the same without reporting a significant change in the pre- and post-oil periods.[22]

Therefore, our findings in this section prove the first part of our hypothesis that oil created a positive impact on Sudanese economy.

---

[18] Despite the huge export earnings from oil, the current account balance has been in deficit at 8 % of GDP on average during 1999–2005. This is partly induced by increased imports of manufactured, machinery and transport equipments and other commodities. The impact of these expenses in the overall balance of payments is subdued by the influx of FDI. In 2004 and 2005, the influx of FDI led to overall surplus in the balance of payments.

[19] See WB-DTIS (2008), p. 4.

[20] See the IMF 'First Review of Performance Under the 2007–2008 Staff-Monitored Program' (June 2008), p. 2, 6.

[21] See the Sudan Ministry of Finance and National Economy 'Sudan Economy in Figures' (2002), Ministry of Finance and National Economy, Macroeconomic Policies and Programme Directorate MEPPD, First Edition (2002), p. 27.

[22] For instance, we find that the significant Chinese investment in the oil sector in Sudan has motivated China to increase very limited technical support for capacity building in Sudan, though the available information implies that direct allocation of Chinese aid to training and education sector is very limited.

2.3.3.2 Oil and the Challenges of Development in Sudan

After explaining the positive impact of oil and the opportunities for development it is useful to elucidate also the negative impact of oil and the challenges for development in Sudan. These include the high uncertainty, volatility and risk of dependence on highly fluctuating oil prices in the international market, unsustained oil revenues; the lack of diversification; Dutch Disease and potential future Sudan-Southern Sudan conflict.

The first challenge related to oil is that the real economic activity is currently high, but the lack of economic diversification raises concerns over longer term sources of growth and sustained development, therefore diversification towards non-oil exports is imperative for long run sustainable development strategies. Sudan has experienced a revival in its exports, but this is largely due to the export of oil. Since 1999 the exploitation of oil resources has led to large increases in national wealth, but it has also complicated macro-economic management with recent pressures toward internal and external imbalances, as well as a heightened concern for balanced growth in the non-oil sectors, which are important for sustainable growth in Sudan. On the external side, the current account has deteriorated since the oil boom and the real exchange rate has appreciated significantly. Therefore, the major challenge created by oil is the need for diversification, although oil has driven the recent surge in real economic growth. To sustain growth and provide broader income opportunities, Sudan will need to pursue a strategy of diversifying its sources of growth, including enhancing its non-oil exports (e.g. traditional agricultural exports that have provided export earnings over the past half century).[23]

Another challenge is that oil earnings enter the economy predominantly through public finance channels, yielding significant volatility for fiscal policy. The expansion in public sector expenditures has crowded out private credit and stressed the financial sector. Oil export earnings now support the majority of public finance (55 % in 2007) and expose fiscal policy to the volatilities of domestic production and international price fluctuations. Significant oil revenue volatility and shortfalls were observed in late 2006 and early 2007 resulting in the highest fiscal deficits since the macro stabilisation of the early 1990s, which accounted for 4.3 % and 3.1 % of GDP in 2006 and 2007 respectively (see Fig. 2.7 below). The volatility in revenue has greatly complicated public expenditure management.[24] Moreover, there is a considerable decline in revenue from 24,707.9 million Sudanese pounds in 2008 to 20,045.6 million Sudanese pounds in 2009, at a rate of 18.9 %, attributed to the great decline in the share of oil in total revenues that declined by about 40.0 %.[25]

One important challenge created by oil is its weak effect in improving social development indicators. For instance, despite the increase in development expenditure from public expenditure, rising from 9 % in 1999 to 31 % in 2004, its share

---

[23] See WB-DTIS (2008), p. viii.

[24] See WB-DTIS (2008), pp. 4–6.

[25] See the Central Bank of Sudan ‘Annual Report’ (2009), p. 84.

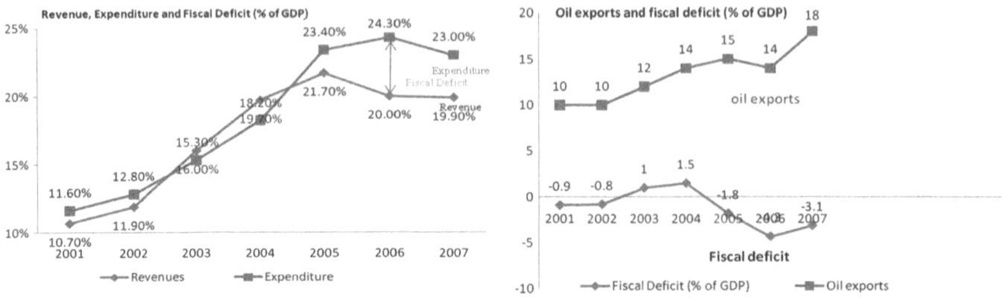

Fig. 2.7 Sudan's fiscal position has deteriorated, while oil earnings have grown, oil exports and fiscal deficit (% of GDP) (Source: The World Development Indicators (WDI)/IMF/World Bank Staff Estimate (2008): Figure 1-3, p. 5)

then declined and flattened out at 24 % from the total public spending over the period 2006–2009. The share of development spending from oil revenues declined from 58 % in 2006 to 34 % in 2008, while the share of current spending from oil revenues increased from 42 % in 2006 to 66 % in 2008.[26] This clearly indicates the bias and deficiency in the use of oil resources on current spending instead of development spending. Despite the high oil revenues and impressive real growth, so far they are not fully utilised and do not prioritise improvement of social development indicators. Consequently, emerging vulnerabilities can be seen from poverty, regional inequalities and a low and deteriorating ranking in the Human Development Index from 147 to 150 and to 154 out of 177 world countries in UNDP-HDI in 2007, 2009 and 2010 respectively.[27,28] The low human development indicators implies that Sudan continued to fall below the Arab states and world average level over the past three decades, for instance, the trend of human development index over the period (1980–2010) implies that Sudan's level in 2010 fell below the Arab states and world average level not only in 2010 but also in 1980 (see Fig. 2.8 below). In addition to high poverty rates, according to Sudan Central Bureau of Statistics Household Survey Report (2009), about 45 % in northern Sudan are estimated to be living below the poverty line of less than US$ 1 a day. Moreover, according to UNDP (2010a), while progress has been made towards several of the Millennium Development Goals (MDG), such as in the area of education, infant and child mortality, access to water and sanitation, Sudan's performance against the MDG indicators demonstrates large inequalities with respect to gender, rural–urban residence, and at the regional and sub-regional

[26] See Sudan Ministry of Finance and National Economy (2002) 'Sudan Economy in Figures', Ministry of Finance and National Economy, Macroeconomic Policies and Programme Directorate MEPPD, First Edition (2002), p. 27.

[27] See WB-DTIS (2008), p. 6.

[28] See UNDP 'Human Development Reports' (2007; 2009; 2010a). See: http://en.wikipedia.org/wiki/Sudan, accessed June 1, 2010.

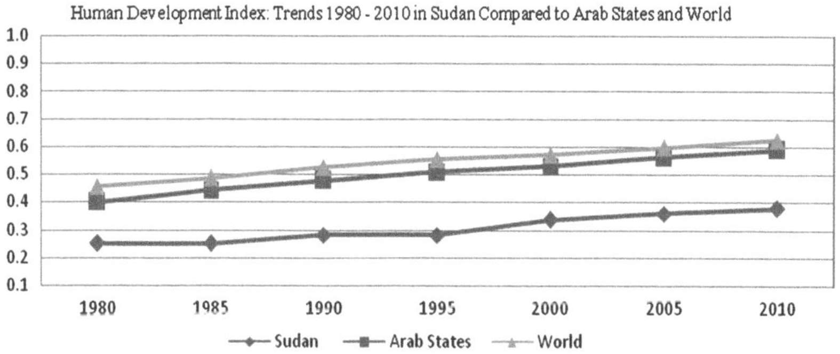

Fig. 2.8 Human development index: trends in (1980–2010) in Sudan compared to Arab States and World (Source: UNDP Sudan Country profile of Human Development Indicators 2010) (See UNDP (2010b): http://hdrstats.undp.org/en/countries/profiles/SDN.html, Accessed on December 22, 2010)

level.[29] The significant regional disparities between regions contributed to growing inequalities and unbalanced development in Sudan (see Tables 2.6 and 2.7 below).

Another challenge related to oil is the potential for Dutch Disease. For instance, the exploitation of oil resources has led the increase in national wealth, but domestic absorption of these large inflows significantly complicates macroeconomic management.[30] There is increasing debate on the potential incidence of the Dutch Disease phenomenon in Sudan's economy. On the one hand, the views in support of the potential incidence of Dutch Disease are based on the argument that

[29] The Millennium Declaration and adoption of the UN MDG in September 2000 implies commitment towards achievement of the eight MDG by 2015. The MDG are: (1) Eradicate extreme poverty and hunger: Halve, between 1990 and 2015, the proportion of people whose income is less than one dollar a day, and halve, between 1990 and 2015, the proportion of people who suffer from hunger. (2) Achieve universal primary education: Ensure that, by 2015, children everywhere, boys and girls alike, will be able to complete a full course of primary schooling. (3) Promote gender equality and empower women: Eliminate gender disparity in primary and secondary education, preferably by 2005, and in all levels of education no later than 2015. (4) Reduce child mortality: Reduce by two thirds, between 1990 and 2015, the under-five mortality rate. (5) Improve maternal health: Reduce by three quarters, between 1990 and 2015, the maternal mortality ratio. (6) Combat HIV/AIDS, malaria and other diseases. (7) Ensure environmental sustainability and (8) Develop a global partnership for development. See UND-HDR 'UN MDGs in Sudan': http://www.sd.undp.org/mdg_sudan.htm, accessed June 1, 2010.

[30] Dutch Disease refers to the experience of the Netherlands in the 1960s, when the economic boom following natural gas discoveries led to a decline in manufacturing and real exchange rate appreciation. In his summary of the literature, Corden defines it as a phenomenon where a boom in one export sector, typically a windfall discovery of a new natural resource, draws factors of production from other sectors of the economy and boosts demand for non-tradeables relative to tradeables, which in turn appreciates the real exchange rate. Traditional exports collapse, due both to the internal reallocation of resources and the real exchange rate appreciation. Corden, W. M., 'Booming Sector and Dutch Disease Economics: Survey and Consolidation', Oxford Economic Papers 36 (November 1984), pp. 360–362.

Table 2.6 The status of MDGs in Northern Sudan in 2008

MDGs/indicators	Indicators	Status in 2004[a]	Current level[b]	Reference year	2015 target
MDG 1 Eradicate extreme poverty and hunger	Estimated poverty incidence (% of total population)	50 %[c]	46.5 %	2009	45 %
	Prevalence of child malnutrition (underweight for age; % under 5)	35 %[c]	31.8 %	2006	16 %
	Prevalence of acute child malnutrition (underweight for weight; % under 5)	16 %[c]			8 %
MDG 2 Achieve universal primary education	Gross primary enrolment ratio	62 %	71.1 %	2009	100 %
	Percentage of cohort completing primary school	21 %			100 %
	Adult literacy rate	65.1 %[b]	77.5 %	2009	25 %[(c)]
MDG 3 Promote gender equality and empower women	Ratio girls to boys in primary education	88 %	53.9–46.1 %	2007	100 %
	Women's literacy rate	62 %	86 %	2009	–
	Percentage of women in National Assembly/Council of States	19 %	25 %	2010	–
MDG 4 Reduce child mortality	Under-5 mortality rate (per 1,000)	105[c]	102	2008	35
	Infant mortality rate (per 1,000 live births)	70[c]	71	2006	–
	One-year-olds immunized against measles	78 %	85 %	2009	–
MDG 5 Improve maternal health	Maternal mortality ratio (per 100,000 live births)	638[c]	534	2006	127
	Birth attended by skilled health staff	57 %[c]	57 %	2006	90 %
MDG 6 Combat HIV AIDS, malaria and other diseases	Contraceptive prevalence (% of women ages 15–49)	7 %	7.6 %	2006	–
	HIV prevalence (% adults ages 15–49)	1.6 %[c]	0.5–1.24[d]	2009	–
	Incidence of TB (per 100,000 per year)	90	120		–
	Children under 5 with fever treated with anti-malarials (%)	54.2 %[c]			–
MDG 7 Integrate the principles of sustainable development into country policies and programmes;	Access to improved drinking water source (% of population)	58.7 %[c]	65 %	2010	85 %
	Access to improved sanitation (% of population)	39.9 %[c]	42 %	2009	67 %

reverse loss of environmental resources						
MDG 8 Develop a global partnership for development	In cooperation with the private sector, make available the benefit of new technologies, especially information and communications	Telephone line per 100 population (% of population)	2 %	(2005)	0.9 %	2009
		Cellular subscribers per 100 population (% of population)	9 %	(2005)	28 %	2009
Internet users per 100 population (% of population)	8.2 % (2009)	10.4 %	2010			

[a]Sudan Millennium Development Goals. Interim Unified Report, 2004 prepared by the UN Resident Coordinator's Support Office, Khartoum, Sudan: http://www.sd.undp.org/mdg_fact.htm, accessed in June 1, 2010.

[b]SPHS-2010, NBHS-2009, SHHS-2006 and administrative data from concerned institutions cited in pp. Sudan MDGs Progress Report 2010 (2011), Sudan National Population council, Khartoum, Sudan, 9–10.

[c]Sudan Health and Household Survey 2006.

[d]0.5 for males and 1.24 for females.

According to United Nations Development Group (2003) the proportion of population below $1 per day is the percentage of the population living on less than $1.08 a day at 1993 international prices. The poverty headcount ratio is the proportion of the national population whose incomes are below the official threshold (or thresholds) set by the national Government. Poverty gap ratio is the mean distance separating the population from the poverty line (with the non-poor being given a distance of zero), expressed as a percentage of the poverty line. Proportion of the population below the minimum level of dietary energy consumption is the percentage of the population whose food intake falls below the minimum level of dietary energy requirements. This is also referred to as the prevalence of under-nourishment, which is the percentage of the population that is undernourished. Prevalence of (moderately or severely) underweight children is the percentage of children under five years old whose weight for age is less than minus two standard deviations from the median for the international reference population ages 0–59 months. Net primary enrolment ratio is the ratio of the number of children of official school age (as defined by the national education system) who are enrolled in primary school to the total population of children of official school age. Primary completion rate is the ratio of the total

(continued)

Table 2.6 (continued)

MDGs/indicators	Indicators	Status in 2004[a]	Current level[b]	Reference year	2015 target

number of students successfully completing (or graduating from) the last year of primary school in a given year to the total number of children of official graduation age in the population. Literacy rate of 15–24 year-olds, or the youth literacy rate, is the percentage of the population 15–24 years old who can both read and write with understanding a short simple statement on everyday life. Ratio of girls to boys in primary, secondary and tertiary education is the ratio of the number of female students enrolled at primary, secondary and tertiary levels in public and private schools to the number of male students. The ratio of literate women to men, 15–24 years old (literacy gender parity index) is the ratio of the female literacy rate to the male literacy rate for the age group 15–24. The proportion of seats held by women in national parliaments is the number of seats held by women expressed as a percentage of all occupied seats. The under-five mortality rate is the probability (expressed as a rate per 1,000 live births) of a child born in a specified year dying before reaching the age of five if subject to current age-specific mortality rates. The infant mortality rate is typically defined as the number of infants dying before reaching the age of one year per 1,000 live births in a given year. The proportion of 1-year-old children immunized against measles is the percentage of children under one year of age who have received at least one dose of measles vaccine. The maternal mortality ratio is the number of women who die from any cause related to or aggravated by pregnancy or its management (excluding accidental or incidental causes) during pregnancy and childbirth or within 42 days of termination of pregnancy, irrespective of the duration and site of the pregnancy, per 100,000 live births. The proportion of births attended by skilled health personnel is the percentage of deliveries attended by personnel trained to give the necessary supervision, care and advice to women during pregnancy, labour and the post-partum period; to conduct deliveries on their own; and to care for newborns. Prevalence of malaria is the number of cases of malaria per 100,000 people. Tuberculosis prevalence is the number of cases of tuberculosis per 100,000 people. HIV prevalence among 15–49 year-old adult is the percentage of adult ages 15–49 whose blood samples test positive for HIV. The proportion of the population with sustainable access to an improved water source, urban and rural, is the percentage of the population who use any of the following types of water supply for drinking: piped water, public tap, borehole or pump, protected well, protected spring or rainwater. Proportion of the urban and rural population with access to improved sanitation refers to the percentage of the population with access to facilities that hygienically separate human excreta from human, animal and insect contact. (See United Nations Development Group 2003.

Table 2.7 Regional disparity in demographic and economic structure and achievements in MDGs in Northern Sudan (2005–2009)

Region			Northern	Khartoum	Central	Kordufan	Darfur	Eastern	Total
Demographic and economic structure (2005–2008)									
Population[a]	2008	Total	1,819	5,274	7,423	4,327	7,516	4,534	30,893
		Share (%)	5 %	13 %	19 %	11 %	19 %	12 %	100 %
Revenues[b]	2005	Total	14,853	15,678	19,267	9546	10,628	25,382	95,354
		Share (%)	16 %	16 %	20 %	10 %	11 %	27 %	100 %
Actual per capita dederal allocation[b]	2005	Total	9,068	8,497	4,872	3,765	2,732	2,553	5,248
Urbanization[b]	2005	Total	27	88	29	29	20	43	39
MDGs[c] (2009)									
Poverty gap ratio	MDG 1.2	Total	9.4	6.4	13.8	23.1	24.6	17.7	16.2
Net enrolment rate in primary education	MDG 2.1	Total	83	85	67	60	62	57	67
Literacy rate of 15–24 years-olds	MDG 2.3.1	Total	88	94	77	69	74	63	77
Literacy rate of 15–24 years-olds	MDG 2.3.2	Men	91	96	84	79	85	68	84
Literacy rate of 15–24 years-olds	MDG 2.3.3	Women	86	92	70	61	64	57	71
Share of women in wage employment in the non-agricultural sector	MDG 3.2	Total	15	19	13	19	22	12	17
Employment ratio to population 15 years old and above	MDG 1.5	Total	35.9	37	41.3	48.5	45.3	39	41.4
Proportion of employed population below poverty line	MDG 1.6	Total	30.1	21	41.5	55.5	58.8	39.2	42.5
Proportion of own account and contributing family workers to total employed	MDG 1.7	Total	39.2	25.3	36	39.8	50	46.7	45

[a] Adapted from Sudan Central Bureau of Statistics Population Census Data (2010): The Fifth Sudan Population and Housing Census (2008).
[b] Elbadawi and Suleiman (2008: 107).
[c] The Sudan Central Bureau of Statistics (2011: 12).

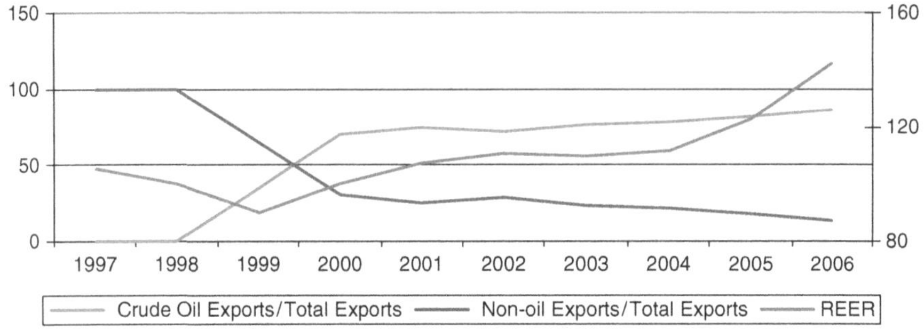

Fig. 2.9 The trend of the share of oil exports and non oil exports in total exports and real effective exchange rate in Sudan (1997–2006) (Source: Elbadawi and Kaltani 2007)

the appreciation of the nominal effective exchange rate and the sustained increases in the general price levels led to the appreciation of real effective exchange rate in recent years. This argument indicates that the inflows through higher levels of government spending put additional pressures on the prices of non-traded goods. Prices of housing, water and electricity grew almost twice as fast as the prices of tradable goods, specifically food, clothing and consumer goods. The real effective exchange rate appreciated by 40 % in 2005–2006, which added to the more fundamental structural rigidities and supply-side constraints already faced by non-oil exporters. This argument indicates that some signs of Dutch Disease are present, though it is difficult to assess the extent of these characteristics, as the country is a relatively new exporter.[31] On the other hand, the views in suspecting of the incidence of Dutch Disease are based on the argument that the agriculture and services sector continues to dominate the economy even after the increasing share of oil in GDP over the period 1990–2009. Moreover, the rise in the share of industry in GDP is mainly attributed to the rise of the share of oil in GDP, while the share of manufacturing in GDP over the period 1999–2006 remained stagnant and the growth rate of manufacturing remained between 1 % and 3 %. This argument implies that it may be too early to confirm any Dutch Disease in Sudanese economy (Fig. 2.9, 2.10, 2.11, and 2.12).[32]

A further challenge related to the dependence on oil is the uncertainty in economic growth as measured by long run GDP and GDP per capita growth rates. This implies that Sudan must implement a strategy to avoid the negative consequences of declining growth rates in GDP and GDP per capita and uncertainty related to a drop in oil reserves by using its oil production. According to UNDP (2010), prior to the global financial crisis, Sudanese economy had been one of the

[31] See WB-DTIS (2008) p. 3. See also Elbadawi and Kaltani (2007).

[32] See Bedawi, W. F. (2007).

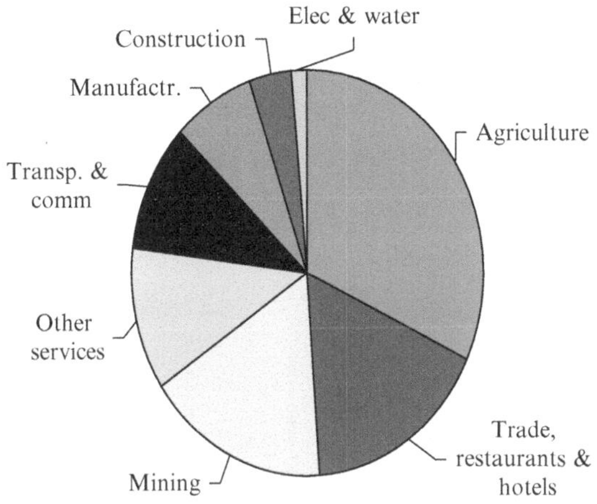

Fig. 2.10 Where is the Dutch Disease? Agriculture and services continue to dominate the economy 2006 GDP by sectors at factor cost (Source: Bedawi 2007)

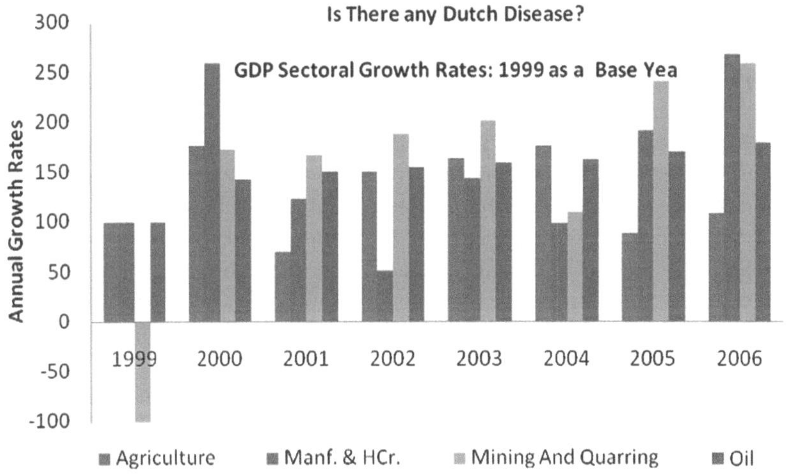

Fig. 2.11 May be too early to say is there any Dutch Disease? Measured by the GDP sectoral growth rate over the period (1999–2006) (Source: Bedawi 2007)

fastest growing in the world, despite United States (US) sanctions. However, the global financial crisis and related shock in 2008 and 2009 resulted in low global oil prices, stagnating domestic oil production and caused a reduction in the GDP

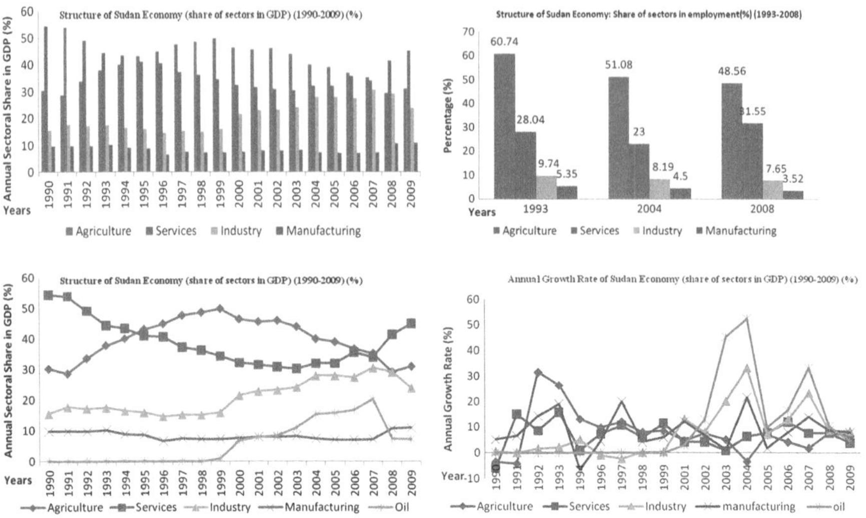

Fig. 2.12 Structure of Sudan economy: share of sectors in GDP, share of sectors in employment and annual sectoral growth rate (share of sectors in GDP) (1990–2009) (%) (Sources: Adapted from the Central Bank of Sudan and Ministry of Finance and National Economy Annual Reports (various issues))

growth rate, dropping from 10.5 % in 2007 to 7.8 % and 5 % in 2008 and 2009 respectively (see Table 2.2 and Fig. 2.1 above).[33]

Another oil related challenge is that oil revenues create other internal problems by increasing internal tensions or conflict related to the desire to maintain control over oil resources and failure to achieve an equitable distribution of oil revenues.[34] The Comprehensive Peace Agreement (2005) states that oil revenues should be shared 50:50.[35] From a political perspective in the short run with the official secession of the Southern Sudan, there is increasing tension and potential conflict

[33] A recent IMF report ranked Sudan as one of the most vulnerable low-income countries in the global financial crisis due to its high vulnerability to trade, aid and remittances shocks. See IMF 'Report on the Implications of the Global Financial Crisis for Low-Income Countries' (March 2009), pp. 48, 50: http://www.imf.org/external/pubs/ft/books/2009/globalfin/globalfin.pdf, Accessed 10 October 2011.

[34] For instance, in the past the exploration and production of Sudan's oil has been a highly controversial issue and is affected by the continuous conflict which involves the war and conflict over controlling oil resources. "Oil has always been an issue in Sudanese conflict. For instance, the organized non-government political activity resisting oil extraction: On 30 August 1999, Sudan's pipeline with a capacity for 100,000 barrels/day filled the first tanker at the supertanker port on the Red Sea. Not one month later, on 20 September, anti-government forces exploded a portion of the pipeline outside the town of Atbara. Moreover, due to conflict, oil exploration has been mostly limited to the central and south-central regions of the country". 'Sudan Factsheet Human Rights and Oil Workshop', (January 31, 2003), pp. 1–2. See also 'Oil Fact Sheet on Sudan' (2006), pp. 1–2.

[35] See for instance, Oil Fact Sheet on Sudan' (September 2006), p. 2.

between Sudan and Southern Sudan that threatens stability and sustainability. This increasing tension is attributed to the fact that the two sides have not reached an agreement on the division of oil revenues after secession. According to official estimates since 70 % of Sudan's crude is pumped in Southern Sudan and since the main oil pipeline, refinery and seaport are located in Sudan, this suggests that Sudanese economy will be affected negatively and lose most of the oil reserves (70 %) and oil revenues (50 %) and Southern Sudan will remain dependent on the main pipeline passing through the north. Even after Southern Sudan's independence, Sudan will remain the former's only export route through a pipeline ending in the seaport of Port Sudan in the Red Sea. This also implies that Sudanese government needs to invest in agriculture and non-oil industries, and that both Sudanese and Southern Sudanese governments need to take measures to counter the negative impacts and ensure their mutual benefit. This demonstrates how oil remains a controversial issue in Sudan-Southern Sudan conflict, and also creates more potential for future conflict between Sudan and Southern Sudan.

Oil has also affected the labour market because the exploration and production of oil leads to the creation of more employment opportunities, although this is difficult to elaborate due to a lack of accurate data.[36] The inflow of FDI and the increased wealth from oil has encouraged migration to Sudan, so migrant workers have increased in the labour market, particularly in the private sector, which may also contribute to the growing unemployment rate. Furthermore, oil has also affected the structure of wages and has led to a wage differential in Sudan; for instance, the results of the comprehensive industrial survey (2005) indicates that the highest salary for workers in the industrial sector is reported in the petroleum refining industry which is 18 times more than the average wage in the industry.[37]

Our findings in this section therefore prove the second part of our hypothesis: that oil has had a negative impact on Sudanese economy.

[36] For instance, of the total labour force estimated at 97,000,000 in 2001, the share and contribution of oil industries in total employment is very minimal and accounted for only 0.0087 % of total employment of labour force and only 0.52 % of total employment in the industrial manufacturing sector in Sudan. It is worthy to note that the contribution of oil industries represent only 0.52 % of total employment and 0.64 % of total number of labour employed in the industrial manufacturing sector in Sudan (2001) but in the meantime oil industries is ranked second in terms of the contribution to industrial value added as it accounts for 11 % of total industrial value added in the manufacturing industries in Sudan; this implies that oil industries tend to use more capital intensive techniques and to be a more capital intensive industry. See for instance, Sudan Central Bureau of Statistics 'Statistical Year Book' (2001), Khartoum, November 2003, for the data on the total number of labor force in Sudan in 2001. See Sudan Ministry of Industry 'Comprehensive Industrial Survey data for 2001' (2005), Tables 12–13; the industrial survey, pp. 72–75, for the data on total number and share of oil in total employment and in employment in the industrial manufacturing sector in Sudan in 2001.

[37] See the 'Executive Summary of Sudan Comprehensive Industrial Survey' (2005), p. 29.

2.4 Structural Problems of Labour in Sudan

Based on the above, in this section it is useful to start by explaining the stylised facts on the characteristics and structural problems of labour market in Sudan. First we explain the relation between the structure of labour market and the demographic structure, participation rates and economic activities, second we show the relation between the structure of labour market and the low skill level and brain drain problems and finally we examine the relation between the structure of labour market and the unemployment and youth unemployment problems in Sudan.

2.4.1 Demographic Structure, Participation Rate and the Declining Productivity of Labour and Economic Growth in Sudan

Before explaining the relation between the structure of labour market and the demographic structure, participation rate and economic activities, it is useful to identify the major stylised facts and characteristics of the labour market in Sudan. For instance, one stylised fact that characterises the labour market in Sudan as in many other Arab and typically developing countries, is the dominance – reflected in the large share – of the public (government) sector in total employment compared to the weakness of the private sector. The organisational structure of the labour market is constrained by weak and inefficient regulations and institutional settings, rigidity and lack of dynamism, deficiency in employment, monitoring, planning and skill upgrading; the high incidence of duality (rural–urban; traditional-modern and formal-informal sectors) and prevalence of high rates of unemployment, especially among youth population and child labour.[38] In addition the labour market is characterised by low participation rates, especially low participation rates for women and the mismatch between educational output (supply) and labour market requirements (demand). These distinctive features of Sudanese and Arab labour markets were caused by such fundamental forces as high population and labour force growth rates, macroeconomic fluctuations caused by oil price instability, and the pervasive role of the state in the region's economic activity. A demographic transition which resulted in rapid population growth, slow down in labour absorption, combined with large-scale shifts of population from rural to urban areas, led to severe pressures on labour markets, especially in urban areas.[39] In addition, there is

[38] Based on the UNDP (2010) definition of child labour as the percentage of children aged 5–14 in the labour market, the UNDP-HDR (2010a) indicates the high rates of child labour in Sudan as the percentage of children aged 5–14 in the labour market accounted for 13 % of children aged 5–14 over the period (1999–2007). See UNDP-HDR (2010a), p. 191.

[39] See for instance, Shaban et al. (1995).

increasing debate that in the Arab oil-dependent countries, the structure of labour market suffers from the impact of the Dutch Disease phenomenon.

One stylised fact of the labour market in Sudan related to demographic structure indicates continuous and rapid increase in total population from 10 million in 1956 to 39 million in 2008, and increase in the growth rate in total population from 2.1 in 1956 to 2.8 in 2008 (see Fig. 2.13 below). According to the Central Bureau of Statistics (2010), data from the 2008 population census indicates that the distribution of population according to the mode of living implies that the majority of Sudan's total population are rural and nomadic (70.5 %) compared to the minority of the urban population (29.5 %). Furthermore, the distribution of population according to gender implies that the majority of Sudan's population are male (51.27 %) compared to female (48.73 %). Moreover, the distribution of population according to age implies that the majority of Sudan's total population are aged 5–24, representing about 47.38 % of Sudan's total population in 2008. In addition, the share of population aged 17 and over (52.85 %) is higher than the share of population aged 16 or less (47.15 %) (see Table 2.8 and Fig. 2.13 below).[40] This structure and distribution of total population by mode of living, gender and age will have several important implications in the structure of labour market, notably, labour force, participation rates, economic activities, skill level, employment and unemployment rates as we will explain in this section below.

For instance, we observe that the first implication and stylised fact in the labour market in Sudan is that the continuous increase in the total population implies continuous increase in the total labour force from 16.5 in 1998 to 22.5 in 2008 and also increasing though low participation rates.[41] The demographic structure, labour force and participation rate in Sudan implies the low share of Sudanese women in the labour force (31.1 %) as compared to Sudanese men (72.2 %) and the total Sudanese labour force (52.4 %), and the low participation rate (for 15–24 year olds) for Sudanese women (6.08 %) as compared to Sudanese men (15 %) and the total Sudanese participation rate (10.08 %). Both crude and adjusted participation rates show continuous rapid increase in the period 1990–1996 in Sudan. Our findings indicate that compared to 1996, in 2008 the crude participation rates for the total population increased to (43.68 %), but they declined for men (28.99 %) and for women (14.69 %), whereas adjusted participation rates declined for men (57.90 %) and increased for women (29.42 %). Both crude and adjusted labour force participation rates in economic activities defined by mode of living and gender, indicate that the participation rates are higher for people (men and women) living in rural areas compared to people (men and women) living in urban areas and participation rates for women are less than men over the period

40 See Sudan Central Bureau of Statistics (2008) 'Central Bureau of Statistics: Sudan Statistics and Statistical Year Book: Sudan (1990–2008)' (2008), p. 3.

41 See for instance, data from: Arab Labour Organization; Sudan Ministry of Labour and Public Service: Migration and Labour Force Surveys 1993 and 1996; and Sudan Central Bureau of Statistics (2010) 'Fifth Sudan Population and Housing Census (2008)'.

Table 2.8 Economically active population according to economic activities (sectoral classification) and employment status defined by gender in Sudan in 1993–2008 (%)

	1993			2004			2008			2008	
Economic activities	M	W	MW	M	W	MW	M	W	MW	M	W
Agriculture, hunting, forestry and fishing	38.79 %	21.94 %	60.74 %	32.6	18.46	51.08	32.57 %	16.00 %	48.56 %	48.13 %	49.48 %
Industry	9.09 %	0.65 %	9.74 %	7.63	1	8.19	6.92 %	0.73 %	7.65 %	10.22 %	2.27 %
Manufacturing	4.81 %	0.54 %	5.35 %	4.04	0.46	4.5	3.09 %	0.43 %	3.52 %	4.57 %	1.32 %
Services	23.51 %	4.53 %	28.04 %	19.61	4	23	24.29 %	7.26 %	31.55 %	35.90 %	22.45 %
Activities not adequately defined or classified	1.07 %	0.41 %	1.49 %	0.64	0.56	1.2	3.89 %	8.34 %	12.24 %	5.75 %	25.80 %
Total	72.46 %	27.54 %	100.00%	60.48 %	24.02 %	83.47 %	67.67 %	32.33 %	100.00 %	100.00 %	100.00 %

Sources: Adapted from Arab Labour Organization (2007), (2) Central Bureau of Statistics – Department of Internal Commerce and Pricing. (3) Own calculation based on Sudan Central Bureau of Statistics Population Census Data (2010): The Fifth Sudan Population and Housing Census (2008)

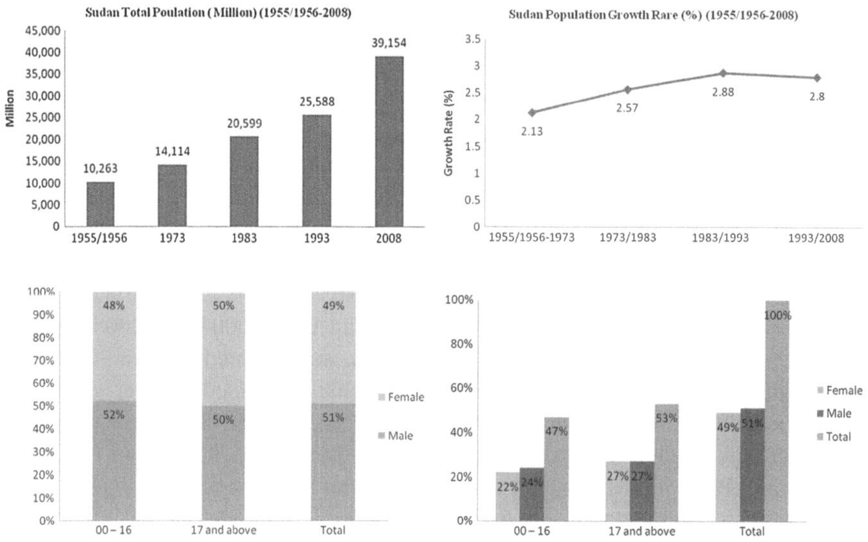

Fig. 2.13 Total population (million) and population growth rare (%) in Sudan over the period (1955/1956–2008) and the distribution of Northern Sudan Total Population defined by age and gender (2008) (Source: Adapted from Sudan Central Bureau of Statistics Population Census Data (2010): The Fifth Sudan Population and Housing Census 2008)

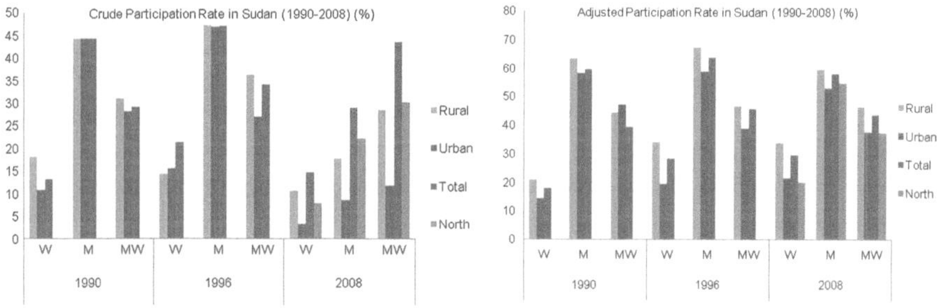

Fig. 2.14 Crude and adjusted participation rate define by gender and mode of living in Sudan (1990–2008) (%). Sources: (1) Figures for 1990 obtained from Ministry of Labour and Administration Reform- Department of Planning and Monitoring and Follow-up. (2) Figures for 1996 obtained from Central Bureau of Statistics – Migration and Labour Force Survey 1996. (3) Central Bureau of Statistics – Department of Internal Commerce and Pricing. (4) Figures for 2008 obtained from own calculation based on Sudan Central Bureau of Statistics Population Census Data (2010): The Fifth Sudan Population and Housing Census (2008), (4) Note: Figures for 1998 from Ministry of Finance and National Economy – Annual Economic Survey 2000, Table 7-2, p. 10

1990–2008 (see Fig. 2.14 above). These findings are consistent with the observed findings from the 1993 population census and 1996 migration and labour force survey. These findings are also consistent with the structure and distribution of population in Sudan in 2008 defined by gender and mode of living as we explained above (see Figs. 2.13 and 2.14 above).

1990–2008 – see Fig. 2.14 above. These findings are consistent with the observed findings from the 1993 population census and 1996 migration and labour force survey. These findings are also consistent with the structure and distribution of population in Sudan in 2008 defined by gender and mode of living as we explained above – see Figs. 2.13 and 2.14 above.

Another stylised fact on the structure of the labour market in Sudan is the inconsistent distribution of economically active population defined according to major economic activities (sectoral classification) and gender. For example, the majority of Sudanese are employed in agriculture sector (51.8 %, 48.56 %), followed by services sector (23 %, 31 %), industry (8.9 %, 7.65 %) and finally few are employed in other activities (1.2 %, 12.24 %) in 2004 and 2008 respectively.[42,43] This structure implies that agriculture is still the predominant activity in Sudan, although its share in employment has gradually declined as other sectors of economic activity have expanded. In the 2008 census almost 48.56 % of the work force were involved in the agriculture sector, compared with 60.74 % in 1993 and 51.08 % in 2004. Services, which included a government work force that grew in terms of employment, emerged as the second largest area of activity, encompassing an estimated 31.55 % of the economically active population in 2008, compared with 28.04 % in 1993 and 23 % in 2004. The industrial sector accounted for 7.65 % in 2008 compared to about 9.74 % in 1993 and 8.19 % in 2004 (see Table 2.8 below). Sudan Central Bureau of Statistics (2010) population census data for 2008 indicates that the distribution of economically active population defined by major economic activities (sectoral classification) and gender implies that the majority of Sudanese men are employed in the agriculture sector (48.13 %), followed by services (35.9 %), industry (10.22 %) and other activities (5.75 %); similarly, the majority of Sudanese women are employed in the agriculture sector (49.48 %), followed by services sector (22.45 %), industry (2.27 %) and other activities (25.8 %). Employed Sudanese men constitute the majority of total employment in all sectors (67.67 %), whereas employed Sudanese women constitute the minority of total employment in all sectors (32.33 %). Sudanese men employed in agriculture, services and industry sectors (32.57 %, 24.29 % and 6.92 % respectively) are higher than Sudanese women employed in these sectors (16 %, 7.26 % and 0.73 % respectively). This implies that Sudanese men employed in agriculture, services and industry sectors are near to twice, near to three times, and near to seven times more than Sudanese women employed in these sectors respectively. These findings are consistent with the results from the 1993 population census and 1996 migration and labour force survey. These findings are also consistent with the structure and distribution of population in Sudan in 2008 defined by gender as we explained above.

---

[42] See for instance, the Arab Labour Organization (2007) for data for 2004 and Sudan Central Bureau of Statistics (2010) population census for data for 2008.

[43] Agriculture sector includes livestock raising, forestry, fisheries, or hunting, services sector includes government work force, wholesale and retail trade, restaurants and hotels, transport, storage and communication, financing, insurance, real estate and business services, community, social and personal services and the industrial sector includes manufacturing, mining, electric power, and construction.

2.4.2 The Low Skill Level and Brain Drain Problem in Sudan

Another stylised fact on the structure of labour market in Sudan can be observed from the skill level defined by occupation (defined by the international definition of major occupational groups classification) and education (defined by school attendance, literacy and education attainment) defined by gender. For example, the definition of skill according to occupation classification indicate that the majority of Sudanese economically active population or workers are medium and low skilled (86 %, 88 %) and minority (14 %, 12 %) are high skilled in 2004 and 2008 respectively.[44] In 2004 only 14 % of men were highly skilled and 86 % are medium and low skilled, and only 15 % of women were highly skilled and 85 % medium and low skilled; women were slightly more skilled than men. In 2008 only 13 % of men were highly skilled and 79 % medium and low skilled, and only 10 % of women were highly skilled and 84 % medium and low skilled; men were slightly more skilled than women. The majority of Sudanese workers were employed in blue collar occupations (70 %, 69.4 %), while the minority were employed in white collar occupations (30 %, 24.4 %) in 2004 and 2008 respectively. In 2004 only 33 % of men were employed in white collar occupations, while 67 % of men were employed in blue collar occupations; only 24 % of women were employed in white collar occupations, while 76 % of women were employed in blue collar occupations. In 2008 only 23.24 % of men were employed in white collar occupations, while 73.42 % of men were employed in blue collar occupations; only 26.9 % of women were employed in white collar occupations, while 61.03 % of women were employed in blue collar occupations (see Table 2.9 below). Moreover, our results from Sudan Central Bureau of Statistics (2010) population census data for 2008 indicate the low skill level and differences in skill level in Sudan that appear in terms of low school attendance, literacy rate and education attainment defined by gender and mode of living. For instance, the distribution of total Sudan population aged 6 years and over according to school attendance and literacy rate, implies that only little above half of Sudan's population aged 6 years and over are currently and/or previously attending school (50.87 %) and are literate (51.59 %), while near to half of Sudan's population aged 6 years and over have never attended school (44.62 %) and are illiterate (45.19 %). The distribution of population aged 6 years and over according to education attainment and currently and/or previously school attendance implies that the majority have less than secondary education and intermediate, primary or less than primary education (76.80 %), this is followed by secondary and post secondary diploma education (13.93 %), and finally followed by a only minority with above secondary education (5.32 %), including university first degree and college education (4.70 %) and post graduate

[44] See for instance, the Arab Labour Organization (2007) for data for 2004 and Sudan Central Bureau of Statistics (2010) population census for data for 2008.

Table 2.9 Economically active population defined according to major occupational groups classification, defined by gender in Sudan (2004–2008) (%)

	2004			2008			2008		
Major occupational groups[a]	M	W	MW	M	W	MW	M	W	North
White Collar high skilled (WCHS)	14.05 %	14.82 %	14.27 %	8.61 %	2.94 %	11.55 %	12.73 %	9.08 %	10.10 %
White Collar low skilled (WCLS)	19 %	9.05 %	16.2 %	7.11 %	5.76 %	12.88 %	10.51 %	17.83 %	6.09 %
Blue Collar high skilled (BCHS)	46.90 %	69.59 %	53.26 %	28.58 %	15.28 %	43.86 %	42.24 %	47.25 %	24.13 %
Blue Collar low skilled (BCLS)	20.04 %	6.53 %	16.25 %	21.10 %	4.46 %	25.56 %	31.18 %	13.78 %	22.03 %
Not stated				2.26 %	3.90 %	6.16 %	3.35 %	12.06 %	6.16 %
Total	100.00	100	100						
White Collar (WC = WCHS + WCLS)	33.05 %	23.87 %	30.47 %	15.72 %	8.70 %	24.42 %	23.24 %	26.90 %	16.19 %
Blue Collar (BC = BCHS + BCLS)	66.94 %	76 %	70 %	49.68 %	19.73 %	69.41 %	73.42 %	61.03 %	46.16 %
High skilled (HS = WCHS)	14 %	15 %	14 %	8.61 %	2.94 %	11.55 %	12.73 %	9.08 %	10.10 %
Medium and low skilled (MLS = WCLS + BCHS + BCLS and not stated)	86 %	85 %	86 %	56.79 %	25.50 %	88.45 %	82.21 %	95.99 %	58.41

Sources: Adapted from the Arab Labour Organization (2007), (2) Central Bureau of Statistics – Department of Internal Commerce and Pricing. (3) Own calculation based on Sudan Central Bureau of Statistics Population Census Data (2010): The Fifth Sudan Population and Housing Census (2008)

[a]The ILO International Standards Classification of Occupations (ISCO) are aggregated in the following way (high skilled includes only the category of WCHS, while medium and low skilled include all other categories: WCLS, BCHS and BCLS):

White-Collar high-skilled (WCHS) includes legislators, senior officials, managers, professionals, technicians and associate professionals.

White-Collar low-skilled (WCLS) includes clerks, services workers, shop and market sales workers.

Blue-Collar high-skilled (BCHS) includes skilled agricultural and fishery workers, craft and related trade workers.

Blue-Collar low-skilled (BCLS) includes plant and machine operators and assemblers and elementary occupations.

diploma, Master's degree and PhD degree (0.62 %).[45] In addition, the distribution of population aged 6 years and over according to current school attendance implies that the majority have less than secondary education and intermediate, primary or less than primary education (73.59 %), followed by secondary and post secondary diploma education (15.40 %), and finally followed by only a minority with above secondary education (6.12 %) – including 5.92 % with university first degree and college education and 0.20 % with Master's and PhD degree education. Moreover, our results from Sudan Central Bureau of Statistics (2010) population census data for 2008 indicate the low skill level and differences in skill level in Sudan that appear in term of low school attendance, literacy rate and education attainment defined by gender and mode of living. For instance, the skill level (defined by school attendance, literacy rate and education attainment) for males is higher than females and for urban are higher than rural population (see Table 2.10 below). These findings are consistent with the structure and distribution of population in Sudan in 2008 defined by gender and mode of living as we explained above.

One stylised fact on the labour market in Sudan is that for a long time Sudan has remained a labour exporting country, especially to the oil rich Arab Gulf countries, and many Sudanese men have worked in other Arab Gulf states; the migration of highly skilled individuals has led to a brain drain problem in Sudan.[46] Based on the

[45] The distribution of population aged 6 years and over according to education attainment and currently and/or previous school attendance implies that about 8.3 % of total Sudan population aged 6 years and over are without educational attainment. It implies that the majority have below primary education (42.58 %), this is followed by primary education (14.84 %), secondary education (12.83 %), intermediate education (4.84 %), university first degree education (4.70 %), post secondary diploma education (1.10 %), postgraduate diploma education (0.29 %), Master's degree (0.24 %) and Ph.D. degree (0.09 %).

[46] "The term human capital flight, more commonly referred to as "brain drain", is the large-scale emigration of individuals with technical skills or knowledge. The reasons usually include two aspects which respectively come from countries and individuals. In terms of countries, the reasons may be social environment (in source countries: lack of opportunities, political instability, economic depression, health risks; in host countries: rich opportunities, comparatively good political system, developed economy, better living conditions). In terms of individual reasons, there are family influences (overseas relatives, and personal preference: preference for exploring, ambition for an improved career, etc.). Although the term originally referred to technology workers leaving a nation, the meaning has broadened into: "the departure of educated or professional people from one country, economic sector, or field for another, usually for better pay or living conditions". Brain drain is usually regarded as an economic cost, since emigrants usually take with them the fraction of value of their training sponsored by the government or other organizations. It is a parallel of capital flight, which refers to the same movement of financial capital. Brain drain is often associated with de-skilling of emigrants in their country of destination, while their country of emigration experiences the draining of skilled individuals. The term brain drain was coined by the Royal Society to describe the emigration of "scientists and technologists" to North America from post war Europe. Another source indicates that this term was first used in the United Kingdom (UK) to describe the influx of Indian scientist and engineers. The converse phenomenon is "brain gain", which occurs when there is a large-scale immigration of technically qualified persons. Brain drain is common amongst developing nations, such as Africa, former colonies of the island nations of the Caribbean, and particularly in centralized economies such as

conventional views in the literature on the incidence of brain drain in typically developing countries, in our view the main reasons for the incidence and continuation of brain drain in Sudan can be perceived from both national and personal perspectives. From the national perspective, the main reasons are related to the internal environment in Sudan due to lack of employment opportunities, political instability and economic instability; in host countries there are rich employment opportunities and better living conditions. From the personal perspective, the main reasons include family influences, overseas relatives and personal preference for an improved career, and better living conditions. The most important reason for the continuation of brain drain in Sudan is that the low standard of economic development has led to low GDP per capita, which implies that high skills are not financially rewarded. The main consequences of the brain drain problem in Sudan is that brain drain is regarded as an economic cost, since emigrants usually take with them the fraction of value of their training sponsored by the government or other organisations. Moreover, brain drain implies that Sudan experiences the draining of skilled individuals and this contributes to vicious circle of underdevelopment in Sudan as a low-income country. This problem of brain drain implies a loss to Sudan that may have amounted to a considerable percentage of its professional and skilled work force. For instance, over the period 2005–2008 the average share of white collar high and white collar Sudanese migrants workers represented about 8.57 % and 11.39 % of total Sudanese migrant workers respectively. Notably, the share of white collar high continuously increased from 5.87 % to 9.42 %, 7.71 % and 9.83 % in 2005, 2006, 2007 and 2008 respectively, and the share of white collar increased from 12.66 % to 13.03 %, 18 % and then declined to 6.6 % in 2005, 2006, 2007 and 2008 respectively (see Table 2.11 below).

In addition, as a result of the brain drain problem which implies a shortage of professional and skilled Sudanese workers, we find the recent phenomena of brain gain of foreign skilled workers which is most probably related to the effects of globalisation and increasing foreign investment, notably foreign investment in the oil sector, which is largely dependent on foreign skills and foreign capital; the easy inflow and employment of foreign workers has caused serious implications because of competition with the local and domestic workers. The presence and high share of skilled foreign workers in total employment of foreign workers means that the

former East Germany and the Soviet Union, where marketable skills were not financially rewarded. Two parties involved in brain drain are developing countries and developed countries. On the left side, because of the disadvantaged social environment (of opportunities, political instability, economic depression, health risks, etc), family influence (overseas relatives, etc), and personal preference (prefer exploring, ambitious to seek brilliant career, etc), many people in developing countries actively choose to migrate. Most of migrations from developing countries are those wealthy or skilled people, whose leaving results in brain drain and slow development of home countries. This contributes to a vicious circle for developing countries (low-income countries). On the other side, the advantaged social environment (rich opportunities, comparatively good political system, developed economy, better living conditions, etc) in developed countries attract talents from other areas, which contribute to brain drain, and finally forms a virtuous circle". See: http://en.wikipedia.org/wiki/Brain_drain, accessed November 14, 2010.

Table 2.10 Sudan-population 6 years of age and over by school attendance, literacy, currently and previously attending school defined by age, gender and mode of living (2008)

	Currently	Previously	Total				Literate	Illiterate	Not stated	
(a) Sudan-population 6 years of age and over by school attendance and literacy defined by age, gender and mode of living										
Group	Attended (currently and previously attended)			Never attended	Not stated	Total	Literacy			Total
	Currently	Previously	Total				Literate	Illiterate	Not stated	
Total	25.50 %	25.38 %	50.87 %	44.62 %	4.51 %	100 %	51.59 %	45.19 %	3.22 %	100 %
Male	14.07 %	14.49 %	28.57 %	19.78 %	2.01 %	50.35 %	29.05 %	19.78 %	1.52 %	50.35 %
Female	11.42 %	10.88 %	22.30 %	24.84 %	2.50 %	49.65 %	22.54 %	25.41 %	1.70 %	49.65 %
Urban	10.44 %	11.89 %	22.33 %	7.02 %	1.22 %	30.57 %	22.56 %	7.21 %	0.80 %	30.57 %
Rural	14.47 %	13.03 %	27.50 %	31.88 %	2.83 %	62.20 %	27.90 %	32.31 %	1.99 %	62.20 %
Nomad	0.59 %	0.46 %	1.04 %	5.72 %	0.47 %	7.23 %	1.13 %	5.68 %	0.43 %	7.23 %
(b) Sudan population 6 years of age and over currently and previously attending school by education attainment, age, gender and mode of living										
Group	Total	Below		secondary: intermediate, primary and below		Secondary and post secondary diploma	Primary and			
				intermediate and secondary	University first degree and colleges		Above university		Not stated	
Total	100 %	76.80 %	13.93 %	94.22 %	4.70 %		0.62 %		3.97 %	
Male	56.16 %	43.29 %	7.80 %	52.94 %	2.52 %		0.39 %		2.15 %	
Female	43.85 %	33.50 %	6.13 %	41.26 %	2.18 %		0.21 %		1.82 %	
Rural	43.89 %	29.56 %	8.52 %	39.70 %	3.51 %		0.47 %		1.83 %	
Urban	54.06 %	45.42 %	5.32 %	52.48 %	1.17 %		0.13 %		2.02 %	
Nomad	2.05 %	1.81 %	0.10 %	2.02 %	0.02 %		0.00 %		0.12 %	
(c) Sudan population 6 years of age and over currently attending school by grade attending, age in single years and sex										
Total	100.00 %	73.59 %	15.40 %	88.68 %	5.92 %		0.20 %		4.89 %	
Male	55.20 %	40.65 %	8.62 %	49.07 %	3.06 %		0.11 %		2.75 %	
Female	44.80 %	32.94 %	6.78 %	39.60 %	2.86 %		0.09 %		2.14 %	
Urban	40.94 %	26.82 %	7.91 %	34.56 %	4.07 %		0.15 %		1.99 %	
Rural	56.76 %	44.94 %	7.26 %	52.06 %	1.82 %		0.04 %		2.71 %	
Nomad	2.30 %	1.84 %	0.24 %	2.07 %	0.04 %		0.00 %		0.19 %	

Source: Own calculation based on Sudan Central Bureau of Statistics Population Census Data (2010): The Fifth Sudan Population and Housing Census (2008) Below secondary includes: intermediate, primary and below primary includes below primary, without educational Attainment, and Khalwa, secondary and less than university includes secondary and post Secondary/Diploma Program, university includes university and colleges, higher than university includes post graduate diploma, Master and Ph.D. degree

Table 2.11 The brain drain: Sudanese working abroad with legal contracts, classified by occupation over the period (2005–2008)

	Total					Share in total (%)				
	2005	2006	2007	2008	2005–2008	2005	2006	2007	2008	2005–2008
Total	8,447	8,302	13,854	22,144	52,747	100 %	100 %	100 %	100 %	100.00 %
White collar high	496	782	1,068	2,177	4,523	5.87 %	9.42 %	7.71 %	9.83 %	8.57 %
White collar low	573	300	1,426	2,481	4,780	6.78 %	3.61 %	10.29 %	11.20 %	9.06 %
White collar	1,069	1,082	2,494	1,363	6,008	12.66 %	13.03 %	18.00 %	6.16 %	11.39 %
Blue collar high	5,878	4,810	6,838	14,416	31,942	69.59 %	57.94 %	49.36 %	65.10 %	60.56 %
Blue collar low	1,500	2,410	4,522	15,652	24,084	17.76 %	29.03 %	32.64 %	70.68 %	45.66 %
Blue collar	7,378	7,220	11,360	18,192	44,150	87.34 %	86.97 %	82.00 %	82.15 %	83.70 %

Source: Sudan Ministry of Labour and Public Service – Annual Economic Reports (various issues)

majority of foreign workers were employed in highly skilled jobs and competed with Sudanese on the available job opportunities in Sudan. Notably, the share of white collar high foreign workers increased from 61 % to 80 %, 89 % and 88 % in 2002, 2003, 2005 and 2006 respectively, and the share of white collar foreign workers increased from 77 % to 90 %, 96 % and 91 % in 2002, 2003, 2005 and 2006 respectively (see Table 2.12 below). The shortage of skilled workers, notably technical skill and mismatch problems in technical jobs also increased the demand for foreign technical workers and this may also contribute to reduce the employment opportunity and therefore increase the unemployment rates for Sudanese workers (see Table 2.13 bclow).

2.4.3 The Unemployment, Youth Unemployment and Skill Mismatch at the Macro Level in Sudan

One stylised fact in the labour market in Sudan is the incidence of a chronically serious unemployment crisis (see Table 2.2 above). Sudan like many other Arab countries not only faces many challenges such as low per capita GDP, low growth of labour productivity and the incidence of a high poverty rate, but also the persistence of a high and rising unemployment rate. This persistent unemployment problem may reflect both a general problem of growth and development and a structural problem of the labour market and inequality, and may lead to several serious implications in hindering the process of development and economic growth. The discussion of unemployment in Sudan is important because of higher and persistent rates of unemployment – now in excess of 20 % (see Table 2.2 above and Fig. 2.15 below). Several studies in the Sudanese literature (cf. Ministry of Labour Report 2004–2005) indicate the problem is due to the demand side, but it is also essential to reflect on the interaction between the supply and demand sides and examine the problem from both perspectives. The UNDP report (2006: 92–94) shows the broad employment trends in Sudan during the 1990s and illustrates a process of jobless growth over that period and highlights the need for employment creation or generation and poverty alleviation in Sudan. Different from the several studies in the Sudanese literature, we explain below four stylised facts on the unemployment problem in Sudan, including the presence of several types of unemployment; the interpretation of unemployment crisis in Sudan from two different endogenous and exogenous perspectives due to endogenous and exogenous causes; the high incidence of unemployment among youth population; and the large mismatch between educational qualifications – supply – and labour market requirements – demand. One advantage of our analysis is that we explain these stylised facts using new data on unemployment based on Sudan Central Bureau of Statistics (2010) population census (2008).

The first stylised fact on the incidence of unemployment in Sudan is the prevalence of several different kinds or types of unemployment including structural, voluntary, involuntary, seasonal, frictional, cyclical, technological, youth, disguised,

Table 2.12 Distribution of foreign workers by occupational classification (%) in Sudan (2002–2006)

Distribution of foreign workers by occupational classification	2002	2003	2004	2005	2006
Total	100 %	100 %	100 %	100 %	100 %
White collar high	61 %	80 %	84 %	89 %	88 %
White collar low	16 %	11 %	12 %	7 %	3 %
White collar	77 %	90 %	96 %	96 %	91 %
Blue collar high	0 %	7 %	0 %	4 %	1 %
Blue collar low	23 %	3 %	4 %	0 %	8 %
Blue collar	23 %	10 %	4 %	4 %	9 %

Source: Adapted from the Statistics of Ministry of Labour

Table 2.13 The share of technician labour in the field of mechanical engineering compared to non-technician and foreign labour

	Technical job/ total job	Non technical work in technical job%	Foreigner in technical jobs%
Cars	59	21.7	19.3
Heat capacity	93.5	6.5	
Heavy machines	86.8	13.2	...
Refrigeration, cooling and conditioning	70.7	25.9	3.4
Welding and blacksmith	53.5	29.3	17.2
Foundries and metal	50.0	50	....
Machining formation	64.1	33.3	2.6
Maintenance of pumps	33.3	66.3	....
Compilation of cooking machine	66.7	33.3	...
Food processing machine	38.5	53.8	7.7
Mineral formation	42.9	28.6	82.6
Electrical steam generators	...	100	...

Source: Prof. Alsheikh Almagzoub (2008) "study of the need in labour market for technician" cited in Sharaf Eldein Ahmed Mohamed (February 2007, 2008) "Globalization and requirements for improving the quality and efficiency of technical education output," paper presented at the third conference for the manager and experts of technical education in states, ministry of education, general department of technical education, June 2008, Table 4, p. 12.

hidden, temporary and open chronic unemployment in Sudan. The presence of seasonal unemployment in Sudan can be perceived from the fact the majority of Sudanese labour force is still hired in the agricultural sector, which is characterised by relative availability of seasonal work in agriculture. In addition the high intensity of labour and family workers in the agriculture sector has probably also caused disguised and hidden unemployment in Sudan. Furthermore, the presence of hidden and disguised unemployment can be perceived from the fact that the public sector is still the main source of job creation in Sudan, it has a limited capacity to hire more workers but the commitment of the government to hire beyond the capacity of the public sector has caused low productivity of workers – at least compared to other Arab countries (see Arab Labour Organization 2007). Furthermore, the presence of frictional unemployment can be perceived from the geographical (temporary)

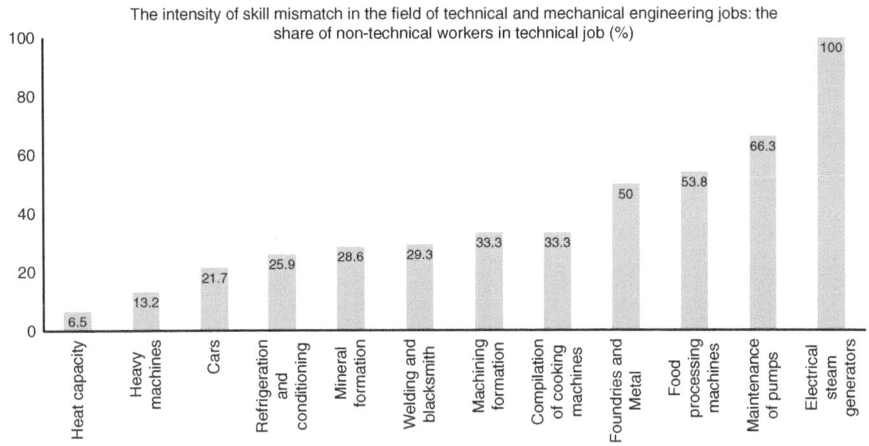

Fig. 2.15 The intensity of skills mismatch in the technical and mechanical engineering jobs: the share of non-technical workers in technical job (%) (Source: Adapted from Almagzoub (2008) and Mohamed (February 2007, 2008))

movement of people (displaced workers and internal refuges). Moreover, the presence of cyclical unemployment can be perceived from the economic crises over the past decades in Sudan. In addition, the presence of technological unemployment can be perceived from the recent expansion in the use of technology, especially ICT in the services, notably the banking sector in Sudan. We observe that the use of new technologies in the banking sector caused displacement and substitution of workers, and contributed to a reduction in the number of employment opportunities. For instance, according to the results of the survey presented in the Central Bank of Sudan (2004) aimed at assessing the impacts of the use of new technologies in Sudanese banking system in 2004, 56.5 % of the respondents indicate that the use of new technologies in the banking system has had some impacts on employment and resulted in the reduction of the number of workers and hence led to an increase in unemployment in Sudan.[47] Furthermore, the presence of youth unemployment can be perceived from the recent information that indicates the rapid increase in unemployment rate in Sudan, especially the youth unemployment rate that reached around 18 % of the total youth population in Sudan; youth unemployment also increased amongst university graduates. In addition, the presence of involuntary unemployment can be perceived from the presence of high unemployment among youth and university graduates, which can be interpreted partly as compulsory unemployment and partly also structural unemployment. The presence of voluntary unemployment can be perceived from the low participation rate, especially for women (housewives). Furthermore, structural unemployment either temporary or open chronically has been persistently present for a long time and can be perceived from the consequences of

[47] See the Central Bank of Sudan (2004), p. 36.

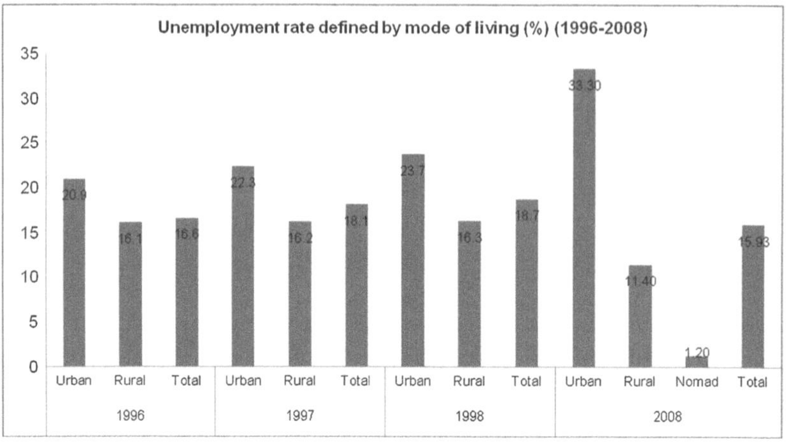

Fig. 2.16 Total unemployment rates in Sudan (%) defined by mode of living (1996–2008) (Sources: (1) Figures for 1996 obtained from Central Bureau of Statistics – Migration and Labour Force Survey 1996. (2) Central Bureau of Statistics – Department of Internal Commerce and Pricing. (3) Figures for 1997/1998 from Ministry of Finance and National Economy – Annual Economic Survey 2000, Table 7-2, p. 10. (4) Figures for 2008 obtained from own calculation based on Sudan Central Bureau of Statistics Population Census Data (2010): The Fifth Sudan Population and Housing Census 2008)

the structural reform and the mismatch between educational qualifications for youth and requirements in the labour market. Moreover, the presence of persistent open chronically long-term unemployment is perceived from the fact that the problem of unemployment existed throughout much of the period 1973–2011. For instance, in Sudan, the rate of unemployment has rapidly increased since 1973 and has more than doubled over the past 15 years (over the period 1993–2008); for instance, the rate of unemployment increased from 10.3 % in 1993 to 20.7 % in 2008, which implies that the unemployment problem remains a chronically persistent one in Sudan (see Table 2.2 above and Fig. 2.16 above).

The second stylised fact is the interpretation of the unemployment crisis in Sudan from two different endogenous and exogenous perspectives due to exogenous and endogenous causes. The exogenous causes include the implications of the internal refuges and migration due to environmental problems, drought and desertification, civil war and conflict, influx of refugees from other neighbouring countries, imbalanced development strategies, globalisation and the use of foreign workers. The exogenous causes also include other factors, for instance, the drop in world oil prices in the 1980s, the Gulf War in the early 1990s and the recent increasing move towards nationalisation of jobs in the Arab Gulf countries, together caused the Arab Gulf states to cut back drastically on their expatriate workers, which resulted in the departure of the thousands of Sudanese workers based in these countries, leaving many of their possessions behind, and leading in turn to increased unemployment in Sudan. Unemployment is also caused by the exogenous environmental problems, for instance, unemployment figures were affected by the severe drought that spread throughout Sudan in the 1980s. In 1983–1984, for example, several million people migrated from the worst hit areas in both the west and the east to Khartoum and other urban areas along the Nile. Many remained in these areas once the drought had eased, living in shantytowns and contributing to unemployment or underemployment in the cities. In addition, more than one million people from the south migrated to the north, as a result of the civil war and famine in these areas. We observe that the lack of political stability and the north–south conflict also contributes to increase unemployment because the civil war not only led to the displacement of many workers and job losses, but also necessitated significant expenditure on defence and security issues rather than prioritising investment in social development and the creation of more employment opportunities. The endogenous causes include the deficient demand caused by deficient macroeconomic policies: privatization; a deficient public sector, a deficient private sector, structural reform, mismatch between educational output and labour market need, unemployment caused by labour market policies, educational policies, and the use of new technologies. In particular, the considerable reduction in the aggregate demand and demand for labour was caused by the liberalisation, structural adjustment programs and privatisation of state owned enterprises during the 1990s that contributed to increased unemployment of the labour force (see Dagdeviren and Mahran 2004). Moreover, the deficiency and low employment of the private sector also contributed to unemployment in Sudan (see Ministry of Labour Report 2004–2005). In addition, we find that the inflation rate is one of the very important endogenous factors that affected the unemployment problem though it does not receive adequate analysis in the Sudanese literature. For this reason, our analysis in this chapter fills this gap in the Sudanese literature and discusses the correlation between inflation rates and unemployment rates. For instance, our findings imply that the increase in unemployment rates seems to be correlated with the increase in inflation rates in Sudan in the period 2000–2008. For instance, when using data and figures on unemployment rates and inflation rates over the period 1990–2008 (presented in Table 2.2 above) and using the ordinary least squares method to examine the correlation between inflation and unemployment rates in Sudan, we find positive and significant correlation between the unemployment rate and inflation rate for the case of Sudan for the period 2000–2008. We find negative significant correlation between inflation rate and unemployment rate over the period 1990–2008, and negative insignificant correlation between inflation rate and unemployment rate over the period 1990–2000. Our findings on the negative correlation between inflation rate and unemployment for the periods 1990–2008 and 1990–2000 are consistent with the studies in the literature in support of the Phillips curve.[48] But our result on the positive correlation between inflation rate and unemployment for the period 2000–2008 is opposite to the findings for the periods 1990–2008 and 1990–2000 and is different from the findings in support of Phillips curve. These contrasting findings for the periods 1990–2000 and 2000–2008 imply

[48] Phillips curve firstly used by Phillips (1958); it indicates a negative correlation between inflation rates and unemployment rates.

Table 2.14 Correlation between unemployment and inflation rates in the Sudan (1990–2008)

Year	SPSS			E-VIEWS		
	Coefficient (*t*-value)	N	R^2	Coefficient (*t*-value)	N	R^2
1990–2008	−0.024* (−2.192)	16	0.256	−0.024* (−2.192)	16	0.256
1990–2000	−0.015 (−1.119)	8	0.173	−0.015 (−1.1197)	8	0.173
2000–2008	0.567* (2.857)	9	0.538	0.567* (2.857)	9	0.538

**Correlation is significant at the 0.5 level of significance

that the correlation between inflation rates and unemployment turned from a negative into a positive correlation in Sudan (see Table 2.14 above and Fig. 2.17 below). The major policy implication from our findings on a significant positive correlation between inflation and unemployment rates for the case of Sudan for the period 2000–2008, implies that an increase in inflation rates have caused an increase in unemployment rates over the period 2000–2008, and so macroeconomic policies aimed at or targeting reducing inflation rates would also contribute to reduce unemployment rates (see Table 2.14 above).

The third stylised fact on the incidence of unemployment in Sudan is the high unemployment among the youth population (see Fig. 2.19 below). Sudan like many other typically developing countries not only suffers from high annual population growth rate (2.8 %) and a very high rate of unemployment (20.7 %), but also the population structure in Sudan as in many other Arab countries – with a high percentage of young people – makes the situation of unemployment even worse and more difficult as most of the population is under 25 years of age; this category of young people represents 22.9 % of Sudan's total population in 2006. Such a population structure has prompted the need to create more job opportunities and is anticipated to put more pressure on the future demands for jobs in Sudan. This situation has led to a very high rate of unemployment among the youth population, for example, according to Arab Labour Organization (2007) data for 2004, the rate of youth unemployment among Sudanese youth is 41.25 %, among females is 43.25 % and males is 36.64 %. The estimated unemployment among Sudanese youth (41.25 %) is among the highest in the world: 2.9 % as much as the international world rate, 2.6 % as much as in Latin America and the Pacific, 2.5 % as much as in South East Asia and 1.4 % as much as in the Arab world (see also Arab Labour Organization (2007); and see Fig. 2.18 below). This situation has ranked Sudan as the fifth worst after Algeria, Iraq, Mauritania and Somalia amongst the Arab countries.[49]

This situation has not only resulted in unemployment but also caused a state of mismatch and underemployment in Sudan as some people were forced to take up jobs for low compensation packages that do not suit their qualifications. (cf. Ministry of Labour Report 2004–2005).[50] The above findings on the high rate of

[49] See for example Arab Labor Organization (2007) and International Labor Organization (2007) recent statistics for 2006.

[50] In this paper the terms Ministry of Labour, Ministry of Labour and Public Service, Ministry of Labour and Administrative Reform are used interchangeably to refer to Ministry of Labour, because the Ministry of Labour is named differently in different regimes.

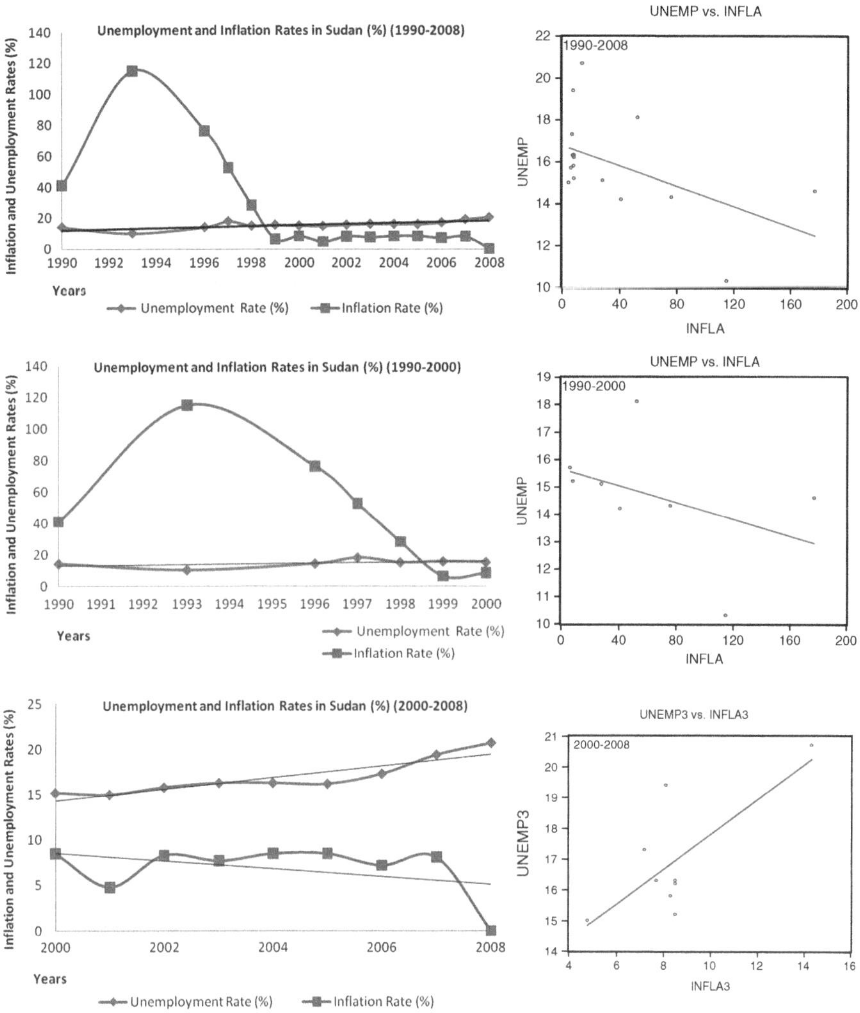

Fig. 2.17 The relation between unemployment and inflation and rates in Sudan (%) (1990–2008) (Source: Own calculation based on data obtained from (1) the Central Bank of Sudan, (2) the central bureau of statistics, (3) Ministry of Finance and National Economy, and (4) Ministry of Labour and public service Annual Economic Reports (various issues))

unemployment for the youth population is also consistent with the findings based on Sudan census data for 1993 and Ministry of Labour and Public Service Migration and Labour Force Surveys 1993 and 1996 that show the presence of persistent unemployment crisis in Sudan and the rising trend since 1979 amongst total population of men and women in rural and urban areas. As for the incidence of unemployment according to age groups in 1996, one should realise that for the total population the highest rate of unemployment is reported amongst the youth

Fig. 2.18 Youth unemployment rates (%) in Sudan compared to world regions (15–24) (2004) (Source: Arab Labor Organization 2007)

population aged 15–24, estimated at 28.4 %, followed by those aged 15 years and over (15.1 %), those aged 25–54 (12.1 %), those aged over 64 (11.4 %), and those aged 55–64 (10.3 %). Therefore, data for 1996 implies that the incidence of open unemployment according to age and gender was higher among the youth population, and notably, young women were more likely to be unemployed compared to young men. Youth unemployment increased from 28.4 % in 1996 to 29 % and 30.8 % in 1997 and 1998 respectively. The distribution of unemployment according to education level indicates that for total population, unemployment is high for primary education (33.8 %), followed by illiterate (29.9 %), illiterate/basic (21.6 %), secondary (11.2 %) and above secondary (3.5 %). The unemployment rate according to education level indicates that for all youth, total unemployment (28.3 %) is high for above secondary (48.7 %), followed by secondary (35.6 %), primary (34.6 %), illiterate/basic (25.59 %) and illiterate (23.4 %). Our findings based on Sudan Central Bureau of Statistics (2010) population census data for 2008, implies that the structure and distribution of the total population and labour force defined by age, gender, mode of living and education attainment have several important implications in the employment rates and unemployment rates. Notably, we find that for the total labour force, employment rates and unemployment rates for men are higher than women and for rural are higher than urban. Moreover, we find that the distribution of unemployment by age groups indicates that the highest unemployment is for the age group 15–24 (32.80 %), followed by the age group 25–39 (32.44 %), age group 10–14 (13.80 %), age group 40–59 (14.57 %) and finally age group 60 and over (4.37 %). These findings are consistent with the findings from the 1993 population census and 1996 migration and labour force survey. These findings are also consistent with the structure and distribution of population in Sudan in 2008 as defined by gender and mode of living as we explain above. Moreover, we find that for the economically active population both employment and unemployment rates

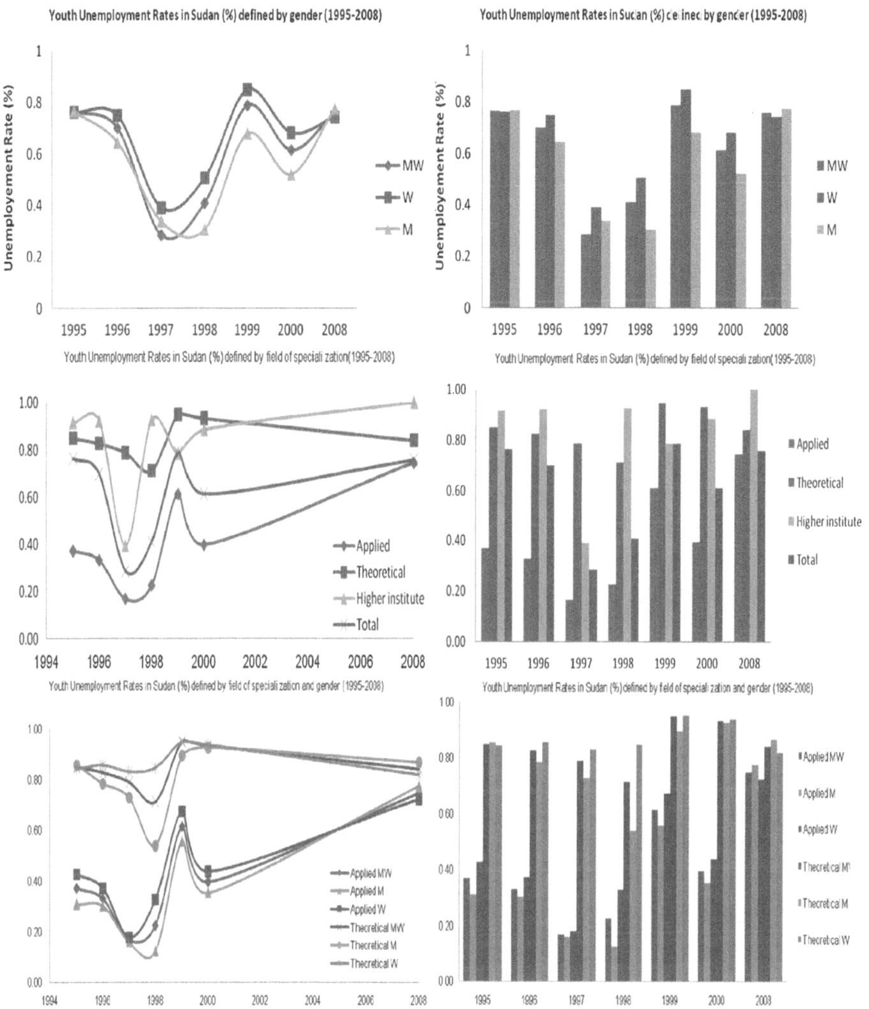

Fig. 2.19 Youth unemployment rates in Sudan (%) defined by gender (1995–2008) (Source: Own calculations based on data from the Sudan federal public service recruitment board – Statistics and Research Administration)

are higher for primary education, followed by secondary and post secondary education respectively. Furthermore, we find that for the economically inactive population the expectations of no hope to find a job (or the potential unemployment) are higher for primary education, followed by secondary and post secondary education respectively (see Table 2.15 below). These findings together imply the importance of education in reducing the incidence of unemployment.

The fourth stylised fact on the incidence on unemployment in Sudan is that the persistence of unemployment (especially among youth) is not only high but also shows a tendency to increase over time in Sudan which is most probably related to the large mismatch between educational qualifications – supply – and labour market requirements – demand – which is perceived from the observed structural change in the demand or changing trends in the share of employment over the period 1988/1989–2008. In particular, on average demand or priority in employment was concentrated among higher secondary school graduates over the period 1988/1989–1996, but the trend changed, and on average priorities in employment turned to become concentrated in applied science colleges followed by social science and art colleges and finally higher institute diploma over the period 2001–2008. This can be attributed to changing trends and priorities from hiring higher secondary school graduates to hiring university graduates, especially applied science college graduates, due to changes in higher educational policies; in particular, the higher education revolution led to expansion in higher educational institutions and increase in student enrolment and graduation during and after the 1990s. This can be interpreted as structural change in the demand for youth labour in favour of university graduates due to structural change in higher educational policies. We observe the changing trend and priorities in reducing employment for higher secondary school graduates to increasing employment for university graduates. However, the structural change in the demand for youth labour in favour of increasing employment for university graduates should not hide the fact that unemployment among university graduates is surprisingly high and continue to increase. Somewhat surprisingly the unemployment crisis was persistent especially among all youth graduates – female and male – in different field of specialisations, even among graduates of applied science colleges. The majority of employment was for graduates in applied science colleges, followed by graduates in social science and art colleges and finally the minority for graduates of higher institute diploma. In general, men were likely to be more employed than women and women were likely to be more unemployed than men.

This fourth stylised fact implies the large mismatch between educational qualifications – supply – and labour market requirement – demand. An important endogenous cause of youth unemployment is the mismatch between educational (qualifications-output) and labour market requirements. Deficiency in educational policies and labour market policies and inadequate planning, assessments and monitoring of policies to create consistency between attained and required education has led to a serious mismatch between educational attainment and labour market requirements. To elaborate the mismatch and unemployment amongst university graduates we utilise the figures on registration and employment obtained from the federal public service recruitment board; we use the figures on registration to refer to supply of university graduates labour; and use the figures on employment to refer to demand for university graduates labour; we calculate the differences between registration and employment to refer to the differences between supply of and demand for university graduates and to define unemployment; then we divide the figures on unemployment by the figures on

Table 2.15 Total population, labour force, economically active population, economically inactive population, employment and unemployment for population 10 years and above defined by age, gender, mode of living, main geographical areas and educational attainment in Sudan (2008)

	Total population (%)	Total labour force (%)	Total economically Active (%)	Total employed (%)	Total unemployed (%)	Total unemployed (%)	Total economically inactive (%)	REASON_REC No hope to find job (%)	REASON_REC Full time student (%)
All Sudan: age groups									
All Sudan	100.00	100.00	43.68	84.07	15.93	100	48.80	10.28	40.70
10–14	18.58	18.58	4.19	7.08	2.52	15.81	13.18	2.78	21.03
15–24	27.98	27.98	9.95	17.56	5.23	32.80	15.89	3.50	18.02
25–39	28.56	28.56	15.98	31.42	5.17	32.44	10.13	2.36	1.70
40–59	17.48	17.48	10.60	21.93	2.32	14.57	5.69	1.74	0.14
60 and over	7.40	7.40	2.96	6.07	0.70	4.37	3.92	0.33	0.01
All Sudan: educational attainment									
No qualifications	4.81		6.08	6.13	5.18	5.18	1.99	0.52	0.77
Primary/junior	11.65		10.89	10.99	9.08	9.08	6.39	0.57	7.20
Secondary	7.61		7.77	7.88	5.76	5.76	3.70	0.37	4.25
Post secondary	3.80		5.32	5.39	4.10	4.10	0.96	0.18	0.91
Khalwa	3.45		4.36	4.32	5.09	5.09	1.34	0.41	0.64
Not stated	73.48		71.66	71.42	75.97	75.97	36.41	8.76	27.71
All Sudan									
Total	100.00	100.00	43.68	84.07	15.93	100	48.80	10.28	40.70
Male	50.07	50.07	28.99	57.23	9.14	57.33	17.28	5.57	22.44
Female	49.93	49.93	14.69	26.84	6.80	42.67	31.52	4.71	18.26
Urban	31.49	31.49	11.83	23.74	3.33	20.93	16.72	2.18	17.69
Rural	61.48	61.48	28.47	53.78	11.40	71.57	29.21	7.24	22.16
Nomad	7.03	7.03	3.38	6.54	1.20	7.51	2.87	0.86	0.85
North									
Total	80.58	80.58	30.10	57.32	11.59	72.73	43.16	6.91	37.07

(continued)

Table 2.15 (continued)

	Total population (%)	Total labour force (%)	Total economically Active (%)	Total employed (%)	Total unemployed (%)	Total unemployed (%)	Total economically inactive (%)	REASON_REC	
								No hope to find job (%)	Full time student (%)
Male	40.41	40.41	22.14	43.60	7.08	44.41	14.64	3.69	20.20
Female	40.17	40.17	7.96	13.71	4.51	28.31	28.52	3.22	16.86
Urban	28.02	28.02	9.71	19.54	2.68	16.83	15.48	1.58	16.56
Rural	45.53	45.53	17.01	31.23	7.71	48.40	24.81	4.46	19.66
Nomad	7.03	7.03	3.38	6.54	1.20	7.51	2.87	0.86	0.85

Source: Own calculation based on Sudan Central Bureau of Statistics Population Census Data (2010): The Fifth Sudan Population and Housing Census (2008)

registration to calculate the unemployment rates for university graduates.[51] We find that unemployment amongst university graduates in all subjects or fields of specialisations is persistent and high for more than two decades over the period 1984/1985–2008. In particular, persistent and high unemployment rates were mostly amongst theoretical, social science and art colleges graduates and technical education high institute (diploma) graduates. The majority of employed graduates were applied science colleges graduates, but this should not hide the fact that unemployment among this category is surprisingly also very high. This implies a mismatch between attained education (educational policies) and required education in the labour market (labour market policies) (see Fig. 2.19 above). We find that the high unemployment is persistent amongst the university graduates with different fields of specialisations over the period 1984/1985–2008, for example, on average the rates of unemployment for all fields of specialization were estimated at: 73 % (1984/1985), 82 % (1985/1986), 78 % (1986/1987), 81 % (1987/1988), 76 % (1995), 69 % (1996), 28 % (1997), 41 % (1998), 78 % (1999), 61 % (2000) and 76 % (2008) respectively. In particular, for applied science colleges unemployment was estimated at: 64 % (1984/1985), 85 % (1985/1986), 75 % (1986/1987), 48 % (1987/1988), 40 % (2000) and 75 % (2008) respectively. As for social science and art colleges, unemployment was estimated at: 88 % (1984/1985), 66 % (1985/1986), 76 % (1986/1987), 90 % (1987/1988), 82 % (1990/1991), 93 % (2000) and 84 % (2008) respectively. As for higher institute diploma, unemployment was estimated at: 74 % (1984/1985), 93 % (1985/1986), 84 % (1986/1987), 75 % (1987/1988), 49 % (1990/1991) and 88 % (2000) respectively.

Over the period 2000–2008, total unemployment for graduates of all fields of specialisation for total, women and men respectively increased from (61.2 %, 68.1 %, 51.9 %) in 2000 to (75.7 %, 74.2 %, 77.5 %) in 2008; unemployment for graduates of applied science colleges and fields of specialisation increased from (39.6 %, 43.7 %, 35.3 %) in 2000 to (74.6 %, 72.1 %, 77.4 %) in 2008; unemployment for graduates of art and social sciences colleges and fields of specialisation decreased from (93.1 %, 93.6 %, 92.4 %) in 2000 to (84 %, 81.7 %, 86.6 %) in 2008; and unemployment for graduates of high institutes (diploma) increased from (88.4 %, 93 %, 82 %) in 2000 to (100 %, 100 %, 100 %) in 2008. Over the period 1995–2000 the distribution of unemployment by gender implies that for all fields of specialisations, applied science colleges and social sciences colleges and fields of specialisation, women are more likely to be unemployed than men. Whereas somewhat surprisingly, the opposite is true in 2008 as the distribution of unemployment by gender implies that for all fields of specialisation, applied science colleges and social sciences colleges and fields of specialisation, men are more likely to be unemployed than women (see Figs. 2.14, 2.15, 2.16, 2.17, 2.18 and 2.19). Therefore, these findings provide further evidence on the serious and

[51] One limitation is that the use of figures on registration and employment to refer to supply of and demand for university graduates labor respectively may be somewhat inaccurate and underestimate the actual figures on supply of and demand for university graduates, because not all university graduates looking for jobs are registered for the federal public service recruitment board and also because figures on employment may include university graduates unregistered for the federal public service recruitment board.

increasing trend of youth unemployment, notably unemployment of university, college and higher institute diploma in Sudan over the period 2000–2008. This implies the urgent need for implementation of sound policies to address the unemployment problem and increase employment opportunities in Sudan.[52]

2.5 Overview on the Recent Restructuring and Challenges for Development in Sudan

The current Sudanese restructuring following the secession of South Sudan has serious implications for development and implies that the challenges are mounting for the economy and technological change in Sudan. From political perspective, the secession of South Sudan has led to political shock, uncertainty and potential future conflict over oil and border regions, particularly because the boundary is not completely delineated. From Northern Sudan's perspective, the demographic and geographical implications of South Sudan separation from North Sudan implies that North Sudan has lost about 21 % of Sudan's population and about 25–30 % of Sudan's area (see Sudan's Central Bureau of Statistics). From Northern Sudan's perspective, the economic implications of South Sudan separation from North Sudan implies that North Sudan has lost about 80 % of Sudan's agricultural and water resources, in addition to the loss of about 75 % of Sudan's proven oil reserves and about 90 % of Sudan's oil and total exports and about 50 % of Sudan's government revenues (see Sudan Central Bureau of Statistics, Central Bank of Sudan and Sudan Ministry of Finance and National Economy). Mainly, the split of the South Sudan means that the North Sudan has lost most of its oil reserves, lost near to two thirds of the country's total oil production of 500,000 barrels a day; this implies the loss of the main source of State revenues. The declining State revenues together with the increasing State expenditure led to a growing budget deficit. Other economic implications are the scarcity of foreign currency, devaluation and loss of near to half value of the Sudanese pound in 1 year that has led to fast rise in prices and near to doubled annual inflation rate that hit almost 28.6 % in April 2012 compared to 16.5 % in April 2011, this implies that the inflation rate is near to doubled during 1 year over the period April 2011 and April 2012 (see Fig. 2.20) (see Sudan central bureau of statistics).[53] Moreover, the political and economic shock implies the lack of political and economic stability, increasing uncertainty for Sudan economy, increasing domestic and external imbalances, difficulties of sustaining macroeconomic stability, limitation on capital inflow and foreign investment and increasing the amount of external debt. The secession of South Sudan resulted in 80 % decline in foreign currency earnings and a 35.6 % reduction in budget revenue. Real GDP grew by 2.8 % in 2011, down from 5 % in 2010. This

[52] See Sudan Federal Public Service Recruitment Board – Statistics and Research Administration (2010).

[53] See http://www.cbs.gov.sd/sites/default/files/Publications/April%20Bulletin.pdf: p. 7, accessed June 03, 2012.

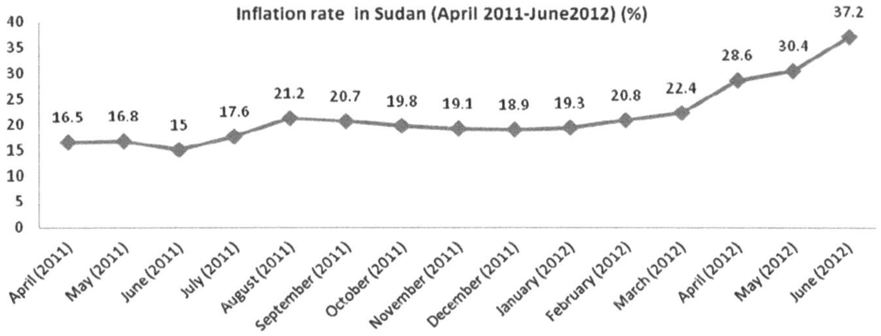

Fig. 2.20 Inflation rates in Sudan (%) (April 2011–April 2012) (Source: Sudan Central Bureau of Statistics (2012): http://www.cbs.gov.sd/sites/default/files/Publications/April%20Bulletin.pdf: p. 7, Accessed June 3, 2012)

slowdown in growth is attributable to the loss of population and oil revenues and GDP growth is estimated to decline further to 2 % in 2012.[54] Deterioration in overall balance of payment, as the balance of payment deficit increased by more than 17 times from US$ 54.2 millions in 2010 to US$ 922.4 millions in 2011.[55] This implies that the economic situation in 2012 had not improved from 2011, and the current economic crisis implies further pressure on the already limited government budget and efforts to allocate more resources for enhancing skill development and local technological capacity building. This implies the importance of implementing sound policies to stabilize Sudan economy and allocate more resources for enhancing skill development and local technological capacity building.

2.6 Conclusions

This chapter explains the general political context and socioeconomic characteristics of Sudan and strategic problems for development in the country, and discusses the impact of oil and the opportunities and challenges for enhancing economic development in Sudan, as well as discussing the strategic problems facing the labour market in Sudan and highlighting the need for skill upgrading and development.

In Sect. 2.2 we started by explaining the general socio-economic characteristics of Sudan's economy. In Sect. 2.3 we then provided a comprehensive analysis using the most recent secondary data, with a view to clarifying the positive and negative effects of oil on Sudan's economic development. We have provided a historical background about the structure of oil investment in Sudan, looking in particular at China's role therein, and explained how oil has created various positive effects and opportunities for development in Sudan. These include the impact of oil in satisfying

[54] See www.africaneconomicoutlook.org/countries/east-africa/sudan/, accessed June 03, 2012.

[55] See the Economic Review June 2012 – Statistical Department – Central Bank of Sudan – Issue No. 06/2012, p. 3.

domestic demand and achievement of self-sufficiency, increasing government resources, revenues and spending, economic growth (GDP growth and composition), foreign trade (volume and structure of exports), balance of trade, balance of payment, FDI and social development in Sudan. We then illustrated the negative impact of oil and the challenges of development in Sudan. These include the volatility and risk of dependence on highly fluctuating oil prices in the international market, unsustainable oil revenues, the lack of diversification, Dutch Disease and the challenges of potential future Sudan-Southern Sudan conflict. Our results in Sect. 2.2 support the stylised fact that oil has had a mixed positive–negative impact on Sudanese economy, arguing that oil is an important resource, particularly in satisfying domestic consumption and the achievement of self-sufficiency by increasing government and public revenues. Although oil has helped to improve economic performance in the country, we find that the recent dependence on oil may spark other problems because it is an exhaustible resource and because instability of oil prices in the international market tend to produce uncertainty in domestic growth. Moreover, the increasing dependence on oil raises the possibilities of the Dutch Disease phenomenon and a lack of diversification, which may aggravate challenges linked to the division of the country and the potential for conflict with newly independent Southern Sudan. Therefore the major policy implication from our findings is that the fulfilment of long run sustainable growth and development strategies in Sudan requires various sources of growth, including revitalising and enhancing non-oil exports, notably traditional agricultural exports.

In Sect. 2.4 we explain several stylised facts on the labour market. First we explain the relation between the structure of the labour market and the demographic structure, participation rates and economic activities. Second, we show the relation between the structure of labour market and the low skill level and brain drain problems and finally we examine the relation between the structure of the labour market and the unemployment and youth unemployment problems in Sudan. We show that the differences in the structure and distribution of the total population defined by age, gender and mode of living have several important implications in the structure of labour market; notably, we find that the labour force, participation rates, economic activities, skill level, employment rates and unemployment rates for men are higher than women, and for rural are higher than urban areas in Sudan.

Different from the several studies in the Sudanese literature we examine in detail four stylised facts on the unemployment problem in Sudan including the presence of several types of unemployment; the interpretation of unemployment crisis in Sudan from two different endogenous and exogenous perspectives due to endogenous and exogenous causes; the high incidence of unemployment among youth population and the large mismatch between educational qualifications – supply – and labour market requirement – demand. Moreover, one advantage of our analysis is that we explain these stylised facts using new data on population, employment and unemployment based on Sudan Central Bureau of Statistics (2010): Fifth Sudan Population and Housing Census 2008.

The major policy implication from our findings indicate that the unemployment crisis is related or linked to the endogenous and exogenous causes explained above, therefore reducing unemployment and enhancement of employment creation are most probably related or linked to several important factors and so policies intervention should deal with these endogenous and exogenous reasons or causes. The solution to the unemployment problem in Sudan not only includes the role of the government and public sector, but also essential roles for the private sector and non-governmental organisations as well as civil society. This implies that the possible policy interventions for reducing unemployment and enhancing employment include increasing employment opportunities in the public and private sectors, improvement of work conditions, employment policies, regulations and legislations, improvement of the federal public recruitment board and improvement of the quality of educational policies to enhance the consistency between the educational attainments and educational requirements in the labour market. One major policy implication from our results implies that an increase in unemployment rates is positively and significantly correlated to an increase in inflation rates over the period 2000–2008, and so macroeconomic policies aimed at targeting reducing inflation rates would also contribute to reduce unemployment rates. Therefore, macroeconomic policies should be used to reduce inflation in order to reduce unemployment in Sudan. Other policies include: reducing the use of foreign workers and the influx of foreign refuges; reducing internal migration by avoiding civil war and conflict and solving political problems and achieving political stabilization; ensuring equity and fairness in the labour market; attracting foreign capital for the creation of new employment opportunities for domestic and local workers and upgrading skill levels; creating more job opportunities for the poor by enhancing small and medium scale enterprises and provide unemployment insurance; enhancing small and family projects; implementing balanced development strategies and improving work conditions and availability of infrastructure and offering incentives to encourage work in the remote states; and finally use of oil revenues to create more and new employment opportunities for domestic workers in Sudan. It is important to realise that the unemployment crisis cannot be managed in a sustainable way through increased employment in an already inflated public sector; productive employment must be generated mostly in the private sector. Dealing with the unemployment crisis and meeting the poverty alleviation challenge requires action in wide-ranging areas of structural reforms to improve the business environment, encourage private sector investment, stimulate productivity growth and enhance efficiency. The implementation of plans simultaneously targeting reducing unemployment and poverty, for instance, provision of more employment opportunities and poverty alleviation, are related to improving infrastructure and facilities of value to the whole society, using labour-intensive methods or schemes to generate employment for large numbers of poor people as well as mobilising small, informal enterprises where many of the poorest workers are concentrated. These strategies are expected to also lead to sustainable job creation and therefore poverty alleviation.

Our findings in this Chapter support our first hypothesis in Chap. 1 that Sudan needs to promote local skills and local technologies in order to implement the five strategies of reducing poverty; achieving economic diversification; reducing unemployment and restructuring the labour market; building local technological capacity and achieving long-term stabilised, sustainable and balanced economic growth and development.

References

Almagzoub, A. (2007/2008, June). Study of the need in labour market for technician. In S. A. Mohamed (Ed.), *Economic globalization and requirements of improving the efficiency and outputs of technical and technological education*. Paper prepared for the Third Conference of Managers and Senior Directors of Technical Education in States. Khartoum: Ministry of Education, General Directorate of Technical Education. Table 4, p. 12.

Arab Labor Organization. (2007). *Statistics on labor market in the Arab region*. http://www.alolabor.org, Accessed 10 March 2009.

Barro, R. J., & Lee, J-W. (2010). A new data set of educational attainment in the World, 1950–2010. (NBER working paper no. 15902), National Bureau of Economic Research, Cambridge, , MA. http://www.nber.org/papers/w15902.pdf?new_window=1. April 2010.

Bedawi, W. F. (2007, July 14–31). *A brief look at Sudan's major economic trends and development challenges*. Paper presented at the Ministry of Finance and National Economy and the World Bank "Macroeconomic and Fiscal Federalism Workshop", Khartoum, Sudan.

Corden, W. M. (1984). Booming sector and Dutch disease economics: Survey and consolidation. *Oxford Economic Papers, 36*(3), 359–380.

Dagdevirena, H., & Ameer Mahran, H. A. (2004). Trade and industry in Sudan: What role for poverty alleviation. Final draft, UNDP project on macroeconomic policy and poverty alleviation in the Arab region. Khartoum: UNDP office in Sudan.

Elbadawi, I. A., & Kaltani, L. (2007, 14–31, July). *The macroeconomics of oil booms: Lessons for SSA, Development Economics Research Group, Economists' Forum, April 19, 2007*. Paper presented at the Ministry of Finance and National Economy and the World Bank "Macroeconomic and Fiscal Federalism Workshop", Khartoum.

Elbadawi, I., & Suleiman, K. (2008). Toward an equitable inter-governmental transfer system for the Sudan. Chapter 3. In S. Abd El Rhman (Ed.), *Generating growth and making fiscal decentralization work in post-conflict Sudan* (p. 107). Khartoum: Ministry of Finance and National Economy. Table 3.

International Monetary Fund (IMF). (2008, June). *Sudan: First review of performance under the 2007–08 staff-monitored program* (IMF Country Rep. No. 08/174). The Middle East and Central Asia Department, International Monetary Fund (pp. 2, 6). Washington, DC: http://www.imf.org/external/pubs/ft/scr/2008/cr08174.pdf. Accessed 1 June 2010.

International Monetary Fund (IMF). (2009, March). *The implications of the global financial crisis for low-income countries*. IMF Report, IMF Multimedia Services Division. Washington, DC: International Monetary Fund. http://www.imf.org/external/pubs/ft/books/2009/globalfin/globalfin.pdf. Accessed 1 June 2010

International Energy Agency. (2005). "International Energy Agency Estimates; International Energy Annual," IPM, cited in Pinaud, C. (2006). UnderstandingSudan.org, 2006. Accessed 20 Dec 2010. (pp. 1–2).

International Labour Organization. (2007). www.ilo.org. Accessed 10 Mar 2009.

Ministry of Labour and Public Service Annual Economic Reports.

Mohamed, S. A. (2008, June). *Economic globalization and requirements of improving the efficiency and outputs of technical and technological education*. Paper prepared for the Third Conference of Managers and Senior Directors of Technical Education in States. Khartoum: Ministry of Education, General Directorate of Technical Education. Table 4, p. 12.

Nour, S. (2011). Assessment of effectiveness of China aid in financing development in Sudan (UNU-MERIT working paper no. 2011–005), Maastricht, The Netherlands. http://www.merit.unu.edu/publications/wppdf/2011/wp2011-005.pdf. Jan 2011

Palma, G. (2003, May 19). *Four sources of de-industrialisation and a new concept of the Dutch Disease*. Paper presented at UNU- INTECH Research Seminar Series 2004-2, p. 21. Unpublished paper.

Phillips, A. W. (1958). The relationship between unemployment and the rate of change of money wages in the United Kingdom 1861–1957. *Economica, 25*(100), 283–299.

Salih, S. H. (2004). *Sudanese oil: The story of the struggle of a nation* (pp. 21–40, 166, 93–94, 91, 182). Khartoum: Sudan Currency Printing Press Company Ltd (in Arabic).
Shaban, R. A., Assaad, R., & Al-Qudsi, S. S. (1995). The challenge of unemployment in the Arab region. *International Labour Review, 134*. www.questia.com/PM.qst?a=o& se=gglsc&d-=5000320794. Accessed 20 Nov 2007.
Sudan Oil fact sheet on Sudan (2006, September). Produced by C. Pinaud for UnderstandingSudan.org, 2006, pp. 1–2
Sudan Central Bureau of Statistics. (2001, November). *Statistical year book*. Khartoum: Sudan Central Bureau of Statistics.
Sudan Central Bureau of Statistics. (2010). *Annul statistical year book and statistical series (1990–2009)* (p. 107). Khartoum: Central Bureau of Statistics, Sudan Ministry of the Cabinet
Sudan Central Bureau of Statistics. (2011). *Sudan national baseline household survey 2009*. Tabulation report, North Sudan. Table 9.1, p. 12.
Sudan Central Bureau of Statistics. (2012). p. 7. http://www.cbs.gov.sd/sites/default/files/Publications/April%20Bulletin.pdf. Accessed 3 June 2012
Sudan Central Bureau of Statistics – Department of Internal Commerce and Pricing.
Sudan Central Bureau of Statistics Population Census Data. (2010). *5th Sudan population and housing census (2008)*. Khartoum: Sudan Central Bureau of Statistics.
Sudan Factsheet Human Rights & Oil Workshop, (2003). cited in Pinaud, C. (2006). UnderstandingSudan.org, 2006. Accessed December 20, 2010. (pp. 1–2).
Sudan Factsheet Human Rights & Oil workshop (2003, January 31). pp 1–2.
Sudan Federal Public Service Recruitment Board – Statistics and Research Administration (2010), Khartoum. Accessed 15 Sept 2010
Sudan Health and Household Survey (2006). Khartoum, Sudan, 2006. cited in the UN Resident Coordinator's Support Office, Khartoum, Sudan: http://www.sd.undp.org/mdg_fact.htm, Accessed on June 01, 2010.
Sudan Ministry of Energy and Mining. (2008). Unpublished Report (2008), Khartoum, Sudan.
Sudan Ministry of Finance and National Economy. (1997–2007: Various Issues). *Annual economic survey report*.
Sudan Ministry of Finance and National Economy. (2000). *Annual economic survey 2000*. Table 7-2, p. 10.
Sudan Ministry of Finance and National Economy. (2002). *The Sudan economy in figure: 1990–2002* (1st ed.). Khartoum: Ministry of Finance and National Economy, Macroeconomic Policies and Programme Directorate MEPPD. p. 27.
Sudan Ministry of Finance and National Economy. (2006). *The performance of Sudan economy 2000–2005* Sudan Ministry of Finance and National Economy, Khartoum, Sudan, 2006 (pp. 16–18).
Sudan Ministry of Finance and National Economy. (2008). Unpublished report.
Sudan Ministry of Foreign Trade and Central Bank of Sudan Annual Foreign Trade Statistical Digest various issues (1992–2010): 2006: p. 20, p. 38. 2005: p. 38, p.20. 2004: p. 20, p. 39. 2002: p. 9, p. 24. 2000: p. 9, p. 24. 1998: pp. 12–13, pp. 33–34. 1997: pp. 12–13, pp. 33–34, 2010.
Sudan Ministry of Industry. (2005). *The comprehensive industrial survey* (Vol. 4). Khartoum, Sudan. www.industry.gov.sd
Sudan Ministry of Industry. (2005, July). *The comprehensive industrial survey (2005)*. Khartoum, Sudan. The executive summary of the comprehensive industrial survey (2005), p. 29. Accessed December 01, 2010
Sudan Ministry of Investment. (2009). Unpublished statistics and data from the feasibility studies (2009). Khartoum
Sudan Ministry of Labour and Administration Reform – Department of Planning and Monitoring and Follow-up, Khartoum.
Sudan Ministry of Labour and Public Service and Central Bureau of Statistics. (1996). Migration and labour force survey (1996). Khartoum.
Sudan Ministry of Labour. (2005). Sudan labour report (2004–2005). Unpublished report, Sudan Ministry of Labour. Khartoum.

Sudan Ministry of Welfare and Social Security and Sudan National Population Council General Secretariat (NPC/GS). (2011). *Sudan millennium development goals progress report 2010*. Khartoum. http://www.sd.undp.org/doc/Sudan%20MDGs%20Report%202010.pdf. Accessed 20 June 2011.

The Central Bank of Sudan. (2004). *The impacts of technologies on the banking sector in Sudan during the period 1999–2003*. Report of the Central Bank of Sudan, Statistical and Research Department Series, Issue No. 6, pp. 13, 36.

The Central Bank of Sudan Annual Report. (2009). p. 84.

The Central Bank of Sudan Annual Report (Various issues: 1999–2009). The central bank of sudan. Khartoum.

The Economic Review. (2012, June). Statistical Department- Central Bank of Sudan – Issue No 06/2012, p.3.

The World Bank (WB-DTS). (2008). *Revitalizing Sudan's non-oil exports: A Diagnostic Trade Integration Study (DTIS) prepared for the integrated framework program (December 2008)*. A Revised version of "Sudan Diagnostic Trade Integration Study (DTS), Poverty Reduction and Economic Management 2- Africa region, Draft: May 2, 2008, Decision Meeting: May 13, 2008": pp. viii–ix; 1–7. http://www.integratedframework.org/files/english/Sudan%20DTIS%20final.pdf. Accessed June 10, 2010.

The World Development Indicators (WDI)/IMF/World Bank Staff Estimate (2008), cited in the World Bank (WB-DTS). (2008). Revitalizing Sudan's non-oil exports: A Diagnostic Trade Integration Study (DTIS) prepared for the integrated framework program (December 2008). A Revised version of "Sudan Diagnostic Trade Integration Study (DTS), Poverty Reduction and Economic Management 2- Africa region, Draft: May 2, 2008, Decision Meeting: May 13, 2008", pp. viii–ix; 1–7.

The World Investment Report. (2002). The united nations conference on trade and development (UNCTAD). The World Investment Report (2002). Geneva: UNCTAD.

United Nations Development Group (2003) "Indicators for Monitoring the Millennium Development Goals: Definitions Rationale Concepts and Sources," September 2003. New York, United Nations Publication. pp. 5, 7, 9, 12, 14, 16, 20, 22, 24, 26, 29, 30, 32, 34, 36, 38, 40, 49, 53, 54, 64, 66: http://www.undp.or.id/mdg/documents/MDG%20Indicators-UNDG.pdf: Accessed on 3 November 2012.

UNDP (United Nations Development Programme). UNDP-HDR. (2007/2008). *Human development report 2007/2008: Fighting climate change: Human solidarity in a divided world* (pp. 231–232, 392–293). New York: UNDP/Palgrave Macmillan.

UNDP (United Nations Development Programme). (2009). Human development report 2009: Overcoming barriers: Human mobility and development. New York: Palgrave Macmillan.

United Nations Development Programme (UNDP). (2010a). *Human development report 2010: The real wealth of nations: Pathways to human development*. Human Development Report 2010 – 20th Anniversary Edition. New York: UNDP/Palgrave Macmillan, Statistical Tables, Statistical Annex, Table 12: Decent Work, p.191.

United Nations Development Programme (UNDP). (2010b). *UNDP Sudan country profile of Human Development Indicators (2010)*. http://hdrstats.undp.org/en/countries/profiles/SDN.html. Accessed 22 Dec 2010.

United Nations Development Programme (UNDP) in Sudan. (2006). Macroeconomic policies for poverty reduction: The case of Sudan. Khartoum, Table 8.1, p. 82.

United Nations Development Programme (UNDP)-The Arab Human Development Report (AHDR). (2003, July). *Arab human development report (2003): Building a knowledge society*. UNDP. New York/Amman: Regional Bureau for Arab States (RBAS)/National Press, p. 56, Table 5.1, p. 102.

United Nations Resident Coordinator's Support Office, Khartoum. http://www.sd.undp.org/mdg_fact.htm, Accessed 1 June 2010.

Wikipedia. http://en.wikipedia.org/wiki/Brain_drain. Accessed 14 Nov 2010.

www.africaneconomicoutlook.org/countries/east-africa/sudan/. Accessed 3 June 2012. UNCTAD 'World Investment Report' (2002).

Part II
Conceptual and Theoretical Framework

Chapter 3
Technological Change and Human Capital: Conceptual Framework, Theoretical and Empirical Literature

Abstract This chapter presents the conceptual and theoretical framework and theoretical and empirical literature that emphasize the positive growth effects of human capital, technological progress and innovation in increasing and sustaining economic growth. We explain that the major difference arise because the exogenous growth theories perceive technical progress and human capital as exogenous variables in growth accounting model, whereas the endogenous growth theory envisages technical progress and human capital as endogenous variables determining the rates and differences of economic growth across countries. We illustrate that the inclusion of human capital and technological change in growth accounting models motivate endogenous growth literature to provide several interesting explanations of the relationship between human capital and technical change. In particular, it stimulates considerable debate about the complementary relationship between human capital and technical progress, skilled biased technical change, the role of technical progress in skill upgrading and the role of skill and improvement in the accumulation of human capital in skill upgrading. Finally, we show the advantages and limitations of several measures of technological change and human capital that have been used in theoretical and empirical literature; some of these measures are used in our analysis.

3.1 Introduction

Technical progress has been essential for the creation, determination, acceleration and improvement of both quantitative and qualitative aspects of economic growth and welfare in any society. Economic growth theories often emphasize the importance of science and technology and the role of technological change in increasing, improving and sustaining the marginal productivity of capital accumulation and the per capita growth rate of the economy. The crucial role of technological change in economic growth has long received particular recognition amongst economists of different schools of thought, from classical, neoclassical, Schumpeterian,

S. Mohamed Nour, *Technological Change and Skill Development in Sudan*, DOI 10.1007/978-3-642-32811-4_3,

evolutionary to new growth theories. However, despite this consonance, both classical and neo-classical economic growth theories view technical progress as an exogenous or unexplained variable. The new growth or endogenous growth theory endogenizes technological progress in economic growth model and explicitly mentions technological progress as the main endogenous factor behind economic growth. Ever since, economists highlight the endogenous role of technical progress in stimulating economic growth and human welfare and identify industrial innovation as the engine of growth (Romer 1990; Freeman and Soete 1997).

Moreover, economic growth literature equally recognizes human capital as an important element for economic growth, and many recent theoretical and empirical studies conducted across countries include some proxies for human capital and emphasize the role of investment in human capital, particularly in the form of education. A higher educational attainment implies more skilled and more productive workers, who in turn contribute to enhancing innovative activities and absorption of advanced technologies. Endogenous growth literature explicitly reveals human capital as one major source of economic growth and acknowledges the endogenous role of human capital accumulation in economic growth. More recent literature finds that various measures of schooling are important determinants of per capita growth: an increase in the quantity of human capital per person leads to higher rates of investment in human capital, and so to higher per capita growth.

In light of this background and the findings in Chap. 2 above, it is therefore reasonable to highlight the need for improvement of education, skill upgrading and technological progress for economic development in the Sudan. Before starting the empirical analysis, it is useful in this chapter to briefly explain the concepts, measures and theoretical and empirical literature in relation to human capital (education), technological change and economic growth. We provide a background for the empirical analysis in the following chapters by surveying the theoretical and empirical literature that emphasizes the positive endogenous growth effects of technical change and human capital in increasing and sustaining economic growth.

The rest of this chapter is organized as follows: in Sect. 3.2 we define the concepts of technological change and human capital; the theoretical and empirical literature on the relationship between technological change, human capital and economic growth are presented in Sect. 3.3. Section 3.4 describes the measures of technological change and human capital; Section 3.5 discusses the role of public policies in supporting endogenous growth, and finally, Sect. 3.6 concludes.

3.2 Conceptual Framework: Technological Change and Human Capital

Before presenting the theoretical and empirical literature, it is useful to begin with the definition of the concepts of technological change and human capital.

Distinction has been made between the term technology, technological change and the effect of technological change. The term technology refers to the branch of knowledge concerned with applied sciences and means the systematic treatment, study, use and application of scientific knowledge for practical purposes, such as in industry. Freeman and Soete (1997) define technology as a body of knowledge about techniques, but frequently used to encompass both the knowledge itself and the tangible embodiment of that knowledge in an operating system using physical production equipment. They use the expression 'technical innovation' or simply 'innovation' to describe the introduction and spread of new and improved products and processes in the economy and 'technological innovation' to describe advances in knowledge.[1]

The rate of technological change is often defined by the rate of increase in the stock of knowledge and relates to the effect it introduces in shifting the production function, leading to either a new shift or an upward shift in the production function. Technological change can be neutral (unbiased) when it does not save any or leads to equal savings of all factors of production, but it can be biased when it results in increased using or saving of one factor rather than others. The classification of technical change into labour (capital) saving technical change implies that it increases the marginal productivity of capital (labour) more than it increases the marginal productivity of labour (capital). Another interpretation indicates that technological change can be equally capital and labour augmenting if it leads to an increase in the production due to either unchanged or equal increases in capital and labour inputs. However, technological change can be purely labour (capital) augmenting, if it leads to increase in effective labour (capital), while effective capital (labour) is constant.

Schumpeter (1934) discusses technological change in the form of innovation including the introduction of new products, services or methods of production; improvement in the quality of existing product or service; development of new markets; exploitation of new sources of supply; and reorganization of methods of operation. Product innovation refers to a substantially new product or an essential improvement to an existing product, while process innovation refers to the introduction of a new or essentially improved method of production.

Human capital refers to health and education that measured by many indicators, for instance, the amount of human capital embodied in people and their respective influence on productivity on the job are determined by skills, ability, education and training (cf. Schultz 1961; Becker 1962). In particular, skill is a broad concept and represents one important form of human capital and tacit knowledge[2]: it refers to acquired and practiced ability or to qualifications needed to perform a job or certain

[1] See Freeman and Soete (1997: 24).

[2] In distinguishing between codified and tacit knowledge Freeman and Soete (1997) argue: "the codified knowledge implies that knowledge is transformed into information which can either be embodied in new material goods (machines, new consumer goods) or easily transmitted through information infrastructure. While, the tacit knowledge refers to that which cannot easily transferred because it has not been stated or measured in an explicit form, one important kind of tacit knowledge is skill, which can be acquired through learning but often of a non-routine kind" (1997: 404, 405).

tasks competently in the labour market. In addition, other indicators such as training, learning by doing and average years of experience are important components in the formation of skills and human capital.

3.3 Theoretical Framework: Technological Change, Human Capital and Economic Growth

Based on the above framework, in this section we show the theoretical and empirical literature on the relationship between technological change, human capital and economic growth. We explain that economic growth theories recognize and provide different perceptions and analytical frameworks for modelling the various effects of technical change, innovation and human capital on economic growth. The major differences arise because exogenous growth theories perceive and model technical progress and human capital as exogenous variables in growth accounting model, while, in contrast, the endogenous growth theory envisages and models technical progress and human capital as endogenous variables in determining growth process.

3.3.1 Economic Growth Theory and Exogenous Technical Change

The classical economists, starting with Adam Smith (1776), observe the importance of the variable of technical progress in the form of invention (discoveries of new goods and methods of production), innovations, increasing specialization of labour and market expansion in the capitalist system. Despite the apparent recognition of the importance of technical progress in the classical growth theory, technical progress is assumed to remain exogenous variable in growth process.

Next, the neo-classical economists place more emphasis on the significance of technological change. For instance, Solow (1957) attributed 90 % of the US growth rate during the period 1909–1949 to technical progress; Abramovitz (1956), Kenderik (1956) and Solow (1957) attributed almost all the change in output per hour worked in 1950s to technological change. Subsequent analysis by Jorgonson et al. (1987) showed the importance of technological change beside the increase in the effective labour force and the effective stock of capital in generating growth in output per worker. The neoclassical growth theory assumes an aggregate production function exhibiting constant returns in labour and capital; the only source of output growth being the increase of capital stock. While the rate of technological change is assumed exogenous variable, represented as a residual factor to measure the growth of TFP, thus in the absence of technological change diminishing returns will eventually cause all economic growth to cease (cf. Solow 1956, 1957; Swan

1956).[3] Therefore, in order to compensate the diminishing returns of capital in the neo-classical framework long run and sustainable growth rate of output per capita assume to equal the continuous advances in the exogenous rates of technological progress in the form of new goods, new markets or new processes. The major limitation of the neoclassical growth theory is that the long run per capita growth is exogenous and determined entirely by the exogenous technical change or residuals factor, which is determined outside the model. This also called the black box problem since the residuals factor includes technical progress beside the contributions of many other variables such as human capital (education), organization, management, knowledge, new machines, etc. Moreover, although, technical progress is included in the neo-classical model, it is not treated as a production factor like capital and labour, and the effect of technical progress is viewed only as a shift in the production function (cf. Solow 1956, 1957).

3.3.2 *Economic Growth Theory and Endogenous Technical Change*

The neo-classical growth theory fails to explain persistent differences in growth rates across countries because it considers the rates of technological progress, which entirely determine the growth rate, as an exogenous variable and fails to deal with increasing returns in a dynamic general equilibrium framework.

The endogenous growth theory contributes to improving understanding of the interaction between technological change and economic growth and fills the gap in the neoclassical theory by recognizing the important endogenous effects of technological progress and innovation for creating and sustaining economic growth. In particular, the endogenous growth theory considers an endogenous technological change and innovation within a dynamic general equilibrium framework and avoids diminishing returns to capital. The endogenous growth theory assumes that technical change and human capital are the major sources of endogenous growth and the presence of increasing returns to scale and externalities prevent diminishing returns to accumulation of capital and so guarantee the steady state of growth in the long run.

An earlier attempt of the endogenous growth theory was made by Schumpeter (1934), who assumed that technological progress, innovation and their diffusion are driving forces and at the centre of the dynamics of the economic system. Schumpeter provides a pioneering theory of innovation that forms the basis for

---

[3] The neo-classical Solow – Swan model assumes a general production function $\boldsymbol{Y(t)}$, in which the flow of output produced at time $\boldsymbol{t}$ and there are only two inputs, physical capital $\boldsymbol{K(t)}$, and labour $\boldsymbol{L(t)}$, the production function takes the form:

$$Y(t) = F\,(K(t), L(t), t)$$

The growth rate of the production function depends on time $\boldsymbol{t}$, which reflects the effect of technological change. The long-run growth rate is determined entirely by exogenous elements such as the saving rate and the level of technology.

the subsequent thinking on the dynamic role of technological innovation in economic growth determination. Schumpeter considers innovation as an activity made by one or more workers (e.g. skilled workers), which produces an economic gain, growth and profit either by reducing costs or creating extra income.

A subsequent attempt by Arrow (1962) indicates that technology improvement and the growth of technical change become endogenous due to an unintended effect of learning by doing. Other earlier endogenous growth models represent the major sources of growth by technical progress, which is viewed as a by-product of production and investment in human capital (cf. Uzawa 1965; Nelson and Phelps 1966). Uzawa (1965) interprets technical progress as representing human capital per worker, assuming that its growth required the use of labour services in the form of educational inputs and analyzed optimal growth paths.[4]

Further efforts by Nelson and Winter (1977, 1982) attempt a search for a useful theory of innovation, and present an evolutionary theory of economic change that assumes economic change partially stems from innovation on the part of the firm. The ensuing attempt by Dosi et al. (1988) contributes to an evolutionary theory of endogenous technical change by investigating the interaction between technical change and economic theory.

Since the mid 1980s, starting with the work of Romer (1986, 1989, 1994), Lucas (1988) and Rebelo (1991), which are based on the work of Arrow (1962) and Uzawa (1965), the endogenous growth theory explicitly recognizes the endogenous role of technical change and distinguishes between labour and human capital. The endogenous growth theory avoids the diminishing returns to the accumulation of capital and highlights the role of increasing returns and assumes that growth may proceed indefinitely due to the presence of human capital and endogenous technical progress. The endogenous growth theory also predicts that, in the long run, economic growth at the aggregate level is determined by endogenous sources of human capital, technical change, learning by doing, spillovers of knowledge, external effect of human capital and R&D.

For instance, Romer (1986) and Lucas (1988) contributed to revitalizing the growth literature using Arrow's (1962) ideas to eliminate the tendency for diminishing returns by assuming that knowledge creation was a side product of investment and a positive effect of experience called learning by doing or learning by investing, but that the rate of technical change remains constant. Next, a major contribution by Romer (1990) presents a pioneering endogenous growth model where technical progress is defined by R&D; assumes that non-homogenous capital consists of a set of different intermediate goods; and new intermediate inputs are discovered when R&D resources are devoted to the search process. Romer (1990) identifies two major sources of increasing returns to capital due to specialization or product differentiation, as in Romer (1987), and research spillovers, in which growth will accelerate indefinitely. Romer (1990) assumes knowledge about technology is a nonrival input and induces spillover effects.

[4] See Aghion and Howitt (1998: 24).

Different from Romer (1990), Aghion and Howitt (1992) present an endogenous growth model that defines technical progress by both R&D and accumulation of technological knowledge through the channel of industrial product and process innovations, which improve the quality of products. Aghion and Howitt's (1990, 1992, 1998) framework differs from earlier models of endogenous growth (Romer 1986, 1990; Lucas 1988) in assuming a model of growth based on Schumpeter's (1942) process of creative-destruction. Where growth results exclusively from technological progress, which has positive and normative implications for growth in creating losses as well as gains, by rendering obsolete skills, goods, markets and manufacturing processes. Innovation consists of 'creative-destructions' rather than just new additions to production, and individual innovations are sufficiently important to affect the entire economy. Aghion and Howitt (1990) follow Romer (1990) in endogenizing technical change in producing endogenous growth, and follow Arrow (1962), Romer (1986) and Lucas (1988) in introducing learning by doing as a second source of growth beside innovation. They assume that the accumulation of learning by doing in the intermediary industry will introduce an increase in productivity in the consumption goods sector, and, in particular, intermediate firms will experience a complete spillover of their learning by doing, which also spills over into the research sector. Different from Romer (1986), the spillover of learning by doing in Aghion and Howitt (1990) leads to private economy growth: they assume that an increase in learning by doing will have a positive direct external effect on the average growth rate. Aghion and Howitt (1992, 1998) assume that a stochastic economic growth is generated by random sequences of product innovations and quality-improving (or vertical) innovations that themselves result from (uncertain) research activities by firms. The average growth rate is determined by the interactions of spillovers or two externalities: positive effect, whereby the knowledge embedded in each innovation can be used by all future researchers to generate growth; and negative effects, namely the business stealing effects.

3.3.3 Human Capital and Economic Growth

Economists have long recognized the importance of human capital to the growth process. For instance, Adam Smith's writings at the beginning of the first industrial revolution recognized that human skills were already becoming more important than raw materials in the designed and manufactured machines. Endogenous growth theory fills the gap in earlier growth theories by assuming the accumulation of human capital is another source of endogenous growth. For instance, both Romer (1986) and Lucas (1988) present endogenous growth models where higher accumulation and an average level of human capital in a context of either increasing or constant returns lead to higher productivity of workers and a higher endogenous growth rate. The endogenous growth theory emphasizes the endogenous role of technology and human capital in economic growth, elaborates on the interaction between them and their central roles in determining the magnitude, speed and difference of growth rates across countries (cf. Abramovitz 1986; Lucas 1988; Romer 1990; Aghion and Howitt 1992). The literature discusses the relationship between human capital and economic growth following the pioneering approaches by Nelson and Phelps (1966) and Lucas (1988) and the contributions of Romer (1986, 1989, 1990) and Aghion and Howitt (1992).

An earlier attempt of the endogenous growth model is the AK model, which assumes a constant return to scale in a broad aggregate capital including physical and human components "K", an improvement in the level of technology, raises the marginal and average product of capital and the growth rate.[5] The AK model has the advantages of inclusion of physical and human capital, elimination of diminishing returns to accumulation of capital and creating endogenous growth; its limitations are the assumption of a fixed level of technology and the failure to explain long run balanced growth rate.

An earlier pioneering approach refers to an important contribution by Nelson and Phelps (1966), which assumes growth rates as being driven by the stock of human capital, which in turn affects a country's ability to innovate or catch up with more advanced countries. Nelson and Phelps (1966) explain differences in growth across countries are primarily due to differences in human capital stocks and the abilities to generate technical progress. They assume that growth is primarily driven by the stock of human capital, but the effects of education and human capital are more important for producing technological change than for producing output under a given technology. Nelson and Phelps (1966) and, more recently, Benhabib and Spiegel (1994) assume that human capital[6] is necessary for innovations (capacity to innovate) and for adapting to new technologies and thereby speeding up technological diffusion throughout the economy. A first implication of the Nelson-Phelps approach is that productivity growth and the rate of innovations should increase with the level of educational attainment, particularly with the enrolment in secondary and higher education, which best reflects the numbers of potential R&D staff in a country. Recent empirical studies verify this result and show the significant impact of secondary and higher educational attainment level on the rate of productivity growth (cf. Barro and Sala-i-Martin 1995; Benhabib and Spiegel 1994). A second implication of Nelson and Phelps (1966) is that the marginal productivity of educational attainment is increasing in or with the rate of technological progress (including both innovation rate and speed of adapting to new technologies). Some studies verify this result and find that education induces a significant impact on productivity growth only when it is explicitly related to the rate of innovations and the speed of technological catch-up (cf. Bartel and Lichtenberg 1987; Benhabib and Spiegel 1994). A third interesting result of Nelson and Phelps (1966) is that education should permit the countries falling behind to

[5] The **AK** production function without diminishing returns and with a fixed level of technology $\boldsymbol{A}$ is defined by: $\boldsymbol{Y = AK}$.

[6] In Nelson and Phelps (1966) approach human capital is referring to education and the highly skilled workers.

learn more from advanced countries and thereby achieve a higher degree of productivity improvement when innovating. Recently, Benhabib and Spiegel (1994) support this result and indicate that the effect of past educational attainment levels on current growth rates is more obvious across countries that fall behind in terms of aggregate productivity, but growth is to be principally driven by technological catch-up. Thus, the inclusion of technical progress beside human capital substantiates the role of human capital in technological catch-up.[7]

A further interesting approach was introduced by Lucas (1988), based on Becker's (1964) theory of human capital, and on the idea that growth is primarily driven by the accumulation of human capital (education).[8] The Lucas (1988) approach is a pioneering contribution to the endogenous growth literature: it regards human capital accumulation as the engine of growth, as an alternative (to technological change) and as a complementary source of sustained growth.[9] Lucas (1988) adapts Uzawa (1965) and Rosen's (1976) formulation and assumes that growth rate is linearly related to human capital level and its accumulation over time. One implication of the Lucas model is that human capital accumulation is a social activity, involving groups of people, in a way that has no counterpart in the accumulation of physical capital. Another implication is that economies with high human capital stock can easily produce more and can thus sustain a high growth rate. On the other hand, an economy beginning with low levels of human and physical capital will remain permanently below an initially better endowed economy.

Hence, in the Lucas model, differences in growth rates across countries are mainly attributable to differences in the rate at which those countries accumulate human capital over time, assuming that the rate of technical progress remains fixed or exogenous, while Nelson and Phelps (1966) explain that differences in growth across countries are primarily due to differences in human capital stocks and the abilities to generate technical progress. Moreover, Lucas (1988) discusses the relationship between productivity growth and the rate of human capital accumulation, whereas Nelson and Phelps (1966) show that productivity growth and the rate of innovations should increase with the level of educational attainment and particularly so with the enrolment in secondary and higher education. Furthermore, Lucas (1988) assumes that the marginal productivity of education is determined and sustained only by the accumulation of human capital, while Nelson and Phelps (1966) assume the marginal productivity of educational attainment is increasing in the rate of technological progress (including both innovation rate and speed of adapting to new technologies).[10]

---

[7] See Aghion and Howitt (1998): 339, 340.

[8] Lucas (1988) defines human capital as general human skills that are produced and acquired by education.

[9] See Aghion and Howitt (1998: 327).

[10] See Aghion and Howitt (1998: 327, 339).

One feature of Lucas' (1988) model is the assumption of constant returns to scale and the accumulation of human capital, which implies that a diminishing return can be avoided when the production function includes both physical and human capital and both these grow at the same rate. Thus, in the steady state, rates of return remain constant and the economy can grow at a constant and sustained rate mainly due to endogenous growth from human capital accumulation, and without the need for external 'engine of growth' or exogenous technological change. Barro and Sala-i-Martin's (1995) results indicate that the presence of human capital (as an alternative to improvements in technology as a mechanism to generate long-term growth) may relax the constraint of diminishing returns to a broad concept of capital and can lead thereby to long term per capita growth in the absence of exogenous technological progress.

Another interesting feature of Lucas' (1988) model is the introduction of human capital with externalities or spillovers effects of education between individuals. Lucas (1988) distinguishes between the internal effects of human capital, i.e. the effects of an individual's human capital on his own productivity, and the external effects of human capital that contributes to the productivity of all factors of production, including his or her own human capital. The external effects of human capital induce more rapid physical than human capital growth; the average skill level of a group of people is assumed to affect the productivity of each individual within the group. Both Lucas (1988) and Romer (1986) highlight the spillover effects or benefits from aggregate human capital, supposing that human capital can be passed down from generation to the next and can therefore grow without bound. Assuming that this special kind of knowledge is only produced as a side effect of other activities, investment in physical capital or investment in schooling respectively, while Romer (1989, 1990) allows this special kind of knowledge to be produced intentionally and not as a side effect. Azariadis and Drazen (1990) find that the existence of threshold externalities in education technology can lead to several steady state growth paths and explain existing continuous and perpetual differences in growth rates across countries due to unequal initial human capital endowments.[11]

Moreover, Lucas (1988) follows the theory of human capital and distinguishes between two main sources of human capital accumulation (or skill acquisition), namely education and learning by doing. Based on theory of human capital, Lucas assumes that the allocation of time over various activities in the current period affect productivity or affect the accumulation of human capital $h(t)$ level in future periods. Lucas identifies both the way the human capital level affects current production and the way the current time allocation affects the accumulation of human capital. Lucas assumes that a worker with skill level h devotes the fraction of $u(h)$ of his non-leisure time to current production and the remaining $1 - u(h)$ to human capital accumulation. So, the human capital equation in the Lucas model is defined by: $h = h(t)\delta(1 - u(t))$, $\delta > 0$ which spells out how current schooling time

[11] See Aghion and Howitt (1998: 331, 333).

$(1 - u)$ affects the accumulation of human capital. If learning by doing rather than education is the primary source of human capital accumulation, the above equation should be replaced by the following equation: $h = \delta hu$.[12]

The subsequent contribution by Romer (1990) presents a growth model of endogenous technical change assuming long run growth is increasing in and driven primarily by both technological change (the accumulation of knowledge) and the stock of human capital[13] rather than the total size of the labour force or the population. He emphasizes the central role of technological change, stock of human capital, externalities and increasing return associated with investments in human capital in the research sector and in determining the rate of growth. He finds that an economy with a larger stock of human capital will experience faster growth. Romer (1990) follows both Romer (1986) and Lucas (1988) in their assumption of external effects arising from knowledge spillover. Romer (1990) assumes that the final output is a function of physical labour, physical capital, human capital and an index of the level of technology. The application of more human capital to research leads to higher rate of production of new designs and stock of knowledge, which increases the productivity of engineers working in the research sector.[14] The output of the design is a linear function of human capital and technology when the other variables are held constant. The marginal productivity of human capital employed in the manufacturing sector grows in proportion to technology. Unlike Lucas (1988), Romer (1990) follows Schumpeter's (1942) assumption that technological change drives growth and provides the incentive for continued capital accumulation, the growth rate is increasing in the stock of human capital. So, both capital accumulation and technological change account for much of the increase in output per worker.

The endogenous growth model proposed by Aghion and Howitt (1992) assumes that capital accumulation includes both physical and human components. They assume that both the average and the variance of the growth rate are increasing functions of the size of innovations, size of skilled labour and the productivity of research, which is measured by the effect of research on the Poisson arrival rate of innovations. They distinguish between three categories of labour: unskilled labour, which can be used only in producing consumption goods; skilled labour, which can be used either in research or in intermediate sector; and specialized labour, which can be used only in research. They assume that skilled labour has two uses: in the manufacture of the latest generation of intermediate goods and research aimed at discovering the next generation of these goods. An expectation of more research in the next period must correspond to an expectation of higher demand for skilled labour in research in the next period. They assume that research produces a random

[12] See Lucas (1988) and Aghion and Howitt (1998: 327, 329).

[13] Romer's (1990) definition of human capital includes activities such as formal education and on the job training.

[14] One implication of Romer's (1990) argument is that, despite having the same amount of human capital, an engineer working at current time has higher productivity than one who worked in the previous century because he acquires the advantages of all additional improvements and accumulation in knowledge since then.

sequence of innovations, and that the Poisson arrival rate of innovations in the economy at any instant is dependant on the current flow of skilled labour used in research. They assume that skilled labour is important factor in research, innovations and economic growth, and that an increase in the endowments of skilled labour increases both the marginal benefit and reduces the marginal costs of research by reducing the wage of skilled labour.

Moreover, several recent empirical studies conducted across countries use many measures of human capital and find that human capital is important determinant of long run economic growth or per capita growth.[15]

3.3.4 *The Relationship Between Technological Progress, Human Capital (Skill) and Skill Upgrading*

In this section, we show that the inclusion of technological change and human capital in growth accounting models motivates endogenous growth literature to postulate several explanations of the relationship between human capital and technical change. In particular, considerable debate arises around four issues regarding the complementary relationship between human capital and technical progress, skilled biased technical change, the role of technical progress in skill upgrading and the role of skill in skill upgrading.

The first hypothesis highlights the complementary relationship between technological progress and human capital. The interpretation of this hypothesis is that the high educated workers can adapt more and easier to changing technologies than the low educated workers. A large endowment of human capital facilitates the fast adaptation of technologies and induces positive impacts on economic growth, and faster technology driven growth in turn can induce more schooling by raising the rate of return on investment in schooling (cf. Nelson and Phelps 1966; Benhabib and Spiegel 1994). Moreover, human capital or skill is found to be more complementary with technology and capital (cf. Goldin and Katz 1998; Mincer 1989). Because the 'embodiment' of technical change in both physical and human capital indicates that the improvement in their quality implies their complementarity with technological change (Bartel and Lichtenberg 1987). In addition, more innovation stimulates human and physical capital accumulation by raising the marginal product of capital, while more capital accumulation stimulates innovation by raising the profit accruing to innovation (Aghion and Howitt 1998). Furthermore, a high proportion of skilled workers in the labour force implies a large market size for skill-complementary technologies and encourages faster upgrading of the productivity of skilled workers (cf. Acemoglu 1998).

[15] For instance, Rebelo (1991), Barro (1991, 1996), Barro and Lee (1993, 1996, 2000a, b, 2010), Benhabib and Spiegel (1994), Barro and Sala-i-Martin (1995), Mankiw et al. (1992), and Kahn and Lim (1998) all find strong positive correlation between schooling and the growth rate or the subsequent growth rate of per capita GDP or TFP.

Several studies use many different indicators to examine the technological progress and human capital complementary hypothesis. For instance, the increasing utilization of higher educated workers shows positive correlation with TFP growth (cf. Kahn and Lim 1998), with physical capital, capital equipment or capital intensity (cf. Griliches 1969; Bartel and Lichtenberg 1987; Goldin and Katz 1998), with R&D (Acemoglu 1998; Machin and Van Reenen 1998) and with the use of new technologies (cf. Acemoglu 1998), especially ICT (cf. Goldin and Katz 1998; Bresnahan et al. 1999; Autor et al. 1998.

The second hypothesis concerns the skill-biased nature of technical change. The rationale for this argument is that technical change has dual implications on employment and demand for skill, which is found biased against low skilled workers and lead to unemployment /job mismatch (cf. Muysken et al. 2002a; b) or crowding out of low skilled workers (cf. Muysken and ter Weel 1998).

Another interpretation is based on the argument that technical changes induce creative-destruction effects on growth and employment. While it enhances productivity growth, stimulates demand and the creation of new jobs, it also destroys jobs because it is primarily labour saving through automation and skill obsolescence (cf. Aghion and Howitt 1992, 1998).

In the recent literature two features have received particular attention: the first issue is that economic debate has become focused on the significant change in the composition of labour demand, particularly the increase in the demand for skilled workers and sharp decline in the demand for low skilled workers. The second issue is focused on the distributional aspects of technical change, particularly the implications of skill-biased technical change (SBTC) on the structure of employment and wages that has shifted against the low skilled workers, leading to either an increase in unemployment of low skilled workers or increasing wages divergence between high skilled and low skilled workers, which leads to greater inequality (cf. Autor et al. 1998; Acemoglu 1998; Bound and Johnson 1992).

The skill-biased technical change (SBTC) hypothesis has been verified both at the macro level across both developed and developing countries (cf. Berman et al. 1998) and at the micro level within industries (cf. Berman et al. 1994). SBTC is related to various measures of technical changes such as TFP growth (cf. Kahn and Lim 1998), R&D (cf. Berman et al. 1994; Acemoglu 1998; Machin, and Van Reenen 1998) and the use of IT or ICT (cf. Bound and Johnson 1992; Berman et al. 1994; Freeman and Soete 1994; Autor et al. 1998).

The third hypothesis explains the role of technical progress in skill upgrading. The interpretation of this hypothesis is that both the technology-human capital complementarity and skill-biased technical change hypotheses imply that a higher rate of technical progress should bring an increase or upgrading in skill level. Several studies in the literature use many indicators to show the role of technical progress (in the form of TFP, R&D, ICT, IT or computer use, etc.) in skill upgrading. For instance, skill upgrading, defined by the increasing incidence of training, increases with the rate of technological change (cf. Mincer 1989; Bartel and Sicherman 1995, 1999),[16] especially in sectors in which the Jorgenson measure of productivity growth was higher (cf. Lillard and Tan 1986) or showing an increasing use of IT or computers (cf. Bresnahan et al. 1999). Skill upgrading, defined by the shift away from unskilled towards skilled employment or increase in the share of white-collar high skilled workers, is also positively correlated to variables related to technological change, such as R&D investment and growth in the number of patent (cf. Colecchia and Papaconstantinou 1996).[17] In addition, skill upgrading, defined by the increase in the wage share of white collar workers, is positively related to two measures of technology: the level of investment in R&D and computers (cf. Berman et al. 1994). Furthermore, skill upgrading, as defined by the change in the share of educated workers in employment and return to schooling or wage rate, is positively correlated with the increase of R&D intensity (cf. Machin, and Van Reenen 1998). Moreover, skill upgrading, as defined by the share of high skilled workers, is positively correlated with TFP (cf. Garcia Cervero 1997) or the use of computers, IT, ICT or computer-intensive industries (cf. Autor et al. 1998; Bresnahan et al. 1999). Skill upgrading, decreasing motor skills and increasing cognitive skills accompany the diffusion of ICT, mainly through occupational change rather than educational improvement, and are also positively correlated with productivity growth (cf. Hwang 2000).

The fourth hypothesis deals with the role of human capital or skill acquisition in skill upgrading. Along with the debate on the relation between technological change and human capital and the positive effects of human capital/education on productivity and economic growth, theoretical and empirical literature highlight the role of human capital/education in skill upgrading through externalities and learning by doing. Educational attainment is important because skills acquisition from formal schooling lead to improvement in training and learning abilities and increase the accumulation of human capital through experience or "learning by doing", which in turn interact together and lead to improvements in workers productivity (cf. Autor 2000).[18] Theoretical literature highlights the role of human capital or skill in skill upgrading through externalities, spillover and learning by doing (cf. Lucas 1988; Romer 1986, 1989, 1990). In addition, the average human capital tends to grow over time as human capital investments have a positive external effect on the human capital of the later cohorts (Stokey 1991). Moreover, recent empirical literature shows that in the developed countries, particularly across the OECD countries, human capital may accumulate at a faster

[16] Bartel and Sicherman (1995) find that on the job training will increase if technological change increases the productivity of human capital, reduces the costs of training or increases the value of time in training relative to work; and that the training gap between the highly educated and the less educated narrows, on average, as the rate of technological change increases.

[17] Colecchia and Papaconstantinou (1996) find that a 1 % point R&D intensity higher than average at the beginning of 1980s has implied about 20 % higher than average upskilling per year.

[18] Autor (2000) argues that: first, training is more productive and therefore valuable to high ability workers; second, workers have some prior information about their ability that is not initially visible to employers; and, third, firms are able to learn about ability through skills training.

rate with the past intensive use of high skilled workers (cf. Colecchia and Papaconstantinou 1996). Furthermore, empirical literature from the developing countries shows that in Singapore and Korea investment in human capital via the expansion and improvement of education and training systems, particularly the development of tertiary, vocational and technical education, leads to an improvement in the overall skill content or skill upgrading of the working population. This appears from the improvement in the educational attainment – defined by highest qualification attained – and skill levels, the rise in the share of high skilled workers, scientists and engineers and the fall in the share of low skilled workers. In addition, upgrading of the occupational structure has resulted from the large/rising share of the supply of high educated, white-collar and non-production workers and the small/falling share of blue-collar workers.[19]

These findings emphasize the importance of the endogenous effects of technical progress and human capital for enhancing economic growth. In particular, these explanations imply that next to the important endogenous effects of technical progress and human capital in economic growth, the complementary relationships between them and between them and skill upgrading are also important for enhancing economic growth.

3.4 Measurement of Technological Change and Human Capital

While it is admitted that technological progress and human capital are difficult to measure, the theoretical and empirical literature use many indicators to approximate their effects. It will be useful to illustrate the advantages and weakness of these measures in order to select some relevant measures for the empirical analysis in the subsequent chapters.

3.4.1 Measurement of Science and Technology Indicators

In recent years, a new economic system has evolved that is characterised by both globalisation and the rise of information and communication technologies. This has driven the need for development in science and technology (S&T), which has become more than simply an element of economic growth and industrial competitiveness, but is now also essential for improving social development, the quality of

[19] See Cheah (1997), Low (1998) and Cheon (1999). In Singapore, the transfer of foreign technology and foreign skills stimulates the acquisition of knowledge and skills from abroad and induces positive spillover in upgrading the skill of domestic workers. This has been accompanied with technological upgrading to promote mechanization, computerization, automation, etc. In the Republic of Korea, the integration into global economy or exposure to foreign competition leads to skill upgrading of domestic workers in the manufacturing sector. Skill upgrading of domestic skills facilitates the adoption of foreign technologies and technological catching-up with the advanced countries.

life and the global environment. For instance, the high level of economic and social development in today's industrialised countries is largely the result of past intensive investment in S&T; similarly, newly industrialised countries are catching up because of their active development of S&T.

> Access to scientific and technological knowledge and the ability to exploit it are becoming increasingly strategic and decisive for the economic performance of countries and regions in the competitive globalized economy. The 50 leading S&T countries have enjoyed long-term economic growth much higher than the other 130 countries of the rest of the world. Between 1986 and 1994 the average growth rate of this heterogeneous group of countries was around three times greater than that of the rest of the world. The average economic wealth per capita of these 50 countries has grown by 1.1 % per year. On the other hand, the per capita income of the group of 130 countries – which perform less well in education, science and technology – has fallen over the same period by 1.5 % per year. These trends prefigure a new division of the global economy, based on access to knowledge and the ability to exploit it. (Organisation for Economic Co-operation and Development (OECD) 1997: ix)

The S&T system is often defined as consisting of all the institutions and organizations essential to the education of scientific people, for example, research and development (R&D) institutions, professional societies and professional organisations linking individual scientists to each other and to their socio-economic environment. The theoretical and empirical literature identifies the important role that S&T plays in promoting economic growth and development in both developed and developing countries.[20]

More recent literature addresses the contribution to S&T performance of the 'national systems of innovation'; a widely used modern term that reflects the link between technical and institutional innovative development, including S&T (e.g. Lundvall 1992; Nelson 1993). Lundvall says this broad definition includes "all parts and aspects of the economic structure and the institutional set-up affecting learning as well as searching and exploring – the production system, the marketing system and the system of finance present themselves as subsystems in which learning takes place" (Lundvall 1992: 12–13). In addition, Freeman and Soete argue:

> The many national interactions (whether public or private) between various institutions dealing with science and technology as well as with higher education, innovation and technology diffusion in the much broader sense, have become known as 'national systems of innovation'. A clear understanding of such national systemic interactions provides an essential bridge when moving from the micro- to the macro-economics of innovation. It is also essential for comprehending fully the growth dynamics of science and technology and the particularly striking way in which such growth dynamics appear to differ across countries (Freeman and Soete 1997: 291).

All the definitions of the systems of innovation share the view that S&T institutions play a vital role in determining or influencing innovation and development. The literature on S&T development often distinguishes between input

[20] For detailed theoretical and empirical literature and assessment studies, see for instance, Freeman and Soete (1997), Dasgupta and David (1994), Foray (1999), Mytelka (2001), Cooper (1991, 1994), Velho (2004). For earlier analyses of S&T in Arab region, see also Qasem (1998a, b), Zahlan (1999a, b), Fergany (1999), ESCWA (1999a, b), ESCWA-UNESCO (1998a, b).

(resources) and output (performance) indicators. For instance, the European Second Report on S&T Indicators (OECD 1997) discusses numerous traditional input and output indicators for S&T development. The input indicators are generally divided into financial and human resources. First financial resource or input indicator includes "R&D expenditure – the most widely accepted indicator for evaluating and comparing S&T efforts in different countries and regions. In the absence of an average measurement to determine R&D within the economic structure and the needs of each country, political decision-makers use indicators such as the intensity of R&D (measured as a percentage of GDP or per capita), R&D area of performance, and origin of funding; change in public spending on education in relation to GDP. [...] In addition to financial resources, human resources are central to research and technological innovation activities". There are also general demographic and human capital indicators, "such as the number of science and technology graduates and the number of scientists and engineers employed in R&D. [...] [There are] four major points relating to human capital: demographic trends, the development of public spending on education, the performance of education systems and researchers and engineers active in R&D". Furthermore, "Human resources in science and technology (HRST) are one of the key resources for economic growth, competitiveness and more general social, economic and environmental improvement" (OECD 1997: 5, 58–59). In addition to total population size and proportion of young people, which represent the human resources potential of each country, educational attainment of the labour force and graduation rates, which show the rate at which newly educated graduates are available at the country level to enter the labour force, particularly the scientific and technological qualifications and doctorate levels, including R&D staff numbers, particularly in S&T fields.

Output indicators, on the other hand, "can be classified according to three parameters: economic, technological and scientific. As to economic outputs, many economists view increases in productivity as a major result of technological investment. [...] The percentage of high-tech exports in total export figures emerges as a potentially useful means of measurement. [...] Clearly not all results are measurable in economic terms. Scientists and engineers often cite the 'learning experience' as one major benefit of engaging in R&D activities. To assess the accumulated knowledge of a given country, its stock of technical knowledge must be quantified. Without doubt, patents and patents applications are the most commonly applied [technological] indicator in this respect and, irrespective of the shortcomings implicit in this approach, they continue to represent a very useful tool". Finally there are direct research outputs or publications, "focusing on the impact of the publication output of a given country or zone and comparing it to the number of publications produced over a certain period of time" (OECD 1997: 79).

3.4.2 Measurement of Technological Change

The literature uses several indicators to measure the role of technical progress in economic growth and particularly distinguishes between input indicators, which include variables such as R&D expenditures and human resources, and output indicators, which comprise variables such as patent, productivity growth, publication, etc. (cf. OECD 1997). A comprehensive approach of technological progress should be based on integration of input and output indicators.

The traditional indicator used in the literature to measure the contribution of technical change in economic progress is represented by total factor productivity (TFP) (cf. Kahn and Lim 1998). It is also known as Solow Residuals, as Solow (1956, 1957) was the one to find that the growth rate of technical progress emerges as the remaining unexplained variable or the residual parameter – defined as the factor of output that can not explained by the input factors.[21] The use of TFP growth measure indicates the high significance of technical progress: for instance, Solow (1957) finds that around 90 % of the growth in US output per worker during the period (1909–1949) was due to the effect of the residual factor, which measures the effect of technical progress. Moreover, Abramovitz (1956), Solow (1957), Kenderik (1961) and Dension (1962) find that about half of the growth of the US economy up to the 1950s was attributed to technical progress measured by TFP. However, the TFP indicator has several drawbacks such as the lack of relevant and adequate data and the inaccuracy and broadness of the concept of TFP, which includes factors other than technological progress such as human capital (education), organization, management, knowledge, new machines, etc. According to Mincer (1989), productivity growth indicates the consequences of technical change, but is not a measure of it; TFP is a useful measure of technological change only if other factors affecting productivity growth are either unimportant or considered in the statistical applications. In addition, TFP growth measure may imply some measurement errors due to business cycles and economies of scale (cf. Mincer 1989: 4).

The major input indicator in measuring scientific and technological progress is often represented by the data relating to R&D expenditures, which have been widely used across the OECD countries due to their consistency and easier computation compared to output indicators in these countries. The R&D expenditure data can be utilized to analyze the comparative development and breakdown of R&D activities according to sector and source of finance. However, R&D expenditure data has several defects: for instance, that the data reflects only the recorded expenditures and the institutionalized aspects of technology aimed at increasing knowledge. It also does not include many activities that contribute to technological knowledge such as design, learning-by-doing, the indirect public spending on R&D, etc. Moreover, R&D data reflects only the effort put into research, and does not reflect the efficiency with which this effort leads to new knowledge, the

[21] Productivity growth is calculated as the differences between the rate of growth of output and (a weighted measure of) the rate of growth of the capital and labour inputs (Mincer 1989), or the differences in growth rates of the social product and the capital and labour production factors.

quality of R&D work undertaken, the quality of the scientists performing the work, the cost of inputs of labour, equipment and materials, etc. Moreover, the definition of R&D expenditure varies substantially across countries and is difficult to measure for a large number of countries, and does not reflect the effects of the international spillover of S&T and variations across countries with respect to R&D performance. In addition, it does not produce immediate results, making it difficult to establish direct relation between R&D performance and indicators of economic growth, because R&D activities lead to knowledge creation, which may lead to improved performance only in the long run.[22]

The major output indicator is defined by the patent indicator, which is utilized in the literature to measure technological capacity and status of a country, sector or company. The literature uses patent data to measure the output of technological activities, to reflect the technological performance over time and across countries, to examine the technological specialization in key sectors of industry and to protect industrial property rights (cf. OECD 1997). The patent indicator is often widely used in the measurement of technological change because it allows for international comparisons over a long period; it provides more accurate and specific analyses by sector and by technology; and allows for more focus on a company, institution and even single inventor or researcher. On the other hand, the patent indicator also has several limitations, such as: a lesser degree of reliability for countries or sectors with a small number of patents; and possible interruptions by reason of having to work with publication rather than priority dates.[23] Further limitations lie in the difficulty in interpreting average annual growth rates per period due to unstable patent numbers at the end of a period, lack of exact measurements and the potential inconsistency between the required and actual measurements, the latter problem admittedly also applicable to other indicators.[24]

Numerous empirical studies use innovation surveys such as the survey of resources (R&D) indicators, survey of direct progress (output) indicators and survey of indirect progress and impact indicators, which can be evaluated by questionnaires to measure technological change. A well-known example of this is the Community Innovation Survey (CIS). Distinction has been made between innovation surveys according to subject and object approaches. The subject approach focuses on the innovator or firm-level innovative activities, identifies both input and output indicators, includes small-scale incremental change, reflects economic indicators and permits for comparisons within industries or inter-firm comparison, but does not allow for comparison between different industries with different outputs. On the other hand, the object approach focuses on significant technological innovations (new product or process) or the objective output of the innovation process, on the technology itself. It allows for an external assessment of the importance of innovation independent of personal judgment and usually

[22] See OECD (1997) Second European Report on Science and Technology Indicators (1997: 37).

[23] The priority date of a patent refers to the date of first filing, whereas the publication date refers to the date on which the patent was published. This leads to a time lag between the priority and publication year. For example, in the European (EPO) system, patents are published 18 months after first application.

[24] See OECD's Second European Report on Science and Technology Indicators (1997: 90, 91).

identified through expert appraisal or through new product announcement in trade journals or other literature. However, it has limitations as it is confined only to major innovations, neglects small incremental innovation and does not include an assessment of the economic significance of innovation.[25]

Some recent studies tend to measure technological change by using an index of use, investment or expenditures on ICT, IT, computers or computer equipment, which are also called the new general purposes technologies. These indicators are relatively easier to calculate and several studies provide strong results when using them to reflect the use and organization of technological innovation (cf. Autor et al. 1998; Bresnahan et al. 1999). However, the utilization of computer use as a measurement of technological progress and innovation may lead to endogenity and measurement problems (cf. Sanders and ter Weel 2000: 26).

For our macro-micro analysis we use R&D, patent and ICT as more relevant measures of technical change at the macro-micro levels. Moreover, at the micro/firm level we use the innovation survey following the subject approach, as it appears more relevant for assessing only small incremental innovation at firm level. Our analysis will not include the object approach since it focuses on big (radical) innovations and seems inappropriate for measuring the small incremental innovations in our case studies in the Sudan. Other measures, such as the TFP measure, are not very relevant for our analysis and will not be included in our study due to a lack of relevant data and information to estimate these at both macro and micro levels. In order to measure these indicators, we will use the available relevant secondary data and information at the macro level and use the firm survey data at the micro level, as we will explain in Chaps. 4 and 6 below.

3.4.3 Measurement of Human Capital (Human Skills)

The most widely used measures of skill in the literature is educational attainment, as measured by the average years of schooling, occupation measure, the share of non-production workers in total employment and the share of non-production workers wages to total wages. In addition, the literature uses other measures of human capital such as school enrolment ratios, adult literacy rates and school quality measures.

School gross and net enrolment ratios reflect current flows of education that accumulates to create the future stocks of human capital and have been used in several studies (cf. Barro 1991; Barro and Lee 1993, 1996).[26] However, they have

[25] See Smith (2000).

[26] The UNESCO definition distinguishes between gross and net enrolment ratios. "Gross enrolment ratio defines the ratio of all persons enrolled in a given level of schooling to the population of the age group that would be enrolled at that level. While, net enrolment ratio modifies the numerator of the gross enrolment ratio to count only the students enrolled within the designated age group i.e. the ratio of students at a given level of schooling in the designated age group to the total population of the same age group. The net enrolment ratios vary between zero and one, whereas the gross enrolment ratios can exceed one" (Barro and Lee 1993: 4). For detailed definition, see also the UNESCO-UIS website: www.uis.unesco.org.

several limitations: for instance, they only measure current flows of schooling and do not reflect the stock of human capital. In addition, they are susceptible to underestimation and overestimation measurement errors[27]: for instance, net enrolment ratios tend to underestimate the actual value of variables on both mortality and migration; gross enrolment ratios introduce errors related to repetition of grades and dropouts, which are widely observed in developing countries. Gross enrolment ratios overestimate the actual value because their calculation are based on annual surveys of educational institutions in each country and reflect registered number of students at the beginning of each school year rather than computing the actual number in attendance. In general, the net enrolment ratio is relatively more appropriate for measuring the accumulation of human capital; however, the gross ratio has been widely used because it is more often available for developing countries.[28]

The adult literacy rates have been frequently used in several studies to estimate the relationship between human capital and economic growth (cf. Barro 1991; Romer 1989). They have an advantage over school enrolment ratios as they reflect the stock of human capital rather than the flow of investment. However, a major problem with adult literacy rates is that they measure only one component of the current stock of human capital or the first stage in the path of human capital formation, but do not reflect the skills that are obtained beyond the most elementary levels of schooling as well as many other important aspects of human capital and various types of technical knowledge, which are important for enhancing labour productivity and economic growth. The use of literacy to measure the stock of human capital implies that education beyond the most elementary level does not contribute significantly to productivity.[29]

The educational attainment or educational level is measured by average years of schooling to reflect the stock of human capital and allow for across countries or international comparison.[30] It has been widely used in the literature (cf. Barro and

[27] Lee and Barro (1997) use an adjusted enrolment ratio to overcome the problems of underestimation in net enrolment ratios and overestimation in gross enrolment ratios.

[28] See Barro and Lee (1993: 4–6).

[29] See Barro and Lee (1993: 6). "The literacy rates have been used in the United Nations Development Programme UNDP, 1990, to construct an index of human capital. Moreover, Barro and Lee (1993) use adult illiteracy rates to proxy for the percentages of adult population who have no school attainment to fill the gap in the availability of census/ survey data" (Barro and Lee 1993: 6, 7). See also the UNESCO-UIS website: www.uis.unesco.org.

[30] The educational attainment is measured by the average years of schooling, which is computed by adding the product of the number of years of schooling times the number of people in each schooling category across school categories, "i.e. defined by the following formula:

$$\textit{Average Years of Schooling} = \Sigma_j\, YR_j.\, H\, Sj$$

Where j is schooling level, YR_j is the number of years of schooling represented by the level j, and H S_j is the fraction of the population for which the jth level is the highest value attained" Barro and Lee (1993: 7). For detailed definition, see also the UNESCO-UIS website: www.uis.unesco.org.

Lee 1993, 1996) as an appropriate and accurate alternative measure to both school enrolment ratios and adult literacy rates. In recent literature, educational attainment is used to reflect the inflows of new school graduates to existing educational stocks across countries (cf. Barro and Lee 2000a, b). The rapid growth in average years of schooling led to double growth in the stock of human capital in the USA (cf. Mulligan and Sala-i-Martin 1995), educational attainment or the average years of schooling has a significant contribution to the growth of total factor productivity (cf. Kahn and Lim 1998). Although the average years of schooling measure is often widely used as the most popular measure of human capital in the new growth literature and comparisons across countries, it has some drawbacks, such as the assumption of constant elasticity of substitution across workers of different group, which implies that always and everywhere workers of each education category are perfect substitutes for workers in other categories. It assumes that productivity differentials among workers with different levels of education are proportional to their years of schooling, that always and everywhere a year of education adds a constant quantity of human capital and delivers the same increase in skill, whether undertaken by a primary pupil or a college student. It implies that always and everywhere a worker with 16 years of schooling is 16 times as productive as worker with 1 year of schooling, irrespective of the wage rate differentials. It does not consider differences in the fields of study and quality of schooling (quality of teachers and education infrastructure) and wage rate across countries. Moreover, the educational attainment does not directly measure the human skills obtained at schools, namely quality of schooling across countries, and it does not reflect the skills and experience gained by individuals after their formal education.[31]

Some studies use schooling quality measure or the quality of educational output to measure the impacts on various dimensions of cognitive skills that affect an individual's productive behaviour, and thereby the quality of the future labour force (cf. Hanushek and Kim 1995; Lee and Barro 1997).[32,33] Although both the quality and the quantity of schooling are important ingredients of human capital, schooling quality measure has several disadvantages as it varies substantially across countries and is difficult to measure for a large number of countries (Lee and Barro 1997: 1).

The occupation measure or classification is based on the definition of employment structure and the relative shares of educated and non-educated workers in total employment. In particular, the ILO International Standards Classification of

[31] See Mulligan and Sala-i-Martin (1995: 2) and Barro and Lee (2000a: 12).

[32] The definition of this measure includes the pupils/teacher ratios, spending per pupil at primary and secondary schools as a percentage of GDP, and also estimates average salaries of primary school teachers to per capita GDP.

[33] Hanushek and Kim (1995) find that cognitive skills are an important component of relevant variations in human capital, reinforcing the attention to school quality found in many countries today. Their results indicate that quality of labour force has a consistent, stable and strong influence on economic growth, the impact of quality indicates that one standard deviation in mathematics and science skills translates into 1 % point in average annual real growth. This growth effect is larger than would be obtained from over eight years in average schooling.

Occupations (ISCO)[34] is a widely used measure for measuring skill composition in the literature (cf. Colecchia and Papaconstantinou 1996; Hwang 2000).[35] According to the ISCO classification, only WCHS is referred to as "high-skilled", with all other groups regarded as low skilled. The advantage of the occupation measure is that the change in occupational distribution of employment provides more information on the skills required and measures the change in skills structure. But it has the drawback that it does not necessarily take into accounts on the job learning and, in particular, skills associated with the use of new technologies.[36] It also fails to account for the changing nature of skills under an occupational title.[37]

The share of non-production workers to total employment measure is defined by the ratio of the non-production workers to total employment and has been usually used in the literature (cf. Kahn and Lim 1998; Cheon 1999). However, it has several limitations: for instance, it does not exactly reflect change in relative demand for non-production workers, and it may over-represent the shift in demand for non-production workers. Moreover, the definition of non-production workers includes a lot of low-skilled service jobs such as janitors, cleaners or simple clerical jobs, and various liberal occupations, while excluding production supervisors, foremen and skilled workers that are of considerable importance in manufacturing sector of developing countries (Cheon 1999: 12, 13).[38]

The share of non-production workers' wages in total wages is measured by the ratio of non-production workers wages to total wages, and has been used in several empirical studies (cf. Kahn and Lim 1998). Its advantage is that the changes in the non-production share in the wage bill provide a better measure of the demand shift toward non-production workers, provided that the elasticity of substitution between production and non-production workers is above one (cf. Berman et al. 1994; Cheon 1999). However, it has several limitations: for instance, the measure is originally based on the definition of non-production and production workers for skill and unskilled workers and may suffer the same measurement errors related to the definition of non-production and production workers as we explained above.

---

[34] The ILO International Standards Classification of Occupations (ISCO) are aggregated in the following way:

White-Collar high-skilled (WCHS) includes legislators, senior officials, managers, professionals, technicians and associate professionals.

White-Collar low-skilled (WCLS) includes clerks, services workers, shop and market sales workers.

Blue-Collar high-skilled (BCHS) includes skilled agricultural and fishery workers, craft and related trade workers.

Blue-Collar low-skilled (BCLS) includes plant and machine operators and assemblers and elementary occupations.

[35] Hwang (2000) finds that skill upgrading, decreasing motor skills and increasing cognitive skills are accompanying the diffusion of ICT, mainly through occupational change rather than educational improvement.

[36] See Colecchia and Papaconstantinou (1996: 8).

[37] See ILO (1998) World Employment Report (1998/1999: 35).

[38] See Kahn and Lim (1998) and Cheon (1999: 12, 13).

Moreover, the wage measure is applicable only when the elasticity of substitution between production and non-production worker is above one. Furthermore, it may be inaccurate to reflect the movement in the stock of human capital when the relative wages change for reasons other than changes in human capital and technological stocks (e.g. due to price change). So, the wage measures may induce some measurement errors (cf. Cheon 1999; Goldin and Katz 1998; Machin and Van Reenen 1998).[39]

For our analysis at the macro level we use the measures of school enrolment ratios, literacy rates, educational attainment, school quality measures and occupational classification to assess skill levels, based on information and data from many relevant sources (e.g. Sudan central bureau of statistics population census data for 2008, the UNESCO, UNDP, etc.). In addition, in our analysis at the micro/firm level, we use two more relevant measures of skill, namely, educational attainment and occupational classification, based on data obtained from the firm survey as we will explain in Chaps. 4, 5 and 6. Our analysis will not include other indicators such as the share of non-production workers in total employment and the share of non-production workers wages in total wages due to a lack of relevant data to estimate these. Instead, we use the share of high-skilled in total employment and the share of high-skilled wages to low-skilled wages according to education and occupation definitions.

3.5 Endogenous Growth and Public Policy

We mentioned in Sect. 3.3 that endogenous growth literature revealed several robust facts and interesting implications that paved the way for growth; it is convenient in this section to explain that it also provided some insights for a possible role for government policy. We explain below the literature and arguments for government intervention to promote the accumulation of technology, human capital and hence growth rate.

The most popular view in the literature is that the rationale for government intervention is basically related to the idea that knowledge (in the form of technical progress or accumulation of human capital) is a public good, which is non-rival and partially excludable. As in Romer (1990) and Barro and Sala-i-Martin (1995), these two features imply an unbounded growth and incomplete appropriability of knowledge and raise the possibility of knowledge spillovers across firms and hence the whole economy. While the feature of spillovers of knowledge supports endogenous growth, it also creates a form of externality and implies that private investments generate a positive external effect and the private returns from investment tend to be lower than the social returns. The outcomes tend not to be Pareto optimal but sub optimal and require government intervention to correct the distortion. The social

[39] See Cheon (1999: 12, 13), Goldin and Katz (1998) and Machin and Van Reenen (1998).

optimum can be achieved by many instruments such as providing subsidies (which can be financed by taxation) to improve the accumulation of technology and human capital, incentives and returns from investment for private investors.

In the endogenous growth literature some studies explicitly model the importance of technology and human capital for endogenous growth, but only implicitly indicate a role for public policy. For instance, while the Lucas (1988) model emphasizes investment in human capital, it only implicitly allows for a role for public policy through subsidies (Haslinger and Ziesemer 1996: 230). Moreover, the Arrow (1962) learning-by-doing and Romer (1986) models imply an indirect intervention: an investment tax credit that increased the accumulation of capital necessarily also increased the accumulation of technology (Romer 1990: S94).

According to Ziesemer's (1987) interpretation, T. W. Schultz (1964) presents a pioneering theoretical justification for a central role for government interference to promote public investment and emphasis their long-run effects on growth and development. Schultz's (1964) theory reveals that the provision of public factors, such as basic education and basic scientific research, is necessary for human capital formation and this would drive technical progress. Therefore, technical progress depends on human capital, and the production of human capital in turn requires public factors such as basic education and basic scientific research. It is assumed that the public goods are financed through a linear income tax rate: the lower the level of public goods and tax rate, the higher is the price of human capital and less human capital is supplied. If the rate of technical progress depends upon human capital, then technological progress is dependent upon taxation and public goods. The contribution by Shell (1967) involves public investment and assumes a public, non-rivalrous stock of technology; a flat-rate income tax is raised to finance the change in the stock of knowledge. Tax has two effects: it increases growth, but also decreases the returns from investment and negatively affects private capital formation.[40]

Ziesemer (1990, 1991, 1995) formulates Schultz's (1964) ideas that public factors – basic education and basic scientific research – are held as necessary for the formation of human capital and the development process. Ziesemer (1990) argues for an essential role of public factors, which are provided by the government and in turn financed by a simple flat rate income tax to introduce an outstanding role of economic policy in economic development. Ziesemer (1991) assumes that, in a growth model with endogenous technical progress, if an externality arises at the firm's level, government intervention is needed to obtain the optimum and perfectly competitive market structure. Hence, a tax subsidy system is introduced to influence the rate of technical progress and brings it to the optimal level of growth. Ziesemer's (1995) model indicates that public factors are used in the formation of

[40] See Ziesemer (1987: 107, 108, 112, 115) and Haslinger and Ziesemer (1996: 230, 232). Ziesemer (1987) summarises T. W. Schultz's (1964) view and indicates that the latter uses Nelson (1959) idea that both basic education and basic scientific research should be viewed as a public good.

human capital and human capital, in turn, is necessary for the production of technical progress. If public factors are financed by a flat-rate income tax, then a higher rate of taxation or shares of public expenditure on education in the GDP will lead to a higher level of public factors, a higher rate of technical progress and will also lead to a higher growth if steady states are stable. In Ziesemer's (1991) model the optimal policy is a technology stock subsidy to reward firms and to provide an incentive for the spillovers of technology formation to the human capital schooling process. In Ziesemer's (1990, 1995) models the optimal policy is a tax financed by government spending on the provision or creation of public knowledge (basic education and basic scientific research). The share of GDP raised and spent on the provision of public factors has an impact on the level of GDP per capita or its growth rate respectively.[41]

One assumption in Romer's (1990) model of endogenous technological change is that technological change arises in response to market incentives, and the latter play an essential role in the model. Romer (1990) assumes that in a growth model with spillover effects, the social optimum can be achieved by subsidizing the accumulation of technology. A subsidy to R&D works to compensate R&D firms for the learning-by-doing and the positive external effects they spillover to other R&D firms; in the absence of R&D, a subsidy creates further incentives for firms to gain entry. Although all the research is embodied in capital goods, a subsidy to physical capital accumulation may be a very poor substitute for direct subsidies that increase the incentives to undertake research. In the absence of feasible policies that can remove the divergence between the social and private returns to research, a second-best policy for a government would be to subsidize the accumulation of human capital. A subsidy to employment in research sector that is financed through lump-sum taxes has the same effects on growth as an increase in the productivity parameter: in the long run, a subsidy will cause an increase in the growth rate (Romer 1990: S72, S74, S93, S98, S99).

Barro and Sala-i-Martin (1992) allow for the effects of fiscal policy on long-term growth and discuss the role of tax policy in various models of endogenous economic growth. In their view, in growth models with learning-by-doing and spillover effects, the social optimum can be attained by financing government consumption purchases with an income tax, and monopoly pricing of new types of capital goods. The tax policies that encourage investment can raise the growth rate and thereby increase the utility of the representative household. In growth models that incorporate public services, the optimal tax policy depends on the characteristics of services. If the public services are publicly provided private goods, which are rival and excludable, or publicly provided public goods, which are non-rival and excludable, then lump-sum taxation is superior to income taxation (Barro and Sala-i-Martin 1992: 645, 660).

In many models of endogenous growth (e.g. Romer 1987, 1990; Grossman and Helpman 1991), technological progress corresponds to an expansion in the number

[41] See Ziesemer (1990: 268–280), (1991: 47–68), (1995: 1–19).

of types of capital goods, inventions require active R&D, and firms are compensated through the retention of monopoly power over the use of their inventions. Therefore, the models involve elements of imperfect competition as the excess of the monopoly price over the competitive one implies that the private rate of return on investment falls short of the social return, and, hence, that the steady-state growth rate is below the socially optimal rate. A common feature in all three types of models – learning-by-doing with spillovers, taxation of income from capital (in models where government services are not subject to congestion) and varieties of capital goods under imperfect competition – is the shortfall of the private rate of return on investment from the social one. This implies that the Pareto optimum can be attained in each model if the government raises the private rate of return on investment to the social one without introducing other distortions. This outcome can be achieved either by subsidizing the purchase of capital goods or by subsidizing the income on capital. Another measure is to subsidize research to raise the private rate of return to the social rate and to provide further incentive to private producers to create new types of capital goods.[42]

Barro and Sala-i-Martin (1995) combine the Arrow (1962) and Romer (1986) assumptions of learning-by-doing and knowledge spillovers. In their model, the social optimum can be attained in a decentralized economy by providing various forms of subsidies that work to raise the private rate of return to investment and thereby tend to eliminate the excess of social over private returns. For instance, the government could induce the private sector to attain the social optimum if it provided subsidies to the purchases of capital goods (an investment-tax credit), financing it through a lump-sum tax. Further options open to the government are: providing subsidies to the purchase of intermediate goods, incentive to expand over time using a lump-sum tax to finance a subsidy, subsidies to final output so that producers receive units of revenue for each unit of good produced, or a direct subsidy to R&D spending to raise the private rate of return on R&D and provide a sufficient incentive for research. Therefore, two policy instruments are needed: one that encourages production of the monopolized intermediates and another that stimulates R&D.[43]

Aghion and Howitt (1998) argue for a role for public intervention to support innovative activities either through the design and use of subsidies (direct targeted or untargeted subsidies) or the design of property rights and patent legislation. They suggest that the R&D investments should be subsidized whenever positive external effects dominate and as a result growth under laissez-faire is suboptimal, but that R&D investment should be taxed if too much "business-stealing" or creative destruction takes place under laissez-faire. They distinguish between targeted and untargeted subsidies; the choice between them depends essentially on availability of information to the government. Targeted R&D subsidies are direct government subsidies that are deliberately aimed at particular sectors (e.g. defence), industries

[42] See Barro and Sala-i-Martin (1992: 651, 652, 654, 655).

[43] See Barro and Sala-i-Martin (1995: 146, 147, 150, 151, 222, 223, 226, 229, 230).

or firms. It may take the forms of public investments in state owned laboratories, research grants, participation in R&D funds, subsidies to enterprises (e.g., input subsidies), credit guarantees, and public investment in high-technology industries. Untargeted subsidies is another important instrument of direct policy intervention in the R&D sector, offered on a non-discriminatory basis according to market decision, without targeting particular firms, industries or projects; untargeted subsidies take the form of research tax credits, tax deductions, credit guarantees, subsidized insurance for risky capital investments, etc. Aghion and Howitt (1998) indicate other forms of government intervention to increase incentives/ subsidies for R&D efforts: through the government's buying up or reducing of the outside investors' share or equity in independent research units and turning it over to R&D firms. Finally, they show that the government has a vital role in enforcing property rights by allowing firms involved to earn monopoly rents as a reward for innovation.[44]

Jones (1998) indicates that many models in the endogenous growth literature have the implication that changes in government policies, such as subsidies to research or taxes on investment, have level effects but no long-run growth effects. For instance, a government subsidy that increases the share of labour in research will typically increase the growth rate of the economy, but only temporarily, as the economy transits to a higher level of income.[45]

Several studies emphasize a role for government intervention and the positive impact of public provision of education and training. For instance, Azariadis and Drazen (1990) suggest a role for government intervention in the education sector (through education subsidies) to avoid low-development traps and thereby promote high sustained growth. Otani and Villanueva (1990), and Barro and Sala-i-Martin (1995) illustrate, for a cross section of countries, a positive impact of government interventions on growth rates coming from the expenditure side (i.e. the share of public expenditure on education of the GDP). Barro and Sala-i-Martin (1995) find that public spending on education has a significant positive effect on growth: a 1.5 % increase of the ratio of public education spending to GDP during the period 1965–1975 would have raised the average growth during the same period by 0.3 % per year. Aghion and Howitt (1998) indicate that centralized funding of education will always favor human capital accumulation and therefore growth in the long run, even though local funding may sometimes be growth-enhancing in the short run.[46] Trostel (2002) suggests that public provision of education through subsidy has the potential to be the most efficient educational policy because it provides incentives and stimulates investment in and accumulation of human capital. A recent report by the UNESCO–UIS/OECD (2003) stresses the role of public finance in enhancing

---

[44] See Aghion and Howitt (1998: 474, 489).

[45] See Jones (1998: 147, 112).

[46] Otani and Villanueva (1990) is cited in Haslinger and Ziesemer (1996: 236). Azariadis and Drazen (1990) and Barro and Sala-i-Martin (1995) respectively are cited in Aghion and Howitt (1998: 333, 328). See also Aghion and Howitt (1998: 338).

investments and returns from education in a number of selected countries. Chatterji (1995) presents a growth model based on Lucas' (1988) model to explore a potential role for government intervention by subsidizing training to compensate private sector for the positive externalities they generate and to provide more incentives for more investment in the accumulation of skills. The model assumes two possible sources of growth: exogenous technical progress and endogenously produced skills growth. The optimal subsidy on training rises with the rise/strength of the externality generated by the average skill level in output production; it also depends on macroeconomic variables such as the extent of productivity growth from other sources in the economy.[47]

One interesting observation by Aghion and Howitt (1998) indicates that the finding with respect to the complementarity between educational attainment and R&D activities has in turn many interesting policy implications. First, it suggests that "macroeconomic policies which affect rates of innovation and investment will affect the relative demand for workers classified by education, and hence the aggregate skill distribution of employment and earning. (Bartel and Lichtenberg 1987)". In other words, governments will increase the average level of education not only directly through education policy, but also indirectly by actively supporting R&D activities. Conversely, government subsidies to education will increase the profitability of research and development activities, and thereby speed up technological progress (Aghion and Howitt 1998: 339–340).

Few studies examine the practical relevance of the models of growth enhancing policies, particularly for the developing countries. For instance, Haslinger and Ziesemer (1996) indicate that most of the models of publicly financed investment in human capital are basically intended for industrialized and not for developing countries. In their view, in the developing countries, raising the publicly financed investment is hampered by the lack of a well-developed institutional setup to use the instruments of taxation, mainly because of substantial engagement in non-monetised activities, a large informal sector, extreme poverty and different effects from the prevalent regressive trade tax (Haslinger and Ziesemer 1996: 240, 241).[48]

Apart from these practical limitations for the developing countries, in the recent years there is a growing body of literature on the role of public policies and government intervention to promote human capital and technological capabilities in the developing countries. For instance, Lall (1999) discusses strategies to develop skills and capabilities in developing countries and argues that there is a valid case for policies to coordinate, guide and subsidize learning; and to develop such factors as skills and technology where externalities and information failures are particularly pervasive. He identifies two broad successful strategies in the developing world to

[47] See Chatterji (1995: 274–282).

[48] In the case of the Sudan, the extent of the 'non-monetised' activities is less clear. However, like most other typically developing countries, in the Sudan both the prevalence of the informal sector and extreme poverty (cf. UNDP 2010)- and the recent structural reform of fiscal and monetary policies and labour market (cf. UNDP 2006) may imply a promising role for government intervention.

promote skills and learning for competitiveness. First, autonomous strategies to accelerate and guide learning by domestic firms by promoting infant industries, coordinating investments in related activities, overcoming externalities, directing credit, and developing specific skills and institutions. Second, foreign direct investment (FDI) dependent strategies that rely on Transnational Corporations (TNCs) to lead export growth and upgrading, which has two subsets of strategies: those based on targeting TNCs and using industrial policy to guide them in more technology intensive activities; and more passive strategies that rely on market forces to attract and upgrade activities. Korea and Taiwan are leading example of national-led strategy, Singapore and Malaysia of the FDI-led targeted strategy, and Mexico and Thailand of the FDI-led market-led strategy (Lall 1999: 9, 10).

3.6 Conclusions

In this chapter we provide a background for the empirical analysis in the following chapters by surveying the theoretical and empirical literature that emphasize the positive growth effects of human capital and technological progress in increasing and sustaining economic growth. We explain that economic growth theories recognized and provided different perceptions and analytical frameworks for modelling the various effects of technical change, innovation and human capital on economic growth. The major differences arise because the exogenous growth theories perceive and model technical progress and human capital as exogenous variables in growth accounting model, whereas the endogenous growth theory envisages and models technical progress and human capital as endogenous variables determining the rates and differences of economic growth across countries. The endogenous growth theory contributes to improve understanding of the interaction between technological change, human capital and economic growth and fills the gap in earlier growth theories by considering the important endogenous effects of human capital, technological progress and innovation. The endogenous growth theory predicts that in the long run economic growth at the aggregate level is determined by endogenous sources of technological change, human capital, learning by doing, spillovers of knowledge and external effects of human capital. The presence of increasing returns to scale and externalities prevent diminishing returns to accumulation of capital, and so ensure the long run steady state of growth within a dynamic general equilibrium framework. We illustrate that the inclusion of human capital and technological change in growth accounting models motivate endogenous growth literature to provide several interesting explanations of the relationship between human capital and technical change. In particular, it stimulates considerable debate about the complementary relationship between human capital and technical progress, skilled biased technical change, the role of technical progress in skill upgrading and the role of skill and improvement in the accumulation of human capital in skill upgrading. These explanations imply that next to the important endogenous effects of technical progress and human capital for economic growth, the complementary relationships between them and between them and skill upgrading are also important for enhancing economic growth. Finally, we show the advantages and limitations of several measures of technological change and human capital that have been used in theoretical and empirical literature; some of these measures are relevant for the empirical analysis in the next chapters according to availability of data.

References

Abramovitz, M. (1956). Resource and output trends in the United States since 1870. *American Economic Review: Papers and Proceedings, 46*, 5–23.

Abramovitz, M. (1986). Catching up, forging ahead and falling behind. *Journal of Economic History, 46*, 385–406.

Acemoglu, D. (1998). Why do new technologies complement skills? Directed technical change and wage inequality. *Quarterly Journal of Economics, 113*(4), 1055–1089.

Aghion, P., & Howitt, P. (1990, January). *A model of growth through creative destruction* (NBER Working Paper Series, No. 3223). Cambridge, MA: National Bureau of Economic Research. http://www.nber.org/papers/w3223.pdf

Aghion, P., & Howitt, P. (1992). A model of growth through creative destruction. *Econometrica, 60*(2), 323–351.

Aghion, P., & Howitt, P. (1998). *Endogenous growth theory*. Cambridge, MA: MIT Press.

Arrow, K. J. (1962). The economic implication of learning by doing. *Review of Economic Studies, 29*, 155–173.

Autor, D. H. (2000, April). *Why do temporary help firm provide free general skills training?* (NBER Working Paper Series, No. 7637, p. 37). Cambridge, MA: National Bureau of Economic Research. http://www.nber.org/papers/w7637.pdf?new_window=1

Autor, D. H., Katz, L. F., & Krueger, A. B. (1998). Computing inequality: Have computers changed the labour market? *The Quarterly Journal of Economics, 113*(4), 1169–1213.

Azariadis, C., & Drazen, A. (1990). Threshold externalities in economic development. *Quarterly Journal of Economics, 105*(2), 501–526.

Barro, R. (1991). Economic growth in a cross section of countries. *Quarterly Journal of Economics CVI, 106*(2), 407–443.

Barro, R. (1996, August). *Determinants of economic growth: A cross- country empirical study* (NBER Working Paper Series, No. 5698). Cambridge, MA: National Bureau of Economic Research. http://www.nber.org/papers/w5698.pdf

Barro, R. J., & Lee, W. (1993). International comparison of educational attainment. *Journal of Monetary Economics, 32*(3), 363–394.

Barro, R. J., & Lee, W. (1996). International measures of schooling years and schooling quality. *American Economic Review, Paper and Proceedings, 86*(2), 218–223.

Barro, R. J., & Lee. W. (2000a, September). *International data on educational attainment updates and implications* (NBER Working Paper Series, No. 7911). Cambridge, MA: National Bureau of Economic Research. http://www.nber.org/papers/w7911.pdf

Barro, R., & Lee, W. -J. (2000b, April). *International data on educational attainment: Updates and implications*. Cambridge, MA: Department of Economics, Harvard University. Data set at International Development at Harvard University (CID), http://www.cid.harvard.edu/cidwp/pdf/042.pdf

Barro, R. J., & Lee. W. (2010, April). *A new data set of educational attainment in the world, 1950–2010* (NBER Working Paper Series, No. 15902). Cambridge, MA: National Bureau of Economic Research. http://www.nber.org/papers/w15902.pdf?new_window=1

Barro, R. J., & Sala-i-Martin, X. (1992). Public finance in models of economic growth. *The Review of Economic Studies, 59*(4), 645–661.

Barro, R. J., & Sala-i-Martin, X. (1995). *Economic growth*. New York: McGraw Hill.

Bartel, A. P., & Lichtenberg, F. (1987). The comparative advantage of educated workers in implementing new technology. *The Review of Economic and Statistics, LXIX*(69), 1–11.

Bartel, A. P., & Sicherman, N. (1995). *Technological change and the skill acquisition of young workers* (Graduate School of Business Paper Series, No. 95–10). New York: Columbia University.

Bartel, A. P., & Sicherman, N. (1999, August) *Technological change and the skill acquisition of young workers* (NBER Working Paper Series, No. 5107). Cambridge, MA. http://www.nber.org/papers/w5107.pdf.

Becker, G. S. (1962). Investment in human capital: A theoretical analysis. *Journal of Political Economy, 70*(1), 9–49.

Becker, G. (1964). *Human capital*. Chicago: The University of Chicago Press.

Benhabib, J., & Spiegel, M. M. (1994). The role of human capital in economic development: Evidence from aggregate cross-country data. *Journal of Monetary Economics, 34*, 143–173. New York.

Berman, E., Bound J., & Griliches Z. (1994). Change in the demand for skilled labour within U.S. manufacturing industries: Evidence from the annual survey of manufacturing. *Quarterly Journal of Economics, CIX*(2), 367–397 (pp. 367–398) (Vol. 109, May).

Berman, E., Bound, J., & Machin, S. (1998). Implications of skill-biased technological change: International evidence. *Quarterly Journal of Economics, 113*(4), 1245–1279.

Bound, J., & Johnson, G. (1992). Change in the structure of wages in the 1980s: An evaluation of alternative explanations. *American Economic Review, 82*(3), 371–392.

Bresnahan, T. F., Brynjolfsson, E., & Hitt, L. M. (1999, May). *Information technology, workplace organization, and the demand for skilled labour: Firm- level evidence* (NBER Working Paper Series, No. 7136). Cambridge, MA: National Bureau of Economic Research. http://www.nber.org/papers/w7136.pdf?new_window=1

Chatterji, M. (1995, September). *Training subsidies, technical progress and economic growth* (pp. 274–282). The Manchester School, *LXIII*(3). Blackwell Publishers Ltd./The Victoria University of Manchester.

Cheah, B. H. (1997). Can government engineer the transition from cheap labour to skill- based competitiveness? The case of Singapore. Chapter 4. In M. Godfrey (Ed.), *Skill development for international competitiveness*. Cheltenham/Brookfield: Edward Elgar.

Cheon, B.-Y. (1999). *Employment, occupations and skills in increased international exposure: The Republic of Korea 1970–1990* (Employment and Training Papers No. 39, pp. 2–3, 12–15, 18, 246–268). ILO

Colecchia, A., & Papaconstantinou, G. (1996). *The evaluation of skills in OECD countries and the role of technology* (STI working paper 1996/6). OECD/GD (96) 183. Paris: OECD.

Cooper. C. (1991). *Are innovation studies on industrialized economies relevant to technology policy in developing countries?* (UNU-INTECH Working Paper Series No. 3). Maastricht: UNU-INTECH.

Cooper. C. (1994). *Science and technology in Africa under conditions of economic crisis and structural adjustment* (UNU-INTECH Working Paper Series No. 4). Maastricht: UNU-INTECH.

Dasgupta, P., & David, P. (1994). Toward a new economics of science. *Research Policy, 23*(5), 487–521. MERIT Research Memoranda Series No. 2/94-003. Maastricht: Maastricht University.

Dension, E. F. (1962). *Sources of growth in the United States and the alternatives before us* (Supplement Paper 13). New York: Committee for Economic Development.

Dosi, G., Freeman, C., Nelson, R. R., Silverberg, G., & Soete, L. (Eds.). (1988). *Technical change and economic theory*. London/New York: Printer Publishers.

ESCWA (Economic and Social Commission for Western Asia). (1999a). *Science and technology policies in the twenty-first century* (pp. 75, 209). New York: United Nations.

ESCWA (Economic and Social Commission for Western Asia). (1999b, March 10–12). *ESCWA proceedings of the expert group meeting on science and technology policies and strategies in the twenty-first century*, Beirut. (E/ESCWA/TECH/1999/8/Rev.1: December 1999). New York: United Nations.

ESCWA (Economic and Social Commission for Western Asia)- UNESCO. (1998a). Research and development system in the Arab States: Development of science and technology indicators. (E/ ESCWA/TECH/ 1998/3). In ESCWA (1999, January) *Review of Science and Technology in ESCWA Member Countries*, Issue, No. 1. 1999 (E/ ESCWA/ TECH/ 1999/Rev. 1), pp. 3–4.

ESCWA-UNESCO (Economic and Social Commission for Western Asia–United Nations Educational, Scientific and Cultural Organization). (1998b). *Research and development systems in the Arab states: Development of science and technology indicators* (E/ESCWA/TECH/1998/ Rev. 1). Cairo: UNESCO.

Fergany, N. (1999). Science and research for development in the Arab region. Chapter 2. In E. Rached & D. Craissati (Eds.), *Research for development in the Middle East and North Africa*. Ottawa: IDRC.

Foray, D. (1999). Science, technology and market. In *World social science report*. Paris: UNESCO/Elsevier.

Freeman, C., & Soete, L. (1994). *Work for all or mass unemployment? Computerised technical change into the twenty-first century*. London: Printer.

Freeman, C., & Soete, L. (1997). The economic of industrial innovation (3rd ed., pp. 2, 24, 404–405). London: Cassell.

Garcia Cervero, S. (1997). *Growth technology, and inequality: An industrial approach* (EUI Working Paper. ECO; EC (471/97/26)). Florence: European University Institute. https://catalogue.lse.ac.uk/Author/Home?author=European%20University%20Institute.%20Economics%20Department.https://catalogue.lse.ac.uk/Record/517760

Goldin, C., & Katz, L. F. (1998). The origin of technology skill complementarity. *The Quarterly Journal of Economics, 113*(3), 693–732.

Griliches, Z. (1969, November). Capital-skill complementarity. *The Review of Economic and Statistics, LI* (69), 465–468.

Grossman, G. M., & Helpman, E. (1991). *Innovation and growth in the global economy*. Cambridge, MA: MIT Press.

Hanushek E. A., & Kim D. (1995, December). *Schooling, labour force quality and economic growth* (NBER Working Paper Series, No. 5399). Cambridge, MA: National Bureau of Economic Research. http://www.nber.org/papers/w5399.pdf?new_window=1

Haslinger, F., & Ziesemer, T. (1996). Endogenous growth and distributional conflict. In C. Gore, G. Köhler, U. Peter Reich, & T. Ziesemer (Eds.), *Questioning development* (pp. 223–240). Marburg: Metroplis Verlag.

Hwang, Gyu-heui. (2000, October 3). *Diffusion of information and communication technologies and changes in skills*. Science and Technology Policy Research (SPRU) (Electronic Working Paper Series, No. 48). Brighton: *SPRU*, University of Sussex. http://www.sussex.ac.uk/Units/spru/publications/imprint/sewps/sewp48/sewp48.pdf

ILO. (1998, September). World employment report 1998/1999: Employability in the global economy how training matters. Chapter 2. In *Globalization, technological change and demand for skilled labour* (p. 35). Geneva: ILO.

Jones, C. (1998). *Introduction to economic growth*. New York/London: W.W. Norton and Company.

Jorgonson, D. W., Gallop, F. M., & Fraumeni, B. M. (1987). *Productivity and U.S. economic growth*. Cambridge, MA: Harvard University Press.

Kahn, J. A., & Lim, J. S. (1998). Skilled labor-augmenting technical progress in U.S. manufacturing. *The Quarterly Journal of Economics, 113*(4), 1281–1308.

Kenderik, J. W. (1956). Productivity trends: Capital and labour. *Review of Economic and Statistics, 38*(3), 248–257.

Kenderik, J. W. (1961). *Productivity trends in the United States*. New York: Princeton University Press.

Lall, S. (1999). *Competing with labour: Skills and competitiveness in developing countries* (Issues in Development Discussion Paper No. 31, p. 17). Geneva: Development Policies Department, ILO.

Lee, W., & Barro, R. J. (1997). *Schooling quality in a cross section of countries* (NBER Working Papers No. 6198). Cambridge, MA: National Bureau of Economic Research, Inc.

Lillard L.A., & Tan, H.W. (1986). *Private sector training: who gets it and what are its effects?* (Santa Monica RAND Corporation Rep. R-3331, DOL-RC, 1986). Santa Monica: Rand Corporation.

Low, L. (1998). *Jobs, technology and skills requirements in a globalized economy: Country study on Singapore* (Employment and Training Paper No. 13). Geneva: Employment and Training Department, ILO.

Lucas, R. E. (1988). On the mechanics of economic development. *Journal of Monetary Economics, 22*, 3–42. North Holland.

Lundvall, B.-Å. (Ed.). (1992). *National systems of innovation: Towards a theory of innovation and interacting learning*. London: Printer Publishers.

Machin, S., & Van Reenen, J. (1998). Technology and changes in skill structure: Evidence from seven OECD countries. *Quarterly Journal of Economics, 113*(4), 1215–1244.

Mankiw, N. G., Romer, D., & Weil, D. (1992). A contribution to the empirics of economic growth. *The Quarterly Journal of Economics, 107*(2), 407–437.

Mincer, J. (1989, December). *Human capital response to technological change in the labor market* (NBER Working Paper Series, No. 3207). Cambridge, MA: National Bureau of Economic Research. http://www.nber.org/papers/w3207.pdf?new_window=1

Mulligan, C. B., & Sala -i- Martin X. (1995, February). *Measuring aggregate human capital* (NBER Working Paper Series, No. 5016, p. 2). Cambridge, MA: National Bureau of Economic Research. http://www.nber.org/papers/w5016.pdf?new_window=1

Muysken, J., & ter Weel, B. (1998). *Overeducation and crowding out of low skilled workers* (MERIT Research Memorandum 98-023). Maastricht: Maastricht University.

Muysken, J., Hoppe, M., & Rieder, H. (2002a). *The impact of education and mismatch on wages: Germany, 1984–2000* (MERIT Research Memoranda 2002-031). Maastricht: Maastricht University.

Muysken, J., Weissbrich, A., & von Restorff, C. H. (2002b). *The impact of education and mismatch on wages: The USA, 1986-1996* (MERIT Research Memoranda 2002-015). Maastricht: Maastricht University.

Mytelka, L. (2001). *Do the least developed countries need science and technology for sustainable development?* http://www.intech.unu.edu/publications/index.php

Nelson, R. R. (1959). The simple economics of basic scientific research. *Journal of Political Economy, 67*, 297–306.

Nelson, R. (Ed.). (1993). *National innovation systems: A comparative study*. New York: Oxford University Press.

Nelson, R., & Phelps, E. (1966). Investment in humans, technological diffusion and economic growth. *American Economic Review: Papers and Proceedings, LVI, 61*, 69–75.

Nelson, R. R. & Winter, S. G. (1977). In search of useful theory of innovation. *Research Policy, 6*(1), *5*, 36–76

Nelson, R. R., & Winter, S. G. (1982). *An evolutionary theory of economic change*. Cambridge, MA: Harvard University Press.

OECD (Organisation for Economic Co-operation and Development). (1997). *The second European report on science and technology indicators (1997)* (pp. 37, 90–91). Paris: OECD.

Otani, I., & Villanueva, D. (1990). Long-term growth in developing countries and its determinants: An empirical analysis. *World Development, 18*(6), 769–783.

Qasem, S. (1998a). Research and development in the Arab states: Development of science and technology indicators. *Review of science and technology in ESCWA member countries, ESCWA* (1999), Issue No. 1. New York: United Nations.
Qasim, S. (1998b). Research and development systems in the Arab states: Development of science and technology indicators, Chapter 4. In *Economic trend in MENA region 2002*. Cairo: Economic Research Forum and the American University in Cairo.
Rebelo, S. (1991). Long-run policy analysis and long-run growth. *Journal of Political Economy, 99*(3), 500–521.
Romer, P. M. (1986). Increasing returns and long run growth. *Journal of Political Economy, 94*(5), 1002–1037.
Romer, P. M. (1987). Growth based on increasing returns due to specialization. *American Economic Review, 77*(2), 56–62.
Romer, P. (1989). *Increasing returns and new development in the theory of growth* (NBER Working Paper Series No. 3098, p. 22).
Romer, P. M. (1990). Endogenous technological change. *Journal of Political Economy, 98*(5, Pt. 2), 71–102.
Romer, P. M. (1994). The origin of endogenous growth. *Journal of Economic Perspectives, 8*, 3–22.
Rosen, S. (1976). A theory of life earnings. *Journal of Political Economy, 84*(4), S45–S67.
Sanders, M., & ter Weel, B. (2000). *Skilled biased technical change: Theoretical concept, empirical problems and a survey of the evidence* (MERIT Research Memorandum 2000-12). Maastricht: Maastricht University.
Schultz, T. W. (1961). Investment in human capital. *American Economic Review, 51*(1), 1–17.
Schultz, T. W. (1964). *Transforming traditional agriculture*. New Haven: Yale University Press.
Schumpeter, J. (1934). *The theory of economic development* (p. 66). Cambridge, MA: Harvard University Press.
Schumpeter, J. (1942). *A capitalism, socialism and democracy*. New York: Harper and Brothers.
Shell, K. (1967). A model of inventive activity and capital accumulation. In K. Shell (Ed.), *Essay on the theory of optimal economic growth* (pp. 67–85). Cambridge, MA: MIT Press.
Smith A. (1776). *An inquiry into the nature and causes of the wealth of nations* (Everyman Edition, 1910). London/New York: J.M. Dent/Random House, 1937.
Smith, K. (2000, August 23–24). *Measurement of innovation in Europe: Concepts, experience and results*. STEP Group. Paper presented to conference on The Measurement of Industrial, Technological Competitiveness for Knowledge-Based Economy, Taipei.
Solow, R. (1956). A contribution to the theory of economic growth. *Quarterly Journal of Economics, 70*, 65–94.
Solow, R. (1957). Technical progress and the aggregate production function. *Review of Economics and Statistics, 39*, 312–320.
Stokey, N. L. (1991). Human capital, product quality and growth. *Quarterly Journal of Economics, 106*(2), 587–616.
Swan, T. W. (1956). Economic growth and capital accumulation. *Economic Record, 32*, 334–361.
Trostel, P. A. (2002). Should education be publicly provided? *Bulletin-of-Economic-Research, 54* (4), 373–391.
UNESCO–UIS/OECD. (2003). *Financing education- investment and returns: Analysis of the world education indicators*. 2002 Edition, UNESCO.
UNDP United Nations Development Programme UNDP. (2010). Human development report 2010: The real wealth of nations: Pathways to human development. In *Human development report 2010* (20th Anniversary Edition). New York: UNDP/Palgrave Macmillan. Statistical Tables, Statistical Annex, Table 12: Decent Work, p.191.
United Nations Development Programme UNDP in Sudan. (2006, June). *Macroeconomic policies for poverty reduction: The case of Sudan* (pp. 92–94). Khartoum: Published for the UNDP in Sudan, Table 8.1, p. 82. UNIDO, reported in Dagdeviren & Mahran (2004) Table 11. Cited in

Trade and Industry – What Role for Poverty Reduction? Chapter 8. In UNDP. (2006). *Macroeconomic policies for poverty reduction: The case of Sudan*, Table 8.1, p. 82.
Uzawa, H. (1965). Optimum technical change in an aggregative model of economic growth. *International Economic Review, 6*(1), 18–31.
Velho, L. (2004). *Science and technology in Latin America and the Caribbean: An overview* (The UNU-INTECH Discussion Paper Series No. 2004-4). Maastricht. http://www.intech.unu.edu/publications/discussion-papers/2004-4.pdf.
Zahlan, A. B. (1999a). *The Arabs and the challenges of science and technology: Progress without change*. Beirut: The Centre for the Study of Arab Unity.
Zahlan, A. B. (1999b, March 10–12). Science policy for the twenty-first century: Mobilization and development. In *ESCWA Proceedings of the Expert Group Meeting on Science and Technology Policies and Strategies for the Twenty-First Century* (pp. 14–16), ESCWA, Beirut.
Ziesemer, T. (1987). *Economic theory of underdevelopment*. Regensburg: Transfer verlag.
Ziesemer, T. (1990). Public factors and democracy in poverty analysis. *Oxford Economic Papers, 42*(Special issue on public economics), 268–280. (Reprinted from *Taxation, private information and capital,* by P. J. N. Sinclair & M. D. E. Slater, Eds., 1991, Oxford: Clarendon Press).
Ziesemer, T. (1991). Human capital, market structure and taxation in a growth model with endogenous technical progress. *Journal of Macroeconomics, 13*(1), 47–68.
Ziesemer, T. (1995). Endogenous growth with public factors and heterogeneous human capital producers. *Finanzarchiv,* Neue Folge, Band 52, 1995, Heft 1, 1–20.

Part III
Empirical Application

Chapter 4
Research Methodology and Methods of Data Collection

Abstract This chapter defines the methods of data collection and explains the motives for performing the macro and firm surveys and selection of a case study, the selection of the sample and composition and operation of the surveys. We explain that the data from the firm and macro surveys provides us with the required information, which is particularly useful from both the analytical and policy perspectives and useful for presenting a macro-micro comparative analysis to identify the causes and consequences of the skills problem, and the plans and policies for skill development from the macro-micro perspectives. One advantage of the macro survey is that it examines the problem after integrating two different perspectives of policy makers and experts and also integrating two different perspectives from both the fields of education and training. Moreover, the data evaluates upskilling efforts, examines the causes and consequences of a deficient educational system, training provision, skills mismatch and transfer of knowledge, and also suggests relevant short and long run plans and policies for the enhancement of skill level, education, training, transfer of knowledge and technological upgrading. The coverage of the firm survey has the advantage of enabling us to compare between firms according to two criteria, i.e. the size of employment and industrial activity.

4.1 Introduction

The previous chapter discussed the theoretical and empirical endogenous growth literature on the relationship between technical progress, human capital (education) and economic growth, which we considered as a useful background for the empirical investigation that follows in the next chapters. Before we go into the empirical analysis, it is appropriate to define the methods of data collection and illustrate the composition and operation of the surveys in this chapter. In Sect. 4.2 we explain the motives for performing the macro and firm surveys and selection of a case study. Next, we show the selection of the sample and composition of the surveys in

S. Mohamed Nour, *Technological Change and Skill Development in Sudan*,
DOI 10.1007/978-3-642-32811-4_4,

Sect. 4.3 and explain the structure and design of the questionnaire in Sect. 4.4. Section 4.5 provides the conclusions, advantages and limitations of the surveys.

4.2 Motivation and Selection of a Case Study

The empirical investigation in this research uses a combination of primary and secondary data covering both the macro and micro levels. We collected our primary data using both macro and firm surveys and interviews. We explain the reasons for undertaking two surveys below, after which we spell out the selection of the case study, the sample, design and composition of the surveys.

The basic objective of conducting the macro and firm surveys was to obtain specific information to provide insights into the factors influencing or the causes and consequences of low skill and technology indicators and to help generate policies to enhance skill and technology. The macro survey examines the causes and consequences of the deficiency of the educational system and the firm survey discusses the implications of the low technology indicators, low skill level and the excessive use of unskilled workers.[1]

We undertook the macro survey due to a lack of sufficient materials to investigate the research problem, assess upskilling efforts and examine the causes and consequences of deficient educational systems, training provision, skills mismatch and transfer of knowledge at the macro level. The survey also aimed to explore, from the perspectives of policy makers and experts, relevant short and long run plans and policies for the enhancement of skill levels, education, training, transfer of knowledge and technological upgrading. The main reason for conducting the firm survey was to fill the information gap due to a lack of relevant, reliable and up-to-date information to allow a more comprehensive analysis and a deeper understanding of the status of skill and technology at the micro level.

The firm survey requested quantitative data to assess technology indicators measured by R&D, patent and ICT and to evaluate skill indicators measured by educational attainment, average years of schooling and experience and occupational level. Additional information was sought to assess the use, level, transfer and dependence on foreign technology, to determine the important factors hindering and those contributing towards, enhancing the provision of training, mechanisms and plans for skill upgrading across firms. Moreover, the survey explains the importance of the industry and examines the important factors hindering, and those contributing toward enhancing the contribution of the industry across firms.

The field research to collect our primary data was held in the period from January to June, 2010, in Sudan as a case study of the Arab countries. The selection

[1] In the appendices, we list the set of questions that were sent out to the sample of firms, policy makers and experts included in the surveys. In addition, we present a brief definition of some of the terms used in the various questions that were provided as a guide to help the respondents.

and focus of our analysis on Sudan was related to the easy accessibility to data and information and facilities for the fulfilment of the fieldwork/surveys, which were offered by the Department of Economics and Faculty of Economic and Social Studies at the University of Khartoum. Moreover, in recent times, has Sudan shown a significant increase in terms of ICT diffusion: for instance, the World Bank (2010) showed the population accessing the Internet as measured by the increase in Internet users (per 100 people) rising from 3.2 % in 2000 to 60.5 % in 2008. In addition, Sudan has shown a rapid increase in the net inflow of FDI, for instance, the World Bank (2010) indicates significant increase in the inflow of FDI from US$ 38 million in 2000 to US$ 159 million in 2008.[2] Moreover, Sudan is the fourth country in Africa and fifth in the Arab regions in terms of attraction of foreign investment, notably, in 2007 Sudan accounts for 1.9 % of total inflow of FDI in the Arab countries.[3]

One stylised fact which we explained in Chap. 2 is that the structure of Sudan economy shows the importance of the agricultural sector (48 %, 31.1 %, 27 %) and services sector (36 %, 45 %, 37 %) compared to the industrial sector (10 %, 23.9 %, 36 %) and manufacturing sector (3.52 %, 10.8 %, 6 %) in terms of the share in total employment (2008), total GDP (2009) and value added as a percentage of GDP (2009) respectively, further to the minor contribution of the industrial sector to total exports in Sudan (17 %) (2001), in addition to the minor contribution of the industry in value added annual growth (15 %) (2005).[4] Moreover, in 2002, Sudan (8.26 % and US$ 28) has been lagging far behind compared to North Africa (15.57 % and US$ 190) and developing countries (20.3 % and US$ 908) in terms of both the manufacturing value added as percentage of GDP and manufacturing value added per capita respectively.[5] Despite the minor contribution of the industrial sector in the Sudanese economy, this research and firm survey focuses on the industrial sector, because of the importance of the industrial sector in achieving both technological development and economic prosperity for any country. Moreover, for the case of Sudan, for many reasons, the industrial sector has the potential to play a significant role in enhancing economic and social development, strengthening and multiplying the capabilities of the national economy and in realising the added

---

[2] See for instance, The World Bank, World Development Indicators Database (2012 a, b) accessed on April 2010 and December 2010.

[3] See for instance, the Arab Investment and Export Credit Guarantee Corporation (2007), p.60.

[4] See for instance, Sudan Central Bureau of Statistics 'Population Census Data' (2010), 'Fifth Sudan Population and Housing Census (2008)', for the data on the distribution and sectoral share in total employment. See also the Central Bank of Sudan 'Annual Report' (2009) for the distribution and sectoral share in total GDP. See also The World Bank, World Development Indicators Database (2010 a, b, 2011) accessed on April 2010, December 2010 and June 2011, for data on the distribution and sectoral share in value added as a percentage of GDP. See also Ministry of Industry 'Comprehensive Industrial Survey' (2005) for data on the share of industry in Sudan's total exports (2001). See also The World Bank, World Development Indicators Database (2005) accessed on April 2005 for data on the share of industry in value added.

[5] See for instance, UNDP (2006), p. 82. Manufacturing Value Added (MVA) per capita is estimated in constant 1995 US$.

value to contribute to the development and improvement of the living conditions of the people in Sudan. Notably, the first reason is that the abundant natural resources including agricultural resources, animal resources, fisheries resources, forests resources, land and water resources, all have the potential to form the basis for the manufacturing industrial development in Sudan. For instance, for some time the agricultural sector has been a leading sector that has often supported the manufacturing industries in Sudan; agricultural crops have often provided the industrial sector with a high ratio of its inputs. For instance, textile industries benefited from the cultivated areas in cotton, and food industries benefited from the high production of oil seeds, particularly vegetable oil. Moreover, the fish and animal resources and their products are being used as inputs in food industries and also animal hides and skins are being used as inputs in the manufacturing of leather and leather products. In addition, the forest natural resources and forestry and its products, form an important input for manufacturing industries related to forestry products, for example, manufacturing of wood and wood products, manufacturing of furniture, manufacturing of other household utensils and building of poles and telephone poles. In addition, the second reason is that the geology of Sudan indicates abundant mineral resources such as iron, copper, chrome, manganese, gold, silicon, limestone, marble, gypsum, mica, and natural gas, all of which have the potential to promote extractive industries in Sudan. The third reason is that the industrial sector is composed of several sub-sectors and as such it has the potential to produce diversified products to cover the diversified needs of the local population, for instance, this includes needs from: food industries; spinning and textile industries; leather industries; chemical and pharmaceutical industries; oil and soap industries; engineering industries; building materials and refractories; and printing and packing. The fourth reason is the relatively high installed capacity of the industrial sector, which enables Sudan to enjoy the third largest industrial basis in Africa after South Africa and Egypt.[6] For instance, the installed capacities in the industrial sector are relatively higher than the national needs, thereby, it has the potential to produce considerable surplus for exports to neighbouring countries if full capacities can be exploited, however, so far the utilised capacities are low, ranging between 20 % and 25 %. Moreover, the fifth reason is that the rate of returns on investment in capital in the industrial sector (34 %) is relatively high compared to other sectors in Sudan economy.[7] Consequently, it is not surprising that the industrial sector (including energy, mining and oil) attracts about 83 % of total foreign investment inflows to Sudan, although the majority of foreign investment is concentrated in extractive oil and energy and mining sector (73 %) and the minority in the manufacturing industry (10 %) as we explained in Chap. 2 above.

[6] See for instance, Sudan Ministry of Industry (2005), 'Comprehensive Industrial Survey Data for 2001'.

[7] See for instance, Sudan Ministry of Industry (2005), 'Comprehensive Industrial Survey Data for 2001'.

The main limitation of our selection was that Sudan does not report the highest performance within the Arab region with respect to both skill and technology indicators. Our analysis in Chap. 5 below illustrates that Sudan shows insufficient and poor performance and is falling behind compared to other Arab countries in terms of skill-technology indicators, technological performance and technology indicators as measured by expenditures on R&D as a percentage of GDP, application to patents, the use of ICT and the use of Internet indicators. We find that Sudan shows poor performance compared to other Arab countries in terms of skill indicators or indices – as measured by the Harbison Myers index, technical enrolment index and engineering enrolment index. Sudan shows poor performance compared to other Arab countries in terms of educational performance as measured by educational enrolment, gross enrolment ratio in tertiary education, school life expectancy, average years of schooling, the share of tertiary students in science, math and engineering, and the share of students enrolled in S&T disciplines. In addition, Sudan has one of the lowest per capita manufacturing output in the Arab region, North Africa and developing countries as we explained above.

The firm survey covers small, medium and large size firms working in four industries in the manufacturing sector: the food, textile, chemical and metal industries.[8] The selection of these industries was based on the following reasons. First, the argument for both upskilling and technological upgrading is promising in these sectors and can be used to reduce the poverty and unemployment problems in the country. Second, the strategic importance of these sectors in creating forward and backward linkages and spin-off effects to other sectors/industries. Third, the food and textile industries represent an agro-industry based on manufactured products that benefit from the comparative advantage of the rich agricultural resources in Sudan, while the chemical and metal sectors have the potential to produce energy intensive use products benefiting from the comparative advantage of the abundant and cheapest mineral and energy resources, particularly petroleum, in Sudan. Fourth, the potential for product diversification in these sectors is promising. Fifth, the important contribution of these sectors in the manufacturing sector, for instance, in terms of gross output, gross value added, capital investment, total labour force, total exports, total imports and total number of industrial establishments in the manufacturing sector. For instance, in 2001, the contribution of these sectors together was around 85.3 % of total gross output, 87.6 % of total gross value added, 66.8 % of capital investment, 76.12 % of total labour force, 77.78 % of total imports and 84.5 % of total number of industrial establishments in the manufacturing sector. In particular, the shares of the food sector accounted for

[8] In the firm survey, the chemical sector includes manufactures of basic industrial chemicals, fertilizers and pesticides, synthetics, resin and related materials, paints, varnishes and lacquers. In addition, petrochemicals, pharmaceuticals, drugs and medicines, soap and cleaning preparations, chemical products, petroleum refineries, miscellaneous petroleum and coal products, tyre and tube industries, rubber products and plastics products are also included in this sector. The metal sector includes basic metal products, fabricated metal products, machinery and equipment and manufacture of electrical machinery and apparatus.

55.3 %, 64.6 %, 46.5 %, 56.32 %, 47.44 % and 70.4 % respectively, the shares of the textile sector accounted for 2.8 %, 2.6 %, 0.8 %, 5.31 %, 0.03 % and 0.2 % respectively, the shares of the chemical sector accounted for 22.6 %, 16.9 %, 9.5 %, 6.95 %, 15.07 % and 1.5 % respectively, while that for the metal sector accounted for 4.6 %, 3.5 %, 10 %, 7.54 %, 14.97 % and 12.4 % respectively.[9] In addition, 90.23 % of total exports of the total manufacturing industries are concentrated in the food (46.21 %) and chemical (44.02 %) industries. The contribution and distribution of these sectors in terms of size of capital and employment implies and enables us to compare between the chemical and metal as capital-intensive sectors and the food and textile sectors as labour-intensive sectors. Table 4.1 below illustrates the major economic indicators defined according to industrial activities for food, textile, chemical and metal manufacturing industries in Sudan in 2001.

In addition, other important reasons for the selection of these industries is that the lack of diversification in manufacturing industries implies the high concentration of manufacturing industries on food and oil industries, which are the two most important sectors with large shares in manufacturing industries in Sudan. The first evidence for this concentration is the high per capita value added for food and oil industries in the manufacturing sector estimated at US$ 26.3 and US$ 4.5 respectively compared to total per capita value added in the total manufacturing sector in Sudan estimated at US$ 40.6. In addition the second evidence for this concentration is the high share in value added for food, oil and chemical (including oil industries) industries that accounted for 6.1 %, 1.1 % and 1.645 % respectively, compared to the share of all other sectors together that accounted for only 2.3 % of total value added in the total manufacturing sector in GDP in Sudan estimated at 9.5 %. This implies that 76 % of total value added in the manufacturing sector in Sudan is concentrated in food and oil industries that show the high share in value added in the manufacturing sector estimated at about 65 % and 11 % respectively; they were followed by the chemical and textile sectors that accounted for 4.1 % and 2.6 % respectively. Moreover, the third evidence for this concentration is the high share in total manufacturing sector exports, for example, 87 % of total manufacturing sector exports is concentrated in food (mainly sugar) and oil industries that show the high share estimated at 46 % and 41 % of total manufacturing sector exports in Sudan respectively. This implies however, the weak structure of industrial exports that still depend basically on exports of raw material products rather than depend on high value manufactured products. In addition, the fourth evidence for this concentration is the high share in total imports, for example, 57.41 % of the total manufacturing sector imports is concentrated in food and chemical industries that show the high share estimated at 47.4 % and 10.01 % of total manufacturing sector imports in Sudan respectively. In addition, the fifth evidence for this concentration is the high share in gross output in the industrial manufacturing, for example, 72 % of gross output in the manufacturing sector is concentrated in food and oil industries that

[9] See for instance, Sudan Ministry of Industry (2005), 'Comprehensive Industrial Survey Data for 2001'.

Table 4.1 Economic indicators defined by activity for food, textile, chemical and metal manufacturing industries in Sudan (2001)

	Gross output		Gross value added		Gross fixed capital formation		No. of labour		Export		Import		No. of establishments	
	Number	%	Number	%	Number	%	Number	%	% of manufacturing in total export	% of sub sector in manufacturing sector	% of manufacturing in total import	% of sub sector in manufacturing sector	Number	%
Food products and beverages	423,637,059	55.3	216,782,220	64.6	15,459,151	46.5	74,058	56.32	5.92	46.21	10.05	47.44	16974	70.4
Textiles	21,696,554	2.8	8,728,836	2.6	260,361	0.8	6,982	5.31			0.06	0.3	58	0.2
Total chemical	172,885,839	22.6	56,599,889	16.9	3,172,453	9.5	9,142	6.95	5.63	44.02	3.2	15.07	343	1.5001
Coke, refined petroleum products and nuclear fuel	128,838,437	16.8	37,248,469	11.1	193,091	0.6	845	0.64	5.33	41.66	0.3	1.4	3	0.0001
Chemicals and chemical products	32,851,260	4.3	13,662,594	4.1	2,409,374	7.2	5,636	4.29	0.3	2.36	2.12	10.01	278	1.2
Rubber and plastics products	11,196,142	1.5	5,688,826	1.7	569,988	1.7	2,661	2.02			0.78	3.66	62	0.3
Total Metal	35,556,940	4.6	11,911,166	3.5	3,330,350	10	9,909	7.54			3.18	14.97	2978	12.4005
Basic metals	11,822,060	1.5	1,363,748	0.4	1,511,996	4.5	973	0.74			1.91	8.99	139	0.6
Fabricated metal products, except machinery and equipment	14,482,531	1.9	6,350,759	1.9	722,067	2.2	7,530	5.73			0.71	3.33	2812	11.7
Machinery and equipment n.e.c	6,354,330	0.8	3,396,428	1	230,504	0.7	461	0.35			0.26	1.25	15	0.1
Electrical machinery and apparatus n.e.c	2,898,019	0.4	800,231	0.2	865,783	2.6	945	0.72			0.3	1.4	12	0.0005
Total sample	653,776,392	85.3	294,022,111	87.6	22,222,315	66.8	100,091	76.12	11.55	90.23	16.49	77.78	20353	84.5006
Total manufactturing	765,429,858	100	335,410,844	100	33,235,336	100	131,506	100	12.8	100	21.19	100	24114	100

Source: Adapted from Sudan Ministry of Industry (2005) the Comprehensive Industrial Survey data for (2001)

show the high share estimated at 55 % and 17 % of total gross output in the manufacturing sector in Sudan respectively. The sixth evidence for this concentration is that food and oil industries constitute 69 % of gross output for large industrial establishments and account for the high share of 48 % and 21 % of gross output for large industrial establishments in Sudan respectively. Finally, further evidence appears from the concentration on food and fabricated metal industries that account for 82 % and 62 % of the total number of manufacturing industrial establishments and total employment in the manufacturing industrial sector in Sudan respectively.

Other important reasons for the selection of the chemical industry are the strategic importance of this industry in the international market, and the large, important and diversified nature of the chemical sector in Sudan, as it includes several important products such as basic chemicals, batteries, gases, matches, medical and pharmaceutical products, paints, plastics, soap, tyres and so on. Therefore, the chemical sector has the potential to satisfy the diversified needs of the local population, for instance, the soap industry's sub-sector is characterised by a relatively high range of manufacturing capacities, which can supply all the demand for the country and also produce a surplus for export. Moreover, the chemical sector is characterised by capital intensity, for instance, in 2001 the refined petroleum products show similar/equal intensity of both labour and capital, whereas, the all chemical sector including the refined petroleum products, petrochemical, chemical and plastic sector shows high intensity of capital and low intensity of labour. It is worthy to note that the contribution of oil industries in total employment represents only 0.5 % but in the meantime it is ranked second in terms of the contribution to industrial value added as it accounts for 11 % of total industrial value added in the manufacturing industries in Sudan; this implies that oil and chemical industries tend to use more capital intensive techniques and to be a more capital intensive industry. Further strategic importance of the oil and chemical industry is the significant contribution of oil in the promotion of oil related industries and the contribution to the industrial exports and total exports of Sudan. For instance, in 2001 the share of the all chemical sector including the refined petroleum products, petrochemical, chemical and plastic sub-sectors together represents 44.02 % of the total manufacturing industrial exports and the balance of trade for the all chemical sector shows a surplus. Whereas, when including only the petrochemical, chemical and plastic sub-sectors and excluding the refined petroleum products sub-sector, the balance of trade for the chemical sector shows a large deficit.[10] Moreover, another reason for the selection of this industry is that oil industry has the highest average labour productivity in the industrial sector, which is 1,353 times above the average for all manufacturing industrial establishments in Sudan; this also implies high level of technology used in the oil sector.[11] In addition, another reason for the selection of this industry is the

[10] See for instance, Sudan Ministry of Industry (2005), 'Comprehensive Industrial Survey Data for 2001'.

[11] See for instance, Sudan Ministry of Industry (2005), 'Comprehensive Industrial Survey Data for 2001'.

high rate of return on investment in capital, for instance, the chemical sector is ranked ninth among the ten highest manufacturing industries in terms of the rates of return on investment in capital, which accounted for 41 %, above the average for all manufacturing industries in Sudan (34 %).[12]

Other main reasons for the selection of the food industry is that the food industry sub-sector is a principal sector in Sudan considering its necessity and its linkages with the agricultural sector, which represents the backbone of the Sudanese economy. Moreover, the food industry is diversified by nature and has relatively high range of manufacturing capacities that can satisfy all the local market demand in Sudan and also produce a surplus for export; for instance, the food industry sub-sector is characterised by relatively high installed and diversified capacity that can deal with seeds, sunflowers, sesame, peanuts, groundnuts and so on, though this also implies the weak structure of manufacturing industrial sector because it is mainly based on agricultural production in Sudan. In addition, another reason for the selection of the food industry is the high rate of return on investment in capital, for instance, the food industry sector is ranked fifth among the ten highest manufacturing industries in terms of the rates of return on investment in capital, which accounted for 49 %, above the average for all manufacturing industries in Sudan (34 %).[13] Moreover, another reason for the selection of food industry is the large share in terms of the total number of industrial manufacturing establishments and total employment, for instance, the sectoral distribution by the number of establishments and employment shows that the food industry is ranked first with high share and accounted for 70 % of total number of industrial manufacturing establishments and accounted for more than half (57 %) of total employment in manufacturing industrial establishments in Sudan. Another reason for the selection of the food industry is that the it has a large share in the total manufacturing industrial output, for example, total output in the food industry represents half of the total output of all manufacturing industrial establishments in 16 states in Sudan. In addition to the large share of the food industry in total gross output of small industrial establishments in Sudan, for instance, food industry is ranked first and accounted for 88 % of total gross output for small industrial establishments in Sudan. Therefore, this implies that since the food industry has a large share in total number of industrial manufacturing establishments, as a result it also has a large share in total output and employment. This also implies that the food industry has the potential to contribute to reduction of the serious unemployment and poverty problems in Sudan as we explained in Chap. 2 above.

Another reason for the selection of the food industry (sugar sub-sector) is the significant contribution of sugar as a sub-sector in food industry. For instance, in 2001, the sugar sub-sector accounted for 33.7 %, 26.4 % and 5 % of total value

[12] See for instance, Sudan Ministry of Industry (2005), 'Comprehensive Industrial Survey Data for 2001'.

[13] See for instance, Sudan Ministry of Industry (2005), 'Comprehensive Industrial Survey Data for 2001'.

added, total output and total number of industrial establishments in the food industry respectively.[14] Notably, the selection of the food industry is based on the important contribution of the sugar industry in terms of total manufacturing employment, as the sugar industry alone accounts for 19.4 % and 28 % of total employment in total manufacturing industrial establishments and total employment in large manufacturing industrial establishments in Sudan respectively. In particular, the sugar industry is a significant industry in Sudan, as it employees 25,460 labourers; all the five sugar factories operating in the sugar industry are large in terms of employment as they all employ more than 100 people, and tend to be more labour intensive and seem more dependent on using labour intensive techniques. Therefore, this implies that the sugar industry has the potential to contribute to reduction of the serious unemployment and poverty problems in Sudan as we explained in Chap. 2. Further to the important share of the sugar industry, it alone contributes by 46 % to the total exports of all manufacturing industrial establishments in Sudan. Moreover, due to availability of natural resources and competent human cadres, Sudan has great potential in terms of sugar production and now it is ranked second in the African continent following South Africa.[15] Notably, today, the Kenana Sugar Company (KSC) is the world's largest producer of white sugar. Prior to the establishment of the KSC and factory, Sudan imported the bulk of its sugar, which was a drain on its limited foreign exchange.[16] Furthermore, the sugar industry could have a wider utility in the region, given the potential increase in the demand for sugar products due to the possible wider use of sugar, owing to the natural characteristics and environmental consideration of potential uses of sugar cane in the production of ethanol products as a clean and sustainable, environmentally friendly energy source in industry to replace/substitute other non-renewable energy sources (such as petroleum). Apart from being a pioneer in the agricultural industry, mainly in sugar production, Kenana is a pioneer in using timber planted for developing new environmentally friendly products and is also a pioneer in using a by-product of sugar production for creating charcoal. Bagasse, Kenana animal feed, with its high nutritional value, has also enjoyed substantial export markets in the Gulf. The inauguration of the Kenana Ethanol factory marked a historic date for Sudan's entrance to the age of green fuel production and implied the establishment of the first Ethanol factory in Africa.[17] Kenana Ethanol factory started off by producing 65 million litres and is set to reach up to 200 million litres by 2011; it aims to put Sudan at an advanced level worldwide in this kind of strategic industry. While the world is seeking to produce other alternative sources

[14] See for instance, Sudan Ministry of Industry (2005), 'Comprehensive Industrial Survey Data for 2001'.

[15] See for instance, the Kenana Sugar Company (KSC) website: www.kenana.com, accessed November 30, 2010.

[16] See for instance, the Kenana Sugar Company (KSC) website: www.kenana.com, accessed November 30, 2010.

[17] The Kenana Ethanol plant factory was inaugurated on 1 June, 2009.

of energy that are environmental friendly and that reduce greenhouse emissions and global warming contributions and so contribute to natural solutions to the present world energy and food crisis, Kenana's new ethanol fuel plant factory will open the door for biofuel and will also be a new addition to Sudan's non-petroleum exports. According to economic reports, Sudan could probably be one of the top producing countries for ethanol gas due to its great potential and capabilities in green ethanol industry field. The inauguration of the ethanol factory in Kenana is regarded as a significant development in the sugar industry and a key addition to the development process in Sudan. With the opening of the ethanol plant, many strategic goals have been achieved by: making maximum use of sugar waste (molasses); supporting the Sudanese economy by producing other energy sources that reduce the carbon dioxide emissions hazard; improving the environment; contributing to securing fuel from various sources, particularly given the expectation of several world experts that ethanol will replace normal fuel within the coming 5 years when it becomes the major energy element in several countries where the traditional energy reservoir is nil. It should be mentioned that ethanol fuel has the advantage that it can be used as a biofuel alternative to gasoline, it is widely used by flex-fuel light vehicles and as an oxygenate to gasoline, because it is easy to manufacture and process and it can be made from very common crops such as sugarcane and corn; moreover, bioethanol, unlike petroleum, is a renewable resource that can be produced from agricultural feedstocks.[18]

Another main reason for the selection of the metal industry is the large share of the metal industry in total number of total industrial manufacturing establishments and total gross output of small industrial establishments in Sudan, for instance, fabricated the metal industry sub-sector is ranked second and accounted for near to 12 % and 5 % of total number of total industrial manufacturing establishments and total gross output of small manufacturing industrial establishments in Sudan respectively. In addition, an important reason for the selection of the metal industry is the large share in terms of employment, for instance, the fabricated metal industry shows high share, ranked third and accounted for near to 7 % of total employment in manufacturing industrial establishments in Sudan. A further reason for the selection of the metal industry is the high rate of return on investment in capital, for instance, the machinery and equipment and fabricated metal sub-sectors are ranked first and sixth among the ten highest manufacturing industries in terms of the rates of return on investment in capital that is estimated at 196 % and 45 % respectively above the average for all manufacturing industries in Sudan (34 %).[19] In addition, machinery and equipment is ranked third in terms of average productivity of labour above the average of the total industrial manufacturing sector in Sudan.

Another main reason for the selection of the textile industry is the importance of it in terms of total capacity considering the availability of raw material (cotton) that

[18] See for instance, Sudan Views website: sudanviews.net, accessed November 30, 2010.

[19] See for instance, Sudan Ministry of Industry (2005), 'Comprehensive Industrial Survey Data for 2001'.

for a long time has supported the emergence of the textile industry in Sudan. Moreover, the textile industry is among the industries that show high performance in terms of average product, as average product in the textile industry is ten times the average product in all manufacturing industries in Sudan.[20] In addition, a further important reason for the selection of the textile industry is the large share in terms of employment in the manufacturing sector in Sudan, for instance, the textile industry shows high share and is ranked fourth, accounting for near to 6 % of total employment in the manufacturing industrial establishments in Sudan. The textile industry contributes highly to employment as it has the tendency to use labour intensive techniques as the majority of the textile firms are among the large size firms in terms of employment and number of labourers. Therefore, this implies that the textile industry has the potential to contribute to the reduction of the serious unemployment and poverty problems in Sudan as we explained in Chap. 2 above.

4.3 The Selection of the Sample and Composition of the Surveys

The sample in the firm survey was drawn from the small, medium and large size firms working in four industries in the manufacturing sector: the food, textile, chemical and metal industries, which are located in Khartoum state.[21] We observe the imbalanced geographical distribution of manufacturing industrial establishments in Sudan. For instance, the majority of the total, large and small manufacturing industrial establishments are concentrated only in three states, notably, Khartoum (19 %, 64 %, 15.6 %), South Darfur (17 %, 5.6 %, 17.7 %) and Al Gezira (13 %, 7.7 %, 13.8 %), and together they constitute the majority of the total, large and small manufacturing industrial establishments in Sudan (49 %, 77.3 % and 47.1 %) respectively.[22] Therefore, the selection of Khartoum state was based on its significant and highest average share in total number of manufacturing industrial establishments, as Khartoum state represents (18.94 %), (64.21 %) and (15.61 %) of the total number of total, large and small factories and manufacturing industrial establishments respectively in the food, textile, chemical and fabricated metal industries in Sudan. Moreover, this implies that most probably Khartoum state also has significant large average share in terms of total employment and capital investment. In addition, Khartoum state represents 13.79 %, 65.52 %, 47.52 % and 26.26 % of the total, large and small factories and industrial establishments in the food, textile, chemical and metal

[20] See for instance, Sudan Ministry of Industry (2005), 'Comprehensive Industrial Survey Data for 2001'. p. 132

[21] For the purpose of this study, firm size is defined by employment size N. The small size firms are firms with $N < 50$, the medium size firms those with $49 < N < 100$, and the large size firms those with $N \geq 100$ workers.

[22] See for instance, Sudan Ministry of Industry (2005), 'Comprehensive Industrial Survey Data for 2001'.

Table 4.2 The total and average share of Khartoum state in total number of establishment in small, medium and large size firms in the food, textile, chemical and metal industries defined by industrial activity and employment size (2001)

Description	Total Sudan				Khartoum	
[Value (000)SD]	Gross total manufacturing output	Value added ('000 SD)[a]	Employment	No. of establishments	No. of establishments	
Food products and beverages						
Large	303.38335	170,616,394	46,451	521	194	37.60
% of large	71.61	78.7	62.7	3	8.29	11.73
Small	120.2586	46,170,721	27,607	16,453	2,147	13.05
% of small	28.39	21.3	37.3	97	91.71	9.56
Share in total manufacturing (%)	55.34	64.63	56.32	70.39	9.71	51.25
Share in total large manufacturing (%)	39.63	50.86	35.32	68.23	0.81	4.25
Share in total small manufacturing (%)	15.71	13.76	20.99	2.14	8.90	47.00
Total	423.64195	216,787,115	74,058	16,974	2,341	13.79
Textiles						
Large	21.622658	8,712,934	6,935	38	18	47.37
% of large	99.66	99.82	99.3	65.5	47.37	1.09
Small	0.073896	15,902	47	20	20	100
% of small	0.34	0.18	0.7	34.5	52.63	0.09
Share in total manufacturing (%)	2.83	2.6	5.31	0.24	0.16	0.83
Share in total large manufacturing (%)	2.82	2.60	5.27	0.16	0.07	0.39
Share in total small manufacturing (%)	0.01	0.00	0.04	0.08	0.08	0.44
Total	21.696554	8,728,836	6,982	58	38	65.52
Chemical						
Large	172.461001	56,418,832	8,594	147	117	79.59
% of large	99.75	99.5	96.8	76	71.78	7.07
Small	0.434629	190,845	548	196	46	23.47
% of small	0.25	0.5	3.2	24	28.22	0.20
Share in total manufacturing (%)	22.59	16.88	6.95	1.42	0.68	3.57
Share in total large manufacturing (%)	22.53	16.82	6.54	0.61	0.49	2.56
Share in total small manufacturing (%)	0.06	0.06	0.42	0.81	0.19	1.01
Total	172.89563	56,609,677	9,142	343	163	47.52
Metal						
Large	28.423257	8,218,761	4,504	102	80	78.43
% of large	79.94	80.9	79.3	37	10.23	4.84

(continued)

Table 4.2 (continued)

Description	Total Sudan				Khartoum	
[Value (000)SD]	Gross total manufacturing output	Value added ('000 SD)[a]	Employment	No. of establishments	No. of establishments	
Small	7.133683	3,692,405	5,405	2,876	702	24.41
% of small	20.06	19.1	20.7	63	89.77	3.13
Share in total manufacturing (%)	4.64	3.55	7.54	12.35	3.24	17.12
Share in total large manufacturing (%)	3.71	2.45	3.42	0.42	0.33	1.75
Share in total small manufacturing (%)	0.93	1.10	4.11	11.93	2.91	15.37
Total	35.55694	11,911,166	9,909	2,978	782	26.26
Total manufacturing						
Large	628.8	280,689,600	90,605	1,654	1,062	4.40
Small	136.7	54,741,145	40,901	22,460	3,506	14.54
Share in total manufacturing (%)	100	100	100	100	100	18.94
Share of large in total manufacturing (%)	82.14	83.68	68.9	6.9	23.25	64.21
Share of small in total manufacturing (%)	17.86	16.32	31.1	93.1	76.75	15.61
Total	765.5	335,430,745	131,506	24,114	4,568	18.94

Source: Adapted from Sudan Ministry of Industry (2005) the Comprehensive Industrial Survey data for (2001)

[a]The value is estimated in Sudanese Dinar (SD)

industries in Sudan respectively (cf. Table 4.2 below).[23] Moreover, the manufacturing industries in Khartoum state is characterised by being more diversified as compared to other states in Sudan. Moreover, the selection of small, medium and large size firms

[23] Khartoum state also has strategic importance in Sudan, for instance, it accounts for more than half of the country's total revenue. "Khartoum accounted for almost 40 % of total revenue collection by states in 1996, its share increased to 50 % in 1999, and is estimated to have increased further [to 71.3 %] by 2001" (see Brixiova et al. 2003, p. 5). In addition, in recent years Khartoum has a thriving economy and has seen significant development driven by Sudan's oil wealth and the concentration of investment in oil; one of Sudan's largest refineries is located in northern Khartoum, and petroleum products are now produced in the north of Khartoum state, providing fuel and jobs for the city. Moreover, Khartoum is a tripartite metropolis with an estimated overall population of over five million people; it accounted for 13.5 % of total population in 2008 (Sudan Central Bureau of Statistics Population Census Data 2010; Fifth Sudan Population and Housing Census 2008). Moreover, it represents a trade and communication center with the highest concentration of economic activity and urban-based services sector in Sudan, such as the construction, telecommunication, infrastructure, banking, health and educational services; Khartoum is the main location for most of Sudan's top educational bodies and accounted for 31 % of total enrolment in higher education in Sudan in 2007 (adapted from the Admission Office cited in Sudan Ministry of Finance and National Economy 'Annual Report' (2007), Table 20-2, p. 24). It accounted for close to 41.23 % of total branches and banking services in Sudan (adapted from the website of the Central Bank of Sudan, Accessed 10 December 2010). http://www.cbos.gov.sd/sites/default/files/banks_spread.pdf.

was based on their shares in total employment; they accounted for 18.94 % of small and large size enterprises working in the manufacturing sector in Khartoum state (14.54 % of small size and 4.4 % of large size enterprises). Our sample drawn from Khartoum state is quite representative, since the coverage of firms in the sample and survey represents 1 %, 16 %, 28 %, and 2 % of the food, textile, chemical and metal industries respectively and 2 %, 7 %, and 1 % of the total firms, medium and large size firms and small size firms respectively.[24,25] We employed the most recent secondary data published by Sudan Ministry of Industry (2005), 'The Comprehensive Industrial Survey (2005)' in selecting a sample of the firms in the survey.[26]

The questionnaire on 'Technological Change and Skill Development' was circulated amongst 100 of the food, textile, chemical and metal small, medium and large size enterprises in Sudan. It aimed at collecting micro qualitative and quantitative data, and covered the small, medium and large size firms engaged in the food, textile, chemical and metal industries in Sudan. Table 4.3 below presents the composition of the firm survey. The response rate varied according to firm size and industrial activity: for the food industry the total response rate was 88 %, and the weighted response rates by employment size were 83 %, 92 % and 88 % for small, medium and large size firms respectively. For the textile industry the total

[24] The distribution of firms in the sample is based on two facts: the great diversity of the food and chemical compared to metal and textile industries and the potential for upgrading skill and technologies in the large compared to small and medium size firms.

[25] The distribution and representation of firms in the sample is reasonable and representative in view of the fact that majority of manufacturing industrial establishments which were included in the comprehensive industrial survey (2005) that was conducted in 2001 seemed to be not working when we conducted the firm survey in Khartoum state over the period January–June 2010.

[26] The Comprehensive Industrial Survey was conducted in 2001 by Sudan Ministry of Industry, Sudan Central Bureau of Statistics, State government and Chambers of Industries with technical support from UNIDO. The Ministry of Industry executed and published the results of the industrial survey in 2005. The report cover sizes of establishments, ownership, sectoral, composition, geographical distribution, employment, wages and salaries, gross output, manufacturing value added, material intensity and import structure of production, export structure and contribution of sectors to the manufacturing trade balance and factor productivity. The industrial survey covered 2,868 manufacturing establishments and the results showed that there are 24,762 industrial establishments in Sudan, 644 (24 %) of them are not working. The survey also showed that 96 % of these establishments belong to the private sector and there are no foreign ones among the minor establishments, whereas there are 25 (1.5 %) foreign among the big establishments and 2.5 % foreign and Sudanese ones. The survey results also showed that the state participated in 113 (0.5 %) of minor industrial establishments and 6.8 % in junior establishments and there are 89 in the state public ownership. The minor establishments contributed with a proportion of 40 % in the transforming industries. The results imply that 80 % of users are concentrated in four industries: foods and drink (57 %), mining products industries (13 %), minerals forming (7 %) and textiles (6 %). Total of manufacturing industries product is estimated at about 765,498 million Sudanese Dinars (equivalent to about US$ 2,958.8 million), the contribution of small establishments is 18 % and the large ones contribution is 82 %, the total of the added value is estimated at about 335,430.8 million Sudanese Dinar (equivalent to about US$ 1,296.6 million). See: www.industry.gov.sd, Accessed 10 December 2011. See also Sudan Ministry of Industry (2005), 'Comprehensive Industrial Survey Data for 2001', pp. 3–7.

Table 4.3 Composition of the firm survey in the Sudan 2010

Activity and size	Employment size	Share of employment in the sample	Share of capital in the sample[a]	Number of the respondent firms	Share of firms In the response rate (%)	Number of firms in the sample	Share of firms in the sample	Response rate (%)
(1) Size								
All firms	Small	17.10 %	99.18 %	21	24 %	25	25	84 %
	Medium	28.04 %	0.06 %	29	33 %	33	33	88 %
	Large	54.86 %	0.76 %	35	40 %	42	42	83 %
	Unknown			2	2 %			2 %
Grand total	Grand total	100 %	100 %	87	100 %	100	100	87 %
(2) Activity								
Chemical	Small	3.18 %	0.0003 %	11	13 %	13	13	85 %
	Medium	5.00 %	0.0169 %	12	14 %	14	14	86 %
	Large	43.75 %	0.3951 %	15	17 %	18	18	83 %
	Unknown			2	2 %			
	Total	51.93 %	0.4123 %	40	46 %	45	45	89 %
Food	Small	12.00 %	99.1762 %	5	6 %	6	6	83 %
	Medium	2.04 %	0.0031 %	11	13 %	12	12	92 %
	Large	7.47 %	0.0296 %	14	16 %	16	16	88 %
	Total	21.51 %	99.2089 %	30	34 %	34	34	88 %
Metal	Small	1.48 %	0.0000 %	4	5 %	5	5	80 %
	Medium	1.87 %	0.0004 %	4	5 %	5	5	80 %
	Large	1.11 %	0.3354 %	4	5 %	5	5	80 %
	Total	4.46 %	0.3359 %	12	14 %	15	15	80 %
Textile	Small	0.44 %		1	1 %	1	1	100 %
	Medium	19.13 %	0.0428 %	2	2 %	2	2	100 %
	Large	2.54 %	0.0001 %	2	2 %	3	3	67 %
	Total	22.10 %	0.0429 %	5	6 %	6	6	83 %
Grand total	Grand total	100 %	100 %	87	100 %	100	100 %	87 %

[a]The share of capital in the sample in this table refers to all respondent firms that respond to the survey question on capital. The high share of food and small size firms in total capital in the sample implies the high response rates of food and small size firms to the survey question on capital, while the low share of chemical, metal, textile and medium and large size firms in total capital in the sample implies the low response rates of chemical, metal, textile and medium and large size firms to the survey question on capital

response rate was 83 %, and the weighted response rates by employment size were 100 %, 100 % and 67 % for small, medium and large size firms respectively. For the chemical industry the total response rate was 89 %, and the weighted response rates by employment size were 85 %, 86 % and 83 % for small, medium and large size firms respectively. For the metal industry the total response rate was 80 %, and the weighted response rates by employment size were 80 %, 80 % and 80 % for small, medium and large size firms respectively. For the total sample, the total response rate was 87 %, and the weighted response rates by employment size were 84 %, 88 % and 83 % for small, medium and large size firms respectively.[27]

The macro survey questionnaire on 'Skill Creation, Human Resources Development and Policy Intervention' was sent to 40 policy makers in the government and experts in the eight public, university, educational, training and research institutions in Sudan. It aimed to collect macro qualitative data to reflect the opinions of policy makers and experts with respect to assessment of skill upgrading efforts and the causes and consequences of the deficient educational system, provision of training, transfer of knowledge and technological upgrading. It was also intended to provide insights to help to generate policies to enhance them by implementation of short- and long-term plans at the macro level. The selection of institutions, policy makers and experts was based on their experience and potential contribution to enhance the upskilling process. The advantage of the macro survey is that it examines these problems after integrating two different perspectives of policy makers and experts and also integrating two different perspectives from both the fields of education (50 %) and training (50 %). Moreover, due to their close association to educational and training institutions, the approached policy makers and experts provided some useful information from both the analytical and policy perspectives. Table 4.4 below presents the composition of the macro survey, and indicates a total response rate of 90 %; the shares of universities, ministries and other public institutions are quite representative and yield different response rates.[28]

The data from the surveys is supported by ten face-to-face interviews with firm managers and five interviews with policy makers and experts. The purpose of these interviews was to obtain more information to support the findings from the macro survey concerning the failure of upskilling efforts, the deficiency in educational and training systems, and the implications on skills mismatch, transfer of knowledge,

[27] For the implementation of the firm survey, a team of part-time researchers from the Sudan Ministry of Industry was hired to make direct personal contact, determine the contact address, handle, distribute and collect the survey from firms. On request from some of the approached firms, an additional copy of the survey was sent by e-mail to accelerate and increase the response rates. The translated Arabic version of the English version of the firm survey was distributed to increase the response rate.

[28] For the implementation of the macro survey, we established direct personal contact to handle and accelerate the distribution and collection of the survey. On request from some of the approached officials and experts, an additional copy of the survey was sent by e-mail to accelerate and increase the response rates. The translated Arabic version of the English version of the macro survey was distributed to increase the response rate.

Table 4.4 Composition of the macro survey in the Sudan, 2010

	Institutions				Individuals			
Representation	Number in sample	Total response	Response rate (%)	Share in total response rate	Number in sample	Total response	Response rate (%)	Share in total response rate
Universities	1	1	100	12.5	5	5	100	14
Ministries	4	4	100	50	24	21	88	58
Other public centres and institutions	3	3	100	37.5	11	10	91	28
Total	8	8	100	100	40	36	90	100

R&D efforts and development of local technologies, as well as the policies and plans of the government and private sector to enhance the upgrading of local skills and local technology, the R&D efforts, networks and transfer of knowledge.

In addition, we conducted the R&D Survey (2010), which is small survey on research and development (R&D) based on 25 face-to-face interviews with the official policy makers and experts in the government and the academic staff in the public and private universities. The main purpose of this survey was to collect primary data to examine the causes and consequences of poor R&D activities, to examine the main factors hindering and those contributing towards the promotion of R&D and then to provide some recommendations to improve R&D and hence S&T development in Sudan.[29]

4.4 Structure and Design of the Questionnaire

We present the general structure and design of the questionnaire of the macro and firm surveys in Table 4.5. The questionnaire in the macro survey was composed of six sections, and the average response rate is ranked in a descending order, i.e. generally higher for the second, third, fifth and fourth sections, moderate for the first section and low for the sixth section.[30] Each of the six sections in the macro survey was designed to request specific information. Section 1 requested general assessment of upskilling efforts from the perspectives of policy makers and experts. Section 2 inquired the relevant policies and important factors contributing to enhance skill upgrading, education, training and transfer of knowledge. Section 3 investigated the causes of low skill, deficient education, training, external effects of schooling, transfer of knowledge and skills mismatch. Section 4 examined both currently implemented plans and suggested future short and long run plans and

[29] The interviews were conducted with the officials and experts (20 %), academics in the public (60 %) and private (20 %) universities. The interviews were conducted with academic staff in the fields of science (36 %), engineering (36 %) and social sciences (8 %) including both males (80 %) and females (20 %). The distribution of the interviewed institutions includes public universities represented by Khartoum University (60 %), private universities represented by University of Medical Sciences and Technology (20 %), Ministry of Science and Technology (12 %) and Ministry of Higher Education and Scientific Research (8 %). For the implementation of the R&D survey, we established direct personal contact to hold direct face-to-face interviews that led to increased response rates. The design of the questionnaire in the R&D survey includes two types of questions: scalar or categories, and open questions. The distribution of question types and their corresponding average response rate are identical, for instance, the majority of questions are of scalar type, which also receives higher average response rate, followed by open questions.

[30] The design of the questionnaire in the macro survey includes three types of questions: nominal (Yes/No), scalar or categories, and open questions. The distribution of question types and their corresponding average response rate are identical, for instance, the majority of questions are of the scalar type, which also receive a higher average response rate, followed by nominal and open questions respectively.

Table 4.5 Structure, aims and average response rate of the questionnaire in the firm and macro surveys in the Sudan, 2010

	Macro survey		Firm survey	
Section no.	Aim of the section	Average response rate (%)	Aim of the section	Average response rate (%)
1	Request general assessment of upskilling efforts	85	Request general background information about the structure, identification and characteristics of the firms	77
2	Investigate the relevant policies to enhance skill upgrading, education, training, transfer of knowledge and building local technological capacity	99	Investigate the use, level, transfer and dependence on foreign technology, value and trend of ICT expenditures. Technology indicators, patent, R&D, ICT and product/process innovations	83
3	Examine the causes of low skill, deficient educational system and the consequences on training provision, external schooling effect/ transfer of knowledge, skills mismatch and low efforts to build local technological capacity	98	Measure human capital (skill) indicators: distribution of workers by skill levels, educational attainment/ average schooling years, occupation level, average years of experience, required qualifications and average wages. Assess the effect of skilled workers and their shortage on firm product, the factors hindering and other enhancing the incidence and transfer of knowledge/ external effects of schooling and firm upskilling plan	89
4	Examine the short and long-term mechanisms for skill upgrading, their effect on technological upgrading and fulfilling socio-economic aims	93	Examine the use of technology and the implications on upgrading skill level and the past and future demand for skilled and unskilled workers. The effect of upskilling on technological upgrading and self-reliance on local skills	87
5	Investigate the incidence of external effects of schooling and transfer of knowledge, their hindering and promoting factors	98	Examine the relative importance of firm training and short and long run skill upgrading mechanisms. The information, coverage, resources and support to firm training, the factors hindering and other promoting the provision of training	76
6	Conclusions	56	Conclusions	91

policies for skill upgrading and determined the role of skill upgrading and technological upgrading in fulfilling the socioeconomic development objectives. Section 5 sought information to check the incidence of external effects of schooling/transfer of knowledge and the important factors hindering and those contributing toward enhancing the external effects and transfer of knowledge. Finally, section 6 requested more recommendations for skill upgrading in Sudan.

The questionnaire in the firm survey was composed of six sections; the average response rate was higher for the sixth section, moderate for the third, fourth and second sections and low for the first and fifth sections.[31] Each of the six sections in the firm survey aimed to request particular information. Section 1 requested general background information about the structure, identification and characteristics of the firms, and it also requested (interval) economic quantitative data on the value and trend of firm production and performance indicators, including: employment, net worth (capital), profit, sale, output and product diversification by sale and employment. Section 2 examined the use of technology, level, transfer and dependence on foreign technologies. It assessed technology indicators, patent applications, spending on R&D, and product and process innovations, the use of ICT, firm production and demand for the highly skilled and also requested quantitative data on the value and trend of ICT expenditure. Section 3 requested quantitative data to measure human capital/skill indicators, defined by the distribution of workers by skill level, educational attainment (average year of schooling), occupational levels, average years of experience, attained and required education and average wages. This section also examined the effect of skilled workers on firm production, the incidence of external effects of schooling, the factors hindering and others contributing toward enhancing the transfer of knowledge at the firm level, firm upskilling plans and their various effects. Section 4 inquired into the implication of technology use on both upgrading skill levels and on the past and future demand for skilled and unskilled workers, and also examined the effect of firm upskilling plans on technological upgrading and self-reliance on local skill. Section 5 investigated the relative importance of the effort of training, the short and long run skill development mechanisms, the coverage, resources and support offered to firm training, the factors hindering and other contribute toward promoting the success of training at the firm level. Finally, Section 6 explained the importance of industry and examines the important factors hindering, and those contributing toward enhancing the contribution of industry and finally requested further recommendations for skill upgrading and technological upgrading in the industrial firms in Sudan.

[31] The design of the questionnaire in the firm survey includes four types of questions: nominal (Yes/No), scalar or categories, open and interval questions. The distribution of the questions and their corresponding average response rates vary. Most of the questions are of the scalar type, followed by nominal, interval and open questions respectively; the corresponding average response rate is higher for nominal questions, followed by scalar, open and interval questions respectively.

4.5 Conclusions

The data from the firm and macro surveys provides us with the required information, which is particularly useful for presenting a macro-micro comparative analysis to identify the causes and consequences of the skills problem, and the plans and policies for skill development from the macro-micro perspectives. The results of the macro survey seem quite representative, since the selection covers government, universities, and educational and training institutions. One advantage of the macro survey is that it examines the problem after integrating two different perspectives of policy makers and experts and also integrating two different perspectives from both the fields of education and training. Another advantage is that, due to their close association to educational and training institutions, the approached policy makers and experts provided some useful information, particularly from both the analytical and policy perspectives. Moreover, the data evaluates upskilling efforts, examines the causes and consequences of a deficient educational system, training provision, skills mismatch and transfer of knowledge, and also suggests relevant short and long run plans and policies for the enhancement of skill level, education, training, transfer of knowledge and technological upgrading. The main limitation of the macro survey is the incomplete and somewhat selective answers offered for some questions, probably because some of the respondent experts lacked adequate information to enable them to contribute to a critical analysis of the causes and consequences of low skill levels.

The results of the firm survey are quite representative, since the selection and coverage of firms in the survey include a broad range of firms working in the food, textile, chemical and metal industries, which provides us with relevant data and information that of considerable use in our analysis. Such coverage also has the advantage of enabling us to compare between firms according to two criteria, i.e. the size of employment and industrial activity.

Moreover, the firm survey presents some background information, which is also quite useful for a further analysis of firms based on other characteristics such as the geographical location, sector (public-private, mixed), net worth (capital), ownership and nationality of owner (government, foreign, mixed) and foreign orientation or affiliation to Multinational Corporation/Transnational Corporation (MNC/TNC). Another advantage of the firm survey is that it presents more specific but also quite comprehensive data and information that allow us to use a wide range of quantitative data and information for measuring skill and technology indicators and the link between them at the micro level/ across firms. In addition, the survey data allows us to approximate and examine the importance of tacit knowledge, and enables us to compare between attained and required education and to measure the skills mismatch across firms. At the micro level, realising the differences in both tacit knowledge and technologies used across food, textile, chemical and metal industries when comparing their effects and interaction, we define tacit knowledge by the share of high skilled workers in total employment and technology by the share or total spending on ICT.

One major limitation with respect to the firm survey is the low response rate for some questions, especially where the answers or data required quantitative measurement. Such problems arose because some of the respondent firms were unwilling to provide complete and reliable quantitative data or some of the respondent firms offered somewhat selective answers. For example, some firms seemed hesitant to provide information about quantitative data on firm performance and also qualitative assessment of training and skill upgrading efforts (see Table 4.5 above). An intensive follow-up with firms often improved the quality and quantity of the response rate. However, the hesitance of some firms compelled us to exclude them when their observations were incomplete, missing and unreliable. Therefore, we used only completed and reliable observations in our estimation and analysis in the next chapters. Apart from this limitation, the data from the firm and macro surveys remains useful from both the analytical and policy perspectives and is suitable to use in the empirical investigation in Chaps. 5, 6, 7 and 8 and also for the policy analysis and recommendations in Chap. 9.

References

Brixiova, Z., Gemayel, E. R., & Said, M. (2003, September). Can fiscal decentralization contribute to poverty reduction? Challenges facing a low income country. *Topics in Middle Eastern and North African Economics, Electronic Journal, 5*, 5. Chicago: Middle East Economic Association and Loyola University Chicago. http://www.luc.edu/publications/academic/

Central Bank of Sudan. (2009). *Central Bank of Sudan annual report (2009)*. Khartoum

Central Bank of Sudan. (2010). *Geographical distribution of the Sudanese banking system*, Khartoum: Central Bank of Sudan. http://www.cbos.gov.sd/sites/default/files/banks_spread.pdf. Accessed 10 Dec 2010

Firm Survey. (2010, June). *Technological change and skill development: A comparative study of chemical, food, metal and textile small, medium and large scale enterprises in the Sudan*. Khartoum

Kenana Sugar Company (KSC). (2010). *KSC achievement report*. Khartoum. http://www.kenana.com. Accessed 30 Nov 2010

Macro Survey. (2010, June). *Skill creation, human resources development and policy intervention: Interviews with policy makers and experts in the Sudan*. Khartoum.

R&D Survey. (2010, June). *Sudan R&D Survey: Efforts for enhancing R&D in the Sudan: Interview with policy makers in the government and academic staff in public and private universities in Sudan*. Khartoum

Sudan Central Bureau of Statistics Population Census Data. (2010). *Fifth Sudan population and housing census (2008)*. Khartoum.

Sudan Central Bureau of Statistics. (2009, December). *Sudan in figures: 2004–2008*. Annual Sudan Statistical Year Book: 2009, Issue No. 9, Khartoum: Sudan Central Bureau of Statistics, Sudan Ministry of the Cabinet, Sudan Currency Printing Press Company Ltd., Table 2, p. 2.

Sudan Ministry of Finance and National Economy. (2007). *Annual economic survey report (2007)*. Khartoum, Table 20-2, p. 24.

Sudan Ministry of Industry. (2005). The comprehensive industrial survey (2005). Khartoum

Sudan Views. sudanviews.net. Accessed 30 Nov 2010

The Arab Investment and Export Credit Guarantee Corporation. (2007). *Investment climate in the Arab countries* (p. 60), Kuwait.

The Kenana Sugar Company (KSC). Web site: www.kenana.com. Accessed 30 Nov 2010
The World Bank, World Development Indicators Database (2010a): http://data.worldbank.org/data-catalog/world-development-indicators. accessed on 10 April 2010.
The World Bank, World Development Indicators Database (2010b): http://data.worldbank.org/data-catalog/world-development-indicators. accessed 1 December 2010.
The World Bank, World Global Development Finance (2010): http://data.worldbank.org/data-catalog/world-development-indicators. accessed on 1 December 2010.
The World Bank, World Development Indicators database (WDI). (2005). Accessed 25 Dec 2005. http://data.worldbank.org/data-catalog/world-development-indicators.
The World Bank, World Development Indicators Database (2011). http://data.worldbank.org/data-catalog/world-development-indicators. Accessed 25 June 2011.
United Nations Development Programme UNDP in Sudan. (2006, June). *Macroeconomic policies for poverty reduction: The case of Sudan*. Khartoum: Published for the UNDP in Sudan, Table 8.1, p. 82, pp. 92–94, UNIDO, reported in Dagdeviren and Mahran (2004), Table 11. Cited in Trade and industry – What role for poverty reduction? Chapter 8 in UNDP (2006) *Macroeconomic policies for poverty reduction: The case of Sudan* Table 8.1, p. 82.

Chapter 5
Assessment of Skill and Technology Indicators

Abstract This chapter uses the results of the macro and firm surveys to show the interaction between the deficient educational system and the high incidence of unskilled workers and their implications.

We find that the major reasons for heavy dependence on foreign technologies at the micro level are low levels of both skill and technology due to the deficient educational system and the high incidence of unskilled workers. We find that the deficient educational system – due to low quality of education – and the excessive share of unskilled workers has led to low skill levels, poor provision of training, serious skills mismatch, weak linkages, lack of a networks and hindered the transfer of knowledge. These factors have interacted with each other and led to poor technology indicators, poor indigenous capability to build the local technology and a heavy dependence on foreign technology. These results prove hypotheses 3.a–3.b in Chap. 1 above concerning the low skill and technology indicators at the micro–macro levels: the serious implications of the interaction between the causes and consequences of the deficient educational system and the high use of unskilled workers. We confirm hypothesis 3.c. in Chap. 1 above that the major causes of low level of local technology are low/a lack of R&D activities due to a lack of skills, transfer of knowledge, networks and collaborations between universities and industry/firms.

5.1 Introduction

In this chapter we use the data and results of the firm and macro surveys set out in Chap. 4 to examine the third hypothesis in Chap. 1 about the serious implications of the interaction between the deficient educational system and the high incidence of unskilled workers and skill mismatch. In particular, we use the results of the macro survey to show the causes of the deficient educational system and consequences on low skill levels, poor provision of training, skills mismatch and low transfer of knowledge at the macro level. In addition, we use the results of the firm survey to illustrate that the high incidence of unskilled workers leads to low skill level, poor

S. Mohamed Nour, *Technological Change and Skill Development in Sudan*,
DOI 10.1007/978-3-642-32811-4_5,

provision of training, skills mismatch, poor technology indicators and a heavy dependence on foreign technologies. The rest of this chapter is organised as follows: Sect. 5.2 shows the causes and consequences of the deficient educational system and the high incidence of unskilled workers and their implications on low skill levels, poor provision of training, skills mismatch, lack of knowledge transfer, low level of local technology and heavy dependence on foreign technology. Section 5.3 presents the micro–macro views about the upgrading of skill and technology and their potential implications. Section 5.4 provides the conclusions.

5.2 Causes and Consequences of Deficient Educational System

The results of the macro survey indicate that, at the aggregate level, the official efforts to promote and upgrade the levels of local skill have been relatively successful only in some sectors.[1] In general, there has been a serious failure and shortcoming of the upskilling process, which is mainly attributed to: (1) inadequate availability and misallocation of resources; (2) the deficient educational system; (3) failure of educated and trained workers to transfer knowledge; and (4) inadequate training provision.[2]

We begin our discussion with the second reason, i.e. the deficient educational system, because we want to argue that both inadequate training provision and the failure of educated and trained workers to transfer knowledge are direct implications of deficient education, while the misallocation of resources is indicated as one cause of both deficient education and the poor provision of training.

5.2.1 Causes of Deficiency in the Educational System

With respect to the second cause, Table 5.1 shows that the deficiencies of the educational system appear in all the basic, technical, secondary and tertiary educational systems. Major causes are the low quality and internal efficiency of the educational system, the lack of infrastructure (due to inadequate investment/public spending on education) and the lack of teachers and mentors.[3] Other important factors are the inadequate assessment and monitoring of educational needs, the lack of modernisation and dynamism and inadequate planning for educational needs.[4]

[1] As reported by 56 % of the respondent policy makers and experts to the macro survey.

[2] As indicated by 100 %, 97 %, 94 % and 92 % of the respondent policy makers and experts respectively.

[3] As reported by 92 %, 92 % and 91 % of the respondent policy makers and experts to the macro survey respectively.

[4] As indicated by 87 %, 86 % and 84 % of the respondent policy makers and experts to the macro survey respectively.

Table 5.1 The causes of deficient educational system in the Sudan, 2010

Causes of deficiency	General educational system (basic + technical + secondary + tertiary) (%)	Basic education (%)	Technical education (%)	Secondary education (%)	Tertiary education (%)
Inadequate assessments and monitoring of educational needs	87	86	94	86	83
Low quality/efficiency of educational system	92	86	92	94	94
Inadequate planning for educational needs	84	81	94	78	81
Lack of flexibility of educational institutions	83	81	78	81	92
Low spending and weak incentives for enrolment in technical education	94	–	94	–	–
Lack of modernization and dynamism	86	78	94	83	89
Low involvement and spending by private sector	79	78	89	72	78
Weak linkages [networks] between universities, colleges, technical and the training institutes	83	–	–	–	83
Lack of infrastructures due to inadequate investment (public spending on education)	92	97	97	92	81
Lack of teachers and mentors	91	94	97	86	86

Source: Own calculation based on the macro survey (2010)

Finally, the lack of flexibility of educational institutions and the weak linkages/networks between universities, colleges, technical and training institutes are also mentioned, but are of somewhat less importance.[5] That also holds good for the low involvement and spending by the private sector.[6] Moreover, the major serious problem for the technical education is the weak incentives for spending and enrolment in technical education.[7] For instance, the share of public spending on education in GDP in Sudan (0.9 %) is low compared to Arab Gulf countries like Saudi Arabia (9.5 %) and other advanced Asian countries such as Korea (3.6 %) and Malaysia (7.9 %) (UNDP 2004) – see our discussion of the supply side of educational policies in Chap. 9. Moreover, according to the twin-peaks analysis in Ziesemer (2004), which compares the distribution of public spending on education across countries, the Sudan's less than 2 % public spending on education falls below the average 4.6 % of public spending on education to improve the accumulation of human capital in 1998. Furthermore, from the official perspective, the other important problems related to basic education level include the lack of interest in using modern technologies, lack of attention to update the curriculum in line with the requirements of the phase, weak financial incentives and motivation for teachers and mentors in basic education, weak laws and regulations under the federal government and the lack of enrolment in the basic education system for some of the population for reasons related to war and displacement, and so on. The other important problems related to technical education level include the lack of priority, facilities, comprehensive government policies for development of technical education, the lack of modernisation in the curriculum of the technical education level, the focus on theoretical aspects of technical education rather than practical aspects of technical education and the negative cultural view towards the technical education level. The other important problems related to secondary education level include the low priority of secondary education, the lack of attention to update the curriculum in line with the requirements of the phase, the lack of interest in using modern technologies, the focus on quantitative aspects of secondary education rather than qualitative aspects of secondary education, the lack of training for teachers, the low commitment to regulations and legislation in secondary education. The other important problems related to tertiary and university education level include the lack of attention to scientific research as a prerequisite of the requirements of university education, the lack of attention to training in and use of modern technical means and ways of education (such as computer, Internet, etc.) learning and self-learning methods in tertiary education. In addition other problems are related to the focus on quantitative aspects of university education rather than qualitative aspects of university education, poor coordination between the agencies of education (academic/technical), the mismatch between the outputs of university education with the requirements and needs of the labour market and employment

[5] As reported by 83 % and 83 % of the respondent policy makers and experts to the macro survey respectively.

[6] As indicated by 79 % of the respondent policy makers and experts to the macro survey.

[7] As reported by 94 % of the respondent policy makers and experts to the macro survey.

opportunities, the lack of periodical review of the contents of educational curriculum in universities and the lack of harmony, uniformity, consistency and match in the system used in different public and private universities.

We observe that, according to 92 % of the respondents to the macro survey, the low quality and efficiency of the educational system appears from the low quality at technical, tertiary, higher secondary and basic education relative to international standards respectively.[8] Important causes are the low rates of accomplishments and motivation at technical, higher secondary and basic education levels relative to international standards, but the problem is somewhat less at the tertiary education level.[9] Other serious problems are the low quality of teachers and mentors, the low public current expenditure per pupil, the low survival rates and high drop-out and the high pupil/teacher ratios,[10] while less important causes include the high repetition rates.[11] Furthermore, from the official perspective, the other important problems related to poor quality of education include the poor curriculum in basic and secondary education, which in some cases do not conform to the students' abilities and seems beyond the capacity of students and their families; for the students the courses include a lot of useless and extra information, including not only basic knowledge but additional redundant courses in family studies, military, agricultural, and so on, therefore, the large number of unrequired courses come at the expense of the basic knowledge. Other problems are related to the subordination of spending on education to localities, poor working conditions for teachers which make the teaching profession unattractive, a lack of proper understanding of the meaning of quality education, a lack of interest in addressing the phenomenon of dropout and leakage from education which is most probably caused by some important reasons including rising poverty, lack of equity and equal opportunities for competition, lack of commitment to offering free access to education and the educational environment, and a lack of follow-up records for students enrolled in the public educational institutions.

5.2.2 *Consequences of the Deficient Educational System*

5.2.2.1 Mismatch Between the Output of Education and the Market Needs

We find that both the deficient basic (primary and secondary) and tertiary educational systems together lead to a serious mismatch between the output of education and the market needs. In particular, about 92 % and 72 % of the respondents to the

[8] As reported by 92 %, 89 %, 89 % and 86 % of the respondent policy makers and experts to the macro survey respectively.

[9] As indicated by 97 %, 89 %, 89 % and 86 % of the respondent policy makers and experts to the macro survey respectively.

[10] As reported by 100 %, 94 %, 89 % and 89 % of the respondent policy makers and experts to the macro survey respectively.

[11] As indicated by about 81 %, of the respondent policy makers and experts to the macro survey.

macro survey reported that the mismatch is mainly attributed to the deficiency of both tertiary and basic education respectively. Moreover, the follow-up interviews with policy makers and experts show that the mismatch is attributed to the deficient educational system,[12] the lack of coordination and planning to meet the critical skill needs and the cultural/social reasons: preference for white collar jobs and bias against technical education and technical jobs. For instance, the deficiency of tertiary educational system is caused by the inconsistent structure: the share of students enrolled in Sudan in 2000 in all social sciences, humanities and art faculties (65 %) was much higher than those of sciences, math and engineering (35 %). The share of students enrolled in sciences, math and engineering in Sudan is also low compared to both Algeria (50 %) and China (53 %) (UNDP 2002, 2003; UNDP-AHDR 2002, 2003) – cf. our discussion on the demand for education in Chap. 9.[13]

5.2.2.2 Lowering Skill Levels

From Sudan population census data (2008) and the educational matrix for the period 1960–2008 set out in Table 5.2, we observe the low skill levels, defined by the educational level of the total population. The share of low educated (99–95 %) is much higher than that of high educated (1–5 %) in the total population over the period 2000–2008. That also indicates a minimal skill upgrading, defined by the relative rise in the share of the high educated population and the relative decline of the share of low educated population during the period 1960–2008.[14] In addition, Sudan Central Bureau of Statistics statistical data (2004–2008) on the distribution of the economically active population by occupational classification (2004–2008) shows the low skill level, defined by occupational levels; for instance, the share of the unskilled population (86–83 %) is much higher than that of the high skilled population (14–12 %). That also implies that rather than improvement there was even a slight deterioration in skill upgrading, defined by the decreasing share of high skilled, despite a falling share of unskilled during the period 2004–2008. Furthermore, as we explain below, the low skill levels at the macro level is consistent with that at the micro level.

[12] As in most other developing countries, the mismatch is attributed to deficiency in the educational system.

[13] These results are also consistent with the findings of El Sabaa (1997), Haan (1999), Nour (2005a, b) and Muysken and Nour (2006) in the UAE. See for instance El Sabaa (1997), pp. 20–21 and Haan (1999), p. 37.

[14] At the aggregate level, the educational matrix implies the distribution of population according to educational level: low level of education refers to illiterate, literate, primary and preparatory school; medium level of education includes secondary, post-secondary and below university; and high level of education includes university and postgraduate levels.

Table 5.2 The Sudan educational matrix: The distribution of population by educational level (1960–2008)

Total population[a,b]		Low	Medium	High	Not stated	Total
1960[a]	Total	0.996000	0.004000	0.000000	0.000000	1
1965[a]	Total	0.994000	0.006000	0.000000	0.000000	1
1970[a]	Total	0.992000	0.007000	0.001000	0.000000	1
1975[a]	Total	0.988000	0.010000	0.002000	0.000000	1
1980[a]	Total	0.981000	0.018000	0.003000	0.000000	1
1985[a]	Total	0.971000	0.027000	0.002000	0.000000	1
1990[a]	Total	0.972000	0.023000	0.004000	0.000000	1
1995[a]	Total	0.966000	0.029000	0.005000	0.000000	1
2000[a]	Total	0.959000	0.034000	0.007000	0.000000	1
2008[b]	Total	0.768000	0.139300	0.053200	0.039700	1
2008[b]		**Low**	**Medium**	**High**	**Not stated**	**Total**
Total Sudan (gender)[b]	Male	0.432900	0.078000	0.029100	0.021500	0.561500
	Female	0.335000	0.061300	0.023900	0.018200	0.438400
	Total	0.768000	0.139300	0.053200	0.039700	1
Total Sudan (mode of living)[b]	Rural	0.295600	0.085200	0.039800	0.018300	0.438900
	Urban	0.454200	0.053200	0.013000	0.020200	0.540600
	Nomad	0.018100	0.001000	0.000200	0.001200	0.020500
	Total	0.768000	0.139300	0.053200	0.039700	1
Total population[b]		0.768000	0.139300	0.053200	0.039700	1
North (gender)[b]	Male	0.375400	0.068200	0.027800	0.021500	0.492900
	Female	0.295300	0.056900	0.023600	0.018200	0.394000
	Total	0.670600	0.125100	0.051300	0.039700	0.886700
South (gender)[b]	Male	0.057700	0.009800	0.001500	0.000000	0.069000
	Female	0.039800	0.004300	0.000500	0.000000	0.044600
	Total	0.097200	0.014100	0.001800	0.000000	0.113100
Total population[b]		0.768000	0.139300	0.053200	0.039700	1
North (mode of living)[b]	Rural	0.268000	0.079300	0.038800	0.018300	0.404400
	Urban	0.384600	0.044900	0.012400	0.020200	0.462100
	Nomad	0.018100	0.001000	0.000200	0.001200	0.020500
	Total	0.670600	0.125100	0.051300	0.039700	0.886700
South (mode of living)[b]	Rural	0.027700	0.005900	0.001100	0.000000	0.034700
	Urban	0.069600	0.008200	0.000700	0.000000	0.078500
	Nomad	0.000000	0.000000	0.000000	0.000000	0.000000
	Total	0.097200	0.014100	0.001800	0.000000	0.113100
Total population[b]		0.768000	0.139300	0.053200	0.039700	1

Note: Figures for 2008 refers to total population with six (6) years old and over
[a]Own calculation from Barro and Lee (2000), cited in Ali Abdel Gadir Ali (2006) "On Human Capital in Post-conflict Sudan: Some Exploratory Results", API/WPS 0602, p. 14
[b]Adapted from Sudan Central Bureau of Statistics Population Census Data (2010): The Fifth Sudan Population and Housing Census (2008)

5.2.2.3 Hampering Transfer of Knowledge

Our results in Table 5.3 show that the low quality of education hindered the easy transfer of knowledge and external schooling effects. According to the macro survey, only 36 % of the respondent policy makers and experts reported that the incidence of

Table 5.3 The factors constrained the transfer of knowledge/external schooling effect in the Sudan, 2010

Factors constrained the transfer of knowledge/external effect of schooling	Officials (%)
Low quality/return from education	89
Low return form/quality of training compared to international standard	92
Prevailing conditions in the Firm conditions do not encourage the external effect	89
Failure of skilled workers to deliver their knowledge and experiences to benefit unskilled workers.	78
Failure of unskilled workers to acquire the knowledge and experience from skilled workers	75
Lack of awareness on the importance of the external effect	92

Source: Own calculation based on the macro survey (2010)

knowledge transfer/external schooling effect is successful, while around 61 % reported that the transfer of knowledge/the external schooling effects are constrained by several factors. The major important factors include: the low quality of training, the lack of awareness about the importance of the external effect of schooling, the low quality of education, the prevailing conditions in the firms do not encourage external effects, failure of skilled workers to deliver knowledge to unskilled workers and failure of unskilled workers to acquire knowledge from skilled workers.[15] Furthermore, from the official perspective, others less important factors or causes hindering the external effect of schooling include the social factors, deficiency in the policy of selection of trainee for training in areas not relevant to their functions and finally, the lack of internal acquisition of education and skills.

In strong contrast to this view, the results of the firm survey show that, at the micro/firm level, the incidence of knowledge transfer/external schooling effect is successful among about 77 % of the respondents firms. It is only unsuccessful within about 23 % of the respondent firms because of the following: the low quality of training, failure of skilled workers to deliver knowledge to unskilled workers and failure of unskilled workers to acquire knowledge from skilled workers, the low quality of education and the prevailing conditions in the firm do not encourage the external effects within firms.[16] From the firm perspective other factors for lowering the external effect of schooling includes the lack of local physical drive, physical motivation, culture, the high illiteracy rate for most of the workers, lack of attention and funding for training.

This contradicting optimistic-pessimistic view at the micro and macro levels regarding the incidence and success of knowledge transfer/external schooling effect implies that the transfer of knowledge/the external schooling effects is probably successful within firms but unsuccessful between firms and within society at large. This is

[15] The transfer of knowledge and external schooling effects refers to knowledge transferred from knowledge holders (high skilled workers/people) to knowledge recipients (low skilled workers/people) – cf. Cowan et al. (2001), p. 9. Knowledge in this sense refers to know how or tacit knowledge embodied in people, and is different from the broad definition of technology, which refers to both embodied and disembodied knowledge.

[16] Another possible explanation for the low transfer of knowledge can be interpreted in relation to the prevailing conditions within private firms. Since within private firms there may be fewer incentives for the incidence of transfer of knowledge from high to low skilled workers.

consistent with our observations from the follow-up interviews that the transfer of knowledge is hindered by both the low quality of education and the lack of cooperation with university sector due to inadequate awareness and lack of social partnership between public sector, private sector, university sector and society. The weak linkages and lack of networks between universities, colleges, technical and training institutes and the productive sectors is mentioned by 83 % of the respondents to the macro survey as factor that constrains the efficiency of educational system – it probably also constrains the transfer of knowledge. An additional factor is that the transfer of knowledge within society at large is probably hindered by the incidence of high illiteracy rate and incidence of high mismatch between educational output of population and labour market. This is mainly due to the excessive share of unskilled workers that probably hindered their sufficient benefit from high skilled workers and population.[17] This is probably also due to a lack of incentives at the aggregate level.[18]

Finally, the macro survey indicates that the contribution of both the educated and trained population to promote the local skills is constrained by several causes. Major causes are the inadequate incentives for trainers, the lack of interaction to market needs (mismatch), the lack of information on educational and training needs in the productive sectors and their demand for graduate students and the uncertainties about the future value of investment in education and training.[19] Other important causes are the uncertainties about future skill needs and the high costs to finance education and training.[20] Relatively less important causes include the lack of a system of certification of skills acquired and risk aversion, in other words the preference of more certain short term returns to available jobs than long-term skill investments.[21] These factors probably also contribute to hinder the transfer of knowledge within society at large.

[17] This result is consistent with the finding of the UAE as reported by El Sabaa (1997), who notes: “It is widely observed that industrial entrepreneurs in the technically advanced projects are strictly against leakage of their technologies outside their factories. Thus, they minimally contribute to developing the technology environment in the country. This adverse impact has been amplified by the unwillingness of foreign as well as local entrepreneurs to employ local manpower, to train them in their factories, either because they doubt their capabilities, or for fear to leaking their technology secret to other competitors. The limited supply of local industrial manpower, coincided with the unwillingness to employ them in both foreign and local industries applying advanced technologies, resulted in constricting the role supposed to be played by expatriate manpower in transferring technology to the industrial sector in the Gulf region. Moreover, the large scale industries despite using more sophisticated advanced technologies, however, they minimally contribute to elevate the technology transfer to the local industrial sector, as they strictly keep their operational and managerial techniques as top confidential secrete and prevent their leaking outside their units. To some extent, the chance of their flow to the rest of the operating factories seems better in the medium size factories”. El Sabaa (1997), pp. 22, 24–25.

[18] The lack of transfer of knowledge can be interpreted as a lack of absorptive capacity, mainly related to deficiencies of education and continued dependence on imported technologies.

[19] As indicated by 97 %, 92 %, 92 % and 92 % of the respondent policy makers and experts respectively.

[20] As reported by 86 % and 86 % of the respondent policy makers and experts.

[21] As indicated by 83 % and 78 % of the respondent policy makers and experts respectively.

5.2.2.4 Poor Provision of Training

Both the deficient educational system and the excess supply of low skilled workers lead to a low skill level and hinder the provision of training. Table 5.4 shows that both the policy makers and experts (officials) and firm managers mentioned the low educational qualifications of workers among the important factors constraining the provision of training. Other important factors are the lack of appreciation/information about training, the lack of finance to cover the high cost of training, the lack of trainers and mentors, inadequate assessment and planning for training programmes and the mismatch problem. Furthermore, from the official perspective other factors for the low provision of training in Sudan include the lack of stimulating environment, the lack of interest in development and continuous training initially as a way of skill formation, the low priority given to training, inadequate incorporation of training in the government policy and the strategic plan or programme and the difficulties of coordination and organisation of training activities due to inadequate resources and facilities offered to the National Council for Training to coordinate and organise the public and private training activities and training institution.[22] In addition to deficiency in the selection of management for training departments, deficiency in determining and setting the goal and the level of requirement of training and deficiency in the selection of learners and trainees for suitable training programmes according to their training needs. In addition, from the firms' perspective the lack of training provision is attributed due to the absence of government support, which is apparent from the lack of regulations from the Ministry of Industry and Ministry of Labour to oblige firm owners and mangers to make the establishment of training units mandatory. Further to the lack of financial resources, only a few firms reported the presence of a budget for the provision of training; for the majority of the firms there is no budget for the provision of training, especially, external training. Moreover, in view of the scarce resources for the industrial firms, the provision of training is most probably hindered by the payment of large number of taxes, fees and levies imposed on the industrial firms, which implies their inability to address skill development issues such as training. In addition from the firms' perspective other problems for the limited provision of training are probably due to: the lack of development; absence of a culture of the importance and need for training given the believe that the nature of the prevailing industrial activities that do not need training; preference for readymade well trained and experienced skilled workers; and lack of important jobs for training for promoting a culture of commitment and affiliation to firms. Moreover, from the perspective of new and relatively new established firms, the lack of the importance of training and R&D over the short-term can be attributed due to their recent establishment; their focus and priority in their early years aims to ensure survival in the market and therefore they believe that both training and R&D are not priorities until after establishment

[22] For instance, the insufficient resources and facilities hindered the ability of the National Council for Training to organise training activities, and most probably lead to existence of chaos and messiness in training activities; there are 195 training institutions and the council does not know about their activities and is unable to organise them due to the lack of necessary resources.

Table 5.4 The factors constrained the provision of training in the Sudan, 2010

Factors constrained the provision of training	Official	All firms	Chemical	Food	Metal	Textile	Large	Medium	Small
Inadequate planning for training programme	94 %	59 %	57 %	55 %	75 %	67 %	81 %	37 %	54 %
Inadequate assessment of training needs.	92 %	65 %	65 %	65 %	63 %	67 %	86 %	47 %	54 %
Mismatch between training programme and changing technical needs	89 %	50 %	48 %	50 %	50 %	67 %	76 %	26 %	38 %
Mismatch between training programmes and changing skill needs	94 %	50 %	43 %	50 %	63 %	67 %	81 %	21 %	38 %
Low quality of trainers and mentors	94 %	50 %	46 %	50 %	78 %	0 %	76 %	33 %	31 %
Low educational qualifications of workers	89 %	67 %	67 %	70 %	63 %	67 %	81 %	55 %	62 %
Lack of trainers and mentors	97 %	60 %	58 %	55 %	88 %	33 %	76 %	40 %	62 %
Lack of appreciation/information about training	94 %	69 %	75 %	70 %	63 %	33 %	86 %	55 %	62 %
Lack of specialized training institutions	86 %	63 %	78 %	50 %	50 %	67 %	71 %	53 %	69 %
Lack of full appropriability of the return from training investment.	92 %	57 %	61 %	60 %	38 %	67 %	62 %	58 %	54 %
Lack of interactions between training institutions and firms	94 %	61 %	61 %	60 %	63 %	67 %	67 %	47 %	77 %
Lack of finance to cover the cost of training	97 %	69 %	74 %	60 %	63 %	100 %	76 %	68 %	54 %
Lack of training materials and equipment	97 %	64 %	71 %	55 %	63 %	67 %	86 %	35 %	69 %
High rate of mobility of trainers to move for better paid jobs after training	89 %	69 %	61 %	75 %	63 %	100 %	76 %	63 %	62 %
Lack of a system of training certification of skills acquired	92 %	50 %	52 %	55 %	38 %	33 %	71 %	37 %	38 %
Total response		56	24	20	9	3	21	21	13

Source: Own calculation based on the macro survey (2010), firm survey (2010)

in the market. Moreover, the results of the firm survey illustrate that the low provision of training appears from the following: (1) The lack of an inhouse training unit – only 27 % of the respondent firms have an inhouse training unit; (2) The complete absence of public financial support: for instance, none of the respondent firms received any government subsidies to support training provision; (3) The selective training provision: in the year 2008, the priority for training among the respondent firms was mostly given to production workers, production engineering staff, management staff and services workers[23]; and for few firms the limited provision of training is also extended to include the provision of training for university students and (4) The limited type of training: most of training provision is focused to on the job training, and on the job and off the job combined, which are preferred by 72 % and 56 % of the respondent firms respectively. The other types of training such as: off the job within the firm (training centre), off the job outside the firm (specialist training centre inside the country) and (outside the country) are very limited[24]; (5) The limited sources of information about training opportunities, as most of the information about training opportunities is provided by private trainers (local and foreign companies) and the chambers of commerce. Few firms find information from government and semi government units, other firms working in the same sector and public educational institutions/universities.[25,26] Few firms

[23] As reported by 51 %, 48 %, 36 % and 21 % of the respondent firms respectively.

[24] As indicated by 42 %, 48 % and 38 % of the respondent firms respectively.

[25] As reported by 46 %, 32 %, 27 %, 26 % and 21 % of the respondent firms respectively.

[26] These results seem consistent with the findings of the earlier studies in the UAE conducted by Nour (2005b), the UAE University (1994, 1997), Gray (1999) and Abdelkarim and Haan (2002). For instance, the UAE Education Assessment Report (1994) shows that both technical and vocational education and training provision are unregulated, uncoordinated and unplanned, while the results of the UAE University (1997) show the limited contribution to private sector training provision by both the public and government sectors. Moreover, the findings of Gray (1999) show that only 30 % of the respondent firms provide systematic training. The provision of training is selective in most cases – focused on some occupational groups, but not others – and for two thirds of the respondent firms, the provision of training was limited to on the job training. The study concluded that the UAE does not have a training-led employment culture. Employers have become used to bringing in their workers from outside the country with readymade skills and replacing them with similarly skilled workers. There has been little incentive to provide skill upgrading except in response to immediate needs such as the introduction of new technology. The local training industry has suffered from the uncoordinated nature of provision and the very limited contribution by public sector organisations and higher education institutions to this variety of provision. Most of training provision has been in low-investment, low-cost and quick-profit areas such as marketing, public relations, sales, computer awareness and management development. The private sector training is relatively undeveloped and uncoordinated, and has limited market due to both limited demand and limited supply. Moreover, the provision of public sector training is constrained by the inadequate involvement of public education institutions. For instance, the Higher Colleges of Technology (HCTs) had little involvement in the important area of adult technical education, including vocational training and retraining. See Gray (1999), pp. 15, 33, 34, 43. Additionally, the findings of Abdelkarim and Haan (2002) show that the UAE public sector training is still limited due to less attention, awareness and resources. See Abdelkarim and Haan (2002), p. 15.

obtained information about training from other sources for example, firms self searching for training, from private Sudanese universities, few public and private institutions and finally foreign expertise.

5.2.3 Consequences of the Low Educational Qualifications of Unskilled Workers at the Micro/Firm Level

In Chap. 2 we explained that one well-known fact about Sudan is the high share of unskilled workers in total employment. In this section we show that next to the consequences of the deficient educational system at the macro level, the high incidence of unskilled workers also causes several serious implications at the micro/firm level.

5.2.3.1 Low Skill Level and Skills Mismatch at the Micro/Firm Level

From the demand perspective, the results of the firm survey can be used to argue that firm demand for low skilled workers leads to an excessive share of low skilled workers.[27] On the other hand, from the supply perspective, our findings from the firm survey show that the excessive share of low skilled workers has direct implications in lowering skill levels at the micro/firm level. For instance, Table 5.5 below shows that across firms the average percentage share of low skilled workers (66 % and 74 %) is much higher than that of high skilled workers (34 % and 26 %), defined by both educational and occupational classifications respectively. Moreover, Tables 5.4 and 5.5 show that the poor educational qualifications of firms workers lead to poor provision of training – see our discussion above – and a serious skills mismatch across firms, as we will explain in detail in the next chapter.[28]

[27] It is convenient in this chapter to briefly indicate the consequences with respect to low skill and skills mismatch at the micro level and to discuss this more fully later in Chap. 7. That serves our aim in this chapter to compare and integrate the macro–micro consequences of low skill level. This brief discussion in this chapter also substantiates the third hypothesis in Chap. 1 above about the interaction between the deficient educational system at the macro level and the high incidence of unskilled workers at the micro level and the serious implications on low skill levels and skills mismatch. It is appropriate to discuss the skills mismatch problem more extensively later in Chap. 7, where we provide a broader, more indepth and coherent analysis of skill problem and the implications of the prevalence of low-skilled workers at the micro/firm level.

[28] We define the mismatch as the differences between the required and actual education. Actual education refers to high (university and above), medium (secondary) and low (below secondary) levels of attained years of education that represent the supply of skills. We define the required education by the required qualifications for each of the occupational classes translated into average years of schooling that represent the demand for skills. We observe that the inconsistency between the required and actual education implies inconsistency between demand for and supply of skills, which we interpreted as skills mismatch (cf. the detailed discussion in Chap. 7 below).

Table 5.5 Main technology and skill indicators defined by firm size and industry in the Sudan, 2010

		Industry				Size		
Indicator	All firms	Chemical	Food	Metal	Textile	Large	Medium	Small
Skill and skills mismatch indicators								
Share of high skilled (education) (%)	34 %	35 %	33 %	31 %	39 %	42 %	32 %	25 %
Share of high skilled (occupation) (%)	26 %	27 %	25 %	21 %	32 %	25 %	26 %	29 %
Share of low skilled (education) (%)	66 %	65 %	67 %	69 %	61 %	58 %	68 %	75 %
Share of low skilled (occupation) (%)	74 %	73 %	75 %	79 %	68 %	75 %	74 %	71 %
Share of firm conducting R&D (%)	20 %	23 %	13 %	17 %	40 %	29 %	10 %	19 %
High skilled wages/low skilled wages	3.5	3.7	3.45	2.96	3.6	4.2	3.1	2.98
Share of firm with skills mismatch (%)								
The high skilled group (%)	40 %	38 %	44 %	22 %	60 %	58 %	36 %	17 %
The medium skilled group (%)	31 %	24 %	38 %	44 %	20 %	30 %	40 %	13 %
The low skilled group (%)	45 %	39 %	39 %	71 %	60 %	43 %	53 %	38 %
Technology indicators								
Share of R&D expenditure/total output (sales value) (%)	0.17	0.12	0.36	0.01	0.11	0.19	0.05	0.225
Average R&D expenditure (× 1 million Sudanese Pound)	0.5	0.65	0.5	0.5	0.1	0.6	0.002	0.5
Number of R&D employees (R)	$1 \leq R \geq 11$	$1 \leq R \geq 11$	$1 \leq R \geq 11$	$1 \leq R \leq 10$	$2 \leq R \geq 11$	$1 \leq R \geq 11$	$1 \leq R \leq 10$	$1 \leq R \geq 11$
Number of full time R&D employees	$1 \leq R \geq 11$	$1 \leq R \leq 10$	$1 \leq R \leq 10$	0	$2 \leq R \geq 11$	$1 \leq R \geq 11$	$1 \leq R \leq 10$	$1 \leq R \leq 10$
Number of full time R&D employees	1–11	1–10	1–10	0	2–11	1–11	1–10	1–10
Number of part time R&D employees	$1 \leq R \geq 11$	$1 \leq R \geq 11$	$1 \leq R \geq 11$	$1 \leq R \leq 10$	0	$1 \leq R \geq 11$	$1 \leq R \leq 10$	$1 \leq R \geq 11$
Number of part time R&D employees	1–11	1–11	1–11	1–10	0	1–11	1–10	2–11
Share of firm applying for patents (%)	6 %	8 %	3 %	8 %	0 %	6 %	8 %	5 %
Share of firm in total spending on ICT (%)	100 %	50 %	30 %	11 %	9 %	37 %	33 %	30 %
Share of firm in total spending on ICT training (%)	100 %	13 %	63 %	13 %	13 %	50 %	25 %	25 %
Share of firm in total spending on ICT (%)	62 %	71 %	52 %	46 %	100 %	57 %	62 %	76 %
Share of firm with spending on ICT training (%)	9 %	3 %	16 %	8 %	20 %	11 %	7 %	10 %

Average spending on training (× 1 million Sudanese Pound	14.42	21.28	2.65	2.00	5.01	18.38	1.06	10.00
Number of training employees	5.77	2.25	2.33	0	25	7	2.5	3.5
The degree of automation/use of sophisticated technologies[a] (%)	54 %	35 %	67 %	61 %	80 %	68 %	55 %	35 %
Dependence on foreign technology[b] (%)	49 %	49 %	58 %	25 %	60 %	51 %	38 %	63 %
Dependence on foreign technology (%)	100 %	100 %	100 %	100 %	100 %	100 %	100 %	100 %
Share of firm providing training (%)	27 %	34 %	15 %	27 %	40 %	34 %	26 %	17 %
Incidence of external schooling effect (%)	77 %	72 %	74 %	91 %	100 %	81 %	74 %	74 %
Dependence on foreign technology-1-q-21	56 %	58 %	56 %	46 %	57 %	57 %	48 %	62 %
Dependence on foreign technology-2-q22	45 %	35 %	46 %	65 %	51 %	45 %	53 %	31 %
Factors constrained the contribution of R&D unit to adaptation of imported technologies								
(a) Shortage of finance	78 %	71 %	83 %	83 %	75 %	82 %	81 %	65 %
(b) Shortage of human resources skilled and qualified workers	74 %	68 %	80 %	75 %	75 %	81 %	67 %	70 %

Source: Own calculation based on the firm survey (2010)

[a]We measure the degree of automation/sophisticated technologies qualitatively, we asked firms about their own appreciation or evaluation of the level of technologies they are using in their production

[b]We measure the dependence on foreign technologies qualitatively, we asked firms if they have an adequate capacity/ability to produce and develop local technologies and if they have purchased equipment, machines and techniques from abroad. Our definition also includes quantitative measurement of the value of imported capital equipment to total capital equipment, the percentage value of capital equipment to total capital equipment that has been build by foreign companies. Finally technology transfer is also an indicator of dependence on foreign technologies – see Table 5.7 below

5.2.3.2 Weak Technology Indicators

From the demand perspective, the results of the firm survey can be used to argue that weak technology indicators lead firms to demand and rely heavily on low skilled workers. On the other hand, from the supply perspective, the excessive share of unskilled workers leads to weak technology input–output indicators across firms. From Table 5.5 above, we observe that a weak R&D indicator appears from the following: (a) The limited R&D activities/efforts performed continuously or occasionally amongst only 20 % of all the respondents firms; the limited prevailing research includes for example the marketing research; (b) The limited R&D activities and expenditures particularly devoted/aimed at improving firm product and to produce a new product; (c) The low R&D expenditures and R&D expenditures as a percentage of total output expenditures. For instance, amongst all the respondent firms, average R&D expenditure was around 0.5 million Sudanese pounds,[29] while the average R&D expenditures as a percentage of total output (sales value) amongst all the respondents firms accounted for only 0.2 %; (d) the low number of both full time and part time R&D employees. For instance, a large majority (91 % and 89 %) of the respondents firms with R&D activities have no or a low number (1–3) of full time research employees and/or part time research employees respectively. Only few (9 % and 11 %) of total firms have more than three full and part-time research employees respectively. Only few firms have more than 11 full and part-time research employees.[30] Only few firms have (6–10) full and part time research employees.[31] Moreover, the contribution of research units in adapting the imported technologies is constrained by a shortage of skilled and qualified workers amongst 74 %, 68 %, 80 %, 75 % and 75 %, 81 %, 67 % and 70 % of all firms, chemical, food, metal, textile, large, medium and small firms respectively (see Table 5.8 below).[32] In addition, the contribution of research units in adapting the imported technologies is constrained by a shortage of finance amongst 78 %, 71 %, 83 %, 83 % and 75 %, 82 %, 82 % and 65 % of all firms, chemical, food, metal, textile, large, medium and small firms respectively. Moreover, the follow up interviews with the officials and firms managers revealed that R&D activities are constrained by several factors such as high costs and low public and private spending, lack of information systems, and the absence of an R&D culture due to inadequate awareness and concern. In addition the very limited cooperation between industrial firms in terms of R&D that appears for example from the only limited cooperation in exchange of information within specific

[29] The value is measured in Sudan's local currency, Sudanese Pounds, which equalled US$ 2.50 Dollars when the survey was held (2010).

[30] As reported by 5 %, 3 % 7 %, 20 %, 9 % and 5 % of the all firms, chemical, food, textile, large and small respondent firms respectively.

[31] As reported by 15 %, 22 %, 10 %, 17 %, 20 %, 14 % and 10 % of the all firms, chemical, food, metal, large, medium and small respondent firms respectively.

[32] We measured the contribution of a research unit to adapt the imported technologies qualitatively, by asking firms how the research unit contributed to adapting imported technologies. In addition to a lack of qualified workers, there may also be other factors such as a lack of incentives or pressure to adapt or master imported technologies as firms can always opt to buy the required technologies or techniques.

limits is most probably limited because of the intense competition between industrial firms. Additional constraints relate to the poor coordination between the institutions engaging in R&D activities, either due to the absence of explicit government policy or the ineffective role of a central body (e.g. the government) to coordinate and promote R&D efforts and motivate collaborative research efforts between the industrial firms and universities. In addition to weak network systems, R&D efforts, in particular, are limited across firms because of weak contact and collaboration with universities; this is probably attributable to the fact that the university sector is lacking resources or concern and interest to conduct joint applied research with industrial firms.[33] Moreover, from the firms' perspective the main problem facing firm strategy to encourage the development of R&D and build of local technology is that research is not a fundamental goal, the primary goal is profit, in addition to the lack of entrepreneur perspective and dominance of commercial perspectives, lack of government policies and support, lack of efficient organisational management and lack of relevant workers. Moreover, from the firms' perspective the development of technology industry in Sudan is hindered by the acute shortage of local raw materials.

Moreover, Table 5.5 above shows the weak technology output indicator as measured by patent applications.[34] For instance, in the year 2008, only 6 % of all respondent firms applied for a patent; the low degree of patenting may be attributable to low R&D efforts.

In addition, a lack of R&D efforts may hinder innovative activities across firms. The increasing uses of technology has encouraged the incidence of product and

[33] This can also be interpreted as a lack of demand-pull since firms can buy all technologies or techniques. Our findings are consistent in some respects with the findings in the UAE as discussed in Nour (2005b) and Haan (1999), who notes "R&D capacity in the UAE is presently very limited. While some research is taking place at the University and other institutes of higher education, it is - as usual in such institutions – more geared towards pure science and tends to have only limited relevance for the productive sector. While the Higher Colleges of Technology place emphasis on more practical training, hardly any R&D takes place, only as a byproduct of the training. Within the private sector in the UAE likewise very few R&D activities are going on. Most manufacturing and other firms tend to rely on imported technologies (both in terms of hardware and software), as well as imported materials and even expatriate manpower. There is very little interest in carrying out research, and the R&D activities are small-scale in nature and mostly only concern minor adaptations to the companies' own products. In all it is estimated that less than 1 % of turnover is used for this purpose. The parastatal sector, in which a number of large basic industries are operating, will do better. There is also some agricultural research ongoing. Without such R&D facilities and efforts, the UAE is almost completely dependent on imported technologies. And without the necessary adaptations to local conditions (e.g. temperatures, effects of dust and sand winds, special cultural aspects, the country's socio-political system, etc.), even these technologies cannot be optimally applied. Moreover, a genuine technology culture to motivate the involvement in R&D and promotion of local technology is now absent in the UAE. The UAE society is geared more predominantly to non-technical education, training and employment. Technical qualifications and occupations are not rated very high by its social values and cultural traditions. The UAE only has a limited industrial tradition (e.g. trade), and lacks explicit policies to stimulate and direct technological development". See Haan (1999), p. 37, 38.

[34] As reported by 6 %, 8 %, 3 %, 8 %, 6 %, 7 % and 5 % of the all firms, chemical, food, metal, large, medium and small respondent firms respectively. This includes five firms: one small chemical, one medium chemical, one medium metal, two large chemical and one large food firms applied for patents.

process innovations, in particular, the incidence of incremental product innovation, namely, improvement of product quality amongst 59 % of all firms. It has also encouraged the incidence of new products, new organisational methods, new combinations of old output, new methods of production, and new processes.[35] It has also encouraged the incidence of great effect on increasing total sales and profits, but it has only slight effect on reducing total costs (see Table 5.6 below).[36]

5.2.3.3 Dependence on Foreign Technology

From the demand perspective, the results of the firm survey can be used to argue that the dependency on foreign technology leads firms to demand and rely heavily on low skilled workers. On the other hand, from the supply perspective, the deficient educational and training system and high supply of low skilled workers lead to low skill levels, lack of knowledge transfer and low capability to build and promote the local technology. The results of the firm survey show that this condition leads to weak technology indicators and dependence on imported technology that appears from the following: (1) The high dependence on the imported equipment, machines and techniques among all of the respondent firms (100 %); (2) The high percentage value of capital equipment to total capital equipment that has been built by foreign companies (56 %) among the respondent firms; (3) The considerable percentage value of imported capital equipment to total capital (45 %) among the respondent firms in the year 2008; (4) The short run plan for 92 % of the respondent firms is based/depends on imported technology[37]; The main reasons for the dependence on foreign technology are the lack of local technology from local suppliers, better quality and better price of foreign technology in that order.[38] Despite the high dependency on imported technologies, it is somewhat surprising that the level of technology used is below international standards amongst the majority of the respondent firms (53 %) and a high level of technology used similar to international standards is limited only within 47 % of all the respondent firms. Moreover, somewhat surprising a high degree of automation through the use of sophisticated and advanced technology is limited only within 54 % of all the respondent firms (see Table 5.5 above). The degree of automation/sophisticated use of advanced technologies is determined by both firm size and industry/activity.[39]

---

[35] As reported by 66 %, 44 %, 43 %, 40 % and 35 % of all respondent firms respectively.

[36] The terms "new product" and "new process" refer to new products and processes intended even just for local firms or for local markets and not necessarily for the international market.

[37] Short, medium and long run refers to next 3 years, next 3–5 years and next 10 years respectively.

[38] As reported by 71 %, 42 % and 14 % of the respondent firms respectively.

[39] These results are consistent with the findings in the UAE as indicated by Nour (2005b) and El Sabaa (1997), who notes "The adoption of different approaches in transferring technology differs according to certain criteria, such as: the scale of industry and its activity. Large size and some specific sectors, namely chemical and petrochemicals industries have better use of sophisticated advanced technologies". See El Sabaa (1997), p. 21, 22.

Table 5.6 The effects of increasing use of technology product and process innovations across firms in the Sudan, 2006–2009

Product/process innovation (2006–2009)	All firms	Firm industry				Firm size		
		Chemical	Food	Metal	Textile	Large	Medium	Small
Improvement of product quality	59 %	69 %	55 %	33 %	60 %	68 %	50 %	53 %
Production of a new method of production	40 %	44 %	30 %	42 %	60 %	49 %	25 %	47 %
Production of a new combination of old output	43 %	50 %	38 %	25 %	60 %	59 %	29 %	37 %
Production of a new process	35 %	22 %	40 %	33 %	100 %	46 %	18 %	37 %
Production of a new product	66 %	50 %	73 %	83 %	100 %	60 %	68 %	74 %
Production of new organizational method	44 %	56 %	31 %	33 %	60 %	53 %	32 %	42 %
Improvement of training within the firm	52 %	57 %	45 %	42 %	80 %	59 %	46 %	44 %
Improvement of communication within the firm	57 %	66 %	52 %	42 %	60 %	65 %	54 %	44 %
Production of more output with low cost	57 %	61 %	50 %	50 %	80 %	74 %	43 %	42 %
Production of the same output with low cost	37 %	44 %	38 %	25 %	0 %	38 %	36 %	32 %
Open of a new market	50 %	58 %	52 %	25 %	40 %	62 %	50 %	32 %
Production of a new service	28 %	33 %	21 %	17 %	60 %	35 %	25 %	21 %
Improvement of process of personal selection	38 %	40 %	38 %	25 %	60 %	50 %	21 %	39 %
Total response	83	36	30	12	5	35	28	19
Reduction in per unit material costs	21 %	22 %	19 %	29 %	25 %	24 %	14 %	28 %
Reduction in per unit energy costs	28 %	36 %	21 %	17 %	20 %	20 %	25 %	46 %
Reduction in total cost	35 %	37 %	29 %	50 %	25 %	25 %	37 %	46 %
Increase in total sales	85 %	92 %	68 %	100 %	100 %	87 %	78 %	93 %
Increase in total profit	81 %	86 %	76 %	72 %	100 %	85 %	76 %	85 %
Total response	64	28	24	7	5	26	24	14

Source: Own calculation based on the firm survey (2010)

The dependence on foreign technologies also appears from the reported information on the transfer of foreign technology that is made through different channels. For instance, Table 5.7 below illustrates that hiring foreign skills/technologically advanced workers/consultants and FDI are more common channels of technology transfer, while strategic alliances, technology licensing and joint ventures are less preferred channels. For a few firms other methods of the transfer of technologies include bringing foreign expertise, the use of Internet and making the self search as needed or according to the nature of the work. The transfer of technology, mainly the transfer of technologically advanced workers/consultants,

Table 5.7 The channels of technology transfer and their effects on firm production and development of local technology across firms in the Sudan, 2010

	All firms	Chemical	Food	Metal	Textile	Large	Medium	Small
(a) Channels of technology transfer (2005–2009)								
Hiring foreign skills/technologically advanced workers/consultants	48 %	56 %	35 %	62 %	40 %	57 %	39 %	50 %
FDI	31 %	28 %	32 %	31 %	40 %	40 %	29 %	20 %
Strategic alliance	18 %	22 %	20 %	8 %	0 %	21 %	11 %	25 %
Licensing	15 %	19 %	13 %	8 %	20 %	18 %	21 %	5 %
Joint ventures	7 %	8 %	7 %	0 %	20 %	9 %	7 %	5 %
Others (e.g. in house technology development by hiring technologically advanced persons)								
Total response (2005–2009)	85	36	31	13	5	35	28	20
(b) The effects of technology transfer in (2005–2009)								
Enhancing firm production	96 %	95 %	96 %	100 %	100 %	100 %	97 %	90 %
Enhancing the capacity to develop the local technologies	85 %	90 %	73 %	93 %	100 %	92 %	82 %	75 %
Total response (2005–2009)	85	37	30	13	5	35	29	20
(c) The effects of technologically advanced workers in								
Enhancing firm production	91 %	95 %	90 %	85 %	100 %	91 %	93 %	90 %
Enhancing the capacity to develop the local technologies	81 %	89 %	70 %	85 %	80 %	92 %	75 %	70 %
Total response (2005–2009)	84	36	30	13	5	35	28	20

Source: Own calculation based on the firm survey (2010)

has induced important effects in enhancing firm production but has slight less effect in enhancing the capacity to develop the local technologies.[40]

[40] These results are consistent in some respects but differ in others with the findings in the UAE as indicated by Nour (2005b), El-Sabaa (1997) and Haan (1999) respectively. "The major channels of technology transfer are: joint ventures, and industrial foreign projects, the latter accounts for the first source of technology transfer. The turn-key projects are preferred channel of technology transfer in the Gulf region mainly because of the keenness to avoid defects of execution and to guarantee the maximum consistency of the project's design, lines of production, quality of the products, facilities of training, etc. But it has very limited role in transferring technology to local industry, because it is confined to their plants, with no minimum leakage allowed. Thus they contribute nothing to implant advanced technologies in the country. Technology transfer to the UAE has obviously contributed to accelerating industrial and economic growth, elevating the standard of national products both quality-wise and quantity-wise. In particular, the transfer of technology contributed to rapid growth of local industrial sector. However, a number of negative factors are still adversely affecting the transfer of technology; the technologies transferred could hardly approach its target of constituting an autonomously developing local technological base, similar to those in the Far East industrial countries. Because of: the inadequate awareness of the end target of technology transfer, the lack of a constitutional framework or comprehensive plan for transferring technology, the insufficient local base of technological data, the lack of qualified local

In the firm survey questionnaire the question on the channels of technology transfer allows for multiple answers, assuming that firms may choose more than one channel to transfer technology.[41] Our results indicate that all respondent firms are less interested in transferring technologies through formal licenses. These may not be often requested, probably because of more liberalised and open market policies that led to considerable presence of foreign capital investment and allowed for foreign and mixed ownership (cf. Table 7.1 in Chap. 7 below).[42]

---

manpower necessary for transferring technology and the contracts of technology transfer". See El Sabaa (1997), pp. 23–26. "The UAE is almost completely dependent on imported technologies. And without the necessary adaptations to local conditions (e.g. temperatures, effects of dust and sand winds, special cultural aspects, the country's socio-political system, etc.), even these technologies cannot be optimally applied". See Haan (1999), p. 38.

[41] Our assumption and respective findings are plausible and consistent with the results in the UAE as indicated by Nour (2005b) and the results of El-Sabaa (1997), which indicate numerous different channels of technology transfer to the UAE, such as: foreign industrial investments, offset programs, training missions, technological imports, industrialisation licenses, patents, technological products, foreign manpower and industrial consulting offices. See El Sabaa (1997), p. 26.

[42] For instance, according to Sudan Ministry of Investment, among the efforts that aim at promoting foreign investment, the government has issued the investment encouragement law, which grants encouraging exemptions to investors and indicates that the investor has the right to operate without a Sudanese partner. In addition, in order to promoting foreign investment the government has established free zones including: Suakin Free Zone and Aljaily Free Zone. The Free Zones and Free Markets Law (1994) represents the legislative framework for the establishment and operation of free zones and markets in Sudan. The rules resultant from this law represent the organisational framework for operating and managing free zones in Sudan. This law provides several advantages of investment in free zones, for instance the industrial, commercial or service investments, which are licensed to be established in the free zones enjoy several advantages. This includes the following: exemption of the projects from profits tax for a period of 15 years, renewable for an extra period dependant on the decision made by the concerned minister commencing from the 1 year period of grace which follows the year of commencement of production; salaries of expatriates working in projects within the free zones will be exempted from the personal income tax; exemption of products imported into the free zone or exported abroad from all customs fees and taxes except service fees and any other fee imposed by the board of Sudan Free Zones Company; real estate establishment inside the free zones area are exempted from all taxes and fees; invested capital and profits are transferable from Sudan to abroad through any bank licensed to operate in the free zone and exemption of products of industrial projects established in the free zones from customs fees; depending on materials used and local costs incurred in production, provided that the value be estimated by a committee assigned for this purpose by the board of Sudanese Free Zones Company. See Sudan Ministry of Investment: http://www.sudaninvest.org/English/Sudan-Invest-FreeZone.htm, accessed January 30, 2011. These results are consistent with the findings in the UAE as indicated in Nour (2005b). For instance, Fasano (2002) indicates that other than Abu Dhabi, the emirates have established free zones that allow 100 % foreign ownership of companies. These zones are particularly important in Dubai, where they have attracted a large number of foreign companies. See Fasano (2002), p.331. El Sabaa (1997) finds that the adoption of open market philosophy, supported by the existence of nine free zones in the seven emirates and the advantage of 100 % foreign ownership and control, encourages foreign industrial investors to set up their projects and to promote technology transfer to the UAE. See El Sabaa (1997), p. 23.

These findings on weak technology indicators and dependence on foreign technologies at the micro level are consistent with those at the macro level and the interaction of these findings lead to a large technological gap (see also our later discussion in Chap. 6 below). Our results in this chapter and Chap. 6 below verify the second hypothesis in Chap. 1 above that, in the short- and medium-term, Sudan is unable to rely on local technologies and remains heavily dependent on foreign technologies.

Our findings from the firm survey show that both skill and technology indicators, product and process innovations, the channels of technology transfer and their respective effects vary enormously across firms and seem determined by both firm size and industry level. For instance, Table 5.5 above shows that skill levels, technology input–output indicators (R&D and patent), the provision of training (upskilling), the dependence on imported technology, the degree of automation and the use of sophisticated and advanced technology vary across firms and increase with firm size and industry level. Moreover, the use of ICT and provision of ICT training increase with firm size, while the transfer of knowledge/external effects of schooling increase with industry level.

5.2.3.4 Shortage of Skilled Workers and Weak Adaptation of Imported Technologies

From the demand perspective, the results of the firm survey can be used to argue that a low supply of skilled workers lead firms to demand and rely heavily on low skilled workers. On the other hand, from the supply perspective, our findings show that the excessive share (supply) of low skilled workers and low supply of high skilled workers lead to some shortage of skilled workers, since firms are experiencing increasing demand for skilled workers (see Fig. 5.1 below). We find that the increasing demand for high skilled workers is related to increasing use of new technologies, which has not only raised the demand for these workers in the past few years, but has also encouraged firms to predict a future/long run increase in this demand. For instance, Fig. 5.1 identifies and compares past and future trends of the demand for skilled workers, which vary enormously across firms according to size and industry level. According to 68 % of the respondent firms, the interpretations of the predicted long run increase in the demand for skilled workers are related to change in the investment policy and project plans to increase the level of existing skills, the expansions of production (increase in existing products, new products and new production line), expansion in firms size, work, capacity, facility, activities and employment, increase in seasonal demand, increasing use of technology, modern techniques and new machines, improvement of financial conditions and the

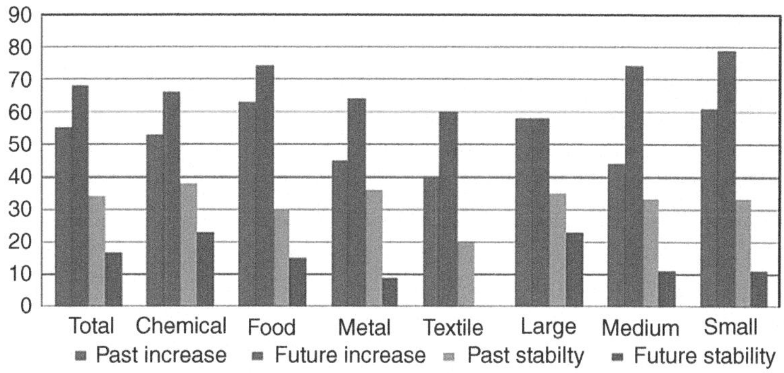

Fig. 5.1 Past and future trends (increase or stability) in the demand for skilled workers across firms (Source: Firm survey 2010)

increasing need for development and for training new workers. In addition the potential future increase is also caused by the economic reasons related to economic recession, deterioration of the industry and increased costs of industrial production, which implies the need for increasing the demand for skilled workers due to increasing motivation to reduce faults, reduce production costs, in addition to increasing industrial awareness for increasing firms efficiency by raising the quality, efficiency and productive skill of local workers. Further to potential vertical expansion, improvement in marketing opportunities and share in the local and global markets, in addition to the increasing need for specialisation and development of manufacturing industries and increasing R&D activities, implementation of new processes, output technologies, advanced control systems and purchases of new machines and equipment. Additional reasons for the rising demand are the increasing need to meet the new requirements of the market and new phase of development despite its slow occurrence, increasing adoption of international standards and enhancement of production, product improvement, diversification and varieties, competition, achieving high standard precision work and improving productivity and quality of work and demand for more specialised skills in IT. On the other hand, the major explanations for the predicted long run stability in the demand for skilled workers across 17 % of all respondent firms is related to the stability in quantity of production, sales, business, demand and market; as well as to the lack of a plan for critical expansion of product operations, stability of seasonal demand, stability of the proportion of machinery, stability of the use of technology, suitability and sufficiency of the available workers, instability of local market conditions, limitations of firms space, inability to add new machines and dependence on the development of the already available products. Moreover, from few firms' perspective the major explanations for the predicted long run decrease in the demand for skilled workers across 15 % of all respondent firms is attributed to upgrading unskilled workers to technicians after selection by the competent authorities which imply that the remaining need for skilled workers will decrease due to increase in the qualifications and training facilities. Other factor for the potential decrease is related to the nature of the industry, for example the decrease in textile industries expected to occur as always in the textile industry it is normal to hire unskilled workers and to train them on the appropriate method of production inside the factory according to firm strategy (in that case the use of skilled workers from other factories is not preferred so as to avoid the transfer of the negative habits from other old firms due to potential conflict to new training strategy in the new firms).

We observe that the expected future rise in the demand for high skilled workers across firms is reasonable since increased use of skilled workers in the past has had significant effects. In particular, there has been increase in firm production, improvements in product quality, improvement in the level of competitiveness in the local market, utilisation of technology and faster adaptation of foreign technology.[43] Further to the introduction of networks, the increasing intensity in the use of modern new technologies, new and more sophisticated machines and machinery, updated equipment, readymade programmes applications and specialised production systems and operating machines related to industrial activities and production all imply the need to hire skilled and trained workers to work with them and that led to increased demand for skilled workers in the firm.[44] In addition, from the firms' perspective the other positive impacts of hiring skilled workers, scientists and engineers include for example, the positive impacts on those Sudanese affected by the positive performance of foreign skilled workers and experts in organisation,

---

[43] As indicated by 83 %, 81 %, 73 %, 73 % and 65 % of the respondent firms respectively.

[44] For instance, from the firms' perspective the increasing need for updating or upgrading the packing machine, the use of communication and information technology and computer technology in production for productive work and the use of production or work-related machines led to increase in production and all has increased the demand for skilled workers. In addition, the introduction of new production lines, the increasing use of modern packaging and covering machines, the use of modern (cut and wrap) machines, the use of modern tagged machines all require a high-technical skilled workers and all led to increase in the demand for skilled workers to facilitate work and increase production. From the firms' perspective for the previous 3 years the important factors for the impact of the use of new technology on causing constant demand for skilled workers related to the use of machines is relatively new, as the introduction of new machinery reduced the number of workers. In addition, from a few firms' perspectives, for the previous 3 years the important factors for the stability in the demand for skilled workers are due to the stability or limited increasing use of new technologies, machines and other technology-related production, that have no effect on increasing the demand for skilled workers and also because the use of the automatic packing machines motivated few firms to reduce employment. Moreover, from a few firms' perspectives, for the previous 3 years the important factors for the impact of the use of new technology on decreasing demand for skilled workers is attributed either due to the decreasing use of machines, equipment and new technology that required the use of high skilled workers for few firms, or because the limited use of new technology replaced the use of workers because the use of new technology is easy to operate in wide areas.

Table 5.8 The shortage of skilled workers and effects across firms in the Sudan, (2006–2009)

Shortage of skilled workers and effects on firm projects	All firms	Chemical	Food	Metal	Textile	Large	Medium	Small
Shortage of skilled workers	41 %	44 %	43 %	36 %	20 %	47 %	39 %	37 %
Effects of shortage of skilled workers on firm projects								
Serious delay of firm project	31 %	39 %	21 %	45 %	0 %	31 %	25 %	42 %
Abolishment/cancellation of firm project	24 %	23 %	21 %	9 %	0 %	22 %	22 %	16 %
Constrained the R&D units to adapt the imported technologies	74 %	68 %	80 %	75 %	75 %	81 %	67 %	70 %
Total response	82	36	29	12	5	32	28	20

Source: Own calculation based on the firm survey (2010)

management and administration. In addition, others positive impacts include the technical update, acquisition of experience in manufacturing of products and the introduction of new industrial activities that were not available in the past, enhancing the local maintenance, educating and learning or teaching of workers new ways to work faster and better and increase the need for continuous development.[45] On the other hand, our results from the firm survey indicate that the relative shortage of skilled workers amongst 41 % of the respondent firms leads to serious delay, slight abolishment/cancellation of project implementation and constrains the R&D units in adapting imported technologies (see Table 5.8 below).[46]

Our analysis of the shortage in skilled workers is based on the economic interpretation and definition of 'skilled shortage' as scarcity or lack of sufficient skilled workers needed, mainly because the supply of skills (as shaped by systems of education and training) has not responded fully to the rising demand across firms. However, managers may have a different interpretation and understand this as a lack of sufficient skilled workers in conjunction with wages constraints, due to limitations on their ability to pay higher wages for the high skilled workers they demand. This may constitute a limitation and appropriate caution should be exercised in interpreting our results with respect to skilled shortage, mainly because the firms answered the questionnaires do not really make it very clear how they have perceived the shortage of skilled workers and their further consequences.

[45] From the firms' perspective the negative impacts of skilled workers are reported by few firms and seems to be limited; this includes for example: the potential increasing unfair and intense competition between workers that probably caused a decline in productivity; the shortage of some skilled workers that probably caused high wages and that probably caused increase in production cost, and the increasing possibility of sagging employment because skilled workers do not continue to work for a long time in the same firm and move to work in other competing firms that probably caused increasing uncertainty and instability of production.

[46] As indicated by 31 %, 24 % and 74 % of the respondent firms respectively.

5.3 The Impacts of Skill Upgrading and Technological Upgrading: Micro–Macro Views

In view of the above findings and our results in Chap. 6 below on poor skill and technology indicators, it is therefore essential to recommend further incentives to upgrade both skill and technology levels at both micro and macro levels. From that perspective, our findings in Table 5.9 below show that at the micro level the upskilling plan amongst 54 % of the respondent firms induced significant effect in enhancing firm production, facilitating the effective utilisation and upgrading of technologies and upskilling national workers in the firm. But it has only a slight effect on reinforcing the employment for the national Sudanese workers, hiring more skilled national workers and reducing the future demand for foreign workers. Moreover, technological upgrading induced significant effects in enhancing firm production and upskilling via raising skill levels and upskilling national workers in the firms, but it has a relatively slight effect on both reinforcing firm ability to promote the local technology and hiring more skilled national workers.

At the micro level, Fig. 5.2 below compares the plans of public and private firms to depend on national skills. It predicts consistency between private and public firms that appears from the potential strong commitment to completely rely on national skills in the short, medium and long run plans in the large public firm. Moreover, our results from the firm survey and Fig. 5.2 below show that managers of private firms have a somewhat optimistic view regarding self-reliance on local skills and the potential role of both technological upgrading and upskilling in reinforcing the self-reliance strategy. For instance only 90 % of the respondent firms have supported the argument in favour of adequate availability of Sudanese national skilled workers and 94 % of the respondent firms have plans to rely on national skilled workers even in restricted fields. Amongst all the respondent firms the plan for completely depending on national skills (by 100 %) over the long run reached 74 % of the total skilled workers. From the firms' perspective since there is no difference between Sudanese workers and foreign workers in the industry, it is better to rely on Sudanese workers, so raising Sudanese workers would imply commitment to work for bringing all the benefits to contribute to their country's progress. This also implies the possibility of bringing in more skilled workers and new foreign engineers for a specific period on a temporary basis provided that they return back after achieving the work.

In strong consistency and similarity to the above view, the macro survey shows that the respondent policy makers and experts are highly optimistic regarding the interactions between technological upgrading and upskilling and their roles in reinforcing economic growth, self-reliance on local skills and restructuring the labour market at the aggregate/macro level. For instance, Table 5.10 below shows that the large majority of all respondent policy makers and experts predict strong linkages between both upgrading of technology and upskilling of labour force to reinforce each other and to have a similar large significant effect on reinforcing the self-reliance strategy and reducing the unemployment rate. The official view

Table 5.9 The effect of upskilling plan and technological upgrading across firms in the Sudan, 2010: micro view

	All firms	Chemical	Food	Metal	Textile	Large	Medium	Small
(a) Upskilling plan and its impacts, self-reliance strategy								
General upskilling plan	54 %	46 %	52 %	82 %	60 %	55 %	59 %	42 %
Self-reliance on national skills-adequate availability of national workers	90 %	89 %	93 %	91 %	80 %	81 %	96 %	94 %
Self-reliance on national skills: plan for dependence on national skill even in restricted fields	94 %	92 %	93 %	100 %	100 %	94 %	96 %	89 %
Self-reliance on national skills – in the short, medium and long run	86 %	85 %	91 %	80 %	92 %	85 %	91 %	92 %
(b)The effect of general upskilling on								
Enhancing firm production	92 %	92 %	89 %	100 %	100 %	94 %	89 %	95 %
Facilitating effective utilization and upgrading of technologies	95 %	97 %	89 %	100 %	100 %	97 %	100 %	82 %
Hiring more skilled national workers	63 %	64 %	59 %	64 %	80 %	69 %	59 %	58 %
Upskilling national workers in the firm	89 %	89 %	81 %	100 %	100 %	94 %	78 %	95 %
Reducing future demand for foreign workers	61 %	53 %	70 %	55 %	80 %	78 %	52 %	42 %
Reinforcing the employment of Sudanese national skill workers	79 %	88 %	67 %	73 %	100 %	81 %	78 %	76 %
(c) The effects of technological upgrading in:								
Enhancing firm production	99 %	100 %	100 %	92 %	100 %	100 %	96 %	100 %
Raising skill level	94 %	92 %	100 %	83 %	100 %	91 %	93 %	100 %
Reinforcing firm ability to promote the local technology	83 %	89 %	80 %	83 %	60 %	80 %	89 %	80 %
Upskilling national workers in the firm	93 %	95 %	97 %	75 %	100 %	97 %	93 %	85 %
Hiring more national skill workers	82 %	78 %	80 %	92 %	100 %	89 %	75 %	80 %
Total response	84	37	30	12	5	35	28	20

Source: Own calculation based on the firm survey (2010)

predicts that the effect of upskilling is stronger than the effect of technological upgrading in both reinforcing economic growth and solving the mismatch and imbalances in the labour market and so reducing the future demand for foreign

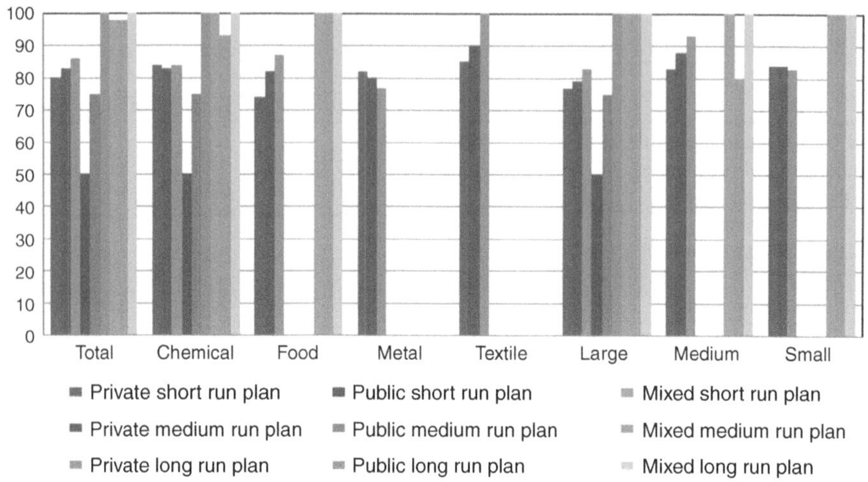

Fig. 5.2 Short, medium and long run plans of self-reliance on national skills across public, mixed and private firms in Sudan (Source: Firm survey 2010)

Table 5.10 The effects of technological upgrading, skill upgrading at the aggregate level in the Sudan 2010: macro view

The effects of technological upgrading	%	The effects of skill upgrading	%
Enhancing/accelerating upskilling	94	Enhancing/accelerating technological upgrading	100
Increasing/reinforcing economic growth	97	Increasing/reinforcing economic growth	100
Reinforcing self-reliance strategy	97	Reinforcing self-reliance strategy	94
Reducing the future demand for foreign skilled workers	92	Reducing the future demand for foreign skilled workers	94
Reducing unemployment rate	94	Reducing unemployment rate	94
Solving the mismatch problem in the labour market: by increasing the match between educational and training output to match with the needs in labour market	97	Solving the mismatch problem in the labour market: by increasing the match between educational and training output to match with the needs in labour market	100

Source: Own calculation based on the macro survey (2010)

skilled workers. Moreover, upskilling is also essential for enhancing productivity, poverty alleviation and for supporting a comprehensive plan on sustainable balanced development for Sudan.

This consistency in optimistic views at the macro and micro levels, regarding the self-reliance on local skills and the role of both technology and upskilling in reinforcing it, imply that the self-reliance strategy is probably not only a preferred government strategy, but also probably favoured by firms. This is consistent with the observation that 92 % of the respondent firms are private firms dominated by low skilled workers and probably realise the negative consequences of the lack the incentives/interests to rely on upskilling workers, mainly due to high costs of training and high salary requirements for high skilled workers. The majority of the respondent firms probably are confronted with the challenge to initiate new plans for skill upgrading, while few of the respondent firms probably faced with the difficulties to maintain the already existing upskilling efforts. For instance, our findings from the firm survey indicate that the lack of finance to cover the high cost of training is mentioned by 69 % of all the respondent firms amongst the important factors hindering the provision of training and hence upskilling of workers across firms at the micro level (see Table 5.4 above). The respondent firms, which are costs minimisers/profits maximisers, are probably willing to continue hiring cheap ready-made skilled workers instead of incurring expensive costs by hiring, training and upskilling unskilled workers. From the firms' perspective, for a few firms skill upgrading is not important for affecting the strategies of firms because few firms are mainly focusing on private profit goals; skill upgrading is not important because the nature of the production in these few firms does not need for raising the skills and the lack of attention to spend on skill upgrading because of the expectation from the past experiences that both the workers and staff are not continuing in the work for a long time period in these few firms.

5.4 Conclusions

This chapter uses the results of the macro and firm surveys to show the interaction between the deficient educational system and the high incidence of unskilled workers and their implications.

Our results confirm a part of the second hypothesis in Chap. 1 above: that in the short- and medium-term, Sudan is unable to rely on local technologies and remain heavily dependent on foreign technologies at the micro level. The major reasons are low levels of both skill and technology due to the deficient educational system and the high incidence of unskilled workers and their implications.

On the one hand, from the demand perspective, the results of the firm survey can be used to argue that firm demand for low skilled workers, weak technology indicators and dependency on foreign technology led firms to demand and rely heavily on low skilled workers. On the other hand, from the supply perspective, our findings in Sect. 5.2 show that the deficient educational system – due to low quality of education – and the excessive share of unskilled workers has led to low skill levels, poor provision of training, serious skills mismatch, weak linkages, lack of a networks and hindered the transfer of knowledge. These factors have interacted with each other and led to poor technology indicators, poor indigenous capability to build the local technology and a heavy dependence on foreign technology. These results prove hypotheses 3.a–3.b in Chap. 1 above concerning the low skill and technology indicators at the micro–macro levels: the serious implications of the interaction between the causes and consequences of the deficient educational system and the high use of unskilled workers. We confirm hypothesis 3.c. in Chap. 1 above that the major causes of low level of local technology are low/a lack of R&D activities due to a lack of skills, transfer of knowledge, networks and collaborations between universities and industry/firms.

Our findings show one surprising contradicting macro–micro view. Notably, contradicting optimistic–pessimistic micro and macro views regarding the incidence and success of knowledge transfer/external schooling effect implies that, probably, the transfer of knowledge/the external effects of schooling is successful within firms but is unsuccessful within society at large. This is probably because the transfer of knowledge is hindered by: low quality of education; the weak linkages and lack of networks between universities, colleges, technical and training institutes and the productive sectors; and the imbalanced structure of population and labour market. We show that the major cause behind the low transfer of knowledge/external schooling effect is low educational qualifications, and deficient educational and training systems. The major consequences are the lack of networks and collaboration between universities and firms, low R&D efforts and low technology indicators. These results prove part of the sixth hypothesis in Chap. 1 above with respect to the failure and the factors hindering the transfer of knowledge/external schooling effect at the macro level. But, on the other hand, our findings surprisingly reject a part of the sixth hypothesis in Chap. 1 above concerning the failure of the transfer of knowledge/external schooling effect at the micro level.

Our results prove part of our hypothesis 8.c. in Chap. 1 above concerning the consistency and similarity optimistic macro and micro views concerning the self-reliance on local skills, and the role of both technological upgrading and upskilling in reinforcing it, implies that the self-reliance strategy is probably not only a preferred government strategy, but probably is also one followed by private firms. Though driven by profit-maximising considerations, private firms are likely to continue in hiring cheap readymade skilled workers rather than in hiring, training and upskilling workers with expensive costs. From these observations, our results accept hypothesis 8.c. in Chap. 1 above about the consistency of upskilling and transfer of knowledge at the macro–micro levels.

References

Abdelkarim, A., & Haan, H. (2002). *Skills and training in the UAE: The need for and the dimensions of institutional intervention* (Policy Research No. 5). Dubai: Centre for Labour Market Research and Information.

Ali, A. A. (2006). *On human capital in post-conflict Sudan: Some exploratory results* (Working Paper Series No. API/WPS 0602). Kuwait: Arab Planning Institute.

Barro, R., & Lee, W. -J. (2000, April). *International data on educational attainment: Updates and implications*. Cambridge, MA: Department of Economics, Harvard University. Data set at International Development at Harvard University (CID). http://www.cid.harvard.edu/cidwp/pdf/042.pdf

Cowan, R., Soete, L., & Tchervonnaya, O. (2001). *Knowledge transfer and the services sector in the context of the new economy* (p. 9) (MERIT-INFONOMIC Research Memorandum No. 2001–021). Maastricht: Maastricht University.

El Sabaa, S. (1997, November) Linking technology to industrial development in the United Arab Emirates. *Trade and Industry Magazine* – Dubai, *22*(263), 16–29

Fasano, G. (2002, October 21). *With open economy and sound policies, U.A.E. has turned oil "Curse" into a blessing* (pp. 330–332). IMF survey. Washington, DC: International Monetary Fund.

Firm Survey. (2010, June). *Technological change and skill development: A comparative study of chemical, food, metal and textile small, medium and large scale enterprises in the Sudan.* Khartoum.

Gray, L. (1999). *Private sector training and employment: Evaluation and intervention strategies* (Specialist Report No. 4). Dubai: The Ministry of Labour and Social Affairs National Strategy for Labour Force Development and Employment Project.

Haan, H. (1999, January–March). *The UAE: Assessment of technology use in some selected sectors – Towards introduction and explaining capital and skill intensive technologies* (Specialist Report No. 6). National Human Resources Development and Employment Strategy Project. Dubai/Amsterdam.

Macro Survey. (2010). *Skill creation, human resources development and policy intervention: Interviews with policy makers and experts in the Sudan.* Khartoum.

Muysken, J., & Nour, S. (2006). Deficiencies in education and poor prospects for economic growth in the Gulf countries: The case of the UAE. *Journal of Development Studies*, *42*(6), 957–980. UK: Taylor and Francis Ltd.

Nour, S. (2005a, September). Science and Technology (S&T) development indicators in the Arab region: A comparative study of Arab Gulf and Mediterranean countries. *The Journal of Science, Technology and Society*, *10*(2), 249–274. SAGE Publications. (Published in *UNU - INTECH Discussion Paper Series* UNU-INTECHDPS2005-3, Maastricht, the Netherlands, August 2005).

Nour, S. (2005b). *Technological change and skill development in the Arab Gulf countries.* Doctoral dissertation, Maastricht University Press, Maastricht, The Netherlands.

Sudan Central Bureau of Statistics Population Census Data. (2010b). *5th Sudan population and housing census (2008).* Khartoum: Sudan Central Bureau of Statistics.

The Arab Human Development Report (AHDR). (2003, July). *Arab human development report (2003): Building a knowledge society.* New York/Amman/Jordan: UNDP/Regional Bureau for Arab States (RBAS)/National Press. P. 56, Table 5.1, p. 102.

The UAE University. (1994). *The UAE education assessment report.* The UAE University, Al-Ain, UA.

The UAE University. (1997). *Private sector employment and labour market in the UAE.* The UAE University, Al-Ain, UA.

UNDP. (2002). *Human development report (2002): Deepening democracy in a fragmental world.* New York/Oxford: UNDP/Oxford University Press.

UNDP. (2003). *Human development report (2003): Millennium development goals: A compact among nations to end human poverty.* New York/Oxford: UNDP/Oxford University Press.

UNDP-AHDR. (2002). *Arab human development report (2002): Creating opportunities for future generations.* Regional Bureau for Arab State (RBAS), United Nations Development Programme (UNDP), New York. Amman: Icons Printing Services.

United Nations Development Programme (UNDP). (2004). *United Nations millennium development goals in Sudan "Sudan millennium development goals interim unified report, 2004".* http://www.sd.undp.org/mdg_sudan.htm. Accessed 1 June.

Ziesemer, T. (2004). Multiple-steady-state growth models explaining twin-peak empirics? In G. Grözinger & A. van Aaken (Eds.), *Inequality: New analytical approaches* (pp. 145–168). Marburg: Metropolis-Verlag.

Chapter 6
Assessment of Science and Technology Indicators and the Need for Science and Technological Development in Sudan

Abstract This chapter investigates the status of S&T indicators in Sudan. We explain that the combination of poor S&T inputs/resources (both financial and human resources) together with an inadequate economic system as a whole, results in Sudan producing poor S&T outputs/performances. Moreover, we find that most R&D and S&T activities and (FTER) employment in Sudan occurs within the public and university sectors, while the private sector and industry make only a minor contribution. We find that the main problems hindering R&D include: the lack of finance from public sector; lack of management and organisational ability; lack of coordination and weak relationships, network and consistency and cooperation between universities and higher education institutions on the one side and the productive sector on the other side; lack of R&D culture; lack of finance from private sector; lack of favourable conditions and the necessary facilities; lack of awareness and appreciation of the economic values of R&D; and lack of human resources (researchers and qualified workers in R&D fields) respectively. Hence, our analysis indicates that in order to improve S&T performance, Sudan needs to invest heavily in both financial and human resources as well as to learn from the lessons of the advanced and developing S&T nations.

6.1 Introduction

Chapter 5 uses the data and results of the firm and macro surveys set out in Chap. 4 to assess skill and technology indicators and to examine the serious implications of the interaction between the deficient educational system and the high incidence of unskilled workers and skill mismatch. To complete our earlier analysis in Chap. 5

Nour, Samia (2012a), "Assessment of science and technology indicators in Sudan," the Journal of Science, Technology and Society, Vol. 17; No. 2; July 2012, pp. 321–352, SAGE Publications, Los Angeles/London/New Delhi/Singapore/Washington DC. DOI: 10.1177/097172181101700206.

DOI 10.1007/978-3-642-32811-4_6, © Springer-Verlag Berlin Heidelberg 2013

above, in this chapter we use the most recent secondary data to discuss science and technology development indicators at the macro level in Sudan, compare the status of Sudan with the rest of the world and highlight the need for technological development and policies to enhance science and technology performance in Sudan. Our analysis in this chapter differs in several ways from the several studies in the literature, which provides an interesting analysis of S&T indicators and performance in the Arab, developing countries and Sudan. First, different from the studies in the Arab literature (Nour 2004, 2005a, b) we provide a more indepth, comprehensive and up to date assessment of S&T input and output indicator by focusing only on Sudan as a new case of the Arab countries. Secondly, we extend our analysis to compare the case of Sudan with other Arab, African and world countries. Given the recent progress of economic globalisation coupled with the emergence of new nations active in S&T in different parts of the world, our analysis in this chapter extends the comparison to include these new countries as well as those in Europe, the United States and Japan, and then draws some policies and recommendations for ways to enhance S&T performance in Sudan. Thirdly, different from the studies in the Sudanese literature we provide a more comprehensive analysis by including both S&T input and output indicators using more up to date data wherever possible. This is so we can help to improve understanding about the urgent need and necessity to stimulate S&T development and support new policies that aim to enhance S&T performance in Sudan and poor countries. Our study highlights recent efforts to create an active Sudanese S&T base but also emphasises the need to improve the quality of resources devoted to S&T development, which will ultimately contribute to and accelerate development in the country. Furthermore, it also helps to encourage the government to provide more incentives for the promotion of S&T indicators in Sudan to obtain the most positive impact possible from technological progress in terms of growth, employment and the wellbeing of all poor Sudanese citizens. Finally, different from the studies in the Sudanese literature, a novel element in our analysis is that we use new survey data based on primary data and 25 face-to-face interviews with the official policy makers and experts in the government and the academics and university staff in the public and private universities to examine the main factors hindering and those contributing towards the promotion of R&D and hence S&T development in Sudan. As mentioned in Chap. 4 above, the main purpose of this survey is to collect primary data to examine the causes of poor R&D activities and then to provide some recommendations to improve R&D and hence S&T indicators in Sudan.

The rest of this chapter is organised as follows: Sect. 6.2 shows the interaction between general socio-economic characteristics of Sudan and S&T. Section 6.3 discusses S&T development indicators in Sudan, including a comparison of the indicators for Sudan with the rest of the world. Finally, Sect. 6.4 draws conclusions and proposes policies to enhance S&T performance in Sudan, based on the results of the Sudan R&D survey and the experiences of other countries.

6.2 General Socio-Economic Characteristics of Sudan and S&T Indicators

S&T performance is often closely related not only to the resources directly devoted to its development but also to the whole economic structure that supports it. Therefore, before assessing S&T performance in Sudan it is useful to explain the interaction between the general socio-economic characteristics of Sudan and S&T indicators. In Chap. 2 above we show the general socio-economic characteristics and lower standards of economic development as measured by GDP per capita and human development index of Sudan as compared to African and Arab countries and the world regions. We explained that after the exploitation of oil in 1999 Sudan economy become increasingly dependent on oil exports, and the economy turned into an oil dependent economy. In recent years the increasing dependence on oil has led to sound economic growth, measured by GDP growth at 6.1 % in 2003 and averaged about 9 % during (2005–2007). However, unfortunately it is only unsustained growth, mainly because of uncertainty and high fluctuation in oil price in the international market, for instance, the recent global financial and economic crisis and related shock in 2008 and 2009 resulted in low global oil prices, lead to significant negative impact on Sudan economy, stagnating domestic oil production and caused reduction in GDP growth rate that dropped from 10.5 % in 2007 to 7.8 % and 5 % in 2008 and 2009 respectively (see Table 2.2 and Fig. 2.2 above).

We are aware of the fact that it may be useful to depart from the analysis of general standardise S&T indicators and to use indepth economic, historical and social evidence to extend our analysis to focus more explicitly on whether the production and export of oil (natural resource-based exports) affected the R&D infrastructure and the growth and development trajectory of Sudan economy. We are aware of the fact that it may be particularly important and interesting to explain the impact of oil in R&D and S&T, but due to practical problems related to availability of adequate and reliable data, unfortunately it will not be possible to discuss this issue in this chapter, so we leave that for a more indepth analysis in our future research. Furthermore, we believe that most probably the impacts of oil in R&D and S&T might be still very limited in view of the very recent start of production and exports of oil just 11 years previous in 1999. Moreover, although oil led to increases in public spending and increases in the share of development expenditure as a percentage of total public expenditure from 9 % in 1999 to around 31 % in 2004, its share declined and sustained at 24 % from the total public spending over the period 2006–2009. Furthermore, the development expenditures include all public spending in development issues including public spending on education, health, etc. Therefore, this implies that it is not at all clear and it is somewhat problematic to distinguish the share and growth of spending on R&D that is mainly attributed to production and export of oil, but it is important to realise that at the macro level the share of spending on R&D as a percentage of GDP most probably remained below the United Nations (UN) (1970) standardised level of spending 1 % of GDP on R&D in the pre- and post-oil periods (see Fig. 6.1 below). In addition, also due to practical problems related to availability of adequate and

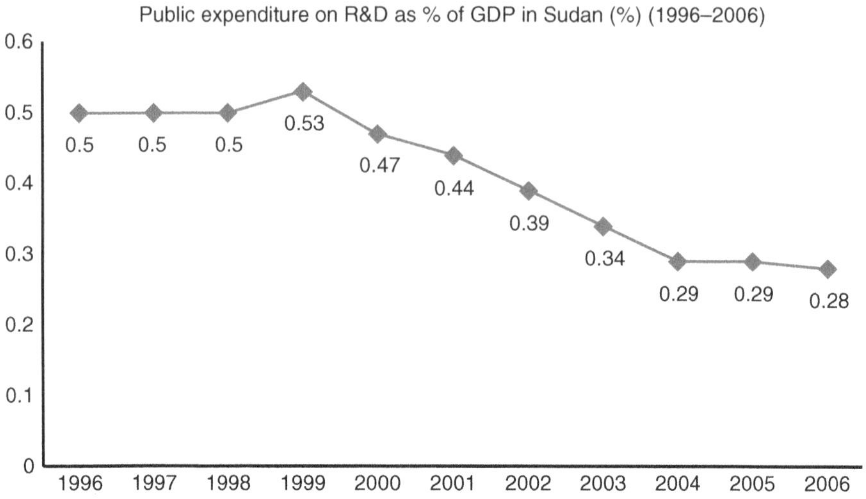

Fig. 6.1 The rate of public expenditure on R&D as a percentage of GDP in Sudan (%) (1996–2006) (Source: UNESCO R&D Statistics (2006))

reliable data unfortunately it will not be possible to give an indepth analysis of the private spending on R&D or the impact of oil companies on R&D at the micro level. So, we hope to cover these issues in our future studies when adequate and reliable data is available. Hence, apart from the limited impact of oil, the next section of this chapter examines whether this economic background affects S&T performance in Sudan.[1]

6.3 S&T Indicators in Sudan

Based on the definition of S&T indicators provided in Chap. 3 above and the explanation of the interaction between the general socio-economic characteristics of Sudan and S&T indicators provided in Sect. 6.2 above, this section extends our assessment of technological indicators at the micro level presented in Chap. 5 above by explaining the technological indicators at the macro level. It is useful to start by explaining the governance of S&T; next we examine input indicators (financial and human resources) and output indicators (scientific and technological performance) required to measure S&T performance at the macro level in Sudan.

[1] One limitation of the comparison in our analysis is that we use data and information from two different local and international sources; the scarcity of data and information covering all indicators limited our attempt to use a unified source.

6.3.1 Governance of S&T

In Sudan the history of S&T governance dates back to the 1970s, when the National Council for Research (NCR) was established in 1970 as a governmental body responsible for formulating policies and plans and coordinating national efforts in this respect. The mandate of NCR was transferred to the Council of Higher Education and Scientific Research in 1991–1992. In S&T education, the government has expanded higher education institutions, there are 85 universities and colleges (private and public), 40 universities and colleges are in the field of applied sciences and about 25 colleges in the fields of engineering and technology. The Sudanese government has also realised the importance of creating high level national science bodies by establishing two important institutions: the National Council for Science and Technology (NCST) and the Ministry of Science and Technology (MOST). The role of the NCST is to formulate the policies of S&T, organise R&D and implement the country's strategies in S&T and to ensure that S&T is utilised in the plans, projects and institutions of the government. A significant development in terms of institutional framework for S&T development in Sudan was the establishment of MOST in 2001. The formation of MOST signified the high priority and importance attached to the promotion of science and technology and to coordinate efforts of national and international links and formulate national strategy for S&T. It led to the centralisation of the public research institutes under the supervision of MOST whereby the public research institutes in the various fields were previously under the jurisdiction of their respective ministries. Scientific research is conducted and governed in three levels: (a) basic research conducted by universities and governed by the Council of Ministry of Higher Education and Scientific Research; (b) R&D research conducted by corporations and centres, governed by MOST, advised by a council and a number of committees; and (c) applied research conducted in some technical department of ministries, administered by the executive authority of each ministry. Given the division among the three sectors, under the new institutional framework, MOST faced the challenges to work as government high coordinating body to coordinating the various diverse fields of research and meeting the needs of the various ministries and industries. MOST includes some specialised research institutes and centres including: Agricultural Research Corporation (ARC); Animal Resources Research Corporation (ARRC); National Centre for Research (NCR); Industrial Research and Consultancy Centre (IRCC); Sudan Atomic Energy Corporation (SAEC); Sudanese Metrology Authority (SMA); Central Laboratories (CL); Sudan Academy of Sciences (SAS); and Social and Economic Research Bureau (SERB).

In terms of S&T planning and in view of the increasingly competitive global environment and rapid advance in technology and increasing importance of S&T in accelerating economic growth and development, the previous comprehensive National Strategy (1992–2002) and current National Quarter Century Strategy (2007–2031) give long-term perspective of S&T development in Sudan. The previous comprehensive National Strategy (1992–2002) provided comprehensive strategies for Science and technology (S&T) development through the preparation of a national plan for scientific research, development of information centres and scientific research as well as the establishment of a national information network, adoption and modification of the important technology system to suit national environment, development of capabilities to invent technology and the maximum utilisation of technology in Sudan. In light of the 25 year long-term strategy, a 5 year strategy was identified and implementation work plan is developed. The 5-year work plan is targeting eight key areas including information, communications and technology and development of scientific research. The plan aims to promote S&T by: promulgating the legislations, laws and regulations conductive to the enhancement of scientific research; recruiting personnel with high abilities and competencies in the fields of scientific research; adopt innovative means to encourage the private sector to participate in scientific research, funding it and benefiting from it; utilising the results of scientific research and modern technology in decision–making and sustainable development planning; developing and disseminating science and knowledge among the people; benefiting from the experience of others in scientific research and also contributing to the advancement of basic sciences.[2] Unfortunately, the implementation of these comprehensive strategies, however, was not fully carried out mainly due to the inadequate financial and human resources needed for S&T development as we explain below.

6.3.2 *Human and Financial Input Indicators*

In terms of both financial and human S&T input/resource indicators there are some differences between Sudan, and the Arab and Sub-Saharan African countries as well as between them and other countries around the world. Table 6.1 shows that both financial and human S&T input indicators in Sudan lag behind the advanced and leading developing countries.

6.3.2.1 Financial Input Indicators

As for the financial resources in S&T, as in most other typically developing countries the Sudanese government seem to afford only a limited budget for S&T. For instance, in 2006, the rate of spending on R&D as a percentage of GDP in Sudan was only 0.2 %, falling behind the standard rate of the world, Arab countries, developing countries, East Asia, Pacific, Latin America and the Caribbean, South Asia, middle income and even low income countries, which spend on R&D as a percentage of GDP about 2.3 %, 0.6 %, 1.0 %, 1.6 %, 0.6 %, 0.7 %, 0.8 %, 0.7 % respectively (see Fig. 6.2 below). The rate of spending on R&D as a percentage of GDP in the developing countries is five times the rate of spending in Sudan. This reflected negatively on the number of researchers and publications, as we will explain below.

[2] See Sudan Ministry of Science and Technology (2008), pp. 3–6.

Table 6.1 S&T resource indicators of the Sudan, Arab and world countries

Country	Public expenditure on education as % of GDP[a]		Public expenditure on education as % of government expenditure [a]		R&D expenditure as % of GDP[a]	Number of scientists and engineers in R&D (per million population) [a]	Number of patents[a, b]	High-technology exports as % of manufactured exports[a]	
	1990	1998–2000	1990	1998–2000	1996–2000	1996–2000	1990–1999	1990	2001
Sudan	0.9	Na	2.8	Na	0.5	225	0	..	7
Gulf countries									
Bahrain	4.2	3.0	14.6	11.4	Na	Na	2	0	0
Kuwait	4.8	Na	3.4	Na	0.2	212	27	4	1
Oman	3.1	3.9	11.1	Na	Na	8	3	11	3
Qatar	3.5	3.6	Na	Na	Na	591	0	0	0
Saudi Arabia	6.5	9.5	17.8	Na	Na	Na	103	0	Na
UAE	1.9	1.9	14.6	Na	Na	Na	15	0	Na
Average Gulf	4.0	4.4	12.3	11.4	0.2	270	25	2.5	1
Mediterranean countries									
Algeria	5.3	Na	21.1	Na	Na	Na	Na	0	4
Egypt	3.7	Na	Na	Na	0.2	493	38	0	1
Lebanon	Na	3.1	Na	11.1	Na	Na	Na	Na	3
Morocco	5.3	5.5	26.1	26.1	Na	Na	Na	0	11
Syria	4.1	4.1	17.3	11.1	0.2	29	3	0	1
Tunisia	6.0	6.8	13.5	17.4	0.5	336	Na	2	3
Average Mediterranean	4.9	4.9	19.5	16.4	0.3	286	20.5	0.4	3.8
Norway	7.1	6.8	14.6	16.2	1.7	4,112	97	8	12
Sweden	7.4	7.8	13.8	13.4	3.8	4,511	285	13	18
UK	4.9	4.5	Na	11.4	1.8	2,666	76	23	31
Korea, Rep. of	3.5	3.8	22.4	17.4	2.7	2,319	931	18	29

(continued)

Table 6.1 (continued)

Country	Public expenditure on education as % of GDP[a]		Public expenditure on education as % of government expenditure [a]		R&D expenditure as % of GDP[a]	Number of scientists and engineers in R&D (per million population) [a]	Number of patents[a, b]	High-technology exports as % of manufactured exports[a]	
	1990	1998–2000	1990	1998–2000	1996–2000	1996–2000	1990–1999	1990	2001
Singapore	Na	3.7	Na	23.6	1.1	4,140	12	39	60
China	2.3	2.1	12.8	Na	0.1	545	793	0	20

[a]UNDP (2003).

[b]United States Patent and Trademark Office (USPTO) website: http://www.uspto.gov. Patent data for Korea, Norway, Singapore, Sweden and the UK obtained from UNDP (2003) and refers to patents granted in 1999 to residents per million people. For China and all Arab countries, patent data was obtained from USPTO during 1991–1999 and refers to the number of registered US patents where the inventor of the patent is resident in the selected countries.

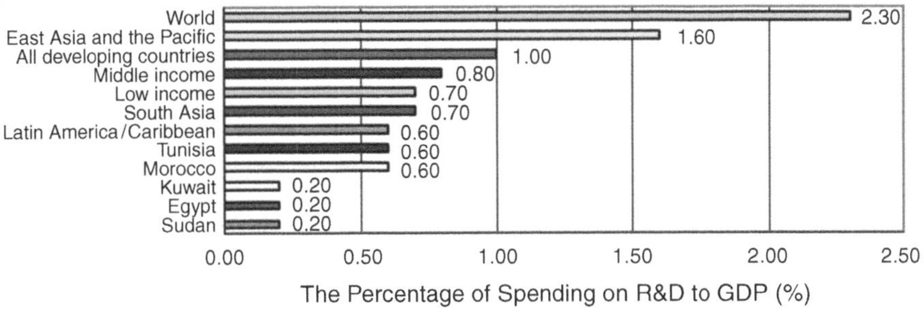

Fig. 6.2 The rate of spending on R&D as a percentage of GDP in Sudan compared to other Arab and world regions (2006) (Source: UNDP, HDR 2007/2008, Table 13: 273–276. P. 240 Arab Human Development Report 2009)

In Sudan the implementation of the comprehensive strategies in the field of S&T, was not fully carried out mainly due to the inadequate financial and human resources. The S&T indicators showed that S&T development was relatively low compared with the average for Arab countries. This was evident as the percentage expenditure in research to total government expenditure in 1998 for Sudan was only 0.04 % compared with the average for seven Arab countries, which was 1.2 %. In terms of expenditure on both education and R&D as percentage of GDP, Sudan performs less than Arab countries. In particular, Table 6.1 shows that the financial resources devoted to S&T, as measured by the percentage share of GDP spent on R&D are poor in Sudan and Arab countries, compared to both advanced and leading developing countries like Singapore and Korea. For instance, in the period 1996–2000, Sudan devoted only 0.1 % compared to Arab countries that devoted an average of only 0.3 % of their GDP to R&D whereas Sweden, one of the leading advanced industrial countries, spent 3.8 % of GDP on R&D. Similarly, spending on education, as measured by percentage of both GDP and total government expenditure, for Sudan was found to be less than Arab countries and the advanced countries.

Comparing S&T indicators between Sudan and other Arab countries, data for 2006 shows that the rate of spending on R&D as a percentage of GDP in Sudan is comparable to the rate of spending in Egypt and Kuwait, but falls behind the rates of both Morocco and Tunisia, notably, the rate of spending on R&D as percentage of GDP in Morocco and Tunisia is three times the rate of spending on R&D in Sudan, Egypt and Kuwait (see Fig. 6.2 above). Moreover, statistics indicate a very high dependence on the public sector on the financial support to S&T (near to 95 % of total financial support to S&T) compared to a very low contribution of the private sector in Sudan (near to only 5 % of total financial support to S&T). There is thus a need to adopt new policies for partnership with the private sector. Investigation of the sectoral distribution of R&D spending by sources of funding in Sudan in 2005 indicates that the public sector is responsible for the majority of R&D activities, accounting for 39.2 % of all Gross Domestic Expenditure on Research & Development (GERD) (see Table 6.2 below). Next to public sector, private sector contributes 33.7 % of GERD; the universities make only a minor contribution,

Table 6.2 Gross domestic expenditure on R&D (GERD) by sector of performance (%) in Sudan (1999–2005)

	Total gross domestic expenditure on R&D (GERD)				GERD by sector of performance (%)		
	Local currency (Sudanese dinar) (000)	PPP$ (000)	As percentage (%) of GDP	Per capita (PPP$)	Business enterprise (%)	Government (%)	Higher education (%)
1999	14,300,000	195,816	0.53	6.0	31.5	38.5	30.1
2000	14,900,000	191,746	0.47	5.7	31.5	38.9	29.5
2001	15,240,000	196,190	0.44	5.8	31.5	39.3	29.2
2002	15,400,000	186,387	0.39	5.4	31.8	39.0	29.2
2003	15,650,000	176,066	0.34	5.0	31.9	39.0	29.1
2004	16,373,000	165,184	0.29	4.6	33.6	38.3	28.1
2005	19,284,000	179,085	0.28	4.9	33.7	39.2	27.1

Source: UNESCO R&D Statistics (2006)

accounting for 27.1 % of GERD. These findings for the case of Sudan seem consistent with the results in Nour (2004, 2005), which implies that in Sudan as in the Gulf and Mediterranean Arab countries the public sector is responsible for the majority of R&D activities and government seems to play a major role in R&D compared to higher education. Moreover, despite the fact that the contribution of the private sector (business enterprises) is near to one third and exceeds the contribution of higher education institutions in Sudan however, this should not hide the fact that business does not seem to play a major role in R&D compared to government. Our findings imply that Sudan is similar to Arab Mediterranean countries, which appear to be more dependent on the public sector than the Arab Gulf countries, reflecting a lack of incentives for private sector institutions to invest in R&D in Sudan and Mediterranean countries compared to the Gulf. The minor contribution of the private sector to R&D activities and spending in Sudan and Arab countries compares poorly to most of the industrialised countries, where more than half of R&D expenditure is financed by industry (OECD1997).

A further problem concerning research funding in Sudan is that not only is Sudan's total GERD is rather fair at about 0.5 % GDP, but also there has been a steady decline in Sudan's total GERD during the 1999–2005 period (see Table 6.2 below). This declining trend implies that the heavy reliance on the limited government and public funding was risky and resulted in poor S&T indicators and inadequate finance for R&D activities that is apparent from the low rates of both the actual received budget relative to approved budget and the approved budget to the proposed budget. For instance, for all institutions of MOST, although the rate of actual received budget relative to approved budget increased from near to 25.7 % in 2003 to 74.7 % in 2009, the actual received budget relative to approved budget covered only 74.7 % of the approved budget in 2009. Implementation of projects is most probably constrained by inadequate finance, for instance, over the period 2003–2009 the average rate of implementation for national ministries and northern states is 60 % (see Table 6.3 below).

6.3.2.2 Human Resources Input Indicators

The human capital for S&T includes human resources in higher education, Master's and doctoral enrolments and the size of the university workforce and research and development personnel. Table 6.1 shows that there are a low number of scientists and engineers in R&D in Sudan and the Arab countries compared to both advanced and leading developing countries. We explain below about human resources in S&T, higher education and research institutions in Sudan.

Beginning with human resources in higher education institutions in Sudan, in the early 1990s, enrolment in both general education and higher education rapidly increased. For instance, during the period 1992–2000 the enrolment rates in both primary (basic) education and in higher secondary education rapidly increased by 54 % and 154 % respectively. As for higher education, following the higher education revolution in the early 1990s, notably 1992/1993, the total number of universities and colleges increased by more than three fold, notably, from 25 in 1993 to 85 in 2008; the number of public government universities increased from 6 universities in 1990 to 14 in 1993 and to 28 in 2008; the private universities and colleges increased from 11 in 1993 to 57 in 2008. The higher education revolution together with the implementation of economic liberalisation and privatisation policies and their related consequences in higher education, led to significant structural change in the share of public and private sectors in higher education institutions in Sudan. For instance, the share of the public government universities declined from 56 % in 1993 to 33 % in 2008, whereas the share of private universities and colleges increased from 44 % in 1993 to 67 % in 2008. The expansion in higher education in the period 1992–2007 led to significant increases in both student enrolment and graduation rates in higher education and universities by 73.78 % and 189.9 % respectively. Student intake jumped from 6,080 in 1989 to 25,018 in 1992/93 and to 43,477 in 2007. The number of female students rose to 40 % of total enrolment in 1995. However, the continued increase in the proportion of female students has not been accompanied by a comparable increase in their representation amongst the faculties: merely 13 % in 1995. The number of students enrolled in private higher education institutions increased nearly nine fold within 4 years: from 2,686 in 1990–1991 to 23,476 in 1994–1995 (see Table 6.4 below). As for Master's and doctoral enrolments, generally, there is remarkable increase in the number of people who participate in postgraduate studies in Sudanese institutions (see Table 6.4 below). The distribution of students enrolled in postgraduate programmes in 24 universities in Sudan indicates that the share of postgraduate students enrolled in 18 universities located in Khartoum state is higher than the share of postgraduate students enrolled in 14 universities located in other Sudanese states. Furthermore, the intensity of students enrolled in the Master's programmes is

Table 6.3 Performance of R&D funding in public research institutions, national ministries and Northern states (2003–2009)

(a) Performance of funding in major public research institutes/ centers

Institutions	Actual received budget/ Approved budget (%) 2003	2004	2005	2007	2008	2009	Approved budget/ Proposed budget (%) 2003	2004	2005	2007	2008
Head of Ministry	...	...	...	...	53.3	70.5	28.4	...	...	...	...
Agricultural Research Corporation (ARC)	21.5	29	44	45.8	74.8	73.7	24.2	17	53	19	40
Animal Resources Research Corporation (ARRC)	22.4	36	42	53.6	77.3	72	34.4	20	32	12	22
National Centre for Research (NCR)	36.9	37	14	46.8	76.6	82	25.4	55	1.23	21	16
Industrial Research and Consultancy Centre (IRCC)	37	...	...	55.5	72.2	83.2	31.7	...	...	...	...
Sudan Atomic Energy Corporation (SAEC)	35.7	...	...	61.2	66.7	73.7	53.8	...	...	...	...
Sudan academy for science	...	...	...	50.6	83.3	86.3	...	...	...	...	...
Social and Economic Research Bureau (SERB)	...	...	...	...	83.3	87.5	...	...	...	...	...
Central Laboratories (CL)	...	...	...	46.9	83.3	83.3	...	...	...	...	...
Sudanese Metrology Authority (SMA)	...	...	...	88.5	...	...	...	...	...	...	...
Total	25.7	...	...	49.1	73.7	74.7	28.2	...	...	...	...

(b) Performance of the national ministries and Northern states

indicators	Year	No of projects	Implemented 100 %	Implementation ongoing	Not implemented	Average performance of area (%)
National ministries	2007	234	42	165	36	63
	2008	165	28	127	10	63
	2009	212	57	149	6	62
Total ministries	2007–2009	620	127	441	52	62

Northern states	2007	19	2	11	6	70
	2008	53	10	27	16	55
	2009	41	13	18	10	54
Total states	2007–2009	113	25	56	32	59
Grand total	2007–2009	733	152	497	84	60
National ministries and states	2007	262	44	176	42	50
	2008	218	38	154	26	52
	2009	253	70	167	16	61
Grand total National ministries and states	2007–2009	733	152	497	84	54

Source: Ministry of science and technology annual reports (2003–2009)

Table 6.4 Growth in higher education institutions and students enrolment in general and higher education in Sudan1992–2008

(a) Higher education institutions 1993–2008

	1993	1994	1996	1998	2000	2002	2003	2004	2005	2008
Total number										
Government Universities	14	23	26	26	26	26	27	27	27	28
Private universities and colleges	11	12	16	18	23	36	37	46	46	57
Other high education institutions	...	...	...	...	...	3	2	2	3	...
Total	25	35	42	44	49	65	66	75	76	85
Share in total (%)										
Government universities	56	66	62	59	53	40	41	36	36	33
Private universities and colleges	44	34	38	41	47	55	56	61	61	67
Other high education institutions	...	...	...	...	...	5	3	3	4	...
Total	100	100	100	100	100	100	100	100	100	100

(b) Enrollment in higher education: % Student enrollment ratio in higher education by field of study (%)

Students enrolled in education	10.9
In arts and humanities and social science	14.7
Enrolled in medical science, (Health and environment)	10.9
In agricultural sciences	6.3
In engineering science	10.7
In basic science	7.1
Share of education, arts and humanities and social science in total enrolment	25.60
Share of agricultural, basic science, engineering science medical science, (Health and environment) in total enrolment	35.00
Share of others in total enrolment	39.40

(c) Enrolment in primary (basic) education: Percentage increase in general education schools (1992–2000)

(continued)

Table 6.4 (continued)

	1992/1993	2000	Increase %
Pre-school education	5,813	8,343	44
Primary education	8,288	11,923	54
Higher secondary education	574	1,457	154

(d) Growth in students enrollment and graduation in higher education and universities (1992–2007)

	1992/1993	2007	Increase %
Enrolled	25,018	43,477	73.78
Graduated	13,183	38,217	189.9

(e) Growth in students enrollment by field of study (%) (1992–2000)

	Number in 1992/1993	Number in 1999/2000	Change (1992–2000)	Growth rate (1992–2000) (%)
Education	4,123	7,269	3,146	76
Humanities and art	4,415	6,412	1,997	45
Social sciences and business administration and law	2,012	12,591	10,579	526
Natural sciences	1,685	3,894	2,209	131
Engineering sciences	2,539	4,545	2,006	79
Agriculture	2,336	4,553	2,217	95
Health and social services	1,760	4,036	2,276	129
Services	148	177	29	20
Total	25,018	43,477	18,459	74

(h) Distribution of postgraduate students in Khartoum state (in 14 universities) and other states (in 18 universities) in Sudan (2006)

Degree	Total number			Share in total (%)				
	Total	Khartoum state (14 universities)	Other states (18 universities)	Khartoum state (14 universities)	Other states (18 universities)	Total	Khartoum state (14 universities)	Other states (18 universities)
Ph.D.	2,885	2,240	645	78	22	14	19	7
M.Sc.	13,340	7,208	6,032	54	45	63	60	67
Higher diploma	4,878	2,613	2,265	54	46	23	22	25
Total	21,103	12,061	8,942	57	42	100	100	100

Source: (a) Sudan Ministry of Higher Education and Scientific Research, the annual higher educational statistics report: Various issues (1993/1994–2009/2010), (b) Sudan Ministry of General r Education, the annual general educational statistics report: Various issues (1993/1994–2009/2010), (c) Elamin (2009)

higher than the intensity of students enrolled in the doctoral and higher diploma programmes in other Sudanese states.[3]

As for human resources for R&D in higher education and universities, many studies indicated a positive relationship between science and technology achievements and the number of engineers and scientists. Despite the significant expansion of higher education and graduate training in the last two decades, the lack of human resources still remains a serious problem that has hindered the promotion of S&T and R&D in Sudan. In particular, despite the presence of 28 public universities and 57 private universities with a capacity of more than 500,000 students, universities produce many more graduates in social sciences than in engineering and science (see Table 6.5 below). Furthermore, many graduates lack the skills to effectively use modern tools and equipment, not to mention develop them. The number of postgraduate and Master's degree graduates in engineering per year is very low, the overall ranking is low, and is continually slipping and consequently, the universities have weak research culture and capabilities.[4] According to the international standard, the number of engineers and scientists per 10,000 people is often used as an international standard indicator of achievement of an acceptable level of research and development. For instance, the presence of less than 10 engineers and scientists per 10,000 people implies weak performance and the presence of gaps in all research sectors; the presence of 15 engineers and scientists per 10,000 people implies a critical level of performance; the presence of 30 engineers and scientists per 10,000 people implies an acceptable performance in science and technology; and the presence of more than 30 engineers and scientists per 10,000 people implies an advanced level in research and development. In Sudan, according to the comprehensive strategy (1992–2002) the standard was 0.02 per 10,000 people. This implies that in Sudan, in order to have satisfactory performance in the science and technology system by applying the international indicator of 30 scientists and engineers per 10,000 people, and based on the last population census (2008), Sudan should have 120,000 scientists and engineers. But the actual number is less than 20,000. This implies that more efforts, resources and time are needed to be equal or near to the international standard. In Sudan the implementation of comprehensive strategies in the field of S&T was not fully carried out mainly due to inadequate financial and human resources. Notably, the ratio of full time researchers in Sudan was 0.2 per 10,000 people in 1990 compared with the average for Arab countries, which was 1.7 per 10,000 people. The ratio of engineers and technicians in 1990 was 1 per 3,000–5,000 people in Sudan, compared with the Arab countries average of 1 per 1,000–2,000 people. In 2008,

[3] See Nkwelo (2008). Naturally, the University of Khartoum – the biggest in Sudan – has the most postgraduate students and one would expect that its science faculties (Engineering and Architecture, Mathematical Sciences, Sciences, Dentistry, Medicine, Medical Laboratory Sciences, Pharmacy, Agriculture, Animal Production, Forestry and Veterinary Science) contribute significantly to the high numbers of postgraduate students (Nkwelo 2008).

[4] See Hassan, A. O. (2009a).

Table 6.5 Size of research and development workforce and total R&D personnel by sector of employment (FTER) in Sudan (1999–2005)

Total R&D Personnel	1999	2000	2001	2002	2003	2004	2005
Total R&D Personnel	18,604	18,808	19,772	23,726	14,923	15,333	16,050
Total researcher	9,100	9,200	9,340	11,208	7,300	7,500	7,850
Total Researchers (HC) per million inhabitants	262	260	258	304	224	225	230
Total Technicians and equivalent staff	3,674	3,714	4,641	5,569	2,947	3,028	3,170
Total Technicians (HC) per million inhabitants	106	105	128	151	90	91	93
Other supporting staff (FTER and HC)	5,830	5,894	5,791	6,949	4,676	4,805	5,030
% Distribution total R&D personnel							
Total researcher	49%	49%	47%	47%	49%	49%	49%
Total researchers (HC) per million inhabitants	1%	1%	1%	1%	2%	1%	1%
Total technicians and equivalent staff	20%	20%	23%	23%	20%	20%	20%
Total technicians (HC) per million inhabitants	1%	1%	1%	1%	1%	1%	1%
Other supporting staff (FTER and HC)	31%	31%	29%	29%	31%	31%	31%
R&D personnel							
Total R&D personnel	18,604	18,808	19,772	23,726	14,923	15,333	16,050
Total male	15,844	16,017	16,730	…	12,982	13,330	13,958
Total female	2,760	2,791	3,042	…	1,941	2,003	2,092
M/F	15,844/2,760	16,017/2,791	16,730/3,042	…	12,982/1,941	13,330/2,003	13,958/2,092
% of female in total	14.8 %	14.8 %	15.4 %	…	13.0 %	13.1 %	13.0 %
Total							
Business enterprise	…	…	180	475	…	…	…
Government	…	…	3,490	4,745	…	…	…
Higher education	…	…	16,102	…	…	…	…
Share in total							
Business enterprise			1 %	9 %			
Government			18 %	91 %			
Higher education			81 %	…			
Researcher							

(continued)

Table 6.5 (continued)

Total R&D Personnel	1999	2000	2001	2002	2003	2004	2005
Total researcher	9,100	9,200	9,340	11,208	7,300	7,500	7,850
Total Male	6,346	6,416	6,510	...	5,746	5,856	6,186
Total Female	2,754	2,784	2,830	...	1,554	1,644	1,664
M/F	6,346/2,754	6,416/2,784	6,510/2,830	...	5,746/1,554	5,856/1,644	6,186/1,664
% of female (%) in total	30.3 %	30.3 %	30.3 %	...	21.3 %	21.9 %	21.2 %
Total							
Business enterprise	...	...	50	224	...	...	...
Government	...	...	1,168	2,242	...	...	...
Higher education	...	...	8,122	8,742	...	...	...
Share in total							
Business enterprise			1 %	2 %			
Government			13 %	20 %			
Higher education			87 %	78 %			

Sources: UNESCO R&D statistics (2006)

the number of researchers per 10,000 people in Sudan was only 0.7, which is very low compared to Arab countries (1.7) and developed countries (75).[5]

When comparing Sudan with the Arab countries, we find that the latter show better performance than Sudan in terms of the total number of scientists and engineers in R&D. In terms of the human resources devoted to R&D (defined by the total number of full time equivalent researchers (FTER)[6] and their distribution within R&D organisations), we find that the majority of researchers (FTER) are employed by the higher education and government public sectors. In Sudan the percentage share of researchers (FTER) in the higher education is estimated to be 87 % and 78 % in 2001 and 2002 respectively. Next to the university sector, it is the public or government sector that has the second largest percentage share of researchers (FTER): at 13 % and 20 % in 2001 and 2002 respectively, while the private sector accounts for only 1 % and 2 % of total researchers (FTER) in 2001 and 2002 respectively in Sudan. These results for the case of Sudan seem consistent with the results in Nour (2004, 2005) regarding distribution of researchers (FTER) by employment institutions, which implies greater dependence in the public sector on the employment of researchers (FTER) and small contribution of the private sector in the employment of researchers (FTER). Again, it is the lack of incentives for private sector institutions to hire that leads to this disparity.

Moreover, despite the growth in the size of the university workforce and research and development personnel and the number of academic staff according to academic professional positions in higher education institutions, data from the Ministry of Higher Education and Scientific Research indicates clearly that at all the institutions, males still strongly dominate positions, with virtually no female representation at some institutions in Sudan. Furthermore, UNESCO information on the numbers of R&D workforce in Sudan from 1999–2005 indicates a very low number of female personnel even though there has been an increase over the years. Moreover, the share of females is not only low but also declined from 14.8 % in 1999 to 13 % in 2005 in total R&D personnel, and from 30.3 % in 1999 to 20.2 % in 2005 in total researchers. Despite the increase in the number of researchers and technicians, their respective shares in total R&D personnel over the period 1999–2005 remained at 49 % and 20 % respectively (see Table 6.5 below). Moreover, the distribution of staff and human resources in some institutions' units in the MOST over the period 2003–2008 indicates that the share of researchers in the workforce increased from 14 % in 2003 to 20 % in 2008, whereas, the share of technicians declined from 31 % in 2003 to 20 % in 2008 and the share of labour increased from 54 % in 2003 to 60 % in 2008. This implies that the majority of the workforce is labour that constitutes about 60 %, whereas the share of both researchers and technicians together constitutes only 40 % of the total workforce employed in MOST over the period 2003–2008 (see Table 6.6 below).

---

[5] See Elamin (2009).

[6] The concept of full time equivalent researchers (FTER) is adopted by UNESCO statistics on R&D personnel.

Table 6.6 Institutions units and staff distribution in the Sudan Ministry of Science and Technology (2003–2008)

(a) Human resources (2008)

		Human resources (2008)								
		Total number (2008)						% share in total (2008)		
		Researchers								
Institution	No. of work force	PhD	MSc.	BSc.	Total	Technician	labour	Researchers	Technician	labour
NCR	746	65	131	57	253	162	331	34 %	22 %	44 %
ARC	3,369	136	241	115	492	251	2,626	15 %	20 %	65 %
ARRC	1,373	78	99	174	351	369	653	28 %	20 %	52 %
IRCC	286	14	65	19	98	36	152	34 %	13 %	53 %
NAEC	311	9	97	–	106	114	91	34 %	37 %	29 %
SAS	112	5	11	–	16	75	21	14 %	67 %	19 %
CL	116	3	9	34	46	26	44	40 %	22 %	38 %
SMA	603	1	7	43	51	370	182	8 %	61 %	30 %
ESRD	94	6	23	2	31	19	44	33 %	20 %	47 %
MOST(HQ)	186	4	22	–	26	10	150	14 %	5 %	81 %
Total	7,196	321	705	444	1,470	1,432	4,294	20 %	20 %	60 %

(b) Human Resources (2003)

	Total number (2008)				% share in total (2008)		
Institutions	Researchers	Technician	labour	Total	Researchers (%)	Technician (%)	Labour (%)
NCR	318	373	238	929	34	40	26
ARC	423	1,434	3,455	5,312	8	27	65
ARRC	282	548	608	1,438	20	38	42
IRCC	82	94	70	246	33	38	28
NAEC	27	48	24	99	27	48	24
ESRC	15	7	18	40	38	18	45
ERC	35	56	27	118	30	47	23
Total	1,182	2,560	4,440	8,182	14	31	54

Source: Ministry of science and technology annual reports: Various issues (2003–2009)

In addition, there are fewer human resources in S&T in Sudan, and both the Gulf and Mediterranean Arab countries compared to more developed countries, shown in Table 6.7 above. Sudan and the Arab countries score poorly compared to Korea and Singapore for the Harbison Myers Index,[7] technical enrolment index, engineering enrolment index, gross enrolment ratio at tertiary education and the share of tertiary students in science, mathematics and engineering.[8] Hence, these findings imply the insufficiency of human resources necessary for the promotion of R&D and S&T in Sudan.

6.3.3 Science and Technology Output Indicator and Impact

As we explained briefly in Sect. 6.2, the literature distinguishes between S&T outputs, which can be measured in terms of publications and patents, and S&T impact, which can be measured in terms of economic growth. This section discusses S&T output as measured by the number of patent filings and scientific publications (in the international literature) but discusses S&T impact as measured only by the share of high-technology manufacturing exports. Owing to limitations concerning data availability it is not possible to address the impact of technological development on economic/productivity growth in much detail.

We integrate the findings in Sect. 6.3, concerning the general economic characteristics of the Sudanese economy, with those of Sect. 4.2 regarding S&T input indicators. Using a systematic approach we assess S&T performance in terms of inputs and the economic system as a whole. Our analysis aims to explain the connection between the S&T system, S&T profile and the economic or productive structure of Sudan. For example, Table 6.1 shows that for both patent numbers and the percentage of high-technology exports, Sudan and the Arab Gulf and Mediterranean countries are substantially lagging behind the advanced and leading developing countries.

In our view, which is backed up by general S&T literature, the weakness of the S&T base in Sudan and the Arab regions should be interpreted not only in terms of a lack of appropriate inputs but also in relation to a poor economic system as a whole. Measuring the strength of the economic and welfare systems using income per capita implies that Sudan shows low per capita income and also exhibits low S&T activity; this seems consistent with the idea that strong S&T is necessary for economic growth and development. Prior to the heavy dependence on oil, the poor economic structure in combination with inadequate resources devoted to

[7] According to Lall (1999): "Harbison Myers Index is the sum of secondary enrolment and tertiary enrolment times five, both as a percentage of age group. Technical enrolment index is tertiary total enrolment (times 1000) plus tertiary enrolment in technical subjects (times 5000), both as a percentage of population. Engineering skills index is the same as the previous index, with tertiary enrolments in engineering instead of enrolment in technical subjects". See Lall (1999), p.52.

[8] See also Muysken and Nour (2006) and UNDP-AHDR (2003).

Table 6.7 Skills indicators in the Sudan Arab and World countries (1992–2000)

	Skill indices (1995)			Gross enrolment ratio (%) at tertiary education	Share tertiary students in science, math and engineering
Country	Harbison Myers Index[a]	Technical enrolment index[a]	Engineering enrolment index[a]	1998[b]	1994–1997[b]
Arab Gulf (GCC)					
Bahrain	Na	Na	Na	25.2	NA.
Kuwait	19.10	36.49	30.57	21.08	23
Oman	8.95	5.35	4.44	NA	30
Qatar	Na	Na	Na	27.66	NA.
Saudi Arabia	13.45	18.96	14.42	20.71	18
UAE	12.20	7.51	5.70	12.10	27
Average Gulf countries	13.425	17.0775	13.7825	21.35	24.5
Arab Mediterranean					
Algeria	11.65	31.14	21.55	15	50 %
Egypt	16.45	16.10	13.87	39	15 %
Lebanon	21.60	46.89	34.60	36	17 %
Morocco	9.55	23.73	11.46	9	29 %
Syria	13.35	23.47	17.67	6	31 %
Tunisia	12.55	24.49	16.15	17	27 %
Average Mediterranean	14.19	27.64	19.22	20.33	28.17 %
Other Arab countries					
Libyan Arab Jamahiriya	Na	Na	Na	56	Na.
Jordan	18.55	39.27	27.64	29[1]	27
Iraq	Na	Na	Na	13	Na
Sudan	2.80	3.50	2.92	7	Na
Yemen	4.45	4.60	4.17	11	6
Mauritania	3.55	5.28	3.74	6	Na
Average all Arab countries	12.01	20.48	14.92	19.636	12.091
Other advanced countries					
Norway	38.85	73.52	60.25	64.83	18 %
Sweden	34.45	64.50	49.94	62.3	31 %
Canada	62.05	103.02	86.01	58.93	Na.
USA	50.25	88.10	68.98	75.66	Na.
UK	37.55	68.69	49.83	58.39	29 %
Australia	50.55	112.70	84.29	63[6]	32 %
Japan	30.05	63.54	63.54	44	23 %
Korea, Republic of	36.10	132.06	113.83	71.69[1]	34 %
Iran	14.30	37.58	30.03	10[1]	36 %

Sources: [a]Lall (1999), [b]UNDP (2002), Human Development Report (2002)
(1) Data refer to the year 1999

S&T development leads to poor S&T performance in Sudan compared to advanced and developing world countries. After the heavy dependence on oil, and despite the growing wealth from oil Sudan still lacks well-defined, targeted plans, policies and proper incentives to promote S&T performance. Sudan needs to benefit from the experience of other countries, for instance, other Arab countries, for while the Gulf countries perform better than the Mediterranean countries in economic terms they lag behind in measurements of S&T performance. Therefore, the big wealth from oil, far from contributing to the improvement of S&T performance in the Gulf may actually hinder it as it masks the need to develop incentives and effective policies to enhance S&T development.

6.3.3.1 Scientific Publications

As for research output and scientific publications, as an output indicator the number of scientific publications depends on input financial and human resources devoted to S&T[9]. The international standard rate is 70–80 researchers for every 10,000 people; currently in Sudan the rate is 0.2. This reflected negatively on the number of publications per researcher per year, which is 0.03 in average compared to the international rate of 2 papers for each researcher.[10]

In terms of research outputs and publications, according to the Institute for Scientific Research, Sudan has produced quite a number of publications between the years 1994–2004, even though the numbers are very low for a country with so many tertiary and research institutions. The publications of selected research institutes involved in R&D as cited by ISI gives the impression that Sudan has a strong inclination towards health related research, followed by agricultural research, and to some extent nuclear related research. Table 6.8 shows that the number of scientific publications for Sudan and the Arab countries grew over the period 1970–1995; Egypt and Saudi Arabia show the largest overall number. Sudan performed better than some Arab countries, but meanwhile, perform less than Egypt, Saudi Arabia, Kuwait, Algeria, Morocco, Tunisia, Jordan and Iraq in terms of the number of scientific publications, which could be a consequence of the superiority of these countries compared to Sudan in terms of most of the S&T indicators: total expenditure on both education and R&D; number of R&D employees; and number of R&D scientists and engineers. Moreover, Table 6.9 indicates that in terms of the average share of Sudan and African countries in world share of Institute for Scientific Information- science and engineering papers

[9] The OECD (1997) report indicates that prizes awarded to individual scientists is an extreme indicator of S&T performance and is one way of measuring R&D output. Of all scientific prizes, the Nobel Prizes for science, which have been awarded to scientists in the fields of chemistry, physics and medicine/physiology since 1901, are probably the most prestigious. The Arab countries have only received one Nobel Prize between them: in 1999 an Egyptian scientist received the Nobel Prize for chemistry.

[10] See Sudan Ministry of Science and Technology (2008), p.65

Table 6.8 Change in R&D spending and the Number of S&T publications (papers published in referred international journals) in Sudan and the Arab countries (1970–1997)

	Enrolment in tertiary[b]	Publications [c]		Share of public spending on education % GDP	Percentage change in GDP Per capita	Percentage change in R&D Spending	Total R&D Spending (US$ Million)[I]	Researchers (FTER)[i]
Country	1998	(1970–1975)[b]	(1990–1995)[c]	(1995–1997)[a]	(1992–1996)[a]	(1992–1996)[a]	1996	1996
Bahrain	21	Na	453	3	−3.7	94.7	3.7	143
Kuwait	21	148	1936	5	32.3	42.2	67.1	1,130
United Arab Emirates (UAE)	10	1	579	1.9	196.4	0.9	10.9	313
Qatar	23	Na	377	3.6	−32.4	27.9	5.5	74
Average (total) high income	18.75	149	(3,345)	3.375	192.6	165.7	87.2	1,660
Oman	7	1	466	4.2	−9.6	83.1	10.8	382
Saudi Arabia	22	126	8,306	9.5	−5.0	49.6	196.1	2,421
Algeria	15	338	1,431	5.3	−13.8	6.0	35.6	2,588
Egypt	38	3,261	12,072	3.7	49.1	45.6	227.5	37,073
Lebanon	45	743	500	2.9	319.7	27.6	7.5	444
Morocco	10	96	2,418	5.1	12.3	5.9	74.9	7,329
Syrian Arab Republic	6	38	471	4	25.5	64.6	24.2	2,105
Tunisia	23	145	1,832	6.8	37.2	75.2	28.9	1,132
Palestine	31	Na	51	Na	Na	Na	Na	Na
Libyan Arab Jamahiriya	58	96	348	2.7	10.3	26.1	16.9	903
Jordan	31	61	1,936	4.6	27.8	36.4	20.6	1,471
Iraq	14	380	931		4.7	−16.6	27.6	2,840
Djibouti	1		Na	3.5	Na	Na	Na	Na
Average (total) medium income	23.155	5,285	(30,762)	4.75455	458.2	403.5	670.6	58,688
Sudan	7	426	690	0.9	−60.3	13.6	10	2,047
Yemen	11	4	155	10	−64.8	56.1	10.3	1,041

(continued)

Table 6.8 (continued)

Country	Enrolment in tertiary[b]	Publications [c]		Share of public spending on education % GDP	Percentage change in GDP Per capita	Percentage change in R&D Spending	Total R&D Spending (US$ Million)[I]	Researchers (FTER)[i]
	1998	(1970–1975)[b]	(1990–1995)[c]	(1995–1997)[a]	(1992–1996)[a]	(1992–1996)[a]	1996	1996
Mauritania	4	Na	27	3.6	Na	Na	4.3	509
Somalia	Na	1	79	Na	Na	Na	Na	Na
Comoros	1		Na	3.8	Na	Na	Na	Na
Average (total) low income	5.75	431	951	4.575	−125.1	69.7	24.6	3,597
Average (total Gulf)	17.333	276	(12,117)	4.5333	178	298.4	294.1	4,463
Average (total) Mediterranean	28.25	4,621	(19,123)	4.3571429	430	224.9	415.5	51,574
Average (total) Arab region	15.885	5,865	(35,058)	4.23485	Na	Na	782.4	63,945

[a]ESCWA –UNESCO (1998)
[b]Qasem (1998)
[c]Zahlan (1999a, b).

Table 6.9 The average share of Sudan and African countries in world share of ISI-listed S&E papers (2003–2007)

	Africa	World share (%)
1.	South Africa	0.372
2.	Egypt	0.272
3.	Tunisia	0.111
4.	Morocco	0.089
5.	Nigeria	0.088
6.	Algeria	0.074
7.	Kenya	0.054
8.	Cameroon	0.029
9.	Tanzania	0.029
10.	Ethiopia	0.026
11.	Uganda	0.024
12.	Ghana	0.019
13.	Senegal	0.018
14.	Zimbabwe	0.016
15.	Burkina Faso	0.012
16.	Cote d'Ivoire	0.012
17.	Botswana	0.011
18.	Malawi	0.011
19.	Madagascar	0.011
20.	Sudan	0.010
	Rest of Africa (33 countries)	0.096
	Total Africa	1.383

Source: Third World Academy of Sciences (TWAS) (May 2008) (See M.H.A. Hassan (2009b)

(ISI-listed S&E papers) over the period 2003–2007, of the African countries, South Africa has the best percentage share of total world scientific publications; it is followed by Egypt, Tunisia and Morocco respectively. However, the average share of Sudan is very low, for instance, Sudan is ranked 20 after Madagascar, and contributed only about 0.01 % of world share of ISI-listed S&E papers over the period 2003–2007.[11] This implies the problem of knowledge gaps even between Sudan and some African countries.

6.3.3.2 Patent Applications

Table 6.1 above shows the low number of patent applications made by Sudan and the Arab countries compared to advanced and leading developing countries like Singapore, Korea and China. In light of our earlier findings, this can be attributed to

[11] The Institute for Scientific Information (ISI) offered bibliographic database services, it maintains citation databases covering thousands of academic journals, indexing service the Science Citation Index (SCI), as well as the Social Sciences Citation Index (SSCI), and the Arts and Humanities Citation Index (AHCI), ISI-listed S&E papers offers citation of science and engineering papers.

Sudan and the Arab countries' low percentage share of GDP spent on R&D and the small number of scientists and engineers in R&D. The low number of patent applications implies a low level of innovative activities in Sudan and the other Arab countries compared to both advanced and developing countries.

Regarding the use of the number of patent applications as a measure for S&T output indicators, Nour (2004, 2005a, b) shows the low number of patent applications over the period 1990–1999 and hence S&T output indicators, across all the Arab countries (168), Arab Gulf countries (150), Arab Mediterranean countries (41) compared to advanced and leading developing countries like Singapore (27), Korea (931) and China (793).[12] Nour (2004, 2005a, b) finds that the poor application to patent rate can be attributed to the low percentage share of spending on R&D to GDP and the number of scientists and engineers in R&D in the Arab countries compared to advanced and developing countries like Singapore, Korea and China. The low patent applications imply the low innovative activities across the Arab countries compared to both advanced and leading developing countries like Singapore, Korea and China. In addition, Table 6.10 below shows the number of patent applications made between 2007–2010 in Sudan and the Arab countries, by residents and non-residents of Sudan and the Arab countries. During that period residents made fewer patent applications than non-residents in all Arab countries. Among the Arab countries in 2007–2010, the highest number of patent applications was filed by residents in Egypt followed by either Saudi Arabia or Morocco, followed by Algeria and Jordan; the highest number of patent applications was filed by non-residents in Egypt followed by Morocco, Algeria, Saudi Arabia and Jordan. In 2007 the lowest numbers of patent applications were filed by residents and non-residents in Yemen and Sudan. The low number of patent applications from residents than those of non-residents of Sudan and all the Arab countries is consistent with the findings in the literature, which indicate that in developing countries patent applications made and patents held by residents of developing countries (domestic applications or patents) are few. Patents are overwhelmingly foreign resident-owned. In most developing countries, domestic applications accounted only for 1–8 % of total applications. Thus, the role of the patent system is less visible to domestic users of the patent system in developing countries. The reason for the low level of patenting in developing countries by their nationals and residents can be explained by a number of reasons, including non-use of the system by universities and local research institutions.[13] The low number of

---

[12] See for example, US Patent and Trademark office website: www.uspto.gov, Accessed 9 September 2005.

[13] See for instance, 'WIPO Patent Agenda Study' by Getachew Mengistie, the Ethiopian Intellectual Property Office, A/39/13 Add.1 available at: http://www.wipo.int/documents/en/document/govbody/wo_gb_ab/doc/a_39_13add1.doc, Accessed 02 February 2008.

Table 6.10 Patents for Sudan compared to selected Arab Countries and selected African countries (1995–2010)

(a) Patents for selected Arab countries (2007–2010)

	(a) Patent applications, nonresidents				(b) Patent applications, residents			
	2007	2008	2009	2010	2007	2008	2009	2010
Algeria	765			730	84			76
Egypt	1,589	1,649	1,452	1,625	516	481	490	605
Jordan	507	535	446	429	59	50	60	45
Morocco	782	834	856	882	150	177	135	152
Saudi Arabia	642			643	128			288
Sudan	13				3			
Yemen	24			55	11			20
Total Arab	4,322	3,018	2,754	4,364	951	708	685	1,186

(c) Patents for selected African countries (pre 1995–2005)

	Pre 1995	1995	1996	1997	1998	1999	2000	2001	2002	2003	2004	2005	All years
South Africa	2200	123	111	101	115	110	111	120	113	112	100	87	3,694
Zimbabwe	42	1	1	0	0	0	0	1	1	1	1	1	53
Mali	25	0	0	0	0	0	0	0	0	0	0	0	25
Tunisia	14	0	0	1	0	0	0	0	1	0	1	1	19
Tanzania	9	0	0	0	0	0	0	1	0	1	0	0	12
Sudan	7	0	0	0	0	0	0	0	0	0	0	0	7
Libya	4	0	0	0	0	0	0	0	0	0	0	0	4

Sources: (a) The World Bank- World Development Indicators Database (2012). (b) The World Bank- World Development Indicators Database (2012). (c) UNESCO (2006)

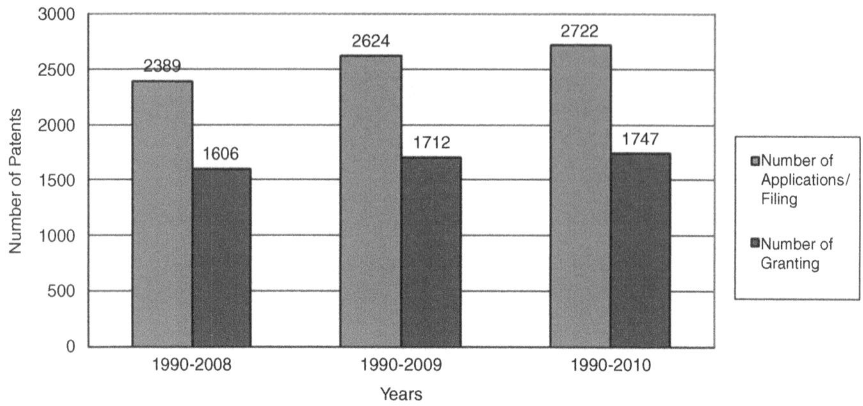

Fig. 6.3 Patents applications (Filing) and granting in Sudan at home level (1990–2010) (Source: Unpublished data and statistics from the General Registrar of IPR Sudan Office (2010))

patents filed by residents of Sudan and the Arab countries can be related to low S&T activity in the countries. The low number of patents recorded by non-residents, however, needs a different interpretation. It is partially because there is a lack of adequate patent legislation, but more importantly it is also due to the lack of an economic structure within which to take advantage of patents. Foreign companies will only register a patent in a country if they fear that a local competitor might exploit their technology without paying for it. Therefore the low number of patents filed by non-residents in Sudan implies that Sudan lacks industries that are internationally competitive, which can also be interpreted in terms of there being a poor economic structure. Moreover, Table 6.10 shows that Sudan and the African countries together have filed far fewer patents than South Africa; the highest numbers of patent applications were made by South Africa, followed by Zimbabwe, Mali, Tunisia, Tanzania, Sudan and Libya. According to the United States Patent and Trademark Office (USPTO) report, Sudan produced only seven patents in about 40 years, with no patents at all in the period 1992–1995 and this puts it much lower than most African countries in terms of patents (see Table 6.10 below).

The low number of patents is probably due to Sudan's insufficient science and technology infrastructure. For instance, Fig. 6.3 indicates the growth in both the filling and granting of patents over the period 1990–2010 at the home level, but this should not hide the fact that the granting of international patents is very limited. For instance, Fig. 6.4 below shows the limited international application for Sudan's application for PCT International patent by residents during the period 2003–2007. In addition, according to IPR-Sudan Profile (2003) patents applications filed and/or registered by ARIPO imply that applications by residents are less than by non-residents in 2001 and 2002 respectively.

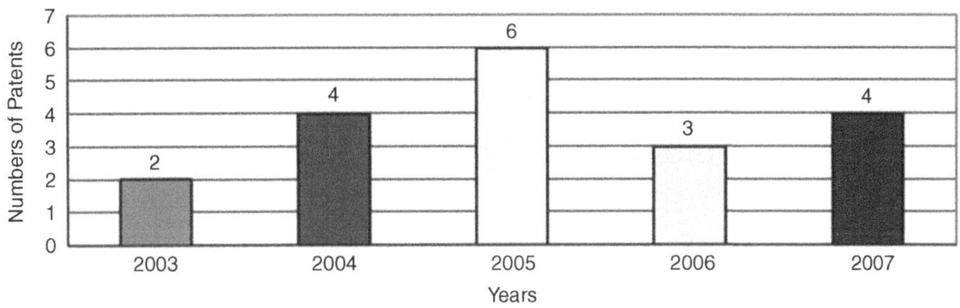

Fig. 6.4 The application for Sudan's application for PCT international patent by residence (2003–2007) (Source: WIPO (2007) Statistics on Applications for PCT)

6.3.3.3 Share of High-Technology Manufacturing Exports

When comparing the average share of exports of high-technology goods manufactured, our findings in Table 6.1 above indicate that in 2001 the highest share of high-technology exports was made by Morocco, followed by Sudan and then other Arab countries. According to Table 6.1 above, Sudan is similar to the Arab countries in having a low share of high-technology manufacturing exports compared to advanced and leading developing countries. In addition, the share of high-technology manufactured goods in Sudan and all the Arab countries in 1995–1997 is well below that of the world average or the corresponding figures for Brazil, Korea, Latin America and the Caribbean, Mexico and Singapore.[14] This can be explained in relation to our earlier findings concerning Sudan's inadequate economic structure, poor spending on R&D, low number of scientists and engineers in R&D and low patent filings.

6.3.3.4 Productivity Growth

In terms of S&T impact as measured by economic growth, Table 6.11 shows significant increase in annual growth rate for average GDP per capita in Sudan during the periods 1975–2001 and 1990–2001 and the average real GDP growth rate during the period 1995–2000 in Sudan is higher than the average for the Arab countries. However, during 1999–2001, Sudan shows a slight drop in the trend of real annual GDP growth rate, whereas the rate of Sudan is higher than the average for developing countries. Sudan is experienced rapid economic growth followed by slight slow down, that most probably due to its heavy dependence on oil (see Fig. 6.1 above).

[14] See for instance, Haddad (2001), Lall (1999).

Table 6.11 Real GDP growth and GDP per capita annual growth rates in the Sudan and Arab countries (1975–2001)

Country	GDP per capita annual growth rate (%)[a]		Real annual GDP growth (%)[b]			
	1975–2001	1990–2001	1995–2000 Average	1999	2000	2001
Sudan	0.8	3.2	6.3	6.9	6.9	5.3
Arab Gulf (GCC[c])						
Bahrain	1.1	1.9	4.3	4.3	5.3	4.8
Kuwait	−0.7	−1.0	3.8	−2.9	2.9	−0.6
Oman	2.3	0.6	3.6	−0.2	5.1	7.3
Qatar	NA	NA	9.4	5.3	11.6	7.2
Saudi Arabia	−2.1	−1.1	1.9	−0.8	4.9	1.2
UAE	−3.7	−1.6	5.7	3.9	5.0	5.1
Total GCC	−0.6	−0.2	4.8	1.6	5.8	4.2
Arab Mediterranean						
Algeria	−0.2	0.1	2.9	2.3	2.8	3.4
Egypt	2.8	2.5	5.3	6.0	5.1	3.3
Lebanon	4.0	3.6	2.3	1.0	−0.5	2.0
Morocco	1.3	0.7	1.9	−0.1	1.0	6.5
Syria	0.9	1.9	3.0	−2.0	0.6	2.7
Tunisia	2.0	3.1	5.1	6.1	4.7	5.0
Total Mediterranean	1.8	2.0	3.4	2.2	2.3	3.8
Arab State	0.3	0.7	3.9	2.4	4.1	3.8
Developing countries	2.3	2.9	5.3	3.9	5.7	4.0

[a]UNDP (2003).
[b]IMF (2002).
[c]GCC Gulf Cooperation Council.

6.3.3.5 Demand for and Supply of Technologies, Technology Infrastructures, and Technology Achievement Index

We measure the demand for and supply of technologies in Sudan using the measurement of demand for and supply of technologies in the Gulf countries discussed in Muysken and Nour (2006).[15] Our results show that on the demand side when using the share of chemicals, manufactured goods, machinery and equipment, transport equipment, petroleum products in total imports as a measure of the demand for imported technology or dependence on foreign technologies, we

[15] According to the UNDP (2001), the technology achievement index (TAI) focuses on four dimensions of technological capacity that are important for reaping the benefits of the network age. TAI includes: (1) Creation of technology as measured by the number of patents granted per capita and receipt of royalty and licenses fees from abroad; (2) Diffusion of recent innovations as measured by diffusion of Internet and export of high and medium technology products as a share of all exports; (3) Diffusion of old innovations as measured by diffusion of telephone and electricity; and (4) Human skills as measured by mean years of schooling and gross enrolment ratio of tertiary students enrolled in science, mathematics and engineering (UNDP 2001).

find heavy dependence on imported technology or dependence on foreign technologies in Sudan. It may be interesting to complement our analysis by also examining the supply side. We measure the supply side by multiplying the manufactures/GDP ratio taken from the Central Bank of Sudan Annual Reports Issues (2000–2002), by value added in machinery and transport equipment as a percentage of value added in total manufactures using World Development Indicators (WDI) (2010) data for 2000 and Sudan Ministry of Industry (2005) Comprehensive Industrial Survey data for (2001); the result is value added in machinery and transport equipment/GDP, which we use as a measure of specialisation in production related to technology.[16] When using this measure, our results show a low technological specialisation in Sudan, which is most probably attributed to lack of both basic and high-technology infrastructure (BTI and HTI) (see Tables 6.12 and 6.13 below).[17] On average both the BTI and HTI for Sudan are poor. Overall, poor BTI is to blame for the low HTI (Rasiah 2002). Consequently, due to lack of both basic and high-technology infrastructure and the low technological specialisation, Sudan shows poor performances in terms of technology achievement index. According to the UNDP (2001) HDI classification of world countries according to technology achievement index, Sudan is classified as being a marginalised adopter of new technologies; amongst marginalised adopter countries in terms of the technology achievement index, Sudan is ranked 71 and placed between Tanzania and Mozambique. Sudan's poor performance lags far behind the world's advanced and leading developing countries which are either leaders or potential leaders in technology. In fact, Sudan also lags behind the countries classified as being dynamic adopters of new technologies in both Arab and African regions, notably: Tunisia (51); Syria (56); Egypt (57); Algeria (58); Zimbabwe (59); Senegal (66); Ghana (67); Kenya (68); and Tanzania (70) (see Table 6.13 below).

[16] Since the recent data from the WDI (2010) is available only for 2000, we therefore use an additional and alternative set of indicators from Sudan Ministry of Industry (2005) 'Comprehensive Industrial Survey Data for 2001'. The observed differences in both measures are most probably because of the differences in the definitions used by both sources. For instance, the only available data from the Sudan Ministry of Industry (2005) 'Comprehensive Industrial Survey Data for 2001' is based on the International Standard Industrial Classification of all economic activities according to ISIC 1968 rather than ISIC Rev 3.

[17] Rasiah (2002) defines basic technology infrastructure (BTI) as weighted proxies representing basic education (enrolment in primary schools), health (physicians per thousand people) and communications (main telephone lines per thousand people), and defines high-technology infrastructure (HTI) as weighted proxies representing R&D investment and R&D scientists and engineers per million people. Rasiah also argues that BTI is an essential but not sufficient condition for economies to achieve advanced technological capacity; the incidence of economies generating innovation is higher when they also have the high-technology support institutions. The lower the BTI implies the lower the capacity and resources for high-technology development.

Table 6.12 Demand for and supply of technologies in the Sudan (1992–2010) (%)

	Demand for technologies			Supply of technologies
	(1)	(2)	(3)	(4)
1992	39	56	84	
1993	47	62	84	
1994	49	56	76	
1995	52	62	78	
1996	51	60	80	
1997	48	59	77	
1998	57	67	80	
1999	50	60	73	
2000	54	64	71	30
2001	54	67	73	8
2002	57	67	72	
2003	58	72	78	
2004	59	77	80	
2005	61	78	83	
2006	61	80	85	
2007	65	82	85	
2008	61	73	80	
2009	62	74	77	
2010	56	67	71	
1992–2010	55	68	78	19

(1) The share of chemicals, manufactured goods, machinery and equipment in total imports (2) The share of chemicals, manufactured goods, machinery and equipment, transport equipment in total imports (3) The share of chemicals, manufactured goods, machinery and equipment, transport equipment, petroleum products in total imports (4) The supply side refers to technological specialization and is measured by the share of value added in machinery and transport equipment/GDP

Source: (a) The demand for technology is calculated from the Sudan Ministry of Foreign Trade and Central Bank of Sudan Annual Foreign Trade Statistical Digest various issues (1992–2010) (b) the supply of technology is calculated from the Central Bank of Sudan Annual Reports Issues (2000–2002), the World Bank-WDI-World Global Development Finance (2010) data for (2000) and the Sudan Ministry of Industry (2005) the Comprehensive Industrial Survey data for (2001)

6.3.3.6 S&T, R&D, Economic Development, Adaptation to Foreign Technologies and Development of Local Technologies

Based on the above findings, this section uses new survey data based on primary data and 25 face-to-face interviews with the official policy makers and experts in the government and academic staff in the public and private universities.[18]

[18] The interviews were conducted with officials and experts (20 %), and academics in the public (60 %) and private (20 %) universities. The interviews were conducted with academic staff in the fields of science (36 %), engineering (36 %) and social sciences (8 %) including both males (80 %) and females (20 %). The distribution of the interviewed institutions includes public universities represented by Khartoum University (60 %), private universities represented by University of Medical Sciences and Technology (20 %), Ministry of Science and Technology (12 %) and Ministry of Higher Education and Scientific Research (8 %).

Table 6.13 Basic and high technology infrastructure and technology achievement index in Sudan (2004–2010)

	2004	2005	2007	2009	2010
(a) Basic Technology Infrastructure (BTI)[a]					
Basic education (enrolment in primary schools)[a]	3,966,944	4,299,737	4,785,952	5,800,000	
Secondary education (enrolment in Secondary schools)[a]	546,305	637,812	636,156	753,000	
Health (Physicians Per thousand people) of Population[a, d]	20	22.6	29.9	30	
Communications (main telephone lines per thousand people)[a, d]	29	18		30	
(b)High Technology Infrastructure (HTI)[b]					
R&D investment: R&D expenditure as % of GDP (2004–2005)[e]	0.29	0.28			
R&D scientists and engineers per million people[b] (1996–2000)[b]	225				
(c)Technology Achievement Index [c]					
(TAI) TAI rank value[c]	0.071				
Diffusion of recent: innovations[c]					
High- and medium technology exports (as % of total goods exports) 1999[c, f]	0.4		0.6	29	
Diffusion of old innovations[c]					
Telephones (mainline and cellular, per 1,000 people)[c, f]	1,029	570	345	370	375
Electricity consumption(kilowatt-hours per capita)[c]	67	79	96	114	
Human skills[c]					
Mean years science of schooling(age 15 and above) 2000[c]–2011 [d]	2.1				3.1
Gross tertiary enrolment ratio (%) (1995–1997)[c]	0.7				5.9

[a]Sudan Central Bureau of Statistics (2010).
[b]UNDP (2003).
[c]UNDP (2001).
[d]UNDP (2011).
[e]UNESCO (2006).
[f]WB-WDI (2012).

The main purpose of this survey is to collect primary data to examine the causes and consequences of poor R&D activities, to examine the main factors hindering and those contributing towards the promotion of R&D and then to provide some recommendations to improve R&D and hence S&T development in Sudan.

As for the importance of R&D the majority of the respondents indicate the importance of R&D in satisfying the needs for economic development, followed by development of local technologies and finally adaptation to imported foreign technologies.[19] As for the contribution of R&D, the majority of the respondents

[19] As indicated by 96 %, 84 % and 76 % of the respondents respectively.

indicate the contribution of R&D in satisfying the needs for economic development, followed by adaptation to imported foreign technologies and finally development of local technologies.[20] When comparing the points of views of the different respondents we find that from the perspective of the private universities, the importance of R&D is viewed with high importance compared to both public universities and officials and experts. However, from the perspective of the private universities, the contribution of R&D is still susceptible, especially with regards to the role of R&D in the development of local technologies; by contrast the views of the public universities, official and experts seem to be somewhat optimistic regarding the role of R&D (see Table 6.14 below).

Regarding the main problems hampering the important contribution of R&D in satisfying the needs for economic development, development of local technologies and adaptation to imported foreign technologies, the majority of the respondents indicate the lack of finance to cover the high costs of R&D as the main problem.[21] Moreover, the lack of human resources (researchers and qualified worker in R&D fields) is also mentioned but of somewhat less importance.[22] When comparing the points of views of the different respondents we find that the views of the private universities, public universities, official and experts seem to be consistent and in agreement with regards to the serious problem of the lack of finance in hampering R&D; from the perspective of both private universities and officials and experts, the importance of the lack of finance in hampering R&D for adaptation to imported foreign technologies is viewed with high importance compared to public universities. However, from the perspective of the private universities, the importance of the lack of human resources seems to be somewhat less compared to the opinions of both the public universities and official and experts (see Table 6.15 below).

As for the main problems hindering R&D, the majority of the respondents indicate the lack of finance from public sector and the weak relationship, network and consistency and cooperation between universities and higher education institutions on the one side and the productive sector (agriculture, industry, services) on the other side.[23] This is followed by the other problems such as the lack of finance from the private sector; the lack of management and organisational ability; and the lack of coordination and R&D culture.[24] Finally the less important factors include the lack of favourable conditions and the necessary facilities; the lack of awareness and appreciation of the economic values of R&D; and the lack of human resources (researchers and qualified workers in R&D fields).[25] When comparing the points of views of the different respondents we find that from the perspective of the public universities, the lack of favourable conditions and the

[20] As indicated by 72 %, 56 % and 48 % of the respondents respectively.

[21] As indicated by 100 %, 100 % and 92 % of the respondents respectively.

[22] As indicated by 88 %, 84 % and 88 % of the respondents respectively.

[23] As indicated by 100 % of the respondents.

[24] As indicated by 96 % of the respondents.

[25] As indicated by 92 %, 92 % and 88 % of the respondents respectively.

Table 6.14 The importance and contribution of R&D to satisfy the economic development in Sudan (2010)

Important	Officials and experts (%)	Private universities (%)	Public universities (%)	All (%)
(a) The importance of R&D				
Satisfying the needs for economic development	100	100	93	96
Development of local technologies	80	100	80	84
Adaptation to imported foreign technologies	80	100	67	76
(b) The contribution of R&D				
Satisfying the needs for economic development	80	40	80	72
Development of local technologies	60	20	53	48
Adaptation to imported foreign technologies	20	40	73	56

Source: Own calculation based on Nour (2010), Sudan R&D Survey 2010

Table 6.15 The main problems hindering the role of R&D and contribution to satisfy the economic development in Sudan (2010)

	Officials and experts (%)	Private universities (%)	Public universities (%)	All (%)
Satisfying the requirements of economic development				
Lack of human resources (researchers and qualified worker in R&D fields)	100	80	87	88
Lack of finance to cover the high costs of R&D	100	100	100	100
Development of local technologies				
Lack of human resources (researchers and qualified worker in R&D fields)	100	60	87	84
Lack of finance to cover the high costs of R&D	100	100	100	100
Adaptation to imported foreign technologies				
Lack of human resources (researchers and qualified worker in R&D fields)	100	60	93	88
Lack of finance to cover the high costs of R&D	100	100	87	92

Source: Own calculation based on Nour (2010) "Sudan R&D Survey 2010"

necessary facilities; the lack of awareness and appreciation of the economic values of R&D; lack of management and organisational ability and the lack of coordination and the lack of R&D culture seems to be the less important problems, whilst from the perspective of the officials and experts the less important problems are the

Table 6.16 The main problems of R&D in Sudan (2010)

	Officials and experts (%)	Private universities (%)	Public universities (%)	All (%)
Lack of finance from public sector	100	100	100	100
Lack of finance from private sector	80	100	100	96
Lack of human resources (researchers and qualified workers in R&D fields)	80	60	100	88
Lack of management and organization ability and lack of coordination	100	100	93	96
Weak relationship, network and consistency and cooperation between universities and higher education institutions on the one side and the productive sector (agriculture, industry, services) on the other side	100	100	100	100
Lack of favorable conditions and the necessary facilities	100	80	93	92
Lack of R&D culture	100	100	93	96
Lack of awareness and appreciation of the economic values of R&D	100	80	93	92

Source: Own calculation based on Nour (2010) "Sudan R&D Survey 2010"

lack of finance from the private sector and the lack of human resources (researchers and qualified workers in R&D fields). Finally, from the perspective of the private universities the less important problems are the lack of favourable conditions and the necessary facilities; the lack of awareness and appreciation of the economic values of R&D; and the lack of human resources (researchers and qualified workers in R&D fields) respectively (see Table 6.16 below).

As for the main suggestions and solutions to improve R&D, the majority of the respondents indicate the availability of sufficient finance from public sector; availability of sufficient finance from private sector; offering incentives and motivation and making availability of sufficient human resources (researchers and qualified workers in R&D fields); improvement of management and organisational ability and coordination; improvement and strengthening the relationship, network and consistency and cooperation between universities and higher education institutions on the one side and the productive sector (agriculture, industry, services) on the other side; and improvement of awareness and appreciation of the economic values of R&D.[26] This is followed by other solutions such as the creation of more favourable conditions and offering all the necessary facilities and improvement of R&D culture.[27] When comparing the points of view of the different respondents we find that the views of the private universities, public universities, officials and experts seem to be consistent and in agreement with regards to the suggestions and solutions for improvement of R&D. However, different from the opinions of both the private

[26] As indicated by 100 % of the respondents.

[27] As indicated by 96 % of the respondents.

Table 6.17 The main suggestions for enhancing the development of R&D in Sudan (2010)

Extremely important moderately important	Officials and experts (%)	Private universities (%)	Public universities (%)	All (%)
Availability of sufficient finance from public sector	100	100	100	100
Availability of sufficient of finance from private sector	100	100	100	100
Offering incentives and motivation and making availability of sufficient human resources (researchers and qualified workers in R&D fields)	100	100	100	100
Improvement of management and organization ability and coordination	100	100	100	100
Improvement and strengthen the relationship, network and consistency and cooperation between universities and higher education institutions on the one side and the productive sector (agriculture, industry, services) on the other side	100	100	100	100
Creation of more favorable conditions and offering all the necessary facilities	100	100	93	96
Improvement of R&D culture	100	100	93	96
Improvement of awareness and appreciation of the economic values of R&D	100	100	100	100

Source: Own calculation based on Nour (2010) "Sudan R&D Survey 2010"

universities and official and experts, and from the perspective of the public universities, the suggestions with regards the creation of more favourable conditions and offering all the necessary facilities and the improvement of R&D culture seems to be less important compared to other suggestions (see Table 6.17 below).

6.4 Conclusions

This chapter shows the status of S&T indicators in Sudan. From our analysis it is clear that Sudan lags behind the world's developed and leading developing countries in terms of the same S&T input and output indicators. We explain that the combination of poor S&T inputs/resources together with an inadequate economic system as a whole, results in Sudan producing poor S&T outputs/ performances. Moreover, we find that most R&D and S&T activities and (FTER) employment in Sudan occurs within the public and university sectors, while the private sector and industry make only a minor contribution.

When comparing the same S&T input and output indicators of Sudan with those of the Arab countries and world's other developed and developing countries, our findings indicate that Sudan lags behind in terms of most S&T input indicators (both financial and human resources). That also holds for the average share of high-technology exports, GDP per capita growth, number of scientific publications, level and share in international publications and number of patent filings.

Our findings indicate that despite the important role of R&D in satisfying the needs for economic development, development of local technologies and adaptation to imported foreign technologies. However, the contribution of R&D seems to be constrained mainly by the lack of finance to cover the high costs of R&D as the main problem, moreover, the lack of human resources (researchers and qualified worker in R&D fields) is also mentioned but is of somewhat less importance. We find that the main problems hindering R&D include: the lack of finance from public sector; lack of management and organisational ability; lack of coordination and weak relationships, network and consistency and cooperation between universities and higher education institutions on the one side and the productive sector (agriculture, industry, services) on the other side; lack of R&D culture; lack of finance from private sector; lack of favourable conditions and the necessary facilities; lack of awareness and appreciation of the economic values of R&D; and lack of human resources (researchers and qualified workers in R&D fields) respectively.

Our results show that the main suggestions to improve R&D include: availability of sufficient finance from public and private sectors; offering incentives and motivation and making availability of sufficient human resources (researchers and qualified workers in R&D fields); improvement of management and organisational ability and coordination; improvement and strengthening of the relationships, network and consistency and cooperation between universities and higher education institutions on the one side and the productive sector (agriculture, industry, services) on the other side and improvement of awareness and appreciation of the economic values of R&D. This is followed by the creation of more favourable conditions and offering all necessary facilities and improvement of R&D culture.

Hence, our analysis indicates that in order to improve S&T performance, Sudan needs to invest heavily in both financial and human resources as well as to learn from the lessons of the advanced and developing S&T nations. Such investment can be more effective if they are made according to targeted, well-defined plans that connect with policies covering industry, science and technology and accompanied by an improvement in the economic system, there is thus a need to adopt new policies for partnership with the private sector. Sudan needs to form a body to formulate a policy on manpower resources for S&T, and suggest measures to minimise brain-drain impacts. Sudan needs to continue building relatively well-developed S&T infrastructure, and a sufficient number of highly qualified university and R&D personnel to put the country in a good position in terms of globally competing in S&T.

So far Sudan does not possess all the human and financial resources necessary to promote S&T. However, Sudan could have a wider range of capabilities to promote S&T if it pooled and integrated its resources. Reforming the economic system, encouraging the private sector, implementing effective S&T cooperation and integration with other Arab and African countries will most likely enhance S&T development and hence long-term harmonious development in Sudan.

Our results in this Chapter at the macro level is consistent with the findings at the micro level presented in Chap. 5 above and support our third and second hypotheses in Chap. 1 above concerning the low skill and technology levels at the macro and micro levels and that in the short- and medium-term, Sudan is unable to rely on local technologies and will remain heavily dependent on foreign technologies respectively. Our findings in this Chapter is consistent with the findings concerning the low skill level and need for skill upgrading presented in Chap. 5 above and also support our first hypothesis in Chap. 1 above that Sudan needs to promote local skills and local technologies in order to implement the five strategies of: reducing poverty; achieving economic diversification; reducing unemployment and restructuring the labour market; building local technological capacity; and achieving long-term stabilised, sustainable and balanced economic growth and development.

References

Elamin, H. B. (2009). *Science, technology and innovation indicators*. UNESCO sub-regional training workshop, Cairo: Sudan Ministry of Science and Technology, September 28–30, 2009.

ESCWA-UNESCO (Economic and Social Commission for Western Asia-United Nations Educational, Scientific and Cultural Organization). (1998). *Research and development system in the Arab states: Development of science and technology indicators* (E/ESCWA/TECH/1998/3). Cairo.

Haddad, M. (2001). *Export competitiveness: Where does the Middle East and North Africa region stand? Background paper to economic trends in the MENA region*. Cairo: Economic Research Forum (ERF). In ERF. (2002). *ERF economic trends in MENA region 2002: Chap. 2: Financial and capital markets, privatization and FDI*. Cairo: AUC.

Hassan, A. O. (2009a). *Sudan National Innovation System*. Paper presented at the Presidential Initiative on the Reform of Science and Technology System in the Sudan, Friendship Hall, January 5–6, 2009.

Hassan, M. H. A. (2009). *Promoting excellence in STI for sustainable development in Africa*. Paper presented at the International Workshop on Science and Technology Systems in Sudan, Khartoum, January 5–6, 2009.

IMF. (2002). *World economic outlook: The Middle East and North Africa regional outlook*. Washington, DC: International Monetary Fund.

Lall, S. (1999). *Competing with labour: Skills and competitiveness in developing countries* (Issues in Development Discussion Paper No. 31, p. 17). Geneva: Development Policies Department, ILO.

Muysken, J., & Nour, S. (2006). Deficiencies in education and poor prospects for economic growth in the Gulf Countries: The case of the UAE. *Journal of Development Studies, 42*(6), 957–980. Taylor and Francis Ltd.: UK.

Nkwelo, M. (2008). *The science and technology system of the republic of Sudan*. UNESCO, 2008.

Nour, S. (2004, March 24–28). *Science and Technology (S&T) development indicators within the Arab Mediterranean countries*. Paper prepared for the Fifth Mediterranean Social and Political Research Meeting, organised by the Mediterranean Programme of the Robert Schuman Centre for Advanced Studies at the European University Institute, Florence, Italy.

Nour, S. (2005a, September). Science and Technology (S&T) development indicators in the Arab region: A comparative study of Arab Gulf and Mediterranean countries. *The Journal of Science, Technology and Society* (SAGE Publications), *10*(2), 249–274 (Published in *UNU-*

INTECH Discussion Paper Series UNU-INTECHDPS2005-3, Maastricht, the Netherlands, August 2005).

Nour, S. (2005b, November). *Technological change and skill development in the Arab Gulf Countries*. Doctoral dissertation, Maastricht University Press, Maastricht, The Netherlands.

Nour, S. (2010, January–June). Firm Survey (2010).*Technological change and skill development: A comparative study of chemical, food, metal and textile small, medium and large scale enterprises in the Sudan*. Khartoum, Sudan.

OECD (Organisation for Economic Co-operation and Development). (1997). *The second european report on science and technology indicators (1997)* (pp.37, 90, 91). Paris: OECD.

Qasem, S. (1998b). Research and development systems in the Arab states: Development of science and technology indicators, Chapter 4. In *Economic Trends in MENA Region 2002*. Cairo: Economic Research Forum and the American University in Cairo.

R&D Survey. (2010, June). *Sudan R&D Survey: Efforts for enhancing R&D in the Sudan: Interview with policy makers in the government and academic staff in public and private universities in Sudan*. Khartoum.

Rasiah, R. (2002, March). *TRIPs and capability building in developing economies*. (UNU/INTECH Discussion Paper Series No. 2002–1).

Sudan Central Bureau of Statistics Population Census Data. (2010c). *5th Sudan population and housing census (2008)*. Khartoum: Sudan Central Bureau of Statistics.

Sudan Ministry of General Education, the annual general educational statistics report: Various issues (1993/1994–2009/2010). Unpublished reports, Sudan Ministry of General Education, Khartoum.

Sudan Ministry of Higher Education and Scientific Research, the annual higher educational statistics report: Various issues (1993/1994–2009/2010). Unpublished reports, Sudan Ministry of Higher Education and Scientific Research, Khartoum.

Sudan Ministry of Industry. (2005, July). *The comprehensive industrial survey*. Khartoum, Sudan, Vol. 4:www.industry.gov.sd

Sudan Ministry of Science and Technology (MOST). (2008). *Status of science and technology in Sudan*. Paper presented in the Commission on Science and Technology for Development, Eleventh session, pp.3–6, 65, 26–30 May 2008.

Sudan Ministry of Science and Technology (MOST), the annual science and technology reports: Various Issues (2003–2009), unpublished reports, Sudan Ministry of science and technology, Khartoum.

The General Registrar of IPR Sudan Office. (2010). *Unpublished data and statistics from the General Registrar of IPR Sudan Office. (2010)*. Khartoum: The General Registrar of IPR Sudan Office. Accessed 10 July 2010.

The International Monetary Fund. (2002). *World Economic Outlook: the Middle East and North Africa Regional Outlook*, Washington, DC: The International Monetary Fund.

The United Nations Development Programme (UNDP)-Arab Human Development Report (AHDR). (2004). *Towards Freedom in the Arab world the UNDP*. New York/Amman: Regional Bureau for Arab States (RBAS)/National Press.

The World Bank- World Development Indicators Database. (2012, April). http://data.worldbank.org/indicator/IP.PAT.NRE. Accessed 5 June 2012.

The World Bank- World Development Indicators Database. (2012, April). http://data.worldbank.org/indicator/IP.PAT.RESD/countries. Accessed 5June 2012.

The World Bank, World Development Indicators database. (2011). Accessed 25 June 2011.

The World Bank, World Development Indicators database (WDI). (2012). Accessed 15 April 2012.

Third World Academy of Sciences (TWAS) (May 2008), cited in Hassan, M. H. A. (2009). *Promoting excellence in STI for sustainable development in Africa*. Paper presented at the International Workshop on Science and Technology Systems in Sudan, Khartoum, 5–6 Jan, 2009.

U.S. Patent and Trade mark office web site: www.uspto.gov

UNDP. (2001). *Summary human development report (2001): Making new technologies work for human development*. New York/Oxford: UNDP/Oxford University Press.
UNDP. (2002). *Human development report (2002): Deepening democracy in a fragmental World*. New York/Oxford: UNDP/Oxford University Press.
UNDP. (2003). *Human development report (2003): Millennium development goals: A compact among nations to end human poverty*. New York/Oxford: Oxford University Press.
UNDP (United Nations Development Programme). (2011). *Human development report 2011: Sustainability and equity: A better future for all*. New York: UNDP, Palgrave Macmillan.
UNDP (United Nations Development Programme). Human development report. (2009). *Human development report 2009: Overcoming barriers: human mobility and development*. New York: Palgrave Macmillan.
UNDP (United Nations Development Programme). (2007/2008). *Human Development report 2007/2008: Fighting climate change: Human solidarity in a divided world*, (pp. 231–232, 292–293). New York: Palgrave Macmillan.
UNDP (United Nations Development Programme). (2010). *Human Development Report 2010: The Real Wealth of Nations: Pathways to Human Development*. Human development report 2010 —20th Anniversary Edition New York, NY: UNDP/Palgrave Macmillan, Statistical Tables, Statistical Annex, Table 12: Decent Work, p.191.
UNESCO (2006): UNESCO R&D Statistics (2006): www.unesco.org
UNESCO Institute of Statistics Data and Statistics. (1996). *UNESCO statistical yearbook: The World Education Indicators on school life expectancy 1992*: www.unesco.org
United Nations. (UN) (1970). *United Nations science and technology for development, proposals for the Second United Nations development decade*. United Nations, New York.
United Nations Development Programme UNDP- The Arab Human Development Report AHDR. (2003). *Arab Human Development Report (2003): Building a Knowledge Society*. UNDP. New York, Regional Bureau for Arab States (RBAS), Amman, Jordan: National Press, July. p. 56; Table 5.1 p.102.
WIPO. (2007). Statistics on Applications for PCT.
WIPO Patent Agenda Getachew Mengistie Study of Ethiopian Intellectual Property Office, A/39/13 Add.1 available at: http://www.wipo.int/documents/en/document/govbody/wo_gb_ab/doc/a_39_13add1.doc
Zahlan, A. B. (1999a). *The Arabs and the challenges of science and technology: Progress without change*. Beirut: The Centre for the Study of Arab Unity.
Zahlan, A. B. (1999b). *Science Policy for the Twenty-First Century: Mobilization and Development*. In *ESCWA Proceedings of the Expert Group Meeting on Science and Technology Policies and Strategies for the Twenty- First Century*, ESCWA, Beirut, 10–12 March 1999, pp. 14–16.

Chapter 7
Relationship Between Skill, Technology and Input–Output Indicators

Abstract This chapter uses the data from the firm survey (2010) to examine skill indicators, their implications and relationships between skill, technology and input-output indicators at the micro/firm level. We illustrate the low skill levels – due to the excessive share of unskilled workers and the implications on skills mismatch, industrial performance indicators and productivity decline across firms. These results are consistent with the micro-macro findings in Chap. 5 above, which indicate the low share of high skilled in total population and employment – measured by both educational and occupational levels – and the serious implications on skills mismatch and the macro-micro duality with respect to upskilling efforts. We find that an increase in skill level and firm size lead to improved relationships between actual and required education, between actual education, experience and wages and between required education, experience and wages. Our results concerning the positive complementary relationships between skill, technology (ICT) and upskilling (ICT training) and between computers, telecommunications and ICT training are consistent with the findings in the new growth literature. We illustrate that an increase in skill level and firm size lead to an improvement in the complementary relationships between skill, upskilling and technology (ICT) at the micro level.

7.1 Introduction

Earlier findings in Chap. 5 show the serious implications of the deficient educational system and excessive use of low educated workers, and illustrate one surprising contradicting macro–micro view regarding the transfer of knowledge/external schooling effects. This surprising result from Chap. 5 motivates our research to attempt a more comprehensive analysis of skill problem and implications of unskilled workers at the micro level/across firms. Hence, the aim of this chapter is to broaden our earlier analysis in Chap. 5 by providing an indepth analysis of skill and technology indicators and the relationship between them and the implications of the prevalence

S. Mohamed Nour, *Technological Change and Skill Development in Sudan*,
DOI 10.1007/978-3-642-32811-4_7, © Springer-Verlag Berlin Heidelberg 2013

of low-skilled workers at the micro level. In addition, we examine the relationships between: skill indicators (education/actual education and occupation/required education respectively and experience) and average wages; between skill, upskilling (spending on ICT training) and technology (spending on ICT); and between technology (spending on ICT) and input–output indicators across firms. We also compare the relevance of our results to the theoretical framework in Chap. 3 and the findings concerning these relationships in the new growth literature.

Prior to investigating the relationships between skill, upskilling, technology and input–output indicators across firms, it is convenient to begin with explaining the importance of the industrial sector across firms, because understanding the importance of the industrial sector from the perspective of the industrial firms can help in supporting the potential contribution of the industrial sector in enhancing economic development in Sudan. Beginning with the importance of the industrial sector for economic development in Sudan, our results from the firm survey (2010) imply that the respondent firms seem to be highly optimistic regarding the potential contribution of the industrial firms in achieving not only the traditional microeconomic aim of maximising private industrial profit but also in achieving the macroeconomic development aims, provided that the appropriate conditions for industrial development is created. For instance, the potential contribution of the industrial sector in: increasing output and income; increasing employment opportunities for present and future labour force (in response to potential population increase); satisfying domestic consumption and achieving self sufficiency by offering the basic and necessary goods for the Sudanese; achieving industrial profit; improving production relationships between workers; and enhancing local technological capability building by adaptation of imported technologies to fit with local needs. This is in addition to: contribution to economic growth through enhancing industrial linkages; reforming the structural imbalances in Sudan economy; decreasing imports; and enhancing optimal and full utilisation of natural resources and local raw materials; enhancing local capability; enhancing development of local technologies to fit with local development needs; supporting development and urbanisation of all regions in Sudan; enhancing local industrialisation using local raw materials; and enhancing economic growth by increasing industrial exports. Finally, meeting the needs and enhancing linkages with other sectors specially agriculture, are also mentioned but are of somewhat less importance (see Table 7.1 below).[1]

The rest of this chapter is organised as follows: Sect. 7.2 defines the variables used in our analysis and the general characteristics of firms; Sect. 7.3 presents our hypotheses and discusses differences in prevalent skill levels and requirements and the implications of low skill levels on skills mismatch, industrial performance

[1] As indicated by 97 %, 91 %, 84 %, 81 %, 81 %, 73 %, 67 %, 67 %, 67 %, 66 %, 63 %, 61 %, 57 %, 49 %, 49 % and 42 % of all the respondent firms respectively.

Table 7.1 The importance of the industrial sector for economic development in Sudan (2008)

Economic development aims	All firms	Industry				Size		
		Chemical	Food	Metal	Textile	Large	Medium	Small
Increasing output and income	97 %	97 %	96 %	100 %	100 %	100 %	96 %	100 %
Increasing employment opportunities for present and future labour force (in response to potential population increase)	91 %	89 %	93 %	90 %	100 %	88 %	92 %	100 %
Satisfying domestic consumption and achievement of self sufficiency by offering basic and necessary goods for Sudanese	84 %	92 %	82 %	60 %	80 %	84 %	85 %	84 %
Achieving industrial profit	81 %	75 %	86 %	100 %	60 %	81 %	85 %	74 %
Creation of improved production relationships between workers	81 %	86 %	75 %	80 %	80 %	81 %	85 %	84 %
Enhancing local technological capability building by adaptation of imported technologies to fit local needs	73 %	78 %	64 %	70 %	100 %	91 %	65 %	63 %
Enhancing economic growth by enhancing industrial linkages	67 %	61 %	71 %	70 %	80 %	69 %	58 %	84 %
Contribution to reform structural imbalances in Sudan economy	67 %	69 %	68 %	50 %	80 %	72 %	62 %	74 %
Contribution to economic growth by decreasing imports	67 %	72 %	54 %	80 %	80 %	59 %	77 %	63 %
Supporting the optimal and full utilisation of natural resources and local raw materials	66 %	69 %	68 %	40 %	80 %	69 %	65 %	63 %
Enhancing local capability	63 %	72 %	50 %	50 %	100 %	78 %	54 %	58 %
Enhancing local technological capability building and reducing technological dependence by development of local technologies	61 %	61 %	57 %	80 %	40 %	72 %	46 %	68 %
Supporting development and urbanisation of all regions in Sudan	57 %	64 %	54 %	60 %	20 %	66 %	54 %	53 %

(continued)

Table 7.1 (continued)

Economic development aims	All firms	Industry				Size		
		Chemical	Food	Metal	Textile	Large	Medium	Small
Enhancing economic growth by local industrialisation of local raw materials that was earlier exported in the form of raw materials	49 %	44 %	54 %	30 %	100 %	63 %	46 %	32 %
Enhancing economic growth by increasing industrial exports	49 %	56 %	46 %	40 %	40 %	53 %	42 %	47 %
Enhancing economic growth by meeting the need and enhancing linkage with other sectors especially agriculture	42 %	44 %	46 %	30 %	20 %	53 %	35 %	37 %
Number of respondents	79	36	28	10	5	32	26	19

Source: Own calculation based on the firm survey (2010)

indicators and productivity decline across firms; Sect. 7.4 examines the relationships between actual and required education, experience and wages; Sect. 7.5 shows the relationships between skill, technology (spending on ICT) and upskilling (spending on ICT training) and between technology (ICT) and input–output indicators; Sect. 7.6 concludes.

7.2 Data, Definition of Variables and General Characteristics of Firms

Before commencing with the empirical analysis, it will be useful to briefly explain the data used in our analysis and the general characteristics of firms.

7.2.1 Data and Definition of Variables

Our analysis in this chapter uses the data from the firm survey (2010), which provides us with three sets of micro variables.[2] The first set includes skill variables,

[2] All data, information and analysis in this chapter are based on the results covering 45 firms obtained from the firm survey (2010).

while the second and third sets include both technology and input–output and performance related variables respectively. We define skill variables by educational attainment, occupational level (measured by the required qualifications/schooling years) and average years of experience.[3] We use the total spending on machinery and equipment to define "old technology" and also we use the total spending on ICT[4] to define "new technology", the share of spending on ICT training as a percentage of total spending on ICT to define "upskilling", total sales value to define "output", total profit and total value added to define "performance", in addition we use economic, productivity, activity and profitability indicators to define industrial performance indicators, and total employment and net worth to define "labour" and "capital" inputs, respectively.[5]

We use the first set of skill variables in Sect. 7.3 to discuss hypotheses 3.b. and 4.a. in Chap. 1 above regarding the implications of unskilled workers across firms. We use input–output and performance indicators to illustrate the decline in industrial performance and productivity indicators and ratios. Next, in Sect. 7.4, we test hypothesis 4.b. in Chap. 1 above about the relationships between actual and required education and experience and wages. In Sect. 7.5, we use the first and second sets of variables including skill, ICT and the share of spending on ICT training to test hypothesis 4.c. in Chap. 1 above regarding the relationship between skill, technology (ICT) and upskilling. Next, we use the second and third sets of technology and input–output variables to test the fifth hypothesis in Chap. 1 above about the relationship between technology (ICT) and input–output indicators.[6]

[3] We classify the educational qualifications of workers into three groups: high skilled (H) with postgraduate, university and diploma degree (more than 12 years of schooling), medium skilled (M) with secondary education (12 years of schooling) and low skilled (L) with less than secondary education (less than 12 years of schooling). We define the occupational status according to five categories, including: white collar high (managers, professionals, management executives, scientists, technicians and engineers); white collar low (clerical and administrative); blue collar high (skilled craftsmen); blue collar low (plant machinery operators, assemblers and elementary occupation); and other workers. We define the required qualifications by required years of schooling including: postgraduate/Ph.D. (19–20 years); professional, MSc/postgraduate (18 years); university graduate (16 years); diploma (14 years); higher secondary schooling (12 years); and less than secondary schooling (less than 12 years). We measure the average wages by average monthly wages (in Sudanese Pounds), and average years of experience by both actual and required average years of experience for both educational and occupational definition respectively.

[4] ICT is the sum of total expenses on computers, telecommunications, Internet, training, maintenance and other related items.

[5] We measure output by the total sales value because the measurement units of sales value is unified (in local currency) across firms, while the measurement units of output in physical terms (tonnes, litres, etc.) varies enormously across firms.

[6] We use few observations in the estimated equations, due to limited availability of reliable data covering these indicators, because some of the respondent firms were particularly reluctant to provide adequate reliable quantitative data covering these indicators.

7.2.2 General Characteristics of Firms

Table 7.2 presents the main general characteristics of firms and economic indicators such as the share of firms in total employment, capital, profit and output (total sales value), and their differences defined by firm size and industry level. We observe that the market size or structure (defined by the share in total employment, raw materials, profit, fixed capital and value added) seems biased toward large size and chemical and food firms respectively. For instance, on average, the large size and chemical firms respectively employ 74 % and 50 % of total workers, absorb 99 % and 73 % of total raw materials, and therefore, it is not surprising that they constitute 99 % and 72 % of total profit. While small size and food industries absorb 99 % and 99 % of total capital, large size and food industries absorb 84 % and 83 % of total fixed capital in the form of machinery and equipment, hence, it is not surprising that they constitute 84 % and 84 % of total value added respectively.[7] In addition, medium size and food industries constitute 63 % and 75 % of total output (total sales value). These differences in market size leads to several implications, as we explain below and in the next sections.

From Table 7.2 we observe the limited contribution of public sector and high share of private sector in the metal, food, chemical and textile industries and medium, small and large size firms respectively. We also note the high share of local ownership and also a limited share of foreign and mixed ownership, which implies the limited dependency on foreign capital and foreign workers. We find that the share of firms in local ownership decreases and so the share in foreign ownership increases with firm size and to some extent with industry level. But despite the presence of foreign capital, there is limited contribution of multinational companies; however, such contribution is diversified as the sources of foreign capital of multinational companies originates from different countries and increases to some extent with industry level and to less extent with firm size. We also observe limited changes in the general structure of firms during the period 2005–2008, which may indicate a lack of dynamism, particularly with respect to the distribution of economic indicators, i.e. total employment, capital and output/sales value across firms. The reported change since establishment in ownership, nationality of main owner and length of years in operation (age) varies across firms and generally increases with firm size and industry level; it was observed only in some of the chemical industries and large and small size firms. In addition, the geographical

[7] We believe that our results should be interpreted carefully, notably when explaining our results related to the share of firms in total capital, which indicate the large share of small size and food industries that absorb 99 % and 99 % of total capital. In particular, we interpret these results due to the relative availability of information and quantitative data covering these financial indicators, notably, due to relatively more response to provide information and quantitative data covering these financial indicators for small size and food industries as compared to other firms, particularly because some of the firms seem to be more reluctant to provide adequate reliable information and quantitative data covering these financial indicators for medium and large size, chemical, metal and textile firms.

Table 7.2 Main characteristics of firms in the Sudan (2005–2008)

Main indicators (2005–2008)[a]		No. of respondent firms	Chemical	Food	Metal	Textile	Large	Medium	Small
Share in employment (%)	2005	85	45 %	37 %	4 %	13 %	71 %	18 %	10 %
	2006	85	52 %	32 %	4 %	13 %	73 %	17 %	10 %
	2007	85	51 %	25 %	4 %	20 %	76 %	16 %	9 %
	2008	85	52 %	22 %	4 %	22 %	77 %	16 %	7 %
	Average 2005–2008	85	50 %	29 %	4 %	17 %	74 %	17 %	9 %
Share in capital (%)	2005	83	0.4 %	99.3 %	0.2 %	0.1 %	0.5 %	0.5 %	99 %
	2006	83	0.4 %	99.1 %	0.4 %	0.1 %	0.5 %	0.5 %	99 %
	2007	83	0.4 %	99.0 %	0.5 %	0.1 %	0.5 %	0.5 %	99 %
	2008	83	0.4 %	99.1 %	0.3 %	0.01 %	0.5 %	0.5 %	99 %
	Average 2005–2008	83	0.4 %	99.2 %	0.3 %	0.1 %	0.5 %	0.5 %	99 %
Share in machinery and equipment (%)	2005	45	2 %	94 %	2 %	2 %	95 %	0.5 %	4.5 %
	2006	45	9 %	79 %	7 %	4 %	84 %	0.5 %	15.5 %
	2007	45	12 %	79 %	9 %	0 %	78 %	1 %	20 %
	2008	45	11 %	80 %	8 %	0 %	80 %	1 %	19 %
	Average 2005–2008	45	9 %	83 %	7 %	2 %	84 %	1 %	15 %
Share in raw materials (%)	2005	45	1 %	96 %	2 %	1 %	97 %	0.5 %	2.5 %
	2006	45	96 %	3 %	1 %	0 %	99 %	0.5 %	0.5 %
	2007	45	95 %	4 %	0 %	0 %	99 %	0.5 %	0.5 %
	2008	45	98 %	2 %	0 %	0 %	100 %	0.5 %	0.5 %
	Average 2005–2008	45	73 %	26 %	1 %	0 %	99 %	0.5 %	0.5 %
Share in profit (%)	2005	45	1 %	90 %	12 %	0 %	99 %	0.5 %	0.5 %
	2006	45	95 %	5 %	0 %	0 %	99 %	0.5 %	0.5 %
	2007	45	95 %	3 %	1 %	0 %	99 %	0.5 %	0.5 %
	2008	45	96 %	2 %	2 %	0 %	99 %	0.5 %	0.5 %
	Average 2005–2008	45	72 %	25 %	4 %	0 %	99 %	0.5 %	0.5 %
Share in output (total sales value) (%)	2005	45	0 %	99 %	1 %	0 %	22 %	78 %	1 %
	2006	45	27 %	72 %	1 %	0 %	36 %	59 %	6 %
	2007	45	30 %	67 %	2 %	0 %	40 %	59 %	1 %
	2008	45	36 %	62 %	2 %	0 %	45 %	54 %	1 %
	Average 2005–2008	45	23 %	75 %	2 %	0 %	36 %	63 %	2 %

(continued)

Table 7.2 (continued)

Main indicators (2005–2008)[a]		No. of respondent firms	Chemical	Food	Metal	Textile	Large	Medium	Small
Share in value added (%)	2005	45	5 %	94 %	1 %	0 %	99 %	0.5 %	0.5 %
	2006	45	2 %	92 %	5 %	1 %	90 %	0.5 %	9.5 %
	2007	45	5 %	77 %	17 %	1 %	76 %	4 %	20 %
	2008	45	8 %	74 %	18 %	1 %	71 %	6 %	22 %
	Average 2005–2008	45	5 %	84 %	10 %	1 %	84 %	3 %	13 %
Share in wage (%)	2005	45	1 %	92 %	2 %	5 %	97 %	0.5 %	2.5 %
	2006	45	4 %	66 %	12 %	18 %	86 %	0.5 %	13.5 %
	2007	45	9 %	77 %	14 %	0 %	80 %	1 %	19 %
	2008	45	10 %	80 %	9 %	0 %	82 %	2 %	17 %
	Average 2005–2008	45	6 %	79 %	9 %	6 %	86 %	1 %	13 %
Share in spending on ICT (%)	Average 2005–2008	54	23 %	53 %	11 %	13 %	48 %	21 %	30 %
Share in spending on ICT training (%)	Average 2005–2008	8	2 %	73 %	12 %	13 %	75 %	18 %	7 %
Share of private firms (%)	2008	87	89 %	94 %	100 %	80 %	89 %	97 %	90 %
Share of ownership (%)[b]	Local – 2008	87	86 %	92 %	68 %	88 %	80 %	84 %	95 %
	Foreign- 2008	87	15 %	8 %	32 %	12 %	20 %	16 %	5 %
Share of main owners (%)	Local – 2008	87	79 %	87 %	62 %	80 %	69 %	83 %	90 %
	Foreign – 2008	87	8 %	3 %	23 %	0 %	9 %	14 %	0 %
	Mixed – 2008	87	13 %	10 %	15 %	20 %	23 %	3 %	10 %
Affiliation to multinational	2008	87	11 %	6 %	8 %	0 %	9 %	10 %	5 %
Change after establishment[c]	2008	87	8 %	0 %	0 %	0 %	6 %	0 %	5 %
Main location (%)	Khartoum – 2008	87	26 %	23 %	38 %	0 %	26 %	24 %	29 %
	Khartoum North-2008	87	58 %	68 %	46 %	80 %	57 %	62 %	71 %
	Omdurman – 2008	87	16 %	10 %	15 %	20 %	17 %	14 %	0 %
Branches other than main location (%)	2008	87	3 %	10 %	15 %	0 %	14 %	3 %	0 %
Average age/operation years	2008	87	18	17	14	16	19	14	17
Average rate of diversification	Sales – 2008	86	1.50	1.47	1.52	1.23	1.57	1.36	1.54
	Employment – 2008	63	1.40	1.31	1.55	1.17	1.47	1.19	1.44

Source: Firm Survey (2010)

[a]All indicators are calculated from the firm survey (2010); some refer to observations over only one year (2008) and others use observations over 4 years (2005–2008).

[b]Some of the respondent firms reported a mixed share of local and foreign ownership.

[c]Change after establishment refers to changes in ownership, management and structure (e.g. expansion; opening new branches or merger with other firms).

distribution of firms indicates that most are clustered in two main locations and only a few of the chemical, food and metal industries and large and medium size firms have branches in cities other than the main location, though the probability of clustering to some extent increases with firm size and industry, and the probability of having branches increases with firm size but to lesser extent increases with industry. Moreover, we realise the limited scope for diversification as measured by sales and employment indices across firms.[8] The average diversification index increases to some extent with firm size but only to a lesser extent increases with industry; this implies that metal and chemical industries and large size firms have more interest in diversification, whereas food and textile industries and medium size firms have less interest in diversification and more interest in concentration and specialisation. As expected, large size firms reported more interest in diversification than medium and small size firms. Somewhat surprising and in contrast to our expectations, the findings across firms indicate that metal firms reported more interest in diversification more than chemical, food and textile firms, moreover, somewhat surprising was that small size firms indicated more interest in diversification than medium size firms.

7.3 Differences in Skill Level and Requirements and the Implications Across Firms

Our earlier findings in Chap. 5 indicate that the share of high skilled workers in total employment, the total number of full time equivalent researchers (FTER), R&D and ICT expenditure, patent, product and process innovations are higher within large size and chemical firms when compared to medium and small size and food, metal and textile firms. Our result with respect to R&D and chemical sector is consistent with the standard classification developed by the OECD in the mid-1980s, which distinguishes between industries in terms of R&D intensity (cf. OECD 1997). For instance, in the mid-1980s, the OECD classification distinguished between industries in terms of R&D intensity, considering pharmaceutical and ICT as high-technology, chemical and vehicle as medium-technology and food and textile as low technology (cf. OECD 1997). Our findings with respect to firm size are consistent with the literature and the Schumpeterian hypothesis, which indicates that large size/market concentration is conducive to R&D investment (cf. Braga and Willmore 1991). For instance, Kumar and Saqib (1994) suggest that the probability of undertaking R&D increases with firm size only up to a certain

[8] We use a modified definition of the diversification index developed by Utton (1979). We define the diversification index by output/ sales diversification $Di = [[P1 + 2\,P2 + 3P3 + 4P4] - 1/\,2]$, where Pi refers to the percentage share of diversified sale product in total sale products within firms. Ranked from large to small, when $Di = 1$, $Di = 4$ and $1 < Di < 4$, it implies complete specialisation, complete diversification and some degree of diversification respectively. We apply the same definition for employment diversification index. Cf. Utton (1979), pp. 15-16, 104–105.

level, while R&D intensity increases with it linearly. However, one should also expect that these results could imply a possibility for reversed causality, mainly because R&D is a fixed cost that requires high financial capacity, which is most likely to be strong amongst large size firms.

In addition to earlier findings, we observe that skill levels and requirements (actual and required education and experience) and skills mismatch are not homogenous across firms and vary with industry and size. As we explained in Sect. 7.3, these findings can be used to test the hypotheses 3.b. and 4.a. in Chap. 1 above that, irrespective of these differences, high skill requirements and low skill levels – due to high share of unskilled workers – lead to skills mismatch and also contribute to industrial performance indicators and productivity decline across firms. In Sects. 7.4 and 7.5, we then examined hypotheses 4.b. and 4.c. in Chap. 1 above that an increase in skill levels and firm size lead to improved relationships between actual and required education and experience, between actual education, experience and wages and between skill, upskilling and technology (ICT). Finally, in Sect. 7.5, we also investigated the fifth hypothesis in Chap. 1 above concerning the relationships between technology (the use of ICT) and input–output indicators at the micro/firm level.

7.3.1 Differences in Skill Level and Requirements (Education and Experience) Across Firms

Prior to investigating the first hypothesis on the extended implications of low skill levels as presented above, it is convenient to begin with explaining differences in skill levels and requirements across firms because understanding why and how they vary with industry and firm size can help in investigating both the first and second hypotheses.

In Figs. 7.1, 7.2, and 7.3 below we explain differences in skill levels and requirements and low skill levels (defined by education and occupation groups) across firms (defined by size and industry).[9] Figures 7.1 and 7.2 show the low share of high skilled – high educated and white collar – workers, differences in skill levels according to education and occupation definitions and differences across firms. For instance, Fig. 7.1 indicates that for 55 % of all respondent firms, the share of high skilled (educated) represents 1–30 % of total employed workers. For a further 20 % of all respondent firms, the share of high skilled (educated) represents 31–50 % of total employed workers, but for the remaining 25 % the share is more than 50 % of the workforce. Figure 7.2 shows, for example, that for 66 % of all respondent firms,

[9] In Figures 7.1–7.3, the horizontal axis defines firms, industry, size (chemical, food, metal and textile, large, medium and small), and skill level (high (H), medium (M) and low (L)). The vertical axis defines the intensity/share of H, M and L across firms. The information in the right margin defines the distribution of workers in Figures 7.1–7.2, and the average required years of education in Figure 7.3.

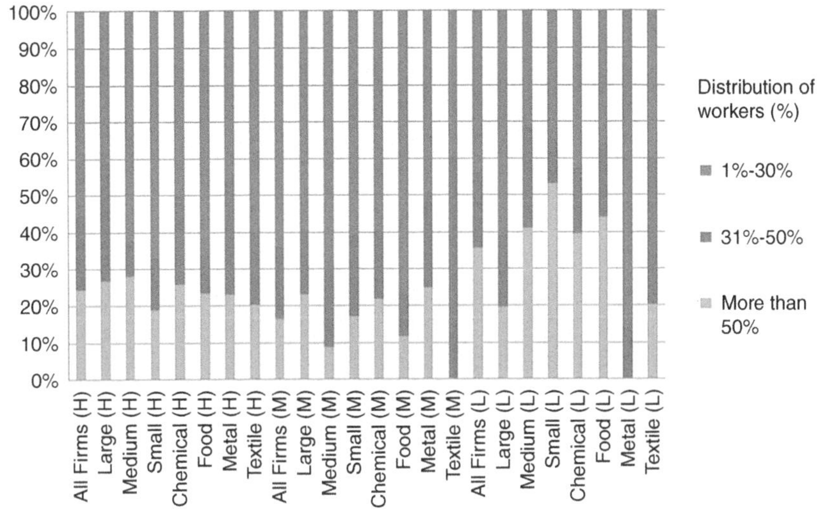

Fig. 7.1 Differences in the distribution of workers by educational level across firms (% share) 2008 (Source: Firm Survey (2010))

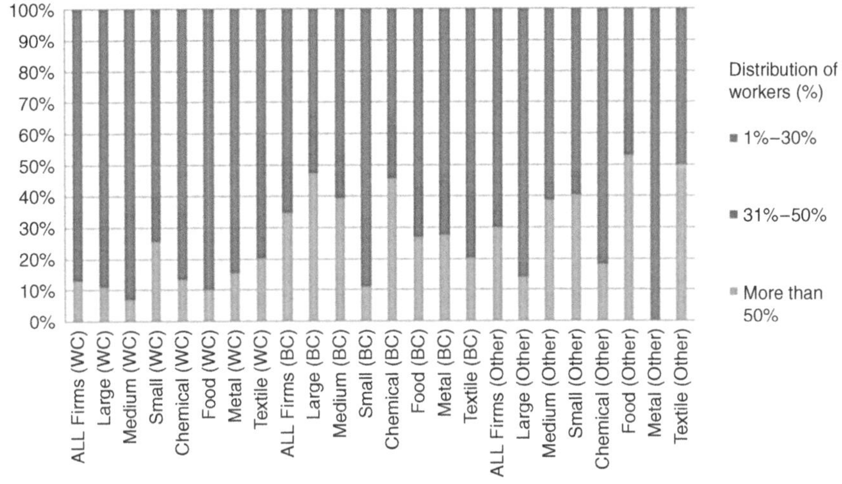

Fig. 7.2 Differences in the distribution of workers by occupational level across firms (% share) 2008 (Source: Firm Survey (2010))

the share of white collar (WC) represents 1–30 % of total employed workers; for 21 % of all respondent firms the WC share is 31–50 % and for 13 % the figure stands at 50 % of total employed workers. The results show that the incidence of high educated and white collar workers constituting more than half of total employment is observed only within 25 % and 13 % of all respondent firms respectively. They also indicate that the share of high skilled – measured by education – is less than

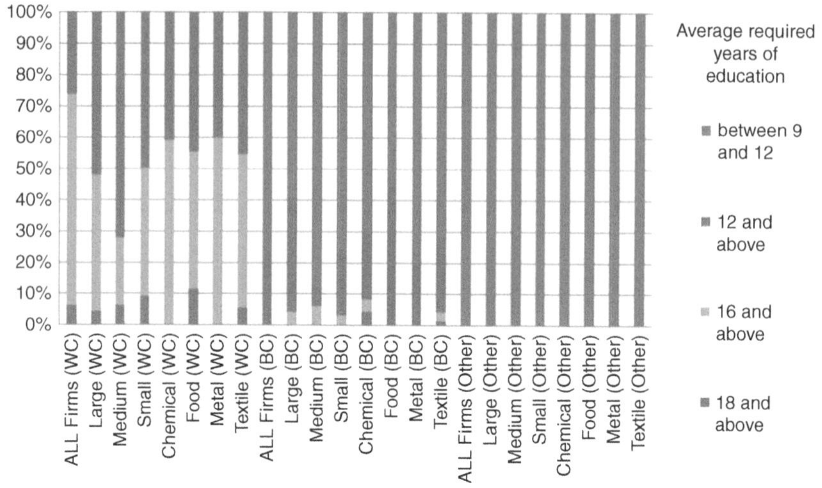

Fig. 7.3 Differences in the educational requirements by occupational level across firms (% share) 2008 (Source: Firm Survey (2010))

one third of total workers for 55 % of all firms and the share of high skilled – white collar measured by occupational level – is less than one third of total workers for 66 % of all firms. That means that across all firms the share of high educated and white collar respectively are less than one third and two thirds; therefore, the majority of employed workers are low and medium skilled.

Figures 7.3 and 7.4 show that skill requirements – average required years of schooling – vary and increase with occupational level across firms.[10] For instance, Fig. 7.3 indicates that for 26 % of all respondent firms the average required years of education for white collar (WC) is 12 and above; 68 % of all respondent firms require an average of 16 years; whilst 6 % of all respondent firms put this figure at 18 and above. Moreover, Fig. 7.4 indicates that for 16 % of all respondent firms the average required years of education for white collar high (WCH) is 14 years (diploma degree), for 47 % the requirement is 16 years (university degree) and for 37 % the requirement is 17–19 years and above (postgraduate degree). The figures show that the university degree is the major preferred required qualification only within the first and second occupational groups, while for the other occupational groups either a diploma or secondary or less than secondary schooling is required.

Figure 7.5 below indicates the variation in skill requirements (required years of experience), defined by educational and occupational levels. For instance, for 36 % of all respondents firms the average required years of experience for high education is 2–5 years; for 39 % the experience requirement stands at 5–10 years, for 17 % the experience requirement stands at 10–15 years and for 8 % the figure is 16 years

[10] White collar (WC) includes white collar high and low. Blue collar (BC) includes blue collar high and low.

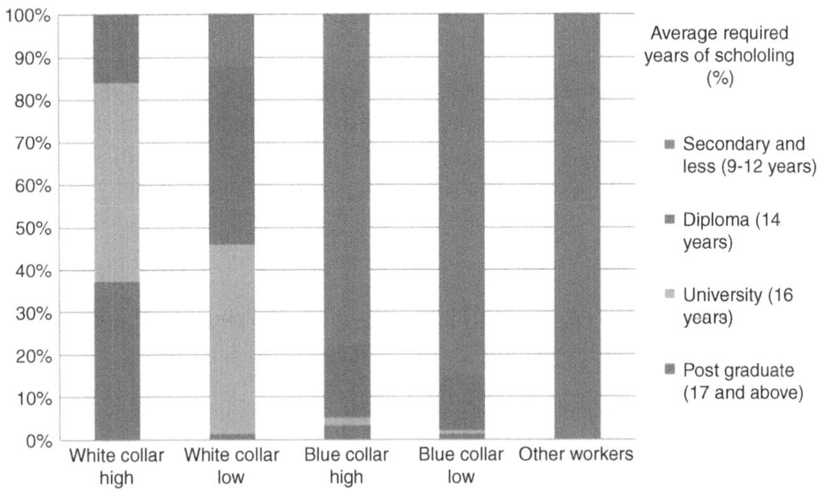

Fig. 7.4 Average required years of schooling defined by occupation classes across firms (2008) (Source: Firm Survey (2010))

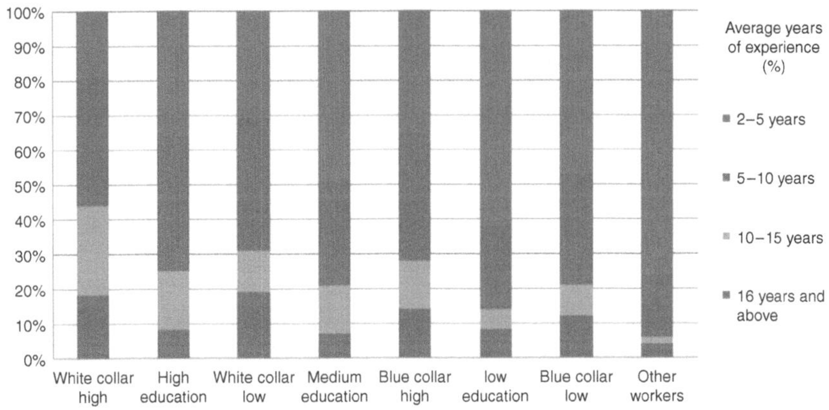

Fig. 7.5 Average years of experience defined by education and occupation classes across firms (2008) (Sources: Firm Survey (2010))

and above. Moreover, for 19 % of all respondent firms the average required years of experience for white collar high (WCH) is 2–5 years, for 37 % the experience requirement stands at 5–10 years, for 26 % the experience requirement stands at 10–15 years and for 18 % the figure is 16 years and above. Figure 7.5 illustrates that average years of experience are increasing in educational and occupational levels respectively. In the next section, we explain the relationships between required education/actual education, occupation/required education and experience and wages across firms.

7.3.2 *The Implications of Low Skill Levels Across Firms*

In this section we examine the first hypothesis that, irrespective of the observed differences in skill levels and requirements and as we explained above, the low skill levels – due to high share of unskilled workers – lead to skills mismatch and probably contribute to industrial performance indicators and productivity decline across firms.

7.3.2.1 Low Skill Levels and Skills Mismatch (Differences in Required and Attained Education)

When comparing the required schooling with the actual/attained schooling, we find that differences in schooling requirements across firms have caused considerable variations between the required and actual/attained schooling for high, medium and low skilled groups. When we interpret the required schooling as the demand for skills and the actual/attained schooling as the supply of skills, we observe that the inconsistency between the required and actual/attained schooling indicates an inconsistency between the demand for and supply of skills, which can be interpreted as skills mismatch.[11] For instance, Fig. 7.6 below illustrates the differences between the required and actual/attained schooling across firms, defined by firm size and industry level and skill levels. We observe that the inconsistency between the demand for and supply of skills, or skills mismatch, is particularly higher/serious within both

[11] Our definition of actual education refers to educational attainment classified under three groups: high (post secondary) educational attainment: university degree and above (16 years of schooling); medium educational attainment: secondary education (12 years of schooling); and low educational attainment: less than secondary education (9 years of schooling). We define the required education by the translated merged required qualifications for each occupation group defined by average years of schooling. The occupational classification includes the following five categories/ groups: (1) managers, professional, management executive, scientific, technical and engineers; (2) clerical and administrative; (3) skilled craftsmen; (4) plant machinery operators, assemblers and elementary occupation; and (5) other workers. We translate the required qualifications associated with each occupational class into average years of schooling and group them in the following way: (1) PhD/postgraduate (19–20 years); (2) professional, MSc/ postgraduates (18 years); (3) university graduates (16 years); (4) Diploma (14 years); (5) higher/ secondary schooling (12 years); and (6) less than secondary schooling (9 years). We then merge the required qualifications into three groups, assuming that the high occupation group includes both the first and second occupation categories, the medium occupation group includes both the third and fourth occupation categories and, finally, the low occupation group includes the fifth occupation category. We then use this definition to compare between the required education for each occupation class and actual/attained education, and we assume that the difference between these indicates the presence of skills mismatch between jobs requirements and educational attainment.

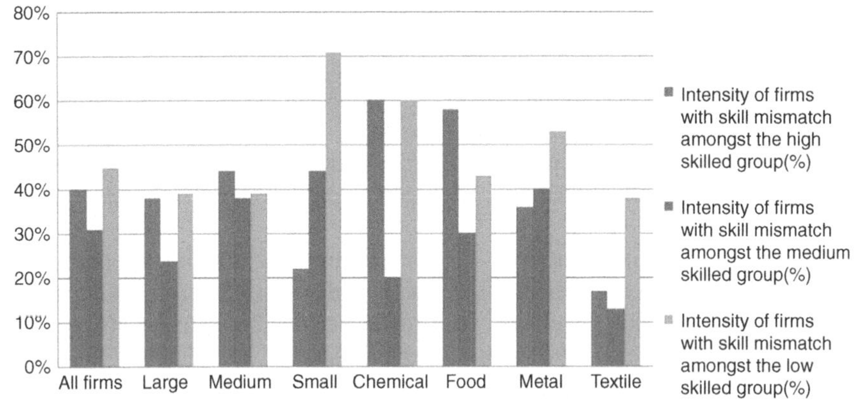

Fig. 7.6 Skills mismatch defined by high medium and low skill levels across firms (%) (2008) (Source: Firm Survey (2010))

high and low skilled groups respectively and across medium, small, chemical, food and metal firms respectively. We find mismatch amongst all employment categories, especially within high, medium and low skilled labour; for instance, we observe that for all firms, on average, the intensity of mismatch for high, medium and low skill groups accounts for 40 %, 31 % and 45 % respectively. This implies that the educational attainment amongst high, medium and low skilled labour does not match the required skills/educational level for high, medium and low skilled jobs across approximately 40 %, 31 % and 45 % of total respondents firms respectively. The mismatch is highest for high, medium and low skills, probably because of both insufficient educational attainment and high educational requirements for high, medium and low skills (see Fig. 7.3 above). Moreover, the intensity of mismatch is more prevalent across small size and medium size and chemical, metal and food firms compared to large size and textile firms. For instance, for medium size firms, on average the mismatch intensity for high, medium and low skill groups accounts for 44 %, 38 % and 39 % respectively, while for small size firms the figures are 22 %, 44 % and 71 % respectively, whereas for large size firms the figures are 38 %, 24 % and 39 % respectively. Moreover, for the chemical industries, on average the mismatch intensity for high, medium and low skill groups accounts for 60 %, 20 % and 60 % respectively, while for food industries the figures are 58 %, 30 % and 43 % respectively, whereas for metal industries the figures are 36 %, 40 % and 53 % respectively, while for textile industries the figures are 17 %, 13 % and 38 % respectively. Hence, our results in this section concerning the presence of serious skills mismatch due to the excessive share of unskilled foreign workers at the micro level are consistent with our earlier findings in Chap. 5 above, which indicates the presence of serious skills mismatch at the macro level.

7.3.2.2 Low Skill Levels and the Declining Trend of Labour Productivity (Output/Labour Ratio)

The low skill levels may contribute to productivity decline across firms.[12] Table 7.3 below illustrates considerable variation in the value and trend of labour productivity (total output/labour ratio) in physical term, in particular, considerable decline in labour productivity (output/labour ratio) for numerous firms over the period 2005–08.[13,14,15]

The declining labour productivity across many firms may not be surprising since the majority of employed workers are low skilled/educated workers – see our result above – and a low skill level may lead to further decline in productivity. For instance, Table 7.3 below shows that over the periods 2005–2006, 2006–2007 and 2007–2008, the declining trend of labour productivity is reversed across 8 out of 37 (22 %) of all respondent firms and the increasing trend continues across 16 out of 37 firms (43 %); however, the increasing trend turns into a declining one across 11 out of 37 firms (30 %), or the declining trend continues across 2 out of 37 (5 %) of all respondent firms. Hence, for the majority 24 out of 37 (65 %) of all respondent firms either the declining trend turns into an increasing one or the increasing trend continues, but for the remaining 13 out of 37 (35 %), i.e. for more than one third of all firms either the increasing trend turns into a declining one or the declining trend continues. For chemical firms over the periods 2005–2006, 2006–2007 and 2007–2008, the declining trend of labour productivity is reversed across 5 out of 18 (28 %) of the chemical firms and the increasing trend continues across 9 out of 18 firms (50 %); however, the increasing trend turns into a declining one across 2 out of 18 firms (11 %), or the declining trend continues across 2 out of 18 (11 %) of the chemical firms. Thus, for the majority 14 out of 18 (68 %) of the chemical firms either the declining trend turns into an increasing one or the increasing trend continues, but for the remaining 4 out of 18 (22 %), i.e. for more than one fifth of the chemical firms either the increasing trend turns into a declining one or the declining trend continues. For food firms over the periods 2005–2006,

---

[12] Productivity is measured in physical terms (tonnes, litres, etc.) for selected firms according to availability of data.

[13] The results from the firm survey (2010) indicate that the declining labour productivity seems to be more sensitive to industry level and less sensitive to firm size as reported by 35 %, 22 %, 42 %, 67 %, 50 %, 47 %, 20 % and 33 % of all firms, chemical, food, metal, textile, large, medium and small size firms respectively.

[14] Due to the small number of observations on the declining trend of labour productivity, our results should be interpreted carefully as probably this may not be the only case; other possible explanations are either the steady or increasing trends amongst the non-respondent firms.

[15] In Table 7.3 we limit our analysis of the productivity decline to compare only the change in labour productivity over the periods 2005–2006, 2006–2007 and 2007–2008 across 42 of the respondent firms. Since our data only reflects skill levels for the year 2008, but does not reflect the change in skill levels over the period 2005–2008. That means we cannot compare the change in productivity with the change in skill levels, so as to attribute the declining trend in productivity over the period 2005–2008 to the declining trend in skill levels.

Table 7.3 Assessment of industrial performance: labour productivity: output/labour ratio measured in physical term across firms (2005–2008)

Variables		Value (2005–2008)				Trend (2005–2008)				Growth rate (2005–2008)			
Average ratio of prod/lab	Group of firms	2005	2006	2007	2008	2005–2006	2006–2007	2007–2008	2005–2008	2005–2006	2006–2007	2007–2008	2007–2008
Chemical	Small	104727.27	119172.41	119172.41	199384.62	+		+	+	14		67	90
		40088.065	59703.806	108426.95	123457.95	+	+	+	+	49	82	14	208
		36000000	32000000	16666667	16666667	−	−		−	−11	−48	0	−54
		2222222.2	2666666.7	3125000	3500000	+	+	+	+	20	17	12	58
		66666.667	71428.571	100000	111111.11	+	+	+	+	7	40	11	67
		57500	63333.333	70000	76666.667	+	+	+	+	10	11	10	33
					25452.475			+	+				
	Medium	1	1	1	8			+	+			700	700
		34379.275	45957.464	33308.435	19848.217	+	−	−	−	34	−28	−40	−42
				18700	36083.333		+	+	+				
		3555.5556	3384.6154	6000	6666.6667	−	+	+	+	−5	77	11	87
	Large		57.915058	61.776062	34.749035	+	+	−	+		7	−44	
				31500	69443.925		+	+	+			120	
		2632.9588	2840.8	2535.7143	3392.3077	+	−	+	+	8	−11	34	29
		78750	70000	120000	133333.33	−	+	+	+	−11	71	11	69
		26557.053	23979.592	20715.596	14434.783	−	−	−	−	−10	−14	−30	−46
			42.857143	41.25	41.257778	+	−	+	+		−4		
				156393.16	179986.3		+	+	+			15	
		78026.433	77137.71	85601.63	90375.527	−	+	+	+	−1	11	6	16
Food	Small	6666.6667	18181.818	36000	1500000	+	+	+	+	173	98	4067	22400
					3055555.6			+	+				
			166.66667	233.33333	400	+	+	+	+		40	71	
		13437500	1.875E + 09	15625000	15000000	+	−	−	+	13853	−99	−4	12
	Medium			60023.543	63433.078		+	+	+			6	
		68432.523	87238.779		17700.333	+	−	+	−	27	−100		−74
		673043.48	673043.48	673043.48	673043.48								
		17.763158	20	24.193548	29.824561	+	+	+	+	13	21	23	68
			4122.75	6369.5918	14983.607	+	+	+	+		54	135	
		7812.5	8531.25	10000	7995.5156	+	+	−	+	9	17	−20	2

(continued)

Table 7.3 (continued)

Variables		Value (2005–2008)				Trend (2005–2008)				Growth rate (2005–2008)			
Average ratio of prod/lab	Group of firms	2005	2006	2007	2008	2005–2006	2006–2007	2007–2008	2005–2008	2005–2006	2006–2007	2007–2008	2007–2008
	Large	172413793	155172414	181034483	65517241	–	+	–	–	−10	17	−64	−62
		96	100.46512	105.36585	109.36709	+	+	+	+	5	5	4	14
		84975.92	97098.257	123589.7	78070.618	+	+	–	–	14	27	−37	−8
		1600	1750	1987.1795	1826.9231	+	+	–	+	9	14	−8	14
Metal	Small				600000			+	+				
		29500	43500	54666.667	5909.0909	+	+	–	–	47	26	−89	−80
		67045455	123456790	79069767	283333333	+	–	+	+	84	−36	258	323
	Medium				538461.54			+	+				
	Large	166.66667	200	235.29412	218.18182	+	+	–	+	20	18	−7	31
Textile	Small	937.5	1187.5	1615.3846	1538.4615	+	+	–	+	27	36	−5	64
	Medium	54110.855	75207.851	47192.489	65621.116	+	–	+	+	39	−37	39	21
	Large			45	450.23333		+	+	+			901	
		251845.94	8436724.6	69963.713	64541.195	+	–	–	–	3250	−99	−8	−74

Source: Own calculation based on the firm survey (2010)

2006–2007 and 2007–2008, the declining trend of labour productivity is reversed across 1 out of 12 (8 %) of the respondent firms and the increasing trend continues across 6 out of 12 firms (50 %); however, the increasing trend turns into a declining one across 5 out of 12 firms (42 %). Therefore, for more than half or the majority 7 out of 12 (58 %) of the food firms either the declining trend turns into an increasing one or the increasing trend continues, but for the remaining 5 out of 12 (42 %), i.e. for more than one third and near to one half of the food firms either the increasing trend turns into a declining one or the declining trend continues. For metal firms over the periods 2005–2006, 2006–2007 and 2007–2008, the declining trend of labour productivity is reversed across 1 out of 3 (33 %) of the metal firms; however, the increasing trend turns into a declining one across 2 out of 3 (67 %) of the metal firms. Hence, for the majority 2 out of 3 (67 %), i.e. for more than two third of the metal firms the increasing trend turns into a declining one, but for the remaining 1 out of 3 (33 %) the declining trend turns into an increasing one. For textile firms over the periods 2005–2006, 2006–2007 and 2007–2008, the declining trend of labour productivity is reversed across 1 out of 4 (25 %) of the textile firms and the increasing trend continues across 1 out of 4 firms (25 %); however, the increasing trend turns into a declining one across 2 out of 4 firms (50 %). Thus, for the first half (2 out of 4 or 50 %), i.e. for one half of the textile firms either the declining trend turns into an increasing one or the increasing trend continues, while for the other half (2 out of 4 or 50 %) the increasing trend turns into a declining one For large size firms over the periods 2005–2006, 2006–2007 and 2007–2008, the declining trend of labour productivity is reversed across 4 out of 15 (27 %) of the large size firms and the increasing trend continues across 4 out of 15 firms (27 %); however, the increasing trend turns into a declining one across 6 out of 15 firms (40 %), or the declining trend continues across 1 out of 15 (7 %) of the large size firms. Thus, for the majority 8 out of 15 (53 %), i.e. for more than one half of the large size firms either the declining trend turns into an increasing one or the increasing trend continues, but for the remaining 7 out of 15 (47 %) either the increasing trend turns into a declining one or the declining trend continues. For medium size firms over the periods 2005–2006, 2006–2007 and 2007–2008, the declining trend of labour productivity is reversed across 3 out of 10 (30 %) of the medium size firms and the increasing trend continues across 5 out of 10 firms (50 %); however, the increasing trend turns into a declining one across 2 out of 10 (20 %) of the medium size firms. Thus, for the majority 8 out of 10 (80 %) of the medium size firms either the declining trend turns into an increasing one or the increasing trend continues, but for the remaining 2 out of 10 (20 %), i.e. for one fifth of the medium size firms the increasing trend turns into a declining one. For small size firms over the periods 2005–2006, 2006–2007 and 2007–2008, the declining trend of labour productivity is reversed across 1 out of 12 (8 %) of the small size firms and the increasing trend continues across 7 out of 12 firms (58 %); however, the increasing trend turns into a declining one across 3 out of 12 (25 %) of the small size firms, or the declining trend continues across 1 out of 12 (8 %) of the small size firms. Thus, for the majority 8 out of 12 (67 %) of the small size firms either the declining trend turns into an increasing one or the increasing trend continues, but for the remaining 4 out of 12 (33 %), i.e. for one third of the small size firms either the increasing trend turns into a declining one or the declining trend continues.

Therefore, our results in this section concerning the declining labour productivity are consistent with our results regarding the declining industrial performance indicators that we measure by three sets of economic-productivity, activity and profitability indicators at the micro level as we show in the next section (see Tables 7.4 and 7.5 below).

7.3.2.3 Low Skill Levels and the Declining Trend of Other Industrial Performance Indicators

The low skill levels may contribute to the decline of industrial performance indicators across firms. The trend of these indicators show considerable variation across firms and in most cases seem to be more sensitive to differences in firm size, industry and sector, in particular, the average performance ratio for different indicators for numerous firms show a considerable decline over the period 2005–2008. Tables 7.4 and 7.5 below illustrate an assessment of the value and trend of industrial performance indicators across firms over the period 2005–2008, which we measure by three different sets of economic and productivity indicators, activity indicators and profitability indicators. Using Al-Quraishi's (2005) definition of industrial performance, first we measure the first set of economic indicators by three indicators including first the degree of industrialisation that is indicated by the ratio of total value added as a percentage of total output measured by total sales value, and second the capital intensity level indicators that we measure by the ratios of capital and fixed capital – measured by total spending in machinery and equipment – as percentages to total labour respectively. We define the third economic indicator by a set of productivity indicators that we measure by: labour productivity indicator measured by the ratio of total value added as a percentage to total labour; capital productivity indicator measured by the ratio of total output measured by total sales value as percentages of total capital; fixed capital productivity indicator measured by the ratio of output measured by total sales value as a percentage of fixed capital or machinery and equipment; wage productivity indicator that we measure by the total output measured by total sales value as a percentage of total wage; and raw materials productivity indicator measured by the ratio of total output measured by total sales value as a percentage to total spending on raw materials. Second, we measure the second set of activity indicators or ratios by first the fixed capital turnover ratio that we measure by the ratio of total sales value as a percentage of fixed capital, and second the capital turnover ratio that we measure by the ratio of total sales value as a percentage of total capital. Third, we measure the third set of profitability indicators by three indicators including first the rate of return on labour that we measure by profit/labour ratio, second the rate of return on capital that we measure by the ratio of profit as a percentage to capital and third profit margin indicator that we measure by the ratio of profit as a percentage to total sales value (Al-Quraishi 2005: 249–277).

Table 7.4 Assessment of the value, trend and growth rates of industrial performance: economic, activity, labour productivity, output/labour and capital/labour ratios and other productivity indicators across firms (2005–2008)

Variables	Group of firms	Value (2005–2008)				Trend (2005–2008)				Growth rate (2005–2008)			
Average ratio of		2005	2006	2007	2008	2005–2006	2006–2007	2007–2008	2005–2008	2005–2006	2006–2007	2007–2008	2005–2008
1. Economic indicators													
The degree of industrialization = value added/output (sale value)	Chemical	20.071384	0.1006944	0.1102117	0.121095	–	+	+	–	−99	9	10	−99
	Food	5.59262	5.7751745	7.4236966	7.7370768	+	+	+	+	3	29	4	38
	Metal	0.0309524	0.0166667	0.1135593	0.0888889	–	+	–	+	−46	581	−22	187
	Textile	0.1262011	0.1229581	0.1379861	0.1255524	–	+	–	–	−3	12	−9	−1
	Large	25.063934	0.0877692	0.0970537	0.1137513	–	+	+	–	−100	11	17	−100
	Medium	7.4804304	7.243592	7.4348101	7.7427771	–	+	+	+	−3	3	4	4
	Small	0.0627458	0.0632965	0.1214039	0.1166114	+	+	–	+	1	92	−4	86
	Private	12.28653	2.6815389	3.1682737	3.3004401	–	+	+	–	−78	18	4	−73
	Mixed	0.0727273	0.0725	0.075	0.0733333	–	+	–	+	0	3	−2	1
	All firms	6.4552894	1.5038734	1.9463634	2.0181533	–	+	+	–	−77	29	4	−69
2. Activityand labourproductivity indicators													
Capital/labour	Chemical	35065602	27590847	20278975	16777192	–	–	–	–	−21	−27	−17	−52
	Food	7.16E + 09	6.10E + 09	6.68E + 09	7.52E + 09	–	+	+	+	−15	9	13	5
	Metal	30876818	74171574	81460349	104421212	+	+	+	+	140	10	28	238
	Textile	10757338	11308723	9320350.8	9304595.7	+	–	–	–	5	−18	0	−14
	Large	18022260	44082391	7016133	11853455	+	–	+	–	145	−84	69	−34
	Medium	716288.06	540146.61	4118685.8	2901363.1	–	+	–	+	−25	663	−30	305
	Small	4.59E + 09	5.01E + 09	5.50E + 09	5.83E + 09	+	+	+	+	9	10	6	27
	Private	2.39E + 09	2.28E + 09	2.08E + 09	2.29E + 09	–	–	+	–	−4	−9	10	−4
	Mixed	153115039	212744727	108623182	117178164	+	–	+	–	39	−49	8	−23
	All firms	1.809E + 09	1.553E + 09	1.697E + 09	1.912E + 09	–	+	+	+	−14	9	13	6
Fixed capital (machinery and equipment)/Labour	Chemical	17324619	14756079	13091702	11310615	–	–	–	–	−15	−11	−14	−35
	Food	296053860	68495853	59997873	63588812	–	–	+	–	−77	−12	6	−79
	Metal	32035906	39615035	41916373	102960921	+	+	+	+	24	6	146	221
	Textile	2866267.3	2503114.8	37533.295	39332.76	–	–	+	–	−13	−99	5	−99
	Large	260069994	51927232	41157115	47809051	–	–	+	–	−80	−21	16	−82
	Medium	84350.355	60784.395	3060039.9	1746163.8	–	+	–	+	−28	4934	−43	1970
	Small	24976660	28926310	27914835	44197977	+	–	+	+	16	−3	58	77
	Private	3438151.1	3258447.2	4053537.4	10617005	–	+	+	+	−5	24	162	209
	Mixed	760056666	283521984	249015472	273580008	–	–	+	–	−63	−12	10	−64
	All firms	87070163	31342520	28760870	44474920	–	–	+	–	−64	−8	55	−49
Raw materials/labour	Chemical	2871098.1	97839439	100277944	107446906	+	+	+	+	3308	2	7	3642
	Food	88134391	9517892.2	26412145	14109863	–	+	–	–	−89	177	−47	−84
	Metal	9684090.9	37673071	15697858	27849382	+	–	+	+	289	−58	77	188

(continued)

Table 7.4 (continued)

Variables	Group of	Value (2005–2008)				Trend (2005–2008)				Growth rate (2005–2008)			
Average ratio of	firms	2005	2006	2007	2008	2005–2006	2006–2007	2007–2008	2005–2008	2005–2006	2006–2007	2007–2008	2005–2008
	Textile	1472742.7	8473466	9432995.2	1311518.5	+	+	–	–	475	11	−86	−11
	Large	78630305	116823438	141977429	153140881	+	+	+	+	49	22	8	95
	Medium	526713.27	653338.89	4091163.3	2391839.4	+	+	–	+	24	526	−42	354
	Small	5617402.1	15395691	8733551.6	10354531	+	–	+	+	174	−43	19	84
	Private	5518075	60875381	64930925	67173818	+	+	+	+	1003	7	3	1117
	Mixed	213506237	6285777.8	54637414	22482133	–	+	–	–	−97	769	−59	−89
	All firms	25540581	38375967	37955236	37679417	+	–	–	+	50	−1	−1	48
Wage/labour	Chemical	629661.54	691335.32	1107005.7	1202982.3	+	+	+	+	10	60	9	91
	Food	19795120	4465405.3	5182546.5	5001834	–	+	–	–	−77	16	−3	−75
	Metal	2795004.4	7412864.2	7561777.4	8655813.8	+	+	+	+	165	2	14	210
	Textile	1123823.6	1174157	10385.944	7701.6136	+	–	–	–	4	−99	−26	−99
	Large	17960782	4505029.2	3964134.6	4244171.8	–	–	+	–	−75	−12	7	−76
	Medium	53597.363	68865.449	416449.87	258538.56	+	+	–	+	28	505	−38	382
	Small	1290975.4	2367999.3	2820244.4	4674367.5	+	+	+	+	83	19	66	262
	Private	1827576.6	1953272.4	2047724.2	2477402.8	+	+	+	+	7	5	21	36
	Mixed	44890563	7038915.8	7394609.9	9456880.8	–	+	+	–	−84	5	28	−79
	All firms	6085902.4	3435940.5	3465428.9	3717082.9	–	+	+	–	−44	1	7	−39
Sales value (output)/labour	Chemical	2965521.1	62897297	50687940	55577289	+	–	+	+	2021	−19	10	1774
	Food	948316463	882596205	729683552	623900429	–	–	–	–	−7	−17	−14	−34
	Metal	65918758	43718037	101317367	120345041	–	+	+	+	−34	132	19	83
	Textile	1667595	3069766.2	956358.61	1035369.3	+	–	+	–	84	−69	8	−38
	Large	143338675	119495935	110775961	116574335	–	–	+	–	−17	−7	5	−19
	Medium	1.15E + 09	893753071	735194023	578678320	–	–	–	–	−22	−18	−21	−50
	Small	9459741.1	168685305	9030811.2	23463531	+	–	+	+	1683	−95	160	148
	Private	337535170	361251148	267733421	250215178	+	–	–	–	7	−26	−7	−26
	Mixed	316187014	60302554	57519586	43154592	–	–	–	–	−81	−5	−25	−86
	All firms	254717084	248070326	220661304	200214532	–	–	–	–	−3	−11	−9	−21
Value added/labour	Chemical	2551317.2	955660.37	1313488.7	1058510.2	–	+	–	–	−63	37	−19	−59
	Food	34483247	5451257.3	5644107.6	4247107.3	–	+	–	–	−84	4	−25	−88
	Metal	1477272.7	1851851.9	7790697.7	23333333	+	+	+	+	25	321	200	1479
	Textile	316276.81	285201.47	444191.11	482803.84	–	+	+	+	−10	56	9	53
	Large	36327013	6253965	6471635.7	4749855.5	–	+	–	–	−83	3	−27	−87
	Medium	3795.4275	4626.0202	1069155.8	762380.47	+	+	–	+	22	23012	−29	1987
	Small	1495991.8	1884495.9	2709645.6	5814842.9	+	+	+	+	26	44	115	289
	Private	1486836.5	859713.02	1576540.1	2743335.1	–	+	+	+	−42	83	74	85
	Mixed	68965517	12500000	12931034	9482758.6	–	+	–	–	−82	3	−27	−86
	All firms	9707028.4	2135992.8	3798121.3	7280438.6	–	+	+	–	−78	78	92	−25

Source: Firm Survey (2010): own calculation from the firm survey (2010)

Table 7.5 Assessment of the value, trend and growth rates of industrial performance: Activity, other productivity and profitability indicators across firms (2005–2008)

Variables	Group of firms	Value (2005–2008)				Trend (2005–2008)				Growth rate (2005–2008)			
Average ratio of		2005	2006	2007	2008	2005–2006	2006–2007	2007–2008	2005–2008	2005–2006	2006–2007	2007–2008	2005–2008
3. Activity and other productivity indicators													
Fixed capital turnover ratio = sale/sale value/fixed capital (machinery and equipment)	Chemical	2.8912123	495.92478	308.19507	366.82629	+	–	+	+	17053	–38	190	12588
	Food	3.1793517	189.03297	146.05162	114.63459	+	–	–	+	5846	–23	–22	3506
	Metal	0.408299	0.8018738	0.648209	0.8518113	+	–	+	+	96	–19	31	109
	Textile	101.48658	60.764981	61.229179	87.570441	–	+	+	–	–40	1	43	–14
	Large	1.8379	673.70374	449.75099	451.41571	+	–	+	+	36556	–33	0	24461
	Medium	3.4770759	2.9621226	3.2838275	134.22139	–	+	+	+	–15	11	3987	3760
	Small	30.012525	160.39301	126.72865	116.94158	+	–	–	+	434	–21	–8	290
	Private	17.22597	353.64104	242.29702	273.28094	+	–	+	+	1953	–31	13	1486
	Mixed	0.1603258	0.3773134	0.4740032	0.496713	+	+	+	+	135	26	5	210
	All firms	26.991361	186.63115	129.03102	142.47078	+	–	+	+	591	–31	10	428
Capital turnover ratio = sale value/capital	Chemical	42.418878	40.067628	3.634055	59.389793	–	–	+	+	–6	–91	1534	40
	Food	412.50789	498.51296	499.94462	255.81401	+	+	–	–	21	0	–49	–38
	Metal	0.4260684	0.4852564	0.4158425	0.6116496	+	–	+	+	14	–14	47	44
	Textile	0.1639614	1.1928782	3.3614254	8.2609911	+	+	+	+	628	182	146	4938
	Large	67.82426	60.273972	6.1373149	44.09251	–	–	+	–	–11	–90	618	–35
	Medium	573.8511	462.21815	444.30504	259.79308	–	–	–	–	–19	–4	–42	–55
	Small	1.0437791	68.812585	0.8258962	0.6976534	+	–	–	–	6493	–99	–16	–33
	Private	167.38012	194.48088	145.13254	111.76126	+	–	–	–	16	–25	–23	–33
	Mixed	2.6348545	0.5323247	3.8765243	11.515423	–	+	+	+	–80	628	197	337
	All firms	113.8792	135.06468	126.83899	81.019111	+	–	–	–	19	–6	–36	–29
Wage productivity ratio = output/(sale value)/wage	Chemical	8.3525744	8.1943495	7.3845158	361.40226	–	–	+	+	–2	–10	4794	4227
	Food	428.41408	552.93182	400.66736	344.83311	+	–	–	–	29	–28	–14	–20
	Metal	4.9500291	5.7200371	5.9344161	12.735474	+	+	+	+	16	4	115	157
	Textile	523.82575	315.64078	506.53952	548.33265	–	+	+	+	–40	60	8	5
	Large	9.4032343	10.088306	9.3580525	9.8078386	+	–	+	+	7	–7	5	4
	Medium	10.118612	14.996222	8.6892371	511.03983	+	–	+	+	48	–42	5781	4950
	Small	496.53461	614.09712	557.16857	538.03343	+	–	–	+	24	–9	–3	8
	Private	242.53064	253.70029	205.28462	379.92113	+	–	+	+	5	–19	85	57
	Mixed	2.4749747	4.802963	4.7386977	3.6878688	+	–	–	+	94	–1	–22	49
	All firms	241.38561	220.62175	230.13145	316.82587	–	+	+	+	–9	4	38	31
Raw materials productivity ratio = output/(sale value)/raw materials	Chemical	1.472792	1.5531448	1.3339254	366.07809	+	–	+	+	5	–14	27344	24756
	Food	6.8706046	1523.3263	157.43672	132.15185	+	–	–	+	22072	–90	–16	1823
	Metal	1.5252941	1.3110383	73.819865	2.4578007	–	+	–	+	–14	5531	–97	61
	Textile	2.0123894	2.0076766	4.1239329	2.1136204	–	+	–	+	0	105	–49	5
	Large	1.8617158	2.8742128	2.3465453	2.0572816	+	–	–	+	54	–18	–12	11
	Medium	2.5102906	2.0460816	154.59166	758.66946	–	+	+	+	–18	7455	391	30122
	Small	5.4970686	1521.3296	18.090003	4.4274845	+	–	–	–	27575	–99	–76	–19
	Private	3.8347064	549.06214	53.093879	255.13267	+	–	+	+	14218	–90	381	6553

(continued)

Table 7.5 (continued)

Variables	Group of	Value (2005–2008)				Trend (2005–2008)				Growth rate (2005–2008)			
Average ratio of	firms	2005	2006	2007	2008	2005–2006	2006–2007	2007–2008	2005–2008	2005–2006	2006–2007	2007–2008	2005–2008
	Mixed	0.6954618	4.7421652	4.0867543	1.1633541	+	–	–	+	582	–14	–72	67
	All firms	2.97027	382.04954	59.178611	125.70034	+	–	+	+	12762	–85	112	4132
4. Profitability indicators													
Rate of return on labour = profit/labour	Chemical	325003.77	97520339	93381463	89415111	+	–	–	+	29906	–4	–4	27412
	Food	16073000	16556605	15371532	8457780.9	+	–	–	–	3	–7	–45	–47
	Metal	25195354	10424630	42266973	45856265	–	+	+	+	–59	305	8	82
	Textile	928346.78	512349.74	226633.83	237267.99	+	+	–	+	–45	–56	5	–74
	Small	14843443	3943959.6	1559333	7393518.3	+	–	+	+	–127	–60	374	–150
	Medium	246042.07	280699.08	261740.38	377443.01	+	–	+	+	14	–7	44	53
	Large	30615658	107746075	119654268	116398986	+	+	–	+	252	11	–3	280
	Private	1343693.4	56024755	59488296	55206716	+	+	–	+	4069	6	–7	4009
	Mixed	62279128	10959241	10342936	8606965	–	–	–	–	–82	–6	–183	–114
	All firms	10166253	30997306	37698334	35872972	+	+	–	+	205	22	–5	253
Rate of return on capital = profit/capital	Chemical	7.202629	6.792908	5.1351797	8.9627444	–	–	+	+	–6	–24	75	24
	Food	2.2030291	10.639528	10.380229	15.224402	+	–	–	–	383	–2	–247	–791
	Metal	0.0336182	0.0365385	0.0194139	0.0901709	+	–	+	+	9	–47	364	168
	Textile	0.358709	2.3015254	0.2610921	0.2539136	–	+	–	–	–742	–111	–197	–171
	Large	10.3743	9.2316877	7.0692848	8.0630912	–	–	+	–	–11	–23	14	–22
	Medium	13.761543	10.120505	9.8050205	7.547615	–	–	–	–	–26	–3	–177	–155
	Small	6.6239792	0.6008153	0.0492431	0.0834129	+	–	+	+	–109	–92	69	–101
	Private	4.3952252	6.8602373	6.0800742	0.1484943	+	–	–	–	56	–11	–102	–103
	Mixed	0.3483543	0.0733435	1.231029	1.2917988	–	–	+	+	–121	1578	–205	271
	All firms	2.4494963	3.7918623	3.9489787	1.6063501	+	+	–	–	55	4	–141	–166
Profit margin = profit/sale	Chemical	4.9863121	4.0622007	2.6593402	2.4372959	–	–	–	–	–19	–35	–8	–51
	Food	4444.0867	4000.932	4000.7488	3334.3731	–	–	–	–	–10	0	–17	–25
	Metal	0.0272396	0.1014987	0.1787782	0.2703077	+	+	–	–	–473	76	–251	892
	Textile	3.7304507	2.0300487	0.0842857	0.2035356	+	+	–	+	–46	–104	–341	–95
	Large	471.82625	0.9103777	0.4437194	0.3052073	–	–	–	–	–100	–51	–31	–100
	Medium	10.019226	6.4150969	5.6685649	4.5480634	–	–	–	–	–36	–12	–20	–55
	Small	3998.7518	4000.0724	3636.4011	3333.2799	+	–	–	–	0	–9	–8	–17
	Private	1741.3817	1430.6505	1178.2029	1054.1797	–	–	–	–	–18	–18	–11	–39
	Mixed	3.8074523	0.0536275	0.0249647	0.0988276	+	–	+	+	–101	–53	296	–103
	All firms	1111.3288	1000.7664	1000.9178	834.08414	–	+	–	–	–10	0	–17	–25

Source: Own calculation based on the firm survey (2010)

Beginning with the first set of economic indicators, we find that for all firms the trend of value and growth rate of the economic indicator as measured by the degree of industrialisation as measured by the value added/sales value (output) ratio, showed a negative decreasing trend over the periods 2005–2006 and 2005–2008 but that again turned into a positive increasing trend over the periods 2006–2007 and 2007–2008. In particular, we find that the economic indicator as measured by the degree of industrialisation as measured by the value added/sales value (output) ratio varied across firms over the period 2005–2008, for instance, either the declining trend continued or the increasing trend turned into a declining trend for metal and textile industries, small size and mixed firms, while by contrast either the increasing trend continued for food industries or the declining trend turned into an increasing trend for all firms, chemical industries and medium size and large size and private firms. Moreover, as for the second economic indicator of capital intensity and productivity indicator as measured by capital/labour productivity indicator or ratio, we find that for all firms the trend of value and growth rate of capital/labour ratio showed a negative decreasing trend over the period 2005–2006 that turned into a positive increasing trend over the periods 2006–2007, 2007–2008 and 2005–2008. In particular, we find that the capital intensity and productivity indicator measured by capital/labour ratio varied across firms over the period 2005–2008, for instance, either the declining trend continued for the chemical industries or the increasing trend turned into a declining trend for textile industries and medium size firms, whereas by contrast either the increasing trend continued or the declining trend turned into an increasing trend for all firms, food and metal industries and small size and large size and private and mixed firms. Moreover, we find that for all firms the trend of value and growth rate of the second economic indicator of capital intensity and productivity indicator measured by fixed capital/labour ratio measured by machinery and equipment/labour ratio showed a negative decreasing trend over the periods 2005–2006, 2006–2007 and 2005–2008 that turned into a positive increasing trend over the period 2007–2008. In particular, we find that the capital intensity and productivity indicator measured by fixed capital/labour ratio measured by machinery and equipment/labour ratio varied across firms over the period 2005–2008, for instance, either the declining trend continued for the chemical industries or the increasing trend turned into a declining trend for the medium size firms, whereas by contrast either the increasing trend continues or the declining trend turned into an increasing trend for all firms, food, metal and textile industries, small size and large size, private and mixed firms. In addition, we find that for all firms the trend of value and growth rate of raw materials/labour ratio showed a positive increasing trend over the periods 2005–2006 and 2005–2008 that turned into a negative decreasing trend over the periods 2006–2007 and 2007–2008. In particular, we find that the raw materials/labour ratio varied across firms over the period 2005–2008, for instance, either the declining trend continues or the increasing trend turned into a declining trend for all firms, food and textile industries and medium size and mixed firms, while by contrast either the increasing trend continued for the chemical industries, large size and private firms or the declining trend turned into an increasing trend for metal industries and small size firms. Moreover, we find that for all firms the trend of value and growth rate of wages/labour ratio showed a negative decreasing trend over the periods 2005–2006 and 2005–2008 that turned into a positive increasing trend over the periods 2006–2007 and 2007–2008. In particular, we find that wages/labour ratio varied across firms over the period 2005–2008, for instance, either the declining trend continued or the increasing trend turned into a declining trend for food and textile industries and medium size firms, while either the increasing trend continued for the chemical and metal industries and small size and private firms, or the declining trend turned into an increasing trend for all firms, large size and mixed firms. Moreover, we find that for all firms the trend of value and growth rate of sales value (output)/labour ratio showed a negative decreasing trend over all the periods 2005–2006, 2006–2007, 2007–2008 and 2005–2008. In particular, we find that sales value (output)/labour ratio varied across firms over the period 2005–2008, for instance, either the declining trend continued for all firms, food industries, medium size, and mixed firms or the increasing trend turned into a declining trend for private firms, while by contrast either the increasing trend continued or the declining trend turned into an increasing trend for the chemical, metal, textile industries and small size and large size firms. Moreover, we find that for all firms the trend of value and growth rate of value added/labour ratio showed a negative decreasing trend over the periods 2005–2006 and 2005–2008 that turned into a positive increasing trend over the periods 2006–2007 and 2007–2008. In particular, we find that the value added/labour ratio vary across firms over the period 2005–2008, for instance, either the declining trend continued or the increasing trend turned into a declining trend for chemical and food industries, medium and large size and mixed firms, while by contrast either the increasing trend continued for metal industries and small size firms or the declining trend turned into an increasing trend for all firms, textile and private firms (see Table 7.4 above). Moreover, we find that for all firms the trend of value and growth rate of other productivity indicators as measured by the wage productivity ratio as measured by sales/wage ratio showed a negative decreasing trend over the period 2005–2006 that turned into a positive increasing trend over all the periods 2006–2007, 2007–2008 and 2005–2008. In particular, we find that the other productivity indicators as measured by the wage productivity ratio as measured by sales/wage ratio varied across firms over the period 2005–2008, for instance, either the declining trend continued or the increasing trend turned into a declining trend for food industries and small size and mixed firms, while by contrast either the increasing trend continued for metal industries or the declining trend turned into an increasing trend for all firms, chemical and textile industries and medium size and large size and private firms. Moreover, we find that for all firms the trend of value and growth rate of other productivity indicators as measured by the raw materials productivity as measured by the sales/raw materials ratio showed a positive increasing trend over the period 2005–2006, that turned into a negative decreasing trend over the period 2006–2007 but that again turned into a positive increasing trend over the periods 2007–2008 and 2005–2008. In particular, we find that the other productivity indicators as measured by the raw materials productivity as measured by the sales/raw materials ratio varied across firms over the period

2005–2008, for instance, either the declining trend continued or the increasing trend turned into a declining trend for food, metal and textile industries and small size and large size and mixed firms, while by contrast either the increasing trend continued or the declining trend turned into an increasing trend for all firms, chemical industries and medium size and private firms. Moreover, we find that for all firms the trend of value and growth rate of the second set of indicators (the activity indicators as measured by fixed capital turnover ratio as defined by the sales/fixed capital (machinery and equipment) ratio) showed a positive increasing trend over the periods 2005–2008 and 2005–2006, that turned into a negative decreasing trend over the period 2006–2007 but that again turned into a positive increasing trend over the period 2007–2008. In particular, we find that the activity and other productivity indicators as measured by the fixed capital turnover ratio as measured by the sales/fixed capital ratio as measured by machinery and equipment varied across firms over the period 2005–2008, for instance, either the declining trend continued or the increasing trend turned into a declining trend for food industries and small size firms, while by contrast either the increasing trend continued for mixed firms or the declining trend turned into an increasing trend for all firms, chemical, metal and textile industries and medium size and large size and private firms. Moreover, we find that for all firms the trend of value and growth rate of activity and other productivity indicators, defined by capital turnover ratio, defined by sales/capital ratio showed a positive increasing trend over the period 2005–2006, that turned into a negative decreasing trend over the periods 2006–2007, 2007–2008 and 2005–2008. Particularly, we find that the activity and other productivity indicators, defined by capital turnover ratio, defined by sales/capital ratio vary across firms over the period 2005–2008; for instance, either the declining trend continued for medium size firms or the increasing trend turned into a declining trend for all firms, food industries and small size and private firms, while by contrast either the increasing trend continued for textile industries or the declining trend turned into an increasing trend for chemical and metal industries and large size and mixed firms (see Table 7.5 below).

As for the third set of profitability indicators from Table 7.5, we find that for all firms the trend of value and growth rate of profitability that we measure by the rate of return on labour or profit/labour ratio showed a positive increasing trend over the periods 2005–2006, 2006–2007 and 2005–2008 that turned into a negative declining trend over the period 2007–2008. In particular, we find that profit/labour ratio varied across firms over the period 2005–2008, for instance, either the declining trend continued for mixed firms or the increasing trend turned into a declining trend for all firms, chemical, food and textile industries, large size and private firms, while by contrast either the increasing trend continues or the declining trend turned into an increasing trend for metal industries, small and medium size firms. In addition, we find that for all firms the trend of value and growth rate of profitability as measured by the rate of return on capital as measured by profit/capital ratio showed a positive increasing trend over the periods 2005–2006 and 2006–2007 that turned into a negative decreasing trend over the periods 2007–2008 and 2005–2008. In particular, we find that profitability as measured by the rate of return on capital measured by profit/capital ratio varied across firms over the period 2005–2008, for instance, either the declining trend continued for medium size firms or the increasing trend turned into a declining trend for all firms, food and textile industries and private firms, while by contrast either the increasing trend continues or the declining trend turned into an increasing trend for chemical and metal industries and small size and large size and mixed firms. Moreover, we find that for all firms the trend of value and growth rate of profitability measured by profit margin that we measure by profit/sales ratio showed a negative decreasing trend over all the periods: 2005–2006, 2007–2008 and 2005–2008. In particular, we find that profitability as measured by profit margin as measured by profit/sales ratio varied across firms over the period 2005–2008, for instance, either the declining trend continued for chemical and food industries, large size, medium size and private firms or the increasing trend turned into a declining trend for all firms, metal and textile industries and small size, while by contrast the declining trend turned into an increasing trend only for mixed firms.

We find that in most cases the trend of these indicators seem to be more sensitive to differences in firm size, industry and sector. In particular, the industrial performance indicators that seem to be more sensitive to differences in firm size, industry and sector include the economic indicator as measured by the degree of industrialisation that we measure by the ratio of total value added as a percentage of total output measured by total sales value. Moreover, other industrial performance indicators that seem to be more sensitive to differences in firm size, industry and sector include three productivity indicators: capital productivity indicator (total output (measured by total sales value)/total capital); the fixed capital productivity indicator (total output (measured by total sales value)/fixed capital (machinery and equipment)); and the wage productivity indicator (total output (measured by total sales value)/total wage). In addition to the activity indicators or ratios measured by fixed capital turnover ratio, measured by the ratio of total sales value as a percentage of fixed capital, and capital turnover ratio measured by the ratio of total sales value as a percentage of total capital, in addition to the profitability indicator measured by the rate of return on capital measured by the ratio of profit as a percentage to capital. We find that the industrial performance indicators that seem to be to some extent sensitive to differences in firm size but less sensitive to industry and sector include the economic or capital intensity level indicator measured by both the ratio of total capital as a percentage to total labour and the ratio of fixed capital or total spending in machinery and equipment as a percentage to total labour. Moreover, we find that the industrial performance indicator that seems to be sensitive to only differences in industry is the raw materials productivity indicator measured by the ratio of total output measured by total sales value as a percentage to total spending on raw material. We find that the industrial performance indictors that seem to be insensitive to differences in firm size, industry and sector include the labour productivity indicator measured by the ratio of total value added as a percentage to the total labour and profitability indicators that we define by profit/labour ratio and profit margin indicator measured by the ratio of profit as a percentage to total sales value. These results imply that in most cases an increase in skill level – share of high skill in total employment – firm size and industry most probably leads to an improvement in most of industrial performance indicators (see Tables 7.4 and 7.5 below).

7.3.2.4 Low Skill Level and Declining Performance of Manufacturing Industrial Firms

The findings from the firm survey (2010) and Table 7.6 below support our argument that the low skill levels may contribute to declining industrial performance indicators: economic, activity, profitability and labour productivity across firms as we explained above. Table 7.6 below shows that the low skill level is indicated by firms among the important problems that are hindering industrial performance and contribution towards economic development in Sudan.[16] For instance, we find that from the perspective of all respondent firms the most important problems are: inadequate finance and inappropriate conditions for industrial development, spread of routine and bureaucracy and slow procedures related to the industrial needs, interruption and inadequate availability and high costs of electricity and water, lack of raw materials, inadequate infrastructure, weak maintenance capability and lack of spare parts, inadequate skills and lack of trained labour force, weak industrial awareness, weak and narrow marketing opportunities, weak and inadequate economic visibility studies, inadequate management and organisational facilities and inadequate transportation equipment respectively (see Table 7.6 below).[17] Moreover, from the firms' perspective other extremely important factors hindering contribution of the industrial sector in economic development in Sudan include the lack of support from Ministry of Industry and the government, and high production costs caused by the imposition of high taxes, fees, levies and customs for clearance of imported raw materials, machines, machinery and equipment imposed on the industrial firms in Sudan.[18] For chemical industries the most important problems are: interruption and inadequate availability and high costs of electricity and water, spread of routine and bureaucracy and slow procedures related to industrial needs, lack of raw materials, inadequate finance and inappropriate conditions for industrial development, inadequate infrastructure, weak industrial awareness, inadequate skill and lack of trained labour force, weak maintenance

[16] For instance, inadequate skills and lack of trained labour force is important problem that reported by 75 %, 76 %, 68 %, 100 %, 60 %, 91 %, 69 % and 60 % of all firms, chemical, food, metal, textile, large, medium and small size firms respectively.

[17] As indicated by 86 %, 85 %, 84 %, 78 %, 76 %, 75 %, 75 %, 73 %, 67 %, 61 %, 57 % and 52 % of all respondents firms respectively.

[18] According to respondent firms 95 % of industrial firms in Khartoum North industrial area are closed due to high production costs.

Table 7.6 The factors constraining improvement of industrial firms performance and economic development in Sudan (2008)

	All firms	Industry				Size		
		Chemical	Food	Metal	Textile	Large	Medium	Small
Inadequate finance and inappropriate conditions for industrial development	86 %	84 %	82 %	100 %	100 %	88 %	88 %	85 %
Spread of routine and bureaucracy and slow procedures related to industrial needs	85 %	89 %	89 %	67 %	60 %	84 %	88 %	85 %
Interruption and inadequate availability and high costs of electricity and water	84 %	89 %	86 %	56 %	80 %	84 %	85 %	85 %
Lack of local raw materials	78 %	86 %	61 %	89 %	100 %	84 %	77 %	75 %
In adequate infrastructure	76 %	81 %	71 %	89 %	40 %	72 %	81 %	80 %
Weak maintenance capability and lack of spare parts	75 %	76 %	75 %	78 %	60 %	91 %	65 %	65 %
Inadequate skill and lack of trained labour force	75 %	76 %	68 %	100 %	60 %	91 %	69 %	60 %
Weak industrial awareness	73 %	78 %	68 %	78 %	60 %	81 %	73 %	65 %
Weak and narrow marketing opportunities	67 %	70 %	64 %	56 %	80 %	66 %	73 %	65 %
Weak and in adequate economic visibility studies	61 %	65 %	61 %	56 %	40 %	59 %	62 %	65 %
Inadequate management and organizational facilities	57 %	65 %	46 %	67 %	40 %	75 %	46 %	45 %
Inadequate transportation equipments	52 %	51 %	46 %	67 %	60 %	59 %	50 %	45 %

Source: Own calculation based on the firm survey (2010)

capability and lack of spare parts, weak and narrow marketing opportunities and inadequate management and organisational facilities respectively.[19] For food industries the most important problems are: spread of routine and bureaucracy and slow procedures related to industrial needs, interruption and inadequate availability and high costs of electricity and water, inadequate finance and inappropriate conditions for industrial development, weak maintenance capability and lack of spare parts, inadequate infrastructure, inadequate skills and lack of trained labour force, weak industrial awareness, weak and narrow marketing opportunities and lack of raw materials respectively.[20] For metal industries the most important

[19] As indicated by 89 %, 89 %, 86 %, 84 %, 81 %, 78 %, 76 %, 76 %, 70 % and 65 % of all respondent chemical firms respectively.

[20] As indicated by 89 %, 86 %, 82 %, 75 %, 71 %, 68 %, 68 % and 64 % and 61 % of all respondent food firms respectively.

problems are: inadequate skills and lack of trained labour force, inadequate finance and inappropriate conditions for industrial development, lack of raw materials, inadequate infrastructure, weak maintenance capability and lack of spare parts, weak industrial awareness, inadequate management and organisational facilities and spread of routine and bureaucracy and slow procedures related to industrial needs respectively.[21] For textile industries the most important problems are: inadequate finance and inappropriate conditions for industrial development, lack of raw materials, interruption and inadequate availability and high costs of electricity and water, weak and narrow marketing opportunities, inadequate skills and lack of trained labour force, spread of routine and bureaucracy and slow procedures related to industrial needs, weak maintenance capability and lack of spare parts and weak industrial awareness respectively.[22] For large size firms the most important problems are: inadequate skills and lack of trained labour force, weak maintenance capability and lack of spare parts, inadequate finance and inappropriate conditions for industrial development, lack of raw materials, interruption and inadequate availability and high costs of electricity and water, spread of routine and bureaucracy and slow procedures related to industrial needs, weak industrial awareness, inadequate management and organisational facilities, inadequate infrastructure and weak and narrow marketing opportunities respectively.[23] For medium size firms the most important problems are: inadequate finance and inappropriate conditions for industrial development, spread of routine and bureaucracy and slow procedures related to industrial needs, interruption and inadequate availability and high costs of electricity and water, inadequate infrastructure, lack of raw materials, narrow marketing opportunities, weak industrial awareness, inadequate skills and lack of trained labour force and weak maintenance capability and lack of spare parts respectively.[24] For small size firms the most important problems are: inadequate finance and inappropriate conditions for industrial development, spread of routine and bureaucracy and slow procedures related to industrial needs, interruption and inadequate availability and high costs of electricity and water, inadequate infrastructure, lack of raw materials, weak maintenance capability and lack of spare parts, weak industrial awareness, weak and narrow marketing opportunities, weak and inadequate economic visibility studies and inadequate skills and lack of trained labour force respectively.[25]

---

[21] As indicated by 100 %, 100 %, 89 %, 89 %, 78 %, 78 %, 67 % and 67 % of all respondent metal firms respectively.

[22] As indicated by 100 %, 100 %, 80 %, 80 %, 60 %, 60 %, 60 %, 60 % and 60 % of all respondent textile firms respectively.

[23] As indicated by 91 %, 91 %, 88 %, 84 %, 84 %, 84 %, 81 %, 75 %, 72 %, and 66 % of all respondent large size firms respectively.

[24] As indicated by 88 %, 88 %, 85 %, 81 %, 77 %, 73 %, 73 %, 69 %, and 65 % of all respondent medium size firms respectively.

[25] As indicated by 85 %, 85 %, 85 %, 80 %, 75 %, 65 %, 65 %, 65 %, 65 % and 60 % of all respondent small size firms respectively.

Hence, our results from Table 7.6 and the firm survey (2010) are consistent with the findings in developing countries and the Sudanese literature that indicate several problems of industrialisation in Sudan (El-Sayed 1998; Abdel-Salam 2006) similar to those in the typically developing countries (Ismail 1994). Different from the studies in the Sudanese literature (El-Sayed 1998; Abdel-Salam 2006) which provide a somewhat general overview concerning the problems of industrialisation in Sudan, an interesting and novel element in our analysis is that our findings are based on recent micro primary data based on the firm survey (2010) and the follow-up interviews with firm managers, and we present a new and more elaborate interpretation of the main problems of industrialisation in Sudan from the perspective of the different industrial firms considering the opinions of a more diversified sample of industrial firms, defined by industry and size as we explained in Table 7.6 below.[26]

Therefore, our findings in this section verify the first hypothesis that high skill requirements and low skill levels – due to high share of unskilled workers – lead to skills mismatch and probably contribute to industrial performance and productivity decline across firms. In the next sections we examine the second and third hypotheses.

7.4 Upskilling, Improving Industrial Performance and Relationships Between Required Education (Occupation), Attained/Actual Education, Experience and Average Wages

Before examining the second and third hypotheses, it is useful to briefly show the importance of upskilling, because explaining this can be used to prevent the decline in labour productivity and industrial performance indicators and to enhance the complementary relationships between skill, technology and upskilling across firms.

7.4.1 Upskilling and Improving Performance of Manufacturing Industrial Firms

The findings from the firm survey (2010) presented in Tables 7.3, 7.4, 7.5, and 7.6 above, support our argument that low skill levels may contribute to the declining of labour productivity and other industrial performance indicators including economic, productivity, activity and profitability indicators across firms as we explained above. These findings imply that improving skill level is an important factor for facilitating improvement of labour productivity and other industrial performance indicators. Table 7.7 below indicates upskilling or improving skill

[26] See for instance, El-Sayed (1998), (pp. 184–188), Abd-Salam (2006), pp. 28–32 and Ismail (1994), pp. 206–209.

Table 7.7 The factors facilitating improvement of industrial firms performance and economic development in Sudan (2008)

	All firms	Industry				Size		
		Chemical	Food	Metal	Textile	Large	Medium	Small
Improving and enhancing adequate availability of finance and appropriate conditions for industrial development	91 %	92 %	89 %	89 %	100 %	97 %	85 %	95 %
Improving and enhancing adequate availability of local raw materials	90 %	92 %	82 %	100 %	100 %	100 %	85 %	85 %
Improving and enhancing adequate availability of industrial awareness	90 %	92 %	86 %	100 %	80 %	97 %	88 %	85 %
Improving and enhancing adequate availability of maintenance capability and spares parts	89 %	92 %	82 %	100 %	80 %	97 %	81 %	90 %
Avoiding of routine and bureaucracy and speed up the procedures related to industrial needs	87 %	86 %	89 %	89 %	80 %	91 %	88 %	85 %
Improving and enhancing adequate availability of infrastructure	86 %	92 %	82 %	89 %	60 %	94 %	81 %	85 %
Improving and enhancing adequate availability of electricity and water with cheap and subsidised price	86 %	84 %	86 %	89 %	100 %	88 %	81 %	95 %
Improving and enhancing adequate availability of skill and trained labour force	85 %	86 %	79 %	100 %	80 %	100 %	81 %	70 %
Improving and enhancing adequate availability of marketing opportunities	85 %	89 %	82 %	78 %	80 %	88 %	81 %	90 %
Improving and enhancing adequate availability of management and organizational facilities	81 %	89 %	68 %	89 %	80 %	100 %	65 %	75 %
Improving and enhancing adequate availability of transportation equipments	75 %	73 %	75 %	89 %	60 %	88 %	62 %	75 %
Improving and enhancing adequate availability of economic visibility studies	72 %	76 %	71 %	67 %	60 %	84 %	58 %	75 %

Source: Own calculation based on the firm survey (2010)

level and adequate availability of skill and trained labour force to be amongst the important factors facilitating improvement of industrial firms performance and contributing towards economic development in Sudan.[27] For instance, we find that from the perspective of all respondent firms the most important factors facilitating improvement are: improving and enhancing adequate availability of finance and appropriate conditions for industrial development, improving and enhancing adequate availability of raw materials, improving and enhancing adequate availability of industrial awareness, improving and enhancing adequate availability of maintenance capability and spare parts and avoidance of routine and bureaucracy and speeding up of the procedures related to industrial needs. In addition to improving and enhancing adequate availability of infrastructure, improving and enhancing adequate availability of electricity and water with cheap and subsidised price, improving and enhancing adequate availability of skill and trained labour force, improving and enhancing adequate availability of marketing opportunities, improving and enhancing adequate availability of management and organisational facilities, improving and enhancing adequate availability of transportation equipment and improving and enhancing adequate availability of economic visibility studies (see Table 7.7 below).[28] Furthermore, from the firms' perspective other extremely important enhancing factors for the development of the performance of the industrial firms include lowering or cancellation of fees, taxes and levies imposed by the government, establishment of databases, reduction of government intervention in the industrial activities and improving and accelerating the procedures for customs clearance of imported raw materials and speeding up of the process of final export of industrial products. From the perspective of chemical firms the most important factors are: improving and enhancing adequate availability of finance and appropriate conditions for industrial development, improving and enhancing adequate availability of raw materials, improving and enhancing adequate availability of industrial awareness, improving and enhancing adequate availability of infrastructure, improving and enhancing adequate availability of maintenance capability and spare parts and improving and enhancing adequate availability of marketing opportunities. In addition to improving and enhancing adequate availability of management and organisational facilities, improving and enhancing adequate availability of skills and trained labour force, avoidance of routine and bureaucracy and speeding up of the procedures related to industrial needs, improving and enhancing adequate availability of electricity and water with

[27] For instance, improving skill level and adequate availability of skill and trained labour force is one important factor facilitating improvement of industrial firms performance and contribution towards economic development that is reported by 85 %, 86 %, 79 %, 100 %, 80 %, 100 %, 81 %, and 70 % of all firms, chemical, food, metal, textile, large, medium and small size firms respectively.

[28] As indicated by 91 %, 90 %, 90 %, 89 %, 87 %, 86 %, 86 %, 85 %, 85 %, 81 %, 75 % and 72 % of all respondent firms respectively.

cheap and subsidised prices.[29] From the perspective of food firms the most important factors are: improving and enhancing adequate availability of finance and appropriate conditions for industrial development, avoidance of routine and bureaucracy and speeding up of the procedures related to industrial needs and improving and enhancing adequate availability of industrial awareness. In addition: improving and enhancing adequate availability of electricity and water with cheap and subsidised prices, improving and enhancing adequate availability of maintenance capability and spare parts, improving and enhancing adequate availability of raw materials, improving and enhancing adequate availability of infrastructure, improving and enhancing adequate availability of marketing opportunities and improving and enhancing adequate availability of skill and trained labour force respectively.[30] From the perspective of metal firms the most important factors are: improving and enhancing adequate availability of skills and trained labour force, improving and enhancing adequate availability of raw materials, improving and enhancing adequate availability of maintenance capability and spare parts, improving and enhancing adequate availability of industrial awareness, improving and enhancing adequate availability of finance and appropriate conditions for industrial development, improving and enhancing adequate availability of infrastructure and avoidance of routine and bureaucracy and speeding up of the procedures related to industrial needs. In addition: improving and enhancing adequate availability of electricity and water with cheap and subsidised prices, improving and enhancing adequate availability of management and organisational facilities and improving and enhancing adequate availability of transportation equipment's respectively.[31] From the perspective of textile firms the most important factors are: improving and enhancing adequate availability of finance and appropriate conditions for industrial development, improving and enhancing adequate availability of raw materials, improving and enhancing adequate availability of electricity and water with cheap and subsidised prices and improving and enhancing adequate availability of skills and trained labour force. In addition to improving and enhancing adequate availability of industrial awareness, improving and enhancing adequate availability of maintenance capability and spares part, improving and enhancing adequate availability of marketing opportunities, improving and enhancing adequate availability of management and organisational facilities, avoidance of routine and bureaucracy and speeding up of the procedures related to industrial needs and improving and enhancing adequate availability of infrastructure respectively.[32] From the perspective of large size firms the most important factors are: improving

[29] As indicated by 92 %, 92 %, 92 %, 92 %, 92 %, 89 %, 89 %, 86 %, 86 % and 84 % of all respondent chemical firms respectively.

[30] As indicated by 89 %, 89 %, 86 %, 86 %, 82 %, 82 %, 82 %, 82 and 79 % of all respondent food firms respectively.

[31] As indicated by 100 %, 100 %, 100 %, 100 %, 89 %, 89 %, 89 %, 89 %, 89 % and 89 % of all respondent metal firms respectively.

[32] As indicated by 100 %, 100 %, 100 %, 80 %, 80 %, 80 %, 80 %, 80 %, 80 % and 60 % of all respondent textile firms respectively.

and enhancing adequate availability of skills and trained labour force, improving and enhancing adequate availability of management and organisational facilities, improving and enhancing adequate availability of raw materials, improving and enhancing adequate availability of finance and appropriate conditions for industrial development, improving and enhancing adequate availability of maintenance capability and spare parts, improving and enhancing adequate availability of industrial awareness and improving and enhancing adequate availability of infrastructure. In addition: avoidance of routine and bureaucracy and speeding up of the procedures related to industrial needs, improving and enhancing adequate availability of electricity and water with cheap and subsidised prices, improving and enhancing adequate availability of marketing opportunities, improving and enhancing adequate availability of transportation equipment and improving and enhancing adequate availability of economic visibility studies respectively.[33] From the perspective of medium size firms the most important factors are: improving and enhancing adequate availability of industrial awareness, avoidance of routine and bureaucracy and speeding up of the procedures related to industrial needs, improving and enhancing adequate availability of finance and appropriate conditions for industrial development, improving and enhancing adequate availability of raw materials and improving and enhancing adequate availability of skills and trained labour force. In addition to improving and enhancing adequate availability of infrastructure, improving and enhancing adequate availability of maintenance capability and spare parts, improving and enhancing adequate availability of electricity and water with cheap and subsidised prices, improving and enhancing adequate availability of marketing opportunities and improving and enhancing adequate availability of management and organisational facilities and availability of transportation equipment.[34] From the perspective of small size firms the most important factors are: improving and enhancing adequate availability of finance and appropriate conditions for industrial development, improving and enhancing adequate availability of electricity and water with cheap and subsidised prices, improving and enhancing adequate availability of maintenance capability and spare parts, improving and enhancing adequate availability of marketing opportunities, improving and enhancing adequate availability of raw materials and improving and enhancing adequate availability of infrastructure. In addition: avoidance of routine and bureaucracy and speeding up of the procedures related to industrial needs, improving and enhancing adequate availability of industrial awareness, improving and enhancing adequate availability of management and organisational facilities, improving and enhancing adequate availability of economic visibility studies, improving and enhancing adequate availability of transportation equipment and

[33] As indicated by 100 %, 100 %, 100 %, 97 %, 97 %, 97 %, 94 %, 91 %, 88 %, 88 %, 88 % and 84 % of large size firms respectively.

[34] As indicated by 88 %, 88 %, 85 %, 85 %, 81 %, 81 %, 81 %, 81 %, 81 %, 65 % and 62 % of medium firms respectively.

improving and enhancing adequate availability of skills and trained labour force respectively.[35]

7.4.2 Relationships Between the Required Education (Occupation), Attained/Actual Education, Experience and Average Wages

Based on the above findings, in this section we examine a part of the second hypothesis that an increase in skill levels and firm size leads to improved relationships between actual and required education, and between actual education, required education, experience and wages across firms.

We begin with the relationship between occupation and education. Using the above definitions of occupation and education/actual and required education respectively, we translate the required qualifications for each of the occupation groups into average years of schooling and use the OLS regression, assuming that the required schooling in each occupation class is dependent on the actual/attained education. Our findings in Table 7.8 and Fig. 7.7 below illustrate that improvement in occupational status (measured by the required education) is positively and significantly correlated with education (measured by actual/attained education) across all firms. In addition, Table 7.8 illustrates that an increase in firm size and industry level leads to improved relationships between required and actual education. For instance, the required education appears to be more sensitive to and increasing in attained/actual education within both large size and chemical and food firms, and more sensitive within all firms. This result is plausible since the skill level – share of high skilled measured by educational attainment – is higher within large size and chemical and food firms compared to metal and textile, medium and small size firms (see Fig. 7.1 above). This is also probably because large size firms are more prevalent in the chemical and food industries (see Table 7.2 above) and may have more consistent recruitment strategies. These results confirm our earlier observations that skill levels and requirements (actual and required education) are non-homogenous across firms and are determined by size and industry.

Concerning the relationship between education, occupation and experience, Table 7.8 above shows that average years of experience are positively correlated and increasing in education (i.e. attained/actual education) and occupation (i.e. required education) respectively. This result is consistent with Fig. 7.5 above, and probably implies that skill indicators – education and experience – are complementing rather than substituting each other.

Table 7.9 below illustrates a considerable variation in the distribution of average wages amongst high, medium and low skill – educational and occupational – levels

[35] As indicated by 95 %, 95 %, 90 %, 90 %, 85 %, 85 %, 85 %, 85 %, 75 %, 75 %, 75 % and 70 % of small firms respectively.

Table 7.8 Required and actual/attained education and experience across firms (2008)

Independent variable		Coefficient (t-value)			R^2	N[a]
Dependent variable	Group of firms and skill	Actual education	Required education	Constant		
Required education. All groups (high, medium and low)	All firms	0.873**		2.101	0.759	74
		(25.172)		(4.691)		
	Large	0. 905**		1.849	0.772	26
		(16.672)		(2.627)		
	Medium	0.864**		2.291	0.766	18
		(14.592)		(2.999)		
	Small	0.825**		2.297	0.742	15
		(11.761)		(2.554)		
	Chemical	0.883**		1.895	0.731	27
		(15.390)		(2.540)		
	Food	0.879**		2.037	0.777	21
		(15.816)		(2.850)		
	Metal	0.814**		3.262	0.793	7
		(9.387)		(2.913)		
	Textile	0.875**		1.749	0.842	4
		(8.338)		(1.316)		
Average years experience	All firms	0.412**		0.767	0.056	73
		(3.469)		(0.505)		
All firms	Large	0.539**		−0.521	0.102	26
		(3.059)		(−0.231)		
	Medium	0.388*		0.777	0.390	18
		(1.576)		(0.245)		
	Small	0.295*		2.232	0.306	15
		(1.429)		(0.852)		
	Chemical	0.274*		3.424	0.023	27
		(1.452)		(1.404)		
	Food	0.617**		−3.065	0.185	21
		(4.010)		(−1.568)		
	Metal	0.371		2.131	0.032	7
		(0.940)		(0.423)		
	Textile	0.068		3.810	0.003	4
		(0.164)		(0.724)		
	All firms		0.641**	−1.810	0.089	71
			(4.260)	(−0.892)		
	Large		0.439**	0.880	0.070	26
			(2.407)	(0.354)		
	Medium		0.566*	−0.697	0.043	18
			(1.606)	(−0.145)		
	Small		1.156**	−8.273	0.288	15
			(4.311)	(−2.369)		
	Chemical		0.465*	1. 528	0.039	27

(continued)

Table 7.8 (continued)

Independent variable		Coefficient (t-value)			R^2	N^a
Dependent variable	Group of firms and skill	Actual education	Required education	Constant		
			(1.875)	(0.458)		
	Food		0.658**	−3. 490	0.148	21
			(3.305)	(−1.304)		
	Metal		1.402**	−12.262	0.373	7
			(3.621)	(−2.290)		
	Textile		0628	−1.047	0.106	4
			(1.034)	(−0.129)		

Correlation is significant * at the 0.05 level (one-tailed); ** at the 0.01 level (one-tailed)
[a]For this regression we use relatively few observations, because some of the respondent firms were particularly reluctant to provide adequate quantitative data on skill indicators. Sometimes we exclude some observations due to inconsistency or unreliability. As we explained in Chap. 4 above, the main problem is the varying response rate for different questions (e.g. to measure education, occupation and wages) across firms. Moreover, the classification of firms into chemical, food, metal and textile industries, small, medium and large size also divided the few observations between them and so allow for only few observations for regression for each group independently.

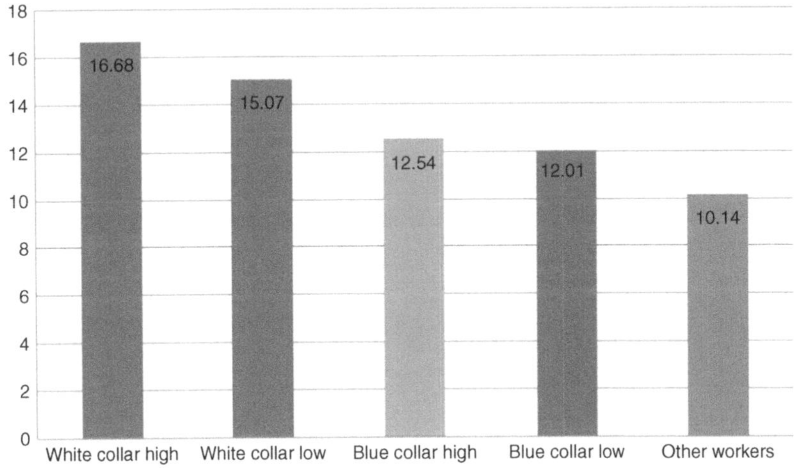

Fig. 7.7 The distribution of occupation classes according to the translated average years of schooling across firms (2008) (Source: Firm Survey (2010))

across firms. When using the occupational rather than the educational definition the distribution of wages shows less fluctuation across firms. Therefore, the effect of occupation/required education on the distribution of average wages across firms

Table 7.9 Differences in the distribution of average wages defined by firm size and industry level and sector (2008)

Characteristics		Industry/activity				Size			Sector		
Wages defined by skill level	All firms	Chemical	Food	Metal	Textile	Large	Medium	Small	Public	Private	Mixed
(a) Skill variables: Education											
High educated/white collar high											
4,001–5,000	1 %	0 %	0 %	13 %	0 %	3 %	0 %	0 %	0 %	2 %	0 %
3,001–4,000	4 %	6 %	0 %	13 %	0 %	7 %	0 %	5 %	0 %	5 %	0 %
2,001–3,000	13 %	19 %	11 %	0 %	0 %	10 %	13 %	16 %	0 %	11 %	40 %
1,001–2,000	46 %	41 %	44 %	50 %	80 %	50 %	48 %	37 %	0 %	45 %	60 %
0,200–1,000	36 %	34 %	44 %	25 %	20 %	30 %	39 %	42 %	100 %	38 %	0 %
Medium educated/white collar low											
1,001–2,000	4 %	6 %	0 %	12 %	0 %	3 %	0 %	11 %	0 %	3 %	20 %
0,200–1,000	96 %	94 %	100 %	88 %	100 %	97 %	100 %	89 %	100 %	97 %	80 %
Low educated/blue collar high											
0,200–1,000	100 %	100 %	100 %	100 %	100 %	100 %	100 %	100 %	100 %	100 %	100 %
(b) Skill variables: occupation											
High educated/white collar high											
4,001–5,000	8 %	6 %	11 %	11 %	0 %	13 %	9 %	0 %	0 %	7 %	20 %
3,001–4,000	11 %	12 %	11 %	11 %	0 %	13 %	9 %	11 %	0 %	10 %	20 %
2,001–3,000	28	21 %	30 %	33 %	60 %	29 %	30 %	26 %	0 %	29 %	20 %
1,001–2,000	34 %	39 %	26 %	33 %	40 %	35 %	26 %	42 %	100 %	32 %	40 %
0,200–1,000	19 %	21 %	22 %	11 %	0 %	10 %	26 %	21 %	0 %	21 %	0 %
Medium educated/white collar low											
2,001–3,000	1 %	3 %	0 %	0 %	0 %	3 %	0 %	0 %	0 %	2 %	0 %
1,001–2,000	24 %	23 %	25 %	33 %	0 %	27 %	26 %	14 %	0 %	22 %	50 %
0,200–1,000	75 %	73 %	75 %	67 %	100 %	70 %	74 %	86 %	100 %	76 %	50 %
Low educated/blue collar high											
1,001–2,000	6 %	4 %	9 %	13 %	0 %	3 %	10 %	7 %	0 %	7 %	0 %
0,200–1,000	94 %	96 %	91 %	88 %	100 %	97 %	90 %	93 %	100 %	93 %	100 %
Blue collar low											
1,001–2,000	3 %	0 %	5 %	13 %	0 %	0 %	9 %	0 %	0 %	3 %	0 %
0,200–1,000	97 %	100 %	95 %	88 %	100 %	100 %	91 %	100 %	100 %	97 %	100 %
Others											
4,001–5,000	2 %	0 %	4 %	0 %	0 %	0 %	5 %	0 %	0 %	2 %	0 %
0,200–1,000	98 %	100 %	96 %	100 %	100 %	100 %	95 %	100 %	100 %	98 %	100 %

Source: Firm Survey (2010)

seems to be less sensitive to differences in firm size and industry. In contrast, when using the educational definition, we observe that the effect of the actual/attained education on the distribution of average wages across firms seems to be more sensitive to differences in firm size and industry. Our interpretation of the observed differences across firms implies the presence of a significant wage differential, the lack of a coherent, homogeneous, unified and sound wage policy and the lack of systematic and consistent recruitment strategies across firms that most probably related to the lack of systematic regulations to organise the labour market in Sudan

The above results are consistent with the OLS regression reported in Table 7.10 below, which indicates that the average wages are positively and significantly correlated with and more sensitive to attained/actual education. For instance, Table 7.10 below illustrates that the average wages are increasing in actual/attained education, experience and its square (cf. Mincer 1974) and therefore, is biased against less educated and experienced workers. These findings support our results from the firm survey, which indicate that wages are increasing in education and biased against low educated workers because the ratio of high skilled to low skilled wages, which can be interpreted as wages/skills premium, exceeds one (see Fig. 7.8 below).[36] These results are consistent with the findings in the new growth literature, particularly skilled biased technical change theorems (cf. Aghion and Howitt 1992, 1998; Acemoglu 1998; Autor et al. 1998). Our results from Table 7.11, which indicate that required education also has significant impact on wages, are plausible and consistent with our expectation in view of the results of the overeducation literature (Hartog 2000; Muysken and ter Weel 1998; Muysken and Ruholl 2001; Muysken et al. 2002a, b; Muysken et al. 2003). We find that the positive correlations between actual education, experience, its square and wages seem more sensitive to firm size and industry level and are particularly significant for large and medium size firms and chemical and food industries, which may not be surprising since these firms have sufficient scope for a coherent wage policy (Nour 2005; Muysken and Nour 2006). This is also probably because large size and medium size firms and chemical and food industries may have more consistent recruitment strategies and high skill levels – share of high skilled workers in total employment (see Fig. 7.1 above and Fig. 7.9 below). These results imply that an increase in skill level/actual education and firm size and industry leads to an improved relationship between actual education, experience and wages.

---

[36] From the firm survey (2010) we find that the proportion of high skilled wages/low skilled wages accounts for 3.5, 3.7, 3.45, 2.96, 3.6, 4.2, 3.1 and 2.98 for all firms, chemical, food, metal, textile, large, medium and small size firms respectively. We find that the wage premium for Sudan in 2010 is less than the wage premium which we estimated for the large and medium size firms active in the chemical and metal industries in the UAE in 2002 (Nour 2005). This result at the micro level is not surprising and it is expected in view of the observed wage differential between Sudan and UAE at the macro level; in particular, this result is consistent with the observed differences in per capita income levels in Sudan and the UAE at the macro level, notably, when using UNDP-HDR (2010) most recent data on per capita income for the year 2008, we realise the low per capita income level in Sudan (US$ 1,353) as compared to high per capita income in the UAE (US$ 56,485) at the macro level.

Table 7.10 Correlation between wages (log) actual and required education and experience (2008) (education definition)

Independent variable Dependent variable: Average wages (log)	Group of firms	Coefficient (t-value) Actual education	Experience	$Experience^2$	Required education[b]	Constant	R^2	N[a]
Average wages (log) high, medium and low skilled	All firms	0.196**				3.948	0.595	73
		(17.478)				(27.570)		
	Large	0.223**				3.663	0.646	26
		(12.532)				(16.114)		
	Medium	0.178**				4.045	0.564	18
		(9.099)				(16.121)		
	Small	0.178**				4.248	0.596	15
		(8.972)				(16.838)		
	Chemical	0.206**				3.860	0.605	27
		(11.742)				(17.156)		
	Food	0.192**				3.896	0.657	21
		(12.131)				(19.349)		
	Metal	0.172**				4.591	0.523	7
		(4.906)				(10.337)		
	Textile	0.197**				3.715	0.724	4
		(5.834)				(8.668)		
	All firms	0.187**	0.021**			3.951	0.625	70
		(16.398)	(3.188)			(27.877)		
	Large	0. 221**	0.010			3.652	0.653	26
		(11.383)	(0.866)			(15.407)		
	Medium	0. 161**	0.033**			4.101	0.675	18
		(9.054)	(3.674)			(18.465)		
	Small	0.175**	0.010			4.225	0.601	15
		(8.582)	(0.780)			(16.575)		
	Chemical	0. 198**	0.023**			3.817	0.634	27
		(11.519)	(2.418)			(17.268)		
	Food	0. 175**	0.027**			3.976	0.689	21
		(10.097)	(2.240)			(19.533)		
	Textile	0.219**	0.040**			3.378	0.912	4
		(9.289)	(2.228)			(11.057)		

All firms	0.187**	0.024	−0.0001		3.951	0.625	70
	(15.360)	(1.249)	(−0.156)		(27.806)		
Large	0.221**	0.009	0.00003		3.652	0.653	26
	(10.626)	(0.263)	(0.18)		(15.302)		
Medium	0.156**	0. 055*	−0.001		4.113	0.678	18
	(8.192)	(1.924)	(−0.792)		(18.192)		
Chemical	0.195**	0.043*	−0.001		3.789	0.636	27
	(10.919)	(1.397)	(−0.687)		(16.818)		
Food	0.176**	0.022	0.000		3.977	0.689	21
	(9.783)	(0.822)	(0.197)		(19.396)		
Textile	0.205**	0.093	−0.004		3.433	0.917	4
	(6.546)	(1.148)	(−0.673)		(10.536)		
All firms	0.153**	0.028*	−0.0004	0.038*	3.868	0.621	69
	(6.019)	(1.360)	(−0.356)	(1.532)	(23.996)		
Large	0.172**	0.030	−0.001	0.047	3.579	0.654	26
	(4.125)	(0.808)	(−0.482)	(1.221)	(13.676)		
Medium	0.126**	0.045*	−0.001	0.037	4.018	0.668	18
	(2.866)	(1.420)	(−0.384)	(0.852)	(15.438)		
Small	0.175**	−0.059	0.004	0.016	4.228	0.614	15
	(3.587)	(−0.989)	(1.076)	(0.314)	(12.664)		
Chemical	0.162**	0.047*	−0.001	0.031	3.791	0.609	27
	(4.064)	(1.360)	(−0.725)	(0.850)	(14.547)		
Food	0.136**	0.024	0.000	0.054*	3.751	0.718	21
	(3.939)	(0.861)	(0.100)	(1.545)	(16.511)		
Textile	0.119*	0.104*	−0.004	0.092	3.291	0.931	4
	(1.486)	(1.305)	(−0.756)	(1.172)	(9.657)		

Correlation is significant * at the 0.05 level (one-tailed); ** at the 0.01 level (one-tailed)

[a]For this regression we use relatively few observations, because some of the respondent firms were particularly reluctant to provide adequate quantitative data on skill indicators. Sometimes we exclude some observations due to inconsistency or unreliability. As we explained in Chap. 4 above, the main problem is the varying response rate for different questions (e.g. to measure education, occupation and wages) across firms. Moreover, the classification of firms into chemical, food, metal and textile industries, small, medium and large size also divided the few observations between them and so allow for only few observations for regression for each group independently.

[b]The required education is not used as a variable in the upper half of this table, because, we want to check the relation with respect to actual/attained education and experience independently and then compare the result when the required education is also included in the regression.

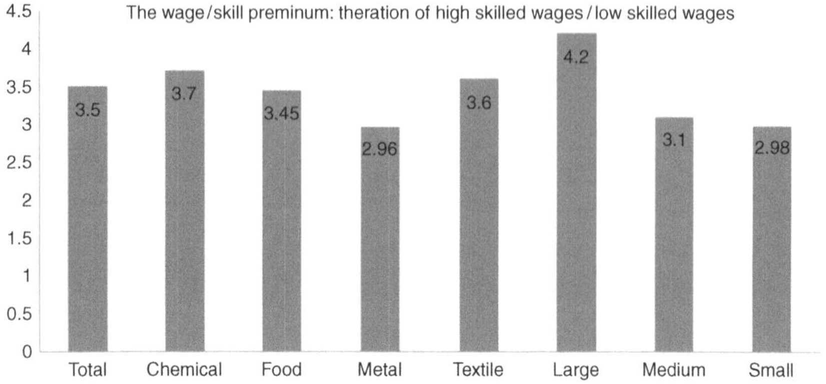

Fig. 7.8 Differences in wage/skill premium (the ratio of high skilled wages/low skilled wages) defined by education levels across firms (2008) (Source: Firm Survey (2010))

One interesting observation from the firm survey data (2010) and the follow-up interviews with firms managers and the results presented in Tables 7.9, 7.10, and 7.11 is that in most cases, the OLS regression results seem to be more significant when using the education definition as compared to occupation definition. This finding seems to be consistent with the observations from Table 7.9 above but seem to be opposite to the observations from the follow-up interviews and the wide belief among firm managers, which probably implies that across the majority of the respondent firms, the structure of wage policy is most likely structured to be more consistent based on occupation definition compared to education definition. This also implies that from the firms' perspective the decision of determining wages levels for workers is most probably determined by the nature of jobs that the workers will do in the firms rather than the years of schooling the workers have already obtained. This also most probably implies the positive but weak return and incentives for additional years of schooling to compensate the costs of additional years of schooling. Another interesting observation is that for all groups of firms when using both education and occupation definitions the OLS regression reported in Tables 7.10 and 7.11 below indicate that the correlations between wages levels and years of education variable are more significant as compared to the correlations between wages levels and average years of experience variable. This result implies that the rate of return to the worker's average years of education is higher and more significant than the average years of experience. This finding is also opposite to the observations from the follow-up interviews and the wide belief among some firm managers which probably implies that across some firms and from some firms' perspective, the decisions of hiring and offering wages are largely determined by experience in the practice of work which is measured by a worker's average years

Table 7.11 Correlation between wages (log) actual and required education and experience (2008) (occupation definition)

Independent variable		Coefficient (t-value)					
Dependent variable: average wages (log)	Group of firms	Actual education[b]	Experience	Experience²	Required education	R² / Constant	N[a]
Average wages (log) high, medium and low skilled	All firms			0.212** (12.176)	3.713 (15.884)	0.429	73
	Large			0.245** (13.152)	3.318 (13.061)	0.678	26
	Medium			0.210** (7.977)	3.749 (10.636)	0.503	18
	Small			0. 204** (7.855)	3.871 (11.500)	0.568	15
	Chemical			0.219** (12.130)	3.639 (15.041)	0.637	27
	Food			0.225** (9.034)	3.551 (10.662)	0.531	21
	Metal			0. 102 (0.904)	5.142 (3.306)	0.036	7
	Textile			0.274** (13.055)	2.972 (11.047)	0.929	4
	All firms	0.034** (4.056)		0.189** (10.315)	3.792 (16.074)	0.472	70
	Large	0.023* (1.930)		0.236** (11.832)	3.289 (12.518)	0.688	26
	Medium	0.029** (3.349)		0.190** (7.998)	3.800 (12.120)	0.627	18
	Small	0.040** (2.991)		0.154** (5.231)	4.252 (12.288)	0.625	15

(continued)

Table 7.11 (continued)

		Coefficient (t-value)					R²	N^a
Independent variable								
Dependent variable: average wages (log)		Group of firms	Actual education[b]	Experience	Experience²	Required education	Constant	
	Chemical		0.025**		0.206**	3.629	0.691	27
			(3.527)		(12.160)	(16.255)		
	Food		0.055**		0.194**	3.641	0.638	21
			(3.683)		(7.620)	(11.365)		
	Textile		0.018*		0.236**	3.376	0.950	4
			(1.611)		(10.982)	(12.444)		
	All firms		0.086**	−0.002*	0.172**	3.840	0.483	70
			(3.148)	(−1.985)	(8.592)	(16.322)		
	Large		0.088**	−0.003*	0.223**	3.255	0.699	26
			(2.202)	(−1.697)	(10.538)	(12.505)		
	Medium		0.066*	−0.001*	0.173**	3.902	0.633	18
			(1.647)	(−0.956)	(5.866)	(11.772)		
	Small		0.069*	−0.001	0.147**	4.259	0.628	15
			(1.464)	(−0.632)	(4.649)	(12.217)		
	Chemical		0.079**	−0.002**	0.188**	3.663	0.710	27
			(3.237)	(−2.307)	(10.297)	(16.795)		
	Food		0.044	−0.000	0.197**	3.631	0.639	21
			(1.025)	(−0.269)	(6.994)	(11.181)		
	Metal		0.271*	−0.009	−0.065**	6.230	0.176	7
			(1.314)	(−0.879)	(−0.439)	(3.585)		
	Textile		0.037	−0.001*	0.233**	3.344	0.951	4
			(0.850)	(−0.455)	(10.042)	(11.360)		
	All firms	0.132**	0.062**	−0.001*	0.068*	3.637	0.519	70

	(3.658)	(2.281)	(−1.344)	(1.964)	(15.526)		
Large	0.193**	0.031	−0.001	0.076**	3.013	0.775	26
	(4.969)	(0.838)	(−0.720)	(2.183)	(12.983)		
Medium	0.111**	0.063*	−0.001	0.076*	3.794	0.663	18
	(2.169)	(1.612)	(−0.979)	(1.449)	(11.698)		
Chemical	0.120**	0.056**	−0.001*	0.097**	3.434	0.753	27
	(3.710)	(2.364)	(−1.559)	(3.243)	(16.206)		
Textile	0.080	0.019	−0.0003	0.163**	3.337	0.960	4
	(1.161)	(0.417)	(−0.145)	(2.519)	(11.611)		

Correlation is significant *at the 0.05 level (one-tailed); **at the 0.01 level (one-tailed)

[a]For this regression we use relatively few observations, because some of the respondent firms were particularly reluctant to provide adequate quantitative data on skill indicators. Sometimes we exclude some observations due to inconsistency or unreliability. As we explained in Chap. 4 above, the main problem is the varying response rate for different questions (e.g. to measure education, occupation and wages) across firms. Moreover, the classification of firms into chemical, food, metal and textile industries, small, medium and large size also divided the few observations between them and so allow for only few observations for regression for each group independently.

[b]The actual/attained education is not used as a variable in the upper half of this table, because, we want to check the relation with respect to required education and experience independently and then compare the result when the actual/attained education is also included in the regression.

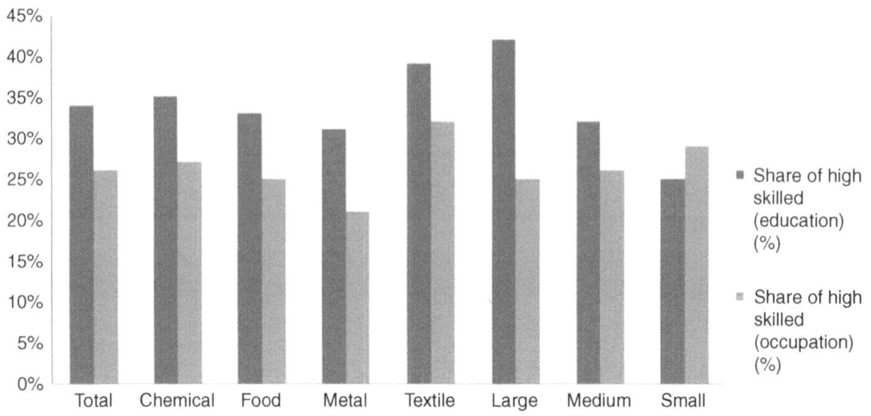

Fig. 7.9 Differences in skill level (share of high skilled) defined by education and occupation classes across firms (2008) (Sources: Firm Survey (2010))

of experience, which is more important than average years of education for some firms that prefer to hire more experienced than educated workers for specific fields.

Therefore, our findings in this section corroborate the first part of the second hypothesis that an increase in skill levels and firm size leads to an improved relationship between actual and required education and experience and between actual education, required education, experience, its square and wages. In the next section we proceed to examine the second part of the second hypothesis that an increase in skill levels and firm size lead to improved relationships between skill, upskilling and technology (ICT). Finally, we test our third hypothesis on the relationship between technology (ICT) and input–output indicators at the micro/firm level.

7.5 Skill, Upskilling (ICT Training), Technology (ICT) and Input–Output Indicators

Based on the above results, in this section we examine the other part of the second hypothesis that an increase in skill levels and firm size leads to improved relationships between skill, upskilling and new technology (ICT) across firms. Before examining this hypothesis, it is useful to briefly show the variations in the use of new technology (spending on ICT) and upskilling (spending on ICT training) across firms, because the observed differences in skill and spending on ICT and ICT training can be used to interpret the complementary relationships between skill, technology and upskilling across firms.

7.5.1 *Skill and the Share of Spending on Technology (ICT) and Upskilling (ICT Training)*

Table 7.12 shows considerable variations in the share and trend of total spending on ICT including computers, telecommunications, training, Internet, maintenance and other items, defined by firm size and industry. The share of telecommunications exhibits a continuous increasing trend for all firms, while that of training shows an opposite declining trend. Table 7.2 above shows that, on average, the share of large size and food and chemical firms represents about 48 %, 53 % and 23 % of total spending on ICT respectively and about 75 %, 73 % and 2 % of total spending on ICT training respectively. However, despite the big share of spending on ICT and ICT training, large size and food firms experienced declining trends of ICT and ICT training (cf. Figs. 7.10 and 7.11). The decline in total ICT spending can be interpreted as being due to a lack of plan for critical expansion in ICT sector or probably due to a general cutback in total spending across food and large size firms. The declining expenses on both ICT training and computers follow the general decline in total ICT spending, which can also be attributed to a lack of plan for critical expansion and a possible change in the strategy of firms that, having already established a sound basis for these components, probably need to shift priority to increase spending on both telecommunications and maintenance.

We now proceed to examine the second part of our second hypothesis that an increase in skill levels and firm size leads to improved complementary relationships between skill, technology (ICT) and upskilling (ICT training) (see Table 7.13 below). For instance, we observe the complementary relationship between the share of high education and the share of expenditure on ICT, which can be seen and understood as complementarity between skill and technology (cf. Goldin and Katz 1998; Acemoglu 1998). We find a complementary relationship between the share of high education and the share of expenditure on ICT training, which can be interpreted as complementarity between skill and upskilling. Tables 7.13 and 7.14 show complementary relationships between the share of expenditure on ICT and ICT training, and between spending on computers, telecommunications, Internet and training, which can be read as complementarity between technology and upskilling (cf. Colecchia and Papaconstantinou 1996; Bresnahan et al. 1999). Our findings, that these complementarities are particularly significant for large size firms, are plausible since these firms have more spending on ICT and ICT training (see Table 7.2 above) and have high skill levels – share of high skilled workers in total employment (see Fig. 7.1 above). These results are consistent with the second part of our second hypothesis that an increase in skill levels and firm size lead to improved complementary relationships between skill, upskilling and technology (ICT) (cf. Acemoglu 1998). The results also imply the importance of a good education/high skill level for the enhancement of skill, technology and upskilling complementarity at the micro level. That also seems consistent with the

Table 7.12 Spending on ICT defined by firm size and industry (2005–2008) (% share in total spending)

Share in total spending in ICT (%) (2005–2008)	Group of Firms/ Years	Industry/activity				Size		
		Chemical	Food	Metal	Textile	Large	Medium	Small
Share in total spending in computer (%)	2005	16 %	48 %	13 %	23 %	28 %	29 %	43 %
	2006	15 %	55 %	14 %	16 %	35 %	22 %	43 %
	2007	32 %	45 %	9 %	14 %	50 %	6 %	45 %
	2008	46 %	42 %	2 %	9 %	53 %	27 %	20 %
Share in total spending in telecommunication (%)	2005	17 %	55 %	24 %	3 %	51 %	5 %	44 %
	2006	13 %	64 %	21 %	2 %	24 %	31 %	44 %
	2007	26 %	52 %	9 %	12 %	50 %	9 %	41 %
	2008	29 %	62 %	3 %	7 %	32 %	42 %	26 %
Share in total spending in training and software development (%)	2005	0 %	100 %	0 %	0 %	97 %	0 %	3 %
	2006	0 %	84 %	16 %	0 %	79 %	16 %	5 %
	2007	3 %	83 %	0 %	15 %	93 %	0 %	7 %
	2008	3 %	82 %	0 %	15 %	82 %	6 %	12 %
	2005–2008	2 %	73 %	12 %	13 %	75 %	18 %	7 %
Share in total spending in internet (%)	2005	25 %	26 %	50 %	0 %	25 %	0 %	75 %
	2006	19 %	24 %	53 %	4 %	19 %	4 %	76 %
	2007	15 %	37 %	26 %	21 %	57 %	0 %	43 %
	2008	10 %	41 %	11 %	38 %	57 %	23 %	21 %
Share in total spending in maintenance services (%)	2005	3 %	3 %	94 %	0 %	3 %	0 %	97 %
	2006	17 %	1 %	81 %	0 %	15 %	21 %	64 %
	2007	6 %	1 %	33 %	61 %	61 %	0 %	39 %
	2008	5 %	13 %	24 %	59 %	59 %	14 %	27 %
Share in total spending in hosting and other relevant ICT services (%)	2005	1 %	99 %	0 %	0 %	100 %	0 %	0 %
	2006	0 %	99 %	0 %	0 %	95 %	0 %	5 %
	2007	0 %	41 %	0 %	59 %	98 %	0 %	2 %
	2008	4 %	28 %	0 %	67 %	94 %	4 %	1 %
Share in total spending in ICT (%)	2005	13 %	57 %	19 %	11 %	47 %	14 %	39 %
	2006	12 %	59 %	22 %	7 %	37 %	22 %	40 %
	2007	22 %	49 %	11 %	18 %	59 %	5 %	36 %
	2008	31 %	52 %	4 %	14	49 %	30 %	21 %
	2005–2008	24 %	53 %	11 %	13 %	49 %	22 %	30 %
Share in average total spending in ICT (%)	2005–2008	23 %	53 %	11 %	13 %	48 %	21 %	30 %
Numbers of respondents	54	27	16	6	5	20	18	16

Source: Firm Survey (2010)

endogenous growth framework and stylised facts concerning the relationships between human capital, technical progress and upskilling (see our theoretical framework in Chap. 3 above).

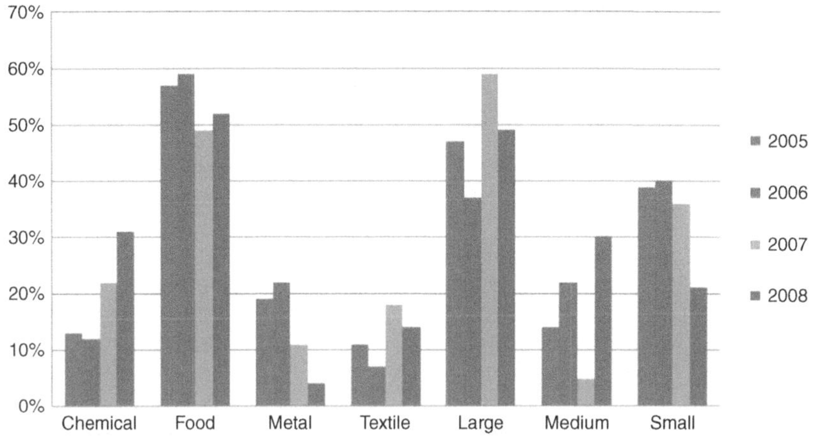

Fig. 7.10 The share and trend of total spending on ICT across firms (2005–2008) (Source: Firm Survey (2010))

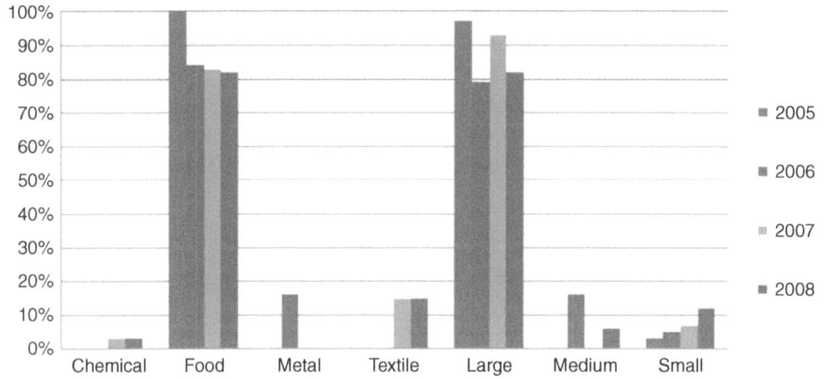

Fig. 7.11 The share and trend of spending on ICT training across firms (2005–2008) (Source: Firm Survey (2010))

7.5.2 The Use of Technology, ICT, Skill and the Demand for Skilled Workers Across Firms

One implication of the above complementary relationship between skill and technology is that the demand for skilled workers has changed in response to the increasing uses of ICT and other technologies. For instance, during the period 2006–2008 the uses of ICT (85 %) increased faster than that of other technologies (70 %); similarly, the corresponding rise in the demand for skilled workers needed for ICT (65 %) was more than that for other technologies (61 %) across all respondents firms (see Fig. 7.12 below). This trend may reflect the fact that the

Table 7.13 The relationship between ICT, skill and upskilling across firms (2008) (2005–2008)

Independent variables		Coefficient (t-value)				
Dependent variables	Group of firms	ICT expenditures	Training expenditures	Constant	R^2	N
High education (linear)	All firms (linear)[a]	0.002* (1.385)		0.344 (14.420)	0.014	34
	All firms (linear)[b]			30.963 (6.770)	0.023	44
	All firms (linear)[b]	0.003** (2.115)		31.724 (10.198)	0.052	82
	Chemical (linear)[b]	0.003* (1.241)		33.111 (7.354)	0.042	36
	Food (linear)[b]	0.002* (1.200)		29.619 (5.387)	0.051	28
	Metal (linear)[b]	0.001 (0.764)		31.684 (3.186)	0.055	11
	Textile (linear)[b]	0. 001** (4.773)		19.140 (3.175)	0.884	4
	Large (linear)[b]	0.006** (2.440)		23.384 (2.740)	0.351	13
	Large (linear)[b]	0.005** (2.349)		26.992 (3.796)	0.283	16
	Medium (linear)[b]	0.004* (1.953)		27.875 (6.103)	0.128	27
	All firms (linear)[a]		0.001** (2.309)	0.3123 (4.765)	0.262	15
ICT(linear: 2005–2008)	All firms (linear)[a]		1.746** (4.827)	15135109 (2.830)	0.608	6
Training expenditures	All firms (linear)[a]	0.349** (4.827)		−1905361 (−0.653)	0.608	6
	All firms (linear)[b]	0.054 (0.485)		4407619.75 (0.680)	0.045	6
	Large (linear)[b]	0.473** (2.301)		−3989610.8 (−0.536)	0.726	3
	Food (linear)[b]	0.036 (0.229)		5907673.03 (0.565)	0.017	4
	All firm (log)[b]	1.018** (5.218)		−1.077 (−0.762)	0.845	6
	Large (log)[b]	1.258** (9.317)		−2.597 (−2.750)	0.977	3
	Food (log)[b]	1.046** (4.627)		−1.399 (−0.851)	0.877	4

Correlation is significant * at the 0.05 level (one-tailed); ** at the 0.01 level (one-tailed)
[a](2005–2008).
[b]2008.

Table 7.14 The relationship between computers, training, internet and telecommunications expenditures across firms (2005–2008)

Independent variables		Coefficient(t-value)						
Dependent variables (All firms)		Computer expenditure	Training expenditure	Telecommunication expenditure	Internet expenditure	Constant	R^2	N
Computer expenditure: All firms	2005			1.587**		748133.60	0.745	17
				(6.836)		(0.980)		
	2005		0.571**			3059520.92	0.783	3
			(2.687)			(0.958)		
	2006			0.367**		241933.61	0.506	20
				(4.410)		(0.333)		
	2006		0.865**			3153599.66	0.713	3
			(2.229)			(0.530)		
	2007		0.147	0.981**		−969394.91	0.674	5
			(0.540)	(2.008)		(−0.241)		
	2008		−0.001	1.129**		−2508705.42	0.832	6
			(−0.002)	(4.445)		(−0.391)		
	2008			1.136**		376032.41	0.579	40
				(7.319)		(0.218)		
	2008		−0.104			15637070.23	0.003	6
			(−0.114)			(1.450)		
	2005–2008			0.763678**		1435747.	0.450	36
				(8.536810)		(2.370763)		
	2005–2008		0.247*	1.023**		165382.95	0.741	6
			(1.523)	(6.353)		(0.063)		
	2005–2008				1.128**	2952505.00	0.169	23
					(3.160)	(2.484)		
	2005–2008			0.780**	0.458*	1312687.00	0.587	23
				(6.971)	(1.684)	(1.493967)		

(continued)

Table 7.14 (continued)

Independent variables		Coefficient(t-value)						
Dependent variables (All firms)		Computer expenditure	Training expenditure	Telecommunication expenditure	Internet expenditure	Constant	R^2	N
Training expenditure: All firms	2005	1.370** (2.687)				−2505285.03 (−0. 439)	0.783	3
	2005			3.054** (22.081)		−676083.74 (−0.973)	0.996	3
	2006	0.824** (2.229)	0.067 (0.044)			138622.51 (0.022)	0.713	3
	2007	0.601 (0.540)				2443767.60 (0.303)	0.236	5
	2005–2008	0.512* (1.523)		−0.416 (−0.986)		5907509.77 (1.708)	0.140	6
	2005–2008	0.001* (1.252)				12.94005 (14.723)	0.125	5
Telecommunication expenditure: All firms	2005	1.379** (4.410)				1752340.80 (1.294)	0.506	20
	2005		0.326** (22.081)			231759.71 (1.046)	0.996	3
	2007	0.584** (2.008)	0.010 (0.044)			2754897.90 (1.023)	0.642	5
	2008	0.510** (7.319)				2032297.91 (1.828)	0.579	40
	2008		−0.092 (−0.125)			16072454.82 (1.845)	0.003	6
	2008	0.737** (4.445)	−0.015 (−0.044)			4553581.73 (0.956)	0.832	6
	2005–2008	0.589516** (8.536810)				758843.9 (1.398)	0.450	36

	2005–2008				0.896**	1864490.00	0.138	23
					(2.936)	(1.921)		
	2005–2008	0.700**	−0.138			2203956.51	0.721	6
		(6.353)	(−0.986)			(1.054)		
	2005–2008	0.645**			0.132	197942.2	0.565	23
		(6.971)			(0.520)	(0.242)		
Internet expenditure: All firms	2005–2008	0.150**				1058449.00	0.169	23
		(3.160)				(2.436)		
	2005–2008	0.122*		0.043		1044110.00	0.174	23
		(1.684)		(0.520)		(2.380526)		
	2005–2008			0.154**		1142708.00	0.138	23
				(2.936)		(2.968)		

Correlation is significant * at the 0.05 level (one-tailed); ** at the 0.01 level (one-tailed)

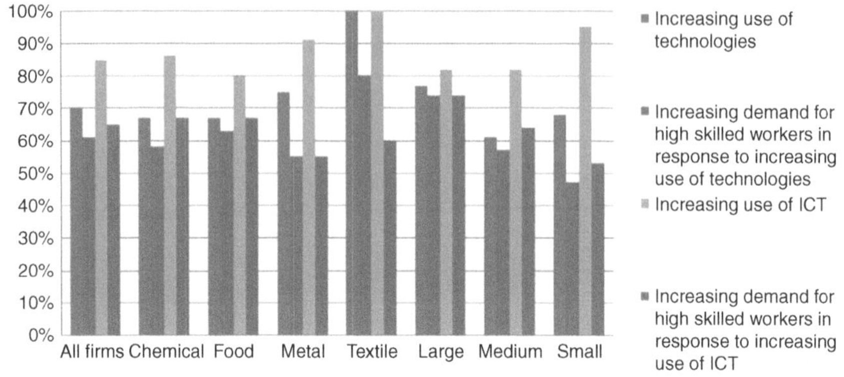

Fig. 7.12 The increasing use of technology, ICT and the demand for high skilled workers across firms, 2006–2008 (Source: Firm survey (2010))

real demand for skilled workers needed for ICT is more than that of other technologies across firms, which may not be surprising given the recent rapid increasing trend of IT diffusion despite the recent history of IT diffusion in Sudan. For instance, according to the World Development WDI database (2005), before 2000 the number of users of both mobile phone and Internet per 1,000 population were zero and up until the year 2000 both were only one; in recent years, Sudan has shown a growing telecommunication network and Internet services but still the highest price/most expensive Internet services as compared to other African and Arab and developing countries.

According to the respondent firms, the increasing use of new technologies caused an increase in both the demand for more skilled workers and the required skill levels of the respective workers involved with them. Table 7.15 indicates that the increasing use of new technologies has important effects on increasing the general skill levels and the demand for skilled workers amongst 88 % and 83 % of the respondent firms respectively.[37] However, it has relatively less important effects on increasing skill levels mainly for unskilled workers, and decreasing and substituting the demand for unskilled workers due to reduction and elimination/substitution of some unskilled jobs. This implies change in the structure of employment/demand for workers in response to the increasing uses of new technologies and is also evidence of skilled-biased technical change theorem.[38]

[37] Firms reported the use of different types of new technologies such as mass chemicals plants, advanced process controls, food processing machines and plants installation, CNC machines, new advanced machines and ICT and computer. In addition to the use of modern new packing and covering machines, modern (cut and wrap) machines and modern tagged machines.

[38] This result is consistent with Skilled Biased Technical Change (SBTC) theorem and our earlier findings indicating that wages are increasing in education and biased against unskilled workers.

Table 7.15 The effects of new technologies on skill level and the demand for workers in the Sudan, 2008

The effects of new technologies in:	All firms	Chemical	Food	Metal	Textile	Large	Medium	Small
Increasing the general skill level	88 %	94 %	92 %	64 %	80 %	91 %	88 %	89 %
Increasing the demand for skilled workers (more educated, trained and experienced workers)	83 %	82 %	81 %	82 %	100 %	84 %	88 %	72 %
Increasing skill level mainly for unskilled workers	79 %	85 %	65 %	82 %	100 %	84 %	76 %	72 %
Reduction in some unskilled jobs	79 %	85 %	80 %	60 %	80 %	88 %	76 %	76 %
Increasing the demand for more professional workers	76 %	76 %	77 %	82 %	60 %	72 %	80 %	78 %
Decreasing the demand for less skilled workers (less educated, trained and experienced workers)	74 %	79 %	65 %	73 %	80 %	81 %	72 %	67 %
Decreasing the demand for production workers	68 %	68 %	69 %	64 %	80 %	84 %	56 %	56 %
Elimination/substitution of some unskilled jobs	65 %	68 %	69 %	50 %	60 %	75 %	52 %	71 %
Substituting the demand for less skilled workers	57 %	53 %	58 %	55 %	80 %	69 %	44 %	50 %
Total response	76 %	34 %	26 %	11 %	5 %	32 %	25 %	18 %

Source: Own calculation based on the firm survey (2010)

Moreover, from the firm survey we find that the increasing use of new technologies has not only raised the demand for high skilled workers in the past years, but also encouraged firms to predict a future/long run increase in the demand for high skilled workers. For instance, for 68 % of the respondent firms the interpretations of the predicted long run increase in the demand for skilled workers are related to planned/expected expansion of production, product diversification, implementation of new process, output technologies, purchases of new machines and equipment and increasing R&D activities.[39] This result seems consistent with the assumption made by Aghion and Howitt (1992) that an expectation of more research in the next period must correspond to an expectation of higher demand for skilled labour in research in the next period.

[39] Moreover, other factors are: the expected increases in market share, turnover, sales, adoption of international standards and enhancement of production, advanced control systems, shortage of manpower, competition, increasing motivation to reduce costs, achieving high standard precision work, improving productivity, quality of work and demand for more specialised skills in IT.

7.5.3 The Share of Spending on ICT and Input–Output Indicators

Finally, in this section we investigate the third hypothesis on the positive relationships between new technology (total expenditures on ICT) and input–output indicators across firms and over time. For instance, when investigating the relationship between ICT and input variables, we find from Table 7.16 that the total spending on ICT is positively correlated and more sensitive to labour (firm size), and industry level throughout the period 2005–2008 and also became sensitive to capital (net worth), notably, throughout the period 2007–2008. Both the total spending on ICT and ICT training (upskilling) are positively and significantly correlated and more sensitive to labour (firm size), and capital (net worth) throughout the period 2005–2008. The relationship between ICT and labour (firm size) is particularly more significant for the large size, chemical and textile firms. The different results across chemical and textile or large size firms is plausible and can be attributed to differences in the skill levels – share of high skilled workers in total employment (see Fig. 7.1 above). This is also because large size firms are more prevalent in the textile and chemical industries, they have high share in total ICT spending, employment, fixed capital, value added and profit (see Table 7.2 above) and probably have more consistent entrepreneurial/organisational strategies.

We examine the relationship between the use of new technology as measured by total spending on ICT, profit and output. Table 7.17 illustrates plausible positive though not significant correlations between the use of new technology (as measured by total spending on ICT) and capital, labour, total output (as measured by total sales value), output diversification (as measured by sales diversification), and productivity (as measured by total sales value/labour ratio) over the period 2007–2008. Moreover, Table 7.17 shows positive significant correlations between the use of new technology as measured by total spending on ICT and total profit and value added over the period 2007–2008.[40] In addition to positive significant correlations between old technology measured by total spending on machinery and equipment and total output measured by total sales value, profit and value added, between value added and old technology measured by total spending on machinery and equipment, spending on raw materials and capital. For old technology measured by total spending on machinery and equipment, the correlation coefficients are more significant than traditional inputs (labour-capital) over the period 2005–2008. These results prove our third hypothesis regarding the positive correlation between ICT and input–output indicators at the micro/firm level. However, our results should be interpreted carefully as they probably have two-ways causality and may leave open the possibility for reversed causality. Mainly because more profit and output would imply more financial capacity that permits more spending on ICT, on the other hand, more spending on ICT implies higher costs and lower profit (see Table 7.17 below).

[40] Except in 2008, where the correlations between labour and profit, labour, capital, productivity and diversification are negative.

Table 7.16 Total spending on ICT, labour and capital across firms (2005–2008)

Independent variables		Coefficient (t-value)				
Dependent variable (ICT expenditures)		Labour	Capital	Constant	R^2	N
ICT expenditures	All firms (2008)	31189.873**		4017773.618	0.090	44
		(2.068)		(1.286)		
	Large	39678.002*		659819.035	0.178	16
		(1.801)		(0.091)		
	Small	851008.625*		−17122964.336	0.132	13
		(1.350)		(−0.871)		
	Chemical	16570.802*		1352016.355	0.077	20
		(1.261)		(0.482)		
	Food	114796.261		2209388.184	0.106	13
		(1.194)		(0.205)		
	Textile	41167.945*		6311339.471	0.505	4
		(1.749)		(0.678)		
ICT expenditures (All firms) (log)[a]	2005	50597.659**	0.00003	200891.316	0.173	23
		(2.047)	(0.624)	(0.051)		
	2006	48260.393*	0.00001	1868612.501	0.132	26
		(1.636)	(1.113)	(0.425)		
	2007	30134.482*	0.00001**	1535671.553	0.266	31
		(1.906)	(2.779)	(0.525)		
	2008	34994.538**	0.00002**	2453825.412	0.525	35
		(2.707)	(5.597)	(0.941)		
ICT expenditures (All firms) (log) (2005–2008)	Total ICT (log)	0.007**	0.002**	10.770	0.166	36
		(3.196)	(3.514)	(24.801)		
	Training (log)	0.03**	0.002**	6.648	0.655	5
		(4.714)	(3.791)	(4.155)		
	Computer (log)	−0.011*	0.002**	11.516	0.167	30
		(−1.649)	(3.227)	(17.591)		
	Telecommunication (log)	−0.004*	0.002**	10.949	0.127	34
		(−1.434)	(3.02)	(24.147)		
	Internet (log)	−0.001	0.001	11.170	0.0394	17
		(−0.145)	(1.134)	(12.641)		
	Maintenance (log)	0.006	0.0004	9.596	0.0271	11
		(0.631)	(0.577)	(7.641)		
	Other, web host (log)	−0.008	0.0002	11.417	0.0426	6
		(−0.745)	(0.222)	(9.707)		

Correlation is significant * at the 0.05 level (one-tailed); ** at the 0.01 level (one-tailed)
[a]Log value for all estimated variables: ICT, labour and capital.

Our findings concerning the significant positive correlations between ICT and profit and value added and the insignificant correlation between ICT and output imply an inconclusive effect at the micro level. These results agree with our

Table 7.17 The correlation between, firm performance, output and profit and labour, capital, total spending on ICT, machinery and equipment and raw materials across firms across firms, (2005–2008)

Independent variables		Coefficient (t-value)							
Dependent variables all firms		Labour	Capital	Total spending on ICT	Total spending on machinery and equipment	Total spending on raw materials	Constant	R^2	N
Total output (total sales value)[a]	2005[a]	0.624 (0.729)	0.196 (1.112)				9.935 (2.024)	0.077	22
	2005[a]			0.097 (0.356)			14.828 (4.497)	0.006	23
	2006[a]	0.748 (0.812)	0.253* (1.526)				8.728 (1.755)	0.110	25
	2006[a]			0.103 (0.464)			14.864 (5.346)	0.008	27
	2007[a]	0.126 (0.167)	0.222* (1.502)				11.509 (2.638)	0.075	30
	2007[a]			0.185 (0.915)			13.847 (5.583)	0.025	34
	2008[a]	0.095 (0.142)	0.193* (1.403)				12.508 (3.091)	0.059	34
	2008[a]			0.154 (0.913)			14.566 (6.965)	0.021	40
	(2005–2008)[a]	0.349 (1.047)	0.337** (5.046)	0.075 (0.845)			8.220 (4.387)	0.244	35
	(2005–2008)[a]	0.003 (1.328)	−0.0005 (−0.869)		0.006** (4.210)		16.195 (36.991)	0.154	35
Profit[a]	2005[a]	0.553 (0.576)	0.416** (2.448)	0.119 (0.488)			4.111 (0.806)	0.419	15

	2006[a]	0.890	0.441	0.040		3.350	0.480	19
		(1.076)	(3.384)	(0.214)		(0.769)		
	2007[a]	0.450	0.433**			5.095	0.392	26
		(0.660)	(3.910)			(1.408)		
	2007[a]			0.273*		11.891	0.055	29
				(1.281)		(4.498)		
	2008[a]	−0.190	0.291**	0.117		9.277	0.312	28
		(−0.359)	(2.270)	(0.818)		(2.905)		
	2008[a]			0.300*		11.740	0.097	33
				(1.854)		(5.820)		
	(2005–2008)[a]	0.001	0.0001	0.0001		14.615	0.013	34
		(0.280)	(0.715)	(0.352)		(29.834)		
	(2005–2008)[a]	0.003	0.0007		0.001**	14.213	0.203	33
		(1.124)	(1.278)		(4.558)	(34.225)		
Value added[a, b]	2005[a]	0.700	0.335**			7.482	0.256	21
		(1.026)	(2.430)			(1.884)		
	2005[a]			0.205		13.820	0.039	20
				(0.883)		(4.806)		
	2006[a]	0.568	0.328**			8.278	0.267	23
		(0.839)	(2.694)			(2.225)		
	2006[a]			0.076		15.249	0.008	24
				(0.426)		(6.948)		
	2007[a]	0.307	0.427**			7.164	0.306	28
		(0.467)	(3.384)			(1.887)		
	2007			0.439**		10.936	0.143	28
				(2.122)		(4.347)		
	2008[a]	0.363	0.399**			7.482	0.286	31
		(0.620)	(3.399)			(2.085)		
	2008[a]			0.275*		12.922	0.071	33
				(1.561)		(5.994)		

(continued)

Table 7.17 (continued)

Independent variables		Coefficient (t-value)							
Dependent variables all firms		Labour	Capital	Total spending on ICT	Total spending on machinery and equipment	Total spending on raw materials	Constant	R^2	N
	(2005–2008)[b]	246591.7	0.02**		0.0154**	0.159**	−82886294	0.981	18
		(0.609)	(2.529)		(4.064)	(13.050)	(−1.441)		
Diversification (sale diversification)	2008[a]	−0.073	−0.013	0.014			0.677	0.056	34
		(−1.137)	(−0.937)	(0.832)			(1.798)		
Productivity (sale/labour)	2005–2008[a]	−0.004	−0.0001	0.0002			12.36	0.027	35
		(−1.450)	(−1.068)	(0.947)			(26.760)		
Total spending on machinery and equipment	(2005–2008)[a]		0.514**				7.631	0.465	32
			(5.195)				(4.437)		
Wage[a]	(2005–2008)[a]	0.395*	0.445**				6.080	0.431	35
		(1.534)	(9.171)				(4.167)		

Correlation is significant * at the 0.05 level (one-tailed); ** at the 0.01 level (one-tailed)
Log value for all estimated variables: ICT, labour and capital
[a]log.
[b]linear.

observations at the aggregate level, which imply that the growing expenditures on ICT in Sudan raises the shares of the population using the Internet, enhances e-business, e-education and e-government. However, despite the growing ICT expenditures, their effects are inconclusive at the aggregate level, probably due to low spending on ICT, high poverty and illiteracy rates, low skill levels and inadequate investment in education.[41] The macro observations are consistent with the recent literature indicating the growing but limited effects of ICT diffusion in the developing countries due to a lack of sufficient investment in the complementary infrastructure such as education, skills and technical skills (cf. Pohjola 2002; Kenny 2002). Therefore, these results prove the third hypothesis in Chap. 1 above about the inconclusive effect of ICT at the micro level.

7.6 Conclusions

In this chapter we use the data from the firm survey (2010) to examine skill indicators, their implications and relationships with average wages, and with upskilling (ICT training) and technology (ICT), ICT and input–output indicators at the micro/firm level.

Our findings in Sect. 7.3 illustrate the low skill levels – due to the excessive share of unskilled workers (Figs. 7.1 and 7.2) – and the implications on skills mismatch (Fig. 7.6), industrial performance indicators and productivity decline across firms (Tables 7.3, 7.4, 7.5, and 7.6). These results are consistent with the micro–macro findings in Chap. 5 above, which indicate the low share of high skilled in total population and employment – measured by both educational and occupational levels – and the serious implications on skills mismatch and the macro–micro duality with respect to upskilling efforts. These findings together with those in Chap. 5 above verify hypotheses 3.b and 4.a in Chap. 1 above regarding the implications of the interaction between the deficient educational system and high use of unskilled workers. These findings then confirm our first hypothesis, which we proved in Chap 2 above, concerning the pressing need for upskilling, particularly within the private sector.

Our results in Sect. 7.4 show positive correlations between actual and required education, experience and average wages (Tables 7.8, 7.9, 7.10, and 7.11). We verify hypothesis 4.b. in Chap. 1 above that an increase in skill level and firm size lead to improved relationships between actual and required education (Table 7.8), between actual education, experience and wages (Table 7.10) and between required education, experience and wages (Table 7.11).

[41] Our attempt to examine the effect of ICT at the macro level in Sudan is constrained by the lack of adequate and reliable data on ICT spending, as the most recent data on the share of spending on ICT relative to GDP (2010) is available only for two years over the period 2007–2008.

In Sect. 7.5 our findings with respect to the positive complementary relationships between skill, technology (ICT) and upskilling (ICT training) and between computers, telecommunications and ICT training (Tables 7.13 and 7.14) are consistent with the findings in the new growth literature. We illustrate and corroborate hypothesis 4.c. in Chap. 1 above that an increase in skill level and firm size lead to an improvement in the complementary relationships between skill, upskilling and technology (ICT).

Taken together, all these results imply the importance of a good education for bridging differences between firms and also for enhancing skill, technology and upskilling complementarity at the micro level. These findings seem consistent with the endogenous growth framework and stylised facts concerning the relationships between human capital, technical progress and upskilling and our theoretical framework in Chap. 3 above.

Finally, our results in Sect. 7.5 indicate positive significant correlations between total spending on ICT and profit and value added, but insignificant correlations between total spending on ICT and output at the micro/firm level (Table 7.17). This result confirms the fifth hypothesis in Chap. 1 above, which implies an inconclusive effect of ICT at the micro level and supports the observations at the macro level in Sudan and the recent literature in the developing countries.

Moreover, our results in Sects 7.4 and 7.5 show the relationships between actual and required education, experience and wages and between skill, technology (ICT) and upskilling (ICT training), defined by firm size and industry level. These results are consistent with our findings in Chap. 5 above, which imply that both skill and technology indicators vary across firms and increase with firm size and industry level.

Therefore, our findings in this chapter are consistent with hypotheses 3.b. and 4. a. in Chap. 1 above with respect to the implications of the excessive use of unskilled workers at the micro level. In addition, our results verify hypotheses 4.b. and 4.c. in Chap. 1 above concerning the relationships between actual and required education and experience and between actual education, required education, experience and wages and the relationships between technology (ICT), skill and upskilling (ICT training). Finally, we corroborate the fifth hypothesis in Chap. 1 above regarding the inconclusive effect of ICT at the micro level.

References

Abdel-Salam, S. A. (2006). *The problems of industrialization in the developing countries, the case of Sudan*. Unpublished partial research paper submitted in the fulfillment of the B.Sc. degree in Economics, Department of Economics, Faculty of Economic and Social Studies, University of Khartoum, Khartoum, pp. 28–32.

Acemoglu, D. (1998). Why do new technologies complement skills? directed technical change and wage inequality. *Quarterly Journal of Economics, 113*(4), 1055–1089.

Aghion, P., & Howitt, P. (1992). A model of growth through creative destruction. *Econometrica, 60*(2), 323–351.

Aghion, P., & Howitt, P. (1998). *Endogenous growth theory*. Cambridge, MA: MIT Press.

Al-Quraishi, M. (2005). *Industrial economy*. Jordan: Dar Wael for publication and distribution.

Autor, D. H., Katz, L. F., & Krueger, A. B. (1998). Computing inequality: Have computers changed the labour market? *Quarterly Journal of Economics, 113*(4), 1169–1213.

Braga, H., & Willmore, L. (1991). Technological imports and technological efforts: An analysis of their determinants in Brazilian firms. *The Journal of Industrial Economics, 39*(4), 421–432.

Bresnahan, T. F., Brynjolfsson, E., & Hitt, L. M. (1999, May). *Information technology, workplace organization, and the demand for skilled labour: Firm- level evidence* (NBER Working Paper Series, No. 7136). Cambridge, MA: National Bureau of Economic Research. http://www.nber.org/papers/w7136.pdf?new_window=1

Colecchia, A., & Papaconstantinou, G. (1996). *The evaluation of skills in OECD countries and the role of technology* (STI Working Paper No. 1996/6). OECD- Paris – OECD/GD (96) 183.

El-Sayed, O. I. (1998, March). Industrial Sector. In *Productive activity Chapter 4 in Sudan economy* (2nd ed., pp. 184–188). Khartoum: Public Enterprise for Printing and Publishing.

Firm Survey. (2010, June). *Technological change and skill development: A comparative study of chemical, food, metal and textile small, medium and large scale enterprises in the Sudan*. Khartoum.

Goldin, C., & Katz, L. F. (1998). The origin of technology skill complementarity. *Quarterly Journal of Economics, 113*(3), 693–732.

Hartog, J. (2000). On returns to education: Wandering along the hills of Oru Land. In H. Heijke & J. Muysken (Eds.), *Education and training in a knowledge based economy*. England: MacMillan Press.

Ismail, M. M. (1994). *Industrial economics* (pp. 206–209). Cairo: Egyptian Universities house for publication.

Kenny, C. (2002, August). *The internet and economic growth in least developed countries* (UNU-WIDER Discussion Paper Series No. DP2002/75). Helsinki: UNU-WIDER.

Kumar, N., & Saqib, M. (1994, January). *Firm size, opportunities for adaptation, and in-house R&D activity in developing countries: The case of Indian manufacturing* (UNU/INTECH Working Paper No.13).

Mincer, J. (1974). *Schooling, experience and earnings*. New York: Columbia University Press.

Muysken, J., & ter Weel, B. (1998). *Overeducation and crowding out of low skilled workers* (Research Memorandum No. 98–023). Maastricht: MERIT, Maastricht University.

Muysken, J., & Ruholl, J. (2001). *The impact of education and mismatch on wages: The Netherlands, 1986–1998* (Research Memoranda No. 2001–030). Maastricht: MERIT, Maastricht University.

Muysken, J., Hoppe, M., & Rieder, H. (2002). *The Impact of education and mismatch on wages: Germany, 1984–2000* (Research Memoranda No. 2002–031). Maastricht: MERIT, Maastricht University.

Muysken, J., Weissbrich, A., & von Restorff, C. H. (2002). *The impact of education and mismatch on wages: The USA, 1986–1996* (Research Memoranda No. 2002–015). Maastricht: MERIT, Maastricht University.

Muysken, J., Kiiver, H., & Hoppe, M. M. (2003). The impact of education and mismatch on wages: Germany, 1984–2000. In F. Büchel, A. de Grip, & A. Mertens (Eds.), *Overeducation in Europe: What do we know?* (pp. 109–132). Cheltenham: Edward Elgar.

Muysken, J., & Nour, S. (2006). Deficiencies in education and poor prospects for economic growth in the Gulf countries: The case of the UAE. *Journal of Development Studies, 42*(6), 957–980. Taylor and Francis Ltd: UK.

Nour, S. (2005, November). *Technological change and skill development in the Arab Gulf countries*. Doctoral dissertation, Maastricht University Press, Maastricht, The Netherlands.

OECD (Organisation for Economic Co-operation and Development). (1997). *The second European report on science and technology indicators (1997)* (pp. 37, 90, 91). Paris: OECD.

Pohjola, M. (2002, July). *New economy in growth and development.* (UNU- WIDER Discussion Paper Series No. DP2002/67).

Utton, M. A. (1979). *"Diversification and competition," The National Institute of Economic and Social Research.* Cambridge, UK: Cambridge University Press. pp. 15, 16, 104, 105.

Chapter 8
The Importance (Impacts) of Knowledge at the Macro–Micro Levels

Abstract This chapter uses the firm survey (2010) data at the micro level and secondary data at the macro level to examine hypothesis 7 in Chap. 1 above concerning the importance/impacts of tacit and codified sources of knowledge at the micro and macro levels respectively in Sudan. Our results prove this hypothesis and show that at the macro level tacit knowledge and codified sources of knowledge are positively and significantly correlated with both schooling years and GDP growth (economic growth rate). At the aggregate level and macro level codified knowledge and the number of FTER show positive correlations with technology (patents) and imply a significant positive complementary relationship between tacit knowledge (measured by the number of FTER) and codified knowledge. At the micro (firm) level, we illustrate the importance of tacit knowledge, which shows positive significant correlations with technology (expenditures on ICT) and upskilling (expenditures on training), output (defined by total sale value), output diversification, productivity and profit. At the micro (firm) level, tacit and codified knowledge show positive significant correlations with total investment, capital, and firm size. This can be interpreted that higher levels of total investment, capital and firm size would correspond to more tacit and codified sources of knowledge across firms.

8.1 Introduction

Our earlier findings in Chap. 5 indicate that the transfer of knowledge is successful within firms, but is somewhat doubtful between firms and universities and within society at large. Our analysis shows that within society at large, the transfer of knowledge is hindered by low skill levels, deficient educational and training systems and the lack of incentives. The transfer of knowledge between universities and firms is hindered by the lack of incentives such as subsidies, and the lack of a networks, information systems, cooperation and interest in conducting joint research between universities and firms and matching the relevance of universities' research to firms needs.

S. Mohamed Nour, *Technological Change and Skill Development in Sudan*,
DOI 10.1007/978-3-642-32811-4_8, © Springer-Verlag Berlin Heidelberg 2013

One implication of our earlier analysis is that Sudan needs to stimulate the incidence and transfer of knowledge at the aggregate level by providing more incentives, for example through subsidies, to education and training to upgrade skill levels, and also by raising spending on R&D and ICT, organisation, coordination and cooperation. Further incentives, such as subsidies, should be provided to stimulate the transfer of knowledge between universities and firms that requires a good knowledge base within firms and further incentives, for example subsidies to education and training to enhance skill levels, and subsidies to R&D, networks organisation, information, coordination and cooperation. In this Chapter we extend our earlier analysis and explain the importance (impacts) of knowledge at both micro and macro levels in Sudan in more detail. In addition, we show the factors contributing to improve the tacit knowledge within firms. Due to the lack of relevant data to assess the transfer of knowledge amongst firms and between firms and universities, we focus only on the impacts of knowledge within the firms and at the aggregate/macro level.

The rest of this chapter is organised as follows: Sect. 8.2 briefly shows the importance and sources of knowledge in the growth literature; Sect. 8.3 presents hypothesis 7 in Chap. 1 above to test some stylised facts about the importance of knowledge and explains the data used to test them; Sect. 8.4 discusses the main findings; and Sect. 8.5 provides the conclusions.

8.2 Definition, Importance, Sources and Measurement of Knowledge in the Growth Literature

Endogenous growth literature recognised the importance of knowledge and its accumulation as a unique source of endogenous technological progress, innovation and economic growth. For instance, in the Lucas (1988) model, knowledge accumulation is vital for the growth process, for knowledge creation, accumulation and acceleration, contribution to scientific and technological progress, innovation, economic growth performance and development.[1]

In defining 'knowledge' the literature makes a distinction between codified and tacit knowledge (Dasgupta and David 1994). "Codified knowledge implies that knowledge is transformed into information which can either be embodied in new material goods (machines, new consumer goods) or easily transmitted through information infrastructure. While, the tacit knowledge refers to that which cannot easily transferred because it has not been stated or measured in an explicit form, skill is an important kind of tacit knowledge"[2] (cf. Freeman and Soete 1997: 404, 405).

In addition, the definition of codified knowledge in the literature is closely related with investment in public spending on education, training, R&D and ICT.

[1] The OECD (1997) confirm that Access to scientific and technological knowledge and the ability to exploit it are becoming increasingly strategic and decisive for the economic performance of countries and regions in the competitive globalized economy.

[2] Disembodied flows of knowledge can be transmitted through movement of people, publications, etc.

Several studies perceive knowledge as a public good, produced through R&D activities that generate spillover and thereby increasing returns (Romer 1994; Grossman and Helpman 1994). Other studies use broader terms to interpret knowledge created and embodied in institutions (cf. Langlois 2001). For instance, Nelson (1993) and Lundvall (1992) emphasise the importance of institutions for the flows of knowledge and information to innovative capability. According to Smith (2002): "R&D is but one component of knowledge and innovation expenditures, and by no means the largest. Because, R&D data tend to either overemphasize the discovery of new scientific or technical innovations, or to exclude a wide range of activities that involve the creation or use of new knowledge in innovation. Thus, innovation rests not only on discovery and R&D but also on learning, external environment (network) of the firm, non-R&D expenditures such as training, market research, design, trail production and tooling up and IPR costs. In addition to capital expenditure, which is a key mode of 'embodied' knowledge spillover from the capital good sector to using industries" (Smith 2002: 14–18).

Moreover, the evolutionary framework developed by Nelson and Winter (1982) makes the nature of knowledge and firms' investment in it a central factor in explaining the size, structure and dynamics of industries. Recent empirical literature (cf. Loof and Heshmati 2002) shows that knowledge capital (defined as the ratio of innovation sales to total sales) is found to be a significant factor contributing to performance heterogeneity and a firm's innovative level. Knowledge capital rises with innovation input, the firm's internal knowledge for innovation and cooperation with domestic universities on matters of innovation. Some empirical studies indicate that survival and growth amongst firms is determined by/or at least influenced by differential rates of investment in knowledge (such as R&D) (cf. Klepper and Simon 1997) or intersectoral differences in the size and R&D intensity of firm (cf. Levin et al. 1985). In addition, Brusoni et al. (2002) and David and Foray (1995) show that an increasing codification of knowledge stock would increase a firm's innovative performance.

In addition, differential in the productivity and growth of different countries is significantly related to improvement in the quality of human capital, technical progress, factors of production and the capacity to create new knowledge and ideas and incorporate them in equipment and people. "Recent growth literature show increasing evidences of the growing relative importance of intangible capital in total productive wealth and the rising relative share of GDP attributable to intangible capital (Abramovitz and David 1996, 1998). Intangible capital largely falls into two main categories: on the one hand, investment geared to the production and dissemination of knowledge (i.e. training, education, R&D, information and coordination); on the other hand, investment geared to sustaining the physical state of human capital (health expenditures). In the US, the current value of the stock of intangible capital (devoted to knowledge creation and human capital) began to outweigh that of tangible capital (physical infrastructure and equipment, inventories, natural resources) at the end of the 1960s. Moreover, since the 1960s annual investment rates in R&D, public education and software have grown steadily at an annual rate of 3 % in the OECD countries" (David and Foray 2001: 1–2).

Furthermore, Drucker (1998: 15) suggests: "knowledge is now becoming the one factor of production, sidelining both capital and labour". In addition, the OECD (1999: 7) has suggested "... the role of knowledge (as compared with natural resources, physical capital and low skill labour) has taken on greater importance".[3] Smith (2002) argues that in recent years, learning and knowledge have attracted increasing attention as a result of the claims that knowledge-intensive industries are now at the core of a growth, knowledge driven economy or even a knowledge society. The role of knowledge as an input to economic processes has fundamentally changed, probably due to rapid technological changes/ advances in ICT; ICT is seen as factor increasing knowledge and increasing the common availability of codified knowledge (David and Foray 1995; Smith 2002). For instance, Van Zon (2001) extends Lucas' (1988) model by incorporating the effects of ICT and capital investment, assuming that ICT has positive influence on growth performance, both by improving the intensity of production and total factor productivity and enhancing the efficiency of knowledge accumulation and learning process.

Moreover, the empirical literature shows that knowledge is positively related to human capital (mainly tacit skill or skill level). For instance, Winter (1987) suggests that tacit and codified knowledge need not be substitutes, but can be seen as complements in the learning process. Brusoni et al. (2002) show a strong positive relationship between the codification of the knowledge base of the industry and its investment in skilled people (high levels of investment in tacit skills) and R&D.

In addition, Cowan et al. (2001: 9) examine knowledge transfer in the services sector "as a process by which knowledge travels from a knowledge holder (a person or organisation possessing the knowledge)" to a knowledge recipient (a person or organisation receiving the knowledge). In their analysis "knowledge holder is important as the "point of departure" of the knowledge being transmitted since they can influence knowledge flows".

Furthermore, the literature indicates a substantial contribution to innovation and therefore to economic growth and public welfare that can be related to an unintended spillover associated with knowledge flows.[4] Distinction has been made between three sources for the flows and transfer of knowledge: for instance, Brusoni et al. (2002) highlight the importance of knowledge sources within the enterprise for innovation among innovative firms in Europe, in particular, the internal divisions (including R&D, design, sales and marketing and senior management). Several other studies have focused on knowledge flows between firms through inter-firm research collaborations (Hagedoorn et al. 2001), user-producer networks (Lundvall 1992), or linkages between competing firms (von Hippel 1988). Yet other studies examine knowledge flows between firms and public research organisations such as universities, public research institutes, government laboratories, and publicly-funded technical institutes (cf. Arundel and Geuna 2001; Mansfield 1991; Mansfield and Lee 1996). At the aggregate level, the transfer of knowledge

[3] See Drucker (1998), p.15 and OECD (1999), p.7 respectively.

[4] Verspagen and Schoenmakers (2000) use patent citations to measure knowledge spillover.

is related to several variables such as the overall quantity of scientific research (publications) and the public research base as measured by the ratio between the total amount of higher education R&D expenditure and the country GDP (cf. Arundel and Geuna 2001: 3, 5).

The notion that knowledge is a public good, produced through education, training and R&D activities that generate spillovers and increasing returns, provides a plausible justification for government intervention to compensate the private sector for the positive externalities they generate and to provide more incentives to support investment and accumulation of knowledge. While Lucas' (1988) model emphasises investment in human capital, it only implicitly allows for a role for public policy through subsidies (Haslinger and Ziesemer 1996: 230). Subsequent studies attempted to fill this gap in Lucas' (1988) model and explicitly indicate a potential role for government intervention and public policies to support the creation and accumulation of knowledge. The main channels are through taxation or subsidisation to the provision of R&D (cf. Romer 1990; Barro and Sala-i-Martin 1992, 1995), public knowledge: basic education and basic scientific research (cf. Ziesemer 1990, 1995) and subsidising training (cf. Chatterji 1995) (see our discussion in Chap. 3 above).

8.3 The Hypothesis, Stylised Facts and Data

Based on the above background, this Section presents hypothesis 7 in Chap. 1 above to test some stylised facts about the importance of knowledge in Sudan and explains the data used to test them.

8.3.1 The Importance (Impacts) of Knowledge at the Micro–Macro Levels in Sudan

In recent times, few studies discuss the status of knowledge in the Arab countries. The UNDP-AHDR (2003), Arab Knowledge report (2009) and Nour (2010) examine the weak status of demand, production and dissemination of knowledge in the Arab states. Aubert and Reiffers (2003) assess the challenges and underline a strategy for the development of knowledge-based economies in the Middle East and North African countries (MENA). All these reports provide significant contribution, but a somewhat general analysis at the aggregate/macro level that refers to all Arab and MENA countries respectively. Since Sudan shows considerable disparity from the other Arab and MENA countries, at least in respect of some indicators such as structure and size of the economy, level of income and structure of labour market, it might be useful to look at it separately. Thus, one obvious advantage of our analysis is that we provide a more specific analysis that focuses only on Sudan as a new case study. Moreover, different from earlier studies, we provide a new empirical investigation of both the importance (impacts) of tacit knowledge at the micro level (see our discussion below) as well as the discrepancy in the transfer of knowledge/external schooling effects at the macro–micro levels (see our discussion in Chap. 5 above).

In this chapter we use the literature presented above to examine hypothesis 7 in Chap. 1 above concerning the importance (impacts) of tacit and codified knowledge at the macro (within society)–micro (inside the firms) levels. In particular, our aim is to test the following stylised facts:

1. At the macro level codified knowledge and tacit knowledge are positively correlated with economic growth (GDP growth rate) and are positively correlated with schooling;
2. At the macro level codified knowledge (the total spending on R&D) and tacit knowledge (the total number of full time equivalent researchers (FTER)) are positively correlated with each other and also with technology (patents);
3. At the micro (firm) level tacit knowledge is positively correlated with technology (ICT), upskilling (training), profit, productivity, output and output diversification;
4. At the micro (firm) level tacit knowledge is positively correlated with market size (firm size; capital; and investment) and firm age.

8.3.2 Definition of Data and Variables

We use the broad definition of knowledge found in the new growth literature that highlights both the tacit and codified components of knowledge. In particular, we define tacit knowledge by the percentage share of high skilled workers in total employment at the micro level.[5] In addition, at the macro level we define tacit knowledge by the share of enrolment in tertiary education; moreover, we use the share of high skilled population with tertiary education in total population and the total number of FTER[6] as other indicators of tacit knowledge at the macro level.[7] We define codified knowledge by the embodied knowledge distributed in many indicators, including the share of spending on R&D, education, training and ICT as

[5] As in Chap. 7 above, our definition of high skilled workers refers to workers with post secondary educational attainment: university degree and above (16 years of schooling).

[6] The concept of full time equivalent researcher (FTER) is adopted by UNESCO statistics on R&D personnel.

[7] The main limitations of our data at the macro/aggregate level are the definition of tacit knowledge by the share/ratios of enrolment in tertiary education (despite their drawback), the adjustment of the variables for different years and the use of unified ratio of education and R&D spending and tacit knowledge and schooling indicators, due to scarcity of data.

percentage of GDP at the macro level.[8] In addition, we use several variables related to knowledge such as patents and schooling years, defined by school life expectancy and mean years of schooling, in Sudan.,[9,10] Table 8.1 below presents the data and variables, which we use in our analysis of the importance (impacts) of knowledge at the macro/aggregate level in Sudan.

As in Chap. 7 above, we obtain our micro/firm data from the firm survey (2010) and use three sets of indicators, including tacit knowledge (technical and non-technical skills), technology and input–output variables. We define tacit knowledge by the share of high skilled/educated workers in total employment; we define codified sources of knowledge by total spending on ICT, R&D and training; we use other definition of codified sources of knowledge, defined by knowledge embodied in machines, defined by total spending on machinery and equipment; we define technology by expenditures on ICT and on machinery and equipment; we define upskilling by expenditure on training; input indicators are labour (employment size) and capital (net worth), output (total sales value), output diversification (sales diversification), productivity and profit.[11]

8.4 The Empirical Results

We use the data presented above and the linear and log linear OLS regression techniques and E-Views and SPSS statistical programmes to test and compare the importance (impacts) of tacit and codified knowledge at the micro and macro levels respectively and compare the relevance of our findings to those in the knowledge literature. Based on Table 8.1 below and Tables 8.2 and 8.3 below we present a panel data analysis reflecting the average for Sudan over the period 1990–2009. Based on data from the firm survey (2010), Tables 8.4, 8.5, 8.6, 8.7, 8.8, and 8.9 reflect the results across firms.

---

[8] At the micro level, the definition of codified knowledge by the relative term or the share of these indicators to total output or sales value does not provide relevant results.

[9] *The concept expected years of schooling (of children) is defined by the number of years of schooling that a child of school entrance age can expect to receive if prevailing patterns of age-specific enrolment rates were to stay the same throughout the child's life, see for instance, UNESCO Institute for Statistics (*2010a*), 'Correspondence on Education Indicators. March 2010. Montreal'.*

[10] *The concept mean years of schooling (of adults) refers to average number of years of education received by people ages 25 and older in their lifetime based on education attainment levels of the population converted into years of schooling based on theoretical durations of each level of education attended, see Barro and Lee (*2010*), 'A New Data Set of Educational Attainment in the World, 1950–2010', NBER Working Paper No. 15902.*

[11] As in Chap. 7 above, we use the same definitions of educational qualifications, ICT, diversification, output, capital, labour (firm's size) and firm's age (total years in operation) - see our definitions in Chap. 7 above. In addition, we obtained information on investment and labour variables from Sudan Ministry of Industry (2005) 'Comprehensive Industrial Survey (2001)'.

Table 8.1 The determinants of knowledge in the Sudan (1990–2009)

Variable/ year	GDP Growth Rate[a, b, c]	Tacit knowledge[d, e, f, g, h] Tertiary[d, e, f,]	Share of high skilled in total population (%)[d, g, h]	Schooling[e, i, j] School Life Expectancy[i, e,]	Mean years of schooling[j]	Codified sources of knowledge: public expenditure on education, R&D, ICT and training as % of GDP (%) Education[k, l, m,]	R&D[n]	ICT[o]	Training[o]	Total spending on education, R&D, ICT and training
1990	5.4	3.0 [d]	0.004 [g]	4.2 [i]	1.5	0.9 [k]	n. a.	n. a.	n. a.	0.9
1991	7.5	3.0 [d]	0.004 [g]	4.2 [i]	1.5	0.6 [l]	n. a.	n. a.	n. a.	0.6
1992	6.5	3.0 [d]	0.004 [g]	4.2 [i]	1.5	0.6 [l]	n. a.	n. a.	n. a.	0.6
1993	4.5	3.0 [d]	0.004 [g]	4.2 [i]	1.5	0.6 [l]	n. a.	n. a.	n. a.	0.6
1994	1	3.0 [d]	0.004 [g]	4.2 [i]	1.5	0.6 [l]	n. a.	n. a.	n. a.	0.6
1995	5.9	3.0 [d]	0.004 [g]	4.2 [i]	1.5	0.6 [l]	n. a.	n. a.	n. a.	0.6
1996	5.9	3.0 [d]	0.004 [g]	4.2 [i]	1.5	0.6 [l]	0.5	n. a.	n. a.	1.1
1997	6.3	3.0 [d]	0.004 [g]	4.2 [i]	1.5	1.4 [m]	0.5	n. a.	n. a.	1.9
1998	6.4	6 [d]	0.004 [g]	4.2 [i]	1.5	1.4 [m]	0.5	n. a.	n. a.	1.9
1999	6.7	6 [d]	0.005 [g]	4.2 [i]	1.5	1.4 [m]	0.53	n. a.	n. a.	1.93
2000	8	6.85 [e]	0.007 [g]	4.4 [i]	2.4	1.4 [m]	0.47	n. a.	n. a.	1.87
2001	6.7	6.85 [e]	0.007 [g]	4.4 [i]	2.4	1.4 [m]	0.44	n. a.	n. a.	1.84
2002	6.5	6.85 [e]	0.007 [g]	4.4 [i]	2.4	1.4 [m]	0.39	n. a.	n. a.	1.79
2003	6	6 [e]	0.007 [g]	4.4 [i]	2.4	1.4 [m]	0.34	n. a.	n. a.	1.74
2004	7.2	6 [e]	0.007 [g]	4.4 [i]	2.4	1.4 [m]	0.29	n. a.	0.0001	1.6901
2005	8	6 [e]	0.007 [g]	4.6 [e]	2.8	1.4 [m]	0.29	n. a.	0.0001	1.6901
2006	10.0	6 [e]	0.007 [g]	4.6 [e]	2.8	1.4 [m]	0.28	n. a.	0.0002	1.6802
2007	10.5	6 [e]	0.007 [g]	4.6 [e]	2.8	1.4 [m]	0.28	0.00039	0.0005	1.68089
2008	7,8	5.9 [f (1)]	0.0532 [h]	4.6 [e]	2.9	1.4 [m]	0.28	0.00018	0.0006	1.68078
2009	6.1	5.9 [f (1)]	0.0532 [h]	4.6 [e]	2.9	1.4 [m]	0.28	n. a.	0.0003	1.6803

		Codified sources of knowledge: total public expenditure on R&D, ICT and training (million pounds)				Share in total R&D Expenditure (%)[n]			Patent [p]		
Year	Total researcher[n]	ICT[o]	Training[o]	R&D[n]	Total R&D, ICT, and training	Business enterprise [n],	Government [n],	Higher education [n]	Resident	Non-resident	Total
1995	n. a.	n. a.	n. a.	n. a.	n. a.	n. a.	n. a.	n. a.	n. a.	n. a.	n. a.
1996	n. a.	n. a.	n. a.	100000	100000	n. a.	n. a.	n. a.	n. a.	n. a.	n. a.
1997	n. a.	n. a.	n. a.	n. a.	n. a.	n. a.	n. a.	n. a.	n. a.	n. a.	n. a.
1998	n. a.	n. a.	n. a.	n. a.	n. a.	n. a.	n. a.	n. a.	n. a.	n. a.	n. a.
1999	9100	n. a.	n. a.	143000	143000	31.5	38.5	30.1	2	4	6
2000	900	n. a.	n. a.	149000	149000	31.5	38.9	29.5	6	16	22
2001	9340	n. a.	n. a.	152400	152400	31.5	39.3	29.2	1	13	14
2002	11208	n. a.	n. a.	154000	154000	31.8	39.0	29.2	2	20	22
2003	7300	n. a.	n. a.	156500	156500	31.9	39.0	29.1	6	11	17
2004	7500	n. a.	7.3	163730	163737.3	33.6	38.3	28.1	4	17	21
2005	7850	n. a.	4.64	192840	192844.64	33.7	39.2	27.1	6	16	22
2006	n. a.	n. a.	5.56	n. a.	5.56	n. a.	n. a.	n. a.	3	13	16
2007	n. a.	9.341	12.00	n. a.	21.341	n. a.	n. a.	n. a.	3	13	16
2008	n. a.	4,60	15.401	n. a.	15.401	n. a.	n. a.	n. a.	n. a.	n. a.	n. a.
2009	n. a.	n. a.	8.559	n. a.	8.559	n. a.	n. a.	n. a.	n. a.	n. a.	n. a.

(1) Data refers to most recent available data between 2001 and 2009). (2) n. a. refers to data not available

Sources: (a) Sudan ministry of Finance and National Economy, (b) Sudan central Bank of Sudan (c) Sudan Central Bureau of Statistics (2009: 39–43), (d) UNDP Human Development Report (HDR) (2002), (e) UNESCO (2011), (f) UNDP HDR (2010), (g) Own calculation from Barro and Lee (2000) and Ali (2006:14), (h) Sudan Central Bureau of Statistics (2010) population census data (2008), (i) UNESCO Institute for Statistics (UIS) (2010a), (j) Barro and Lee (2010), (k) UNDP HDR (2004), (l) UNESCO–UIS (2003), (m) UNESCO–UIS (2000), (n) UNESCO R&D Statistics (2006), (o) Sudan National Council for Strategic Planning- General Secretariat (2010: 493, 497), (p) World Development Indicators database (2011)

Table 8.2 The impacts of tacit and codified sources of knowledge on schooling and GDP in Sudan (1990–2009)

	Coefficient (t-value)						
	Constant	Tacit knowledge (share of enrolment in tertiary education)	Tacit knowledge (Share of high skilled population with tertiary education in total population (%))	Codified knowledge (share of public spending in education, R&D, training and ICT in GDP)	Codified knowledge (share of public spending in education in GDP)	R^2	N
Schooling: school life expectancy	3.994	0.073**				0.485	20
	(43.856)	(4.113)					
	4.283		6.620**			0.330	20
	(109.358)		(2.977)				
	4.110			0.171**		0.293	20
	(43.709)			(2.731)			
	4.014				0.296**	0.428	20
	(41.714)				(3.669)		
Schooling: mean years of schooling	0.690	0.279**				0.582	20
	(2.399)	(5.006)					
	1.830		22.600**			0.313	20
	(13.161)		(2.866)				
	1.142			0.654**		0.347	20
	(3.602)			(3.092)			
	0.806				1.105**	0.485	20
	(2.515)				(4.115)		
Tacit knowledge (share of enrolment in tertiary education)	1.329			2.557**		0.707	20
	(2.284)			(6.586)			
	0.622				3.784**	0.758	20
	(1.035)				(7.504)		
Tacit knowledge (Share of high skilled population with tertiary education in total population (%))	0.001			0.007		0.057	20
	(0.101)			(1.043)			
	−0.004				0.012*	0.096	20
	(−0.347)				(1.382)		

Growth of GDP	4.108			1.783**	0.243	20
	(3.613)			(2.334)		
	6.559	3.163			0.0003	20
	(11.294)	(0.073)				
	3.6120		0.609**		0.261	20
	(2.837)		(2.448)			
Codified knowledge (share of public spending on education,	0.045	0.276**			0.707	20
	(0.205)	(6.586)				
R&D, training and ICT in GDP)	1.315		8.677		0.057	20
	(8.965)		(1.043)			

Correlation is significant * at the 0.05 level (one-tailed), ** at the 0.01 level (one-tailed)

Table 8.3 The impacts of FTER and codified sources of knowledge on each other and patent in the Sudan (1990–2009)

	Coefficient (t-value)							
	Constant	Tacit knowledge (total number of FTER)	Codified knowledge (total public spending on R&D)	R&D (share of government in total R&D spending)	R&D (share of business in total R&D spending)	R&D (share of higher education institutions in total R&D spending)	R^2	N
Tacit knowledge (total number of FTER)	5,392.608		0.014				0.005	8
	(0.381)		(0.157)					
Codified knowledge (total public spending on R&D)	156,116.6	0.351					0.005	8
	(8.525)	(0.157)						
	−2,857,670			39,172.90	31,894.35	16,115.27	0.962	8
	(−0.828)			(1.093)	(0.941)	(0.478)		
Technology (Patents)	−43.323	0.0008*[(1)]					0.779	3
	(−1.283)	(1.876)						
	−13.0895		0.0002* [(1)]				0.277	7
	(−0.585)		(1.383)					

Correlation is significant * at the 0.05 level (one-tailed) ** at the 0.01 level (one-tailed), Notes: N = 6

[1]correlation between tacit knowledge (FTER) codified knowledge (total spending on R&D) and total patent by resident and non-resident

Table 8.4 The significance of tacit knowledge for firms performance across firms, 2010

All equations for 2008		Coefficient (t-value) Constant	Tacit knowledge (share of high skilled in total employment)	R^2	N
Total profit	All firms	15.455	0.211	0.0002	40
	(log)	(24.002)	(0.137)		
Total output (total sales value)	All firms	16.867	0.303	0.0003	44
	(semi log)	(26.663)	(0.194)		
Productivity (total sales value per workers)	All firms	13.003	0.292	0.005	44
	(log)	(21.077)	(0.813)		
Output diversification (sales diversification)	All firms	1.164	0.104	0.035	44
	(log)	(4.190)	(1.253)		
Value added	All firms	−47,932,040.9	8,748,233.6	0.092	13
	(linear)	(−0.163)	(1.101)		
Technology (total expenditures on ICT)	All firms	31.724	0.0000004**	0.052	82
	(linear)	(10.198)	(2.115)		
	All firms	6,270,092.7	53776.9	0.006	44
	(linear)	(1.412)	(0.499)		
	Large	40.404	0.0000003	0.036	32
	(linear)	(6.934)	(1.079)		
	Medium	27.875	0.0000004*	0.128	27
	(linear)	(6.103)	(1.953)		
	Chemical	33.111	0.0000003	0.042	36
	(linear)	(7.354)	(1.241)		
	Food	29.619	0.0000002	0.051	28
	(linear)	(5.387)	(1.200)		
	Metal	1.282	0.060	0.325	4
	(log)	(5.224)	(1.202)		
	Textile	19.140	0.000001**	0.884	4
	(linear)	(3.175)	(4.773)		
Skill upgrading (total expenditure on ICT Training)	All firms	993,061.5	179433.8	0.177	4
	(linear)	(0.090)	(0.802)		
	Large	43.281	0.000001	0.142	3
	(linear)	(2.260)	(0.576)		
	Food	37.543	0.000002	0.087	4
	(linear)	(2.109)	(0.536)		
Skill upgrading (total expenditures on general training)	All firms	14.381	2.726	0.118	9
	(log)	(5.897)	(1.034)		
Skill upgrading (training employees)	All firms	1.917	0.527	0.115	9
	(log)	(3.671)	(1.022)		

Correlation is significant * at the 0.05 level (one-tailed) ** at the 0.01 level (one-tailed)

Tables 8.2, 8.3, 8.4, 8.5, 8.6, 8.7, and 8.8 present our results, which indicate the importance (impacts) of tacit and codified sources of knowledge at the macro (aggregate) and micro (firm) levels respectively. Some of these results are

Table 8.5 The significance of codified sources of knowledge and firms performance across firms, 2010

All equations for 2008		Coefficient (t-value) Constant	Codified knowledge (total spending on R&D, ICT and training)	Codified knowledge (total spending on machinery and equipment)	R^2	N
Total output (total sales value)	All firms (log)	12.733 (6.798)	0.282** (2.062)		0.098	40
	Large (log)	12.808 (3.917)	0.326* (1.568)		0.149	15
	Medium (log)	9.162 (2.482)	0.513* (1.711)		0.226	11
	Chemical (log)	12.619 (6.494)	0.354** (2.350)		0.225	20
	Food (log)	13.965 (2.874)	0.054 (0.150)		0.002	10
	Metal (log)	8.362 (2.237)	0.745** (3.075)		0.825	3
Output diversification (sales diversification)	All firm (log)	0.078 (0.511)	0.017* (1.612)		0.050	50
	Large (log)	0.088 (0.285)	0.018 (0.920)		0.047	18
	Medium (log)	-.178 (−0.701)	0.030* (1.571)		0.150	15
	Small (log)	0.217 (0.773)	0.012 (0.540)		0.020	15
	Chemical (log)	−0.140 (−0.705)	0.038** (2.620)		0.222	25
	Food (log)	−0.093 (−0.352)	0.023* (1.250)		0.115	13
Total profit	All firms (log)	11.042 (6.078)	0.329** (2.479)		0.161	33
	Large (log)	11.293 (2.934)	0.301 (1.215)		0.102	14
	Medium (log)	6.099 (1.644)	0.739** (2.308)		0.432	8
	Small (log)	12.089 (4.625)	0.283* (1.351)		0.186	9
	Chemical (log)	11.111 (4.617)	0.365* (1.985)		0.180	19
	Food (log)	12.404 (4.970)	0.091 (0.474)		0.036	7
	Metal (log)	10.461 (4.873)	0.590** (4.706)		0.957	2

(continued)

Table 8.5 (continued)

All equations for 2008		Coefficient (t-value) Constant	Codified knowledge (total spending on R&D, ICT and training)	Codified knowledge (total spending on machinery and equipment)	R^2	N
Value added	All firms (log)	12.826 (3.331)	0.222 (0.810)		0.062	11
	Large (log)	13.263 (0.825)	0.131 (0.140)		0.010	3
	Medium (log)	−1.074 (−0.429)	1.287** (6.066)		0.948	3
	Small (log)	15.946 (6.592)	0.185 (0.964)		0.317	3
	Chemical (log)	14.728 (2.929)	0.072 (0.200)		0.013	4
	Food (linear)	−114,881,069.48 (−0.341)	17.939** (2.845)		0.730	4
Technology (total expenditures on ICT)	All firms (log)	5.560 (2.060)		0.384** (2.340)	0.142	34
Value added	All firms (linear)	27,460,225.450 (0.738)		0.031** (16.118)	0.956	13
Total profit	All firms (log)	7.682 (2.925)		0.474** (2.945)	0.224	31
Raw materials	All firms (log)	7.161 (2.826)		0.560** (3.614)	0.284	34
Total output (total sales value)	All firms (log)	9.456 (3.436)		0.447** (2.662)	0.172	35
Output diversification (sales diversification)	All firms (linear)	1.485 (17.037)		0.00008 (1.168)	0.039	10

Correlation is significant * at the 0.05 level (one-tailed), ** at the 0.01 level (one-tailed)

consistent with the findings in the literature (cf. Abramovitz and David 1996, 1998; David and Foray 2001; Loof and Heshmati 2002). Our results in Tables 8.2 and 8.3 illustrate the importance of knowledge at the aggregate/macro level. In support of our expectations, the findings in Table 8.2 indicate that at the macro level both codified knowledge (defined by total public spending on education, training, R&D and ICT) and tacit knowledge (defined by tertiary school enrolment ratios and the share of high skilled population with tertiary education in total population) show positive significant correlation with GDP growth rate and positive significant correlation with schooling years (defined by school life expectancy and mean years of schooling). In addition, we find that codified knowledge (defined by the

Table 8.6 The significance of the components of codified sources of knowledge and firm performance across firms, 2010

			Coefficient (t-value)					
All equations for 2008	2008	Constant	Codified knowledge (R&D expenditure)	Codified knowledge (total spending on R&D, ICT and training)	Codified knowledge (general training expenditure	Training employees staff	R^2	N
Total profit	All firms (linear)	−573799690.1 (−1.242)	2.264** (8.181)				0.905	8
Total output (total sales value)	All firms (linear)	1587124792.4 (0.844)	4.493** (3.988)				0.694	8
Codified knowledge (total spending on machinery and equipment)	All firms (linear)	9575800441.6 (0.708)	28.729 (0.258)				0.013	6
Value added	All firms (log)	10.091 (1.118)	0.412 (0.735)				0.350	2
Raw materials	All firms (linear)	879325977.81 (0.596)	3.390 (0.279)				0.015	6
Output diversification (sales diversification)	All firms (linear)	1.676 (8.790)	−0.00007 (−0.584)				0.037	10
Dependence on the import of foreign technology (value of imported capital equipment to total capital equipment (%))	All firms (linear)	0.416 (5.953)	0.00001** (2.468)				0.404	10
	All firm (linear)	0.550 (13.267)		0.000087* (1.449)			0.040	52
Dependence on the import of foreign technology (value of capital equipment to total capital equipment built by foreign companies (%))	All firms (linear)	0.311 (3.052)	0.000001** (2.259)				0.422	8
	All firm (linear)	0.424 (6.905)		0.000001* (1.746)			0.101	28
General Training employees	All firms (linear)	6.146 (1.562)			0.00004 (0.081)		0.001	9
Tacit knowledge (share of high skilled in total employment)	All firms (log)	−1.141 (−3.037)			0.219 (1.022)		0.115	9
	All firms (log)	−1.300 (−2.402)				0.043 (1.034)	0.118	9

Correlation is significant * at the 0.05 level (one-tailed) ** at the 0.01 level (one-tailed)
(1) The logarithm (log) value for all estimated variables is taken. (2) correlation between Codified knowledge (share of R&D expenditure) and labour

Table 8.7 The correlation between tacit and codified sources of knowledge across firms, 2010

All equations for 2008		Constant	Coefficient (t-value) Tacit knowledge (share of high skilled in total employment)	Codified knowledge (spending on R&D, ICT and training)	Codified knowledge (spending on machinery and equipment)	R^2	N
Codified knowledge (total spending on R&D, ICT, training)	All firms	14.290	0.560			0.017	50
	(log)	(13.735)	(0.930)				
	Large	16.804	0.988			0.086	19
	(log)	(12.364)	(1.298)				
	Medium	−3,487,389.092	33,619,426.961*			0.127	27
	(linear)	(−.521)	(1.945)				
	Small	130,726.714	35,922,633.83*			0.089	19
	(linear)	(0.014)	(1.326)				
	Chemical	13.331	0.158			0.002	25
	(log)	(10.035)	(0.224)				
	Food	4,526,245.979	31,159,057.065*			0.071	28
	(linear)	(0.495)	(1.432)				
	Metal	16.324	3.137			0.035	4
	(log)	(1.640)	(0.332)				
	Textile	−22,553,770.978	11,838,6918.5**			0.941	4
	(linear)	(−2.921)	(6.896)				
	Textile	20.030	4.537**			0.712	4
	(log)	(9.936)	(2.721)				
Tacit knowledge	All firms	0.318			0.000003	0.027	35
	(linear)	(7.729)			(0.964)		
	All firms	−1.822		0.031		0.017	50
	(log)	(−3.859)		(0.930)			
	Large	−2.646		0.087		0.086	19
	(log)	(−2.469)		(1.298)			

(continued)

Table 8.7 (continued)

All equations for 2008		Constant	Coefficient (t-value) Tacit knowledge (share of high skilled in total employment)	Codified knowledge (spending on R&D, ICT and training)	Codified knowledge (spending on machinery and equipment)	R^2	N
	Medium	0.279		0.004*		0.127	27
	(log)	(6.103)		(1.945)			
	Small	0.252		0.003*		0.089	19
	(linear)	(5.131)		(1.326)			
	Chemical	−1.622		0.013		0.002	25
	(log)	(−2.001)		(0.224)			
	Food	0.288		0.002*		0.071	28
	(linear)	(5.210)		(1.432)			
	Metal	−1.149		0.011		0.035	4
	(log)	(−2.357)		(0.332)			
	Textile	0.203		0.002**		0.941	4
	(linear)	(4.904)		(6.896)			
	Textile	−3.454		0.157**		0.712	4
	(log)	(−3.890)		(2.721)			
Codified knowledge (total spending on machinery and equipment)	All firms	−148,468,236.04	8,158,430,084.89			0.027	35
	(linear)	(−0.043)	(0.964)				
	All firms	12.231		0.303**		0.123	34
	(log)	(6.528)		(2.154)			
	Large	13.688		0.178		0.051	12
	(log)	(3.995)		(0.769)			
	Medium	13.835		0.070		0.004	9
	(log)	(2.745)		(0.170)			
	Small	9.923		0.617**		0.574	11
	(log)	(4.788)		(3.672)			

	Chemical	13.906	0.159		0.049	19
	(log)	(6.429)	(0.964)			
	Food	9.287	0.513*		0.223	8
	(log)	(1.870)	(1.418)			
	Metal	8.536	0.839**		0.982	2
	(log)	(6.087)	(7.366)			
	Textile	11.491	0.277		0.224	2
	(log)	(1.566)	(0.537)			
Codified knowledge (spending on R&D, ICT and training)	All firms	19,582,066.79		0.001*	0.080	35
	(linear)	(2.163)		(1.721)		
	All firms	6.117		0.406**	0.123	34
	(log)	(1.971)		(2.154)		
Codified knowledge (share of R&D expenditure)	All firms	−5.167		0.112*	0.180	9
	(log)	(−3.505)		(1.326)		
Codified knowledge (total R&D expenditure)	All firms	12.158		0.174	0.037	6
	(log)	(1.763)		(0.439)		

Correlation is significant * at the 0.05 level (one-tailed), ** at the 0.01 level (one-tailed)

Table 8.8 The increased use and effect of skilled workers, scientists and engineers across firms in the Sudan, 2008 (measured in % as indicated by respondents)

The increased use of skilled workers and their effects	All firms	Chemical	Food	Metal	Textile	Large	Medium	Small
(a) Increased use of skilled workers (2006–2008)	50 %	46 %	52 %	55 %	60 %	52 %	41 %	58 %
(b) The effects of increased use of skilled workers								
Increase in firm production	83 %	83 %	78 %	91 %	100 %	87 %	78 %	84 %
Improve product quality	81 %	80 %	78 %	82 %	100 %	77 %	78 %	89 %
Improve the level of competitiveness in the local market	73 %	69 %	74 %	91 %	60 %	65 %	70 %	89 %
Effective utilization of technologies	73 %	80 %	59 %	73 %	100 %	74 %	74 %	68 %
Faster adaptation of technologies	65 %	69 %	52 %	73 %	100 %	74 %	63 %	53 %
Improve the level of competitiveness in the international market	21 %	23 %	19 %	27 %	0 %	32 %	19 %	0 %
Total response	78	35	27	11	5	31	27	19
(c) The effects of scientists and engineers on firm production and acquisition of knowledge								
The effects of scientist and engineers	All firms	Chemical	Food	Metal	Textile	Large	Medium	Small
Shorten development time	89 %	95 %	82 %	91 %	80 %	93 %	89 %	79 %
Add technical, scientific or marketing knowledge to areas where firms already had expertise	80 %	81 %	79 %	82 %	80 %	84 %	78 %	74 %
Add new technical, scientific or marketing knowledge to areas where firms lacked expertise	67 %	69 %	64 %	73 %	60 %	78 %	64 %	52 %
Total response	80	36	28	11	5	32	28	19

Source: Own calculation based on the % as indicated by respondent firms to firm survey (2010)

share of public spending on education as a percentage of GDP) shows positive significant correlation with tacit knowledge (defined by tertiary school enrolment ratios and the share of high skilled population with tertiary education in total population) and with schooling years (defined by school life expectancy and mean years of schooling). These results imply that the enhancement of codified knowledge (defined by increasing public spending on education, training, R&D and ICT) can be used as important mechanism to enhance tacit knowledge (defined by tertiary school enrolment ratios and the share of high skilled population with tertiary education in total population) and to enhance schooling years (defined by school life expectancy and mean years of schooling). In addition, we observe that when excluding the share of public expenditure on training, R&D and ICT relative to GDP from the definition of codified knowledge and limiting the definition of

Table 8.9 The correlations between labour, capital, age, tacit and codified sources of knowledge across firms, 2008

All firms (2005–2008)		Coefficient (t-value)					
Independent variables	Dependent variable	Labour	Capital	Age	Constant	R^2	N
Tacit knowledge (share of high skilled in total employment)	All firms (linear)	1.036 (1.006)			95.529 (2.250)	0.023	44
	All firms (log)		0.006 (0.146)		−1.574 (−2.414)	0.001	35
	All firms (log)			−0.080 (−0.515)	−1.212 (−3.071)	0.006	44
Codified knowledge (total spending on R&D, ICT and training) [1]	All firms (Log)	1.677** (2.372)	0.320** (2.163)		−0.051 (−0.012)	0.210	34
	Large (Log)	1.933 (1.005)	0.634* (1.653)		−6.741 (−0.590)	0.343	10
	Small (Log)	4.227* (1.652)	0.720** (3.461)		−16.684 (−1.792)	0.606	12
	Chemical (Log)	0.723 (0.663)	0.449* (1.513)		1.102 (0.148)	0.148	16
	Food (Log)	3.830** (2.031)	0.226 (1.112)		−6.738 (−0.731)	0.438	8
	Textile (Log)	1.459 (1.013)	0.238 (0.393)		4.184 (0.300)	0.340	4
	All firms (log)	1.864** (2.866)			4.713 (1.623)	0.170	41
	Chemical (Log)	1.393* (1.568)			5.993 (1.471)	0.115	20
	Metal (Log)	4.912 (0.994)			−5.429 (−0.287)	0.248	4
	All firms (log)		0.246* (1.593)		8.273 (3.091)	0.071	34
	Chemical (log)		0.411* (1.439)		4.797 (0.990)	0.121	16
	Metal (Log)		1.138** (7.858)		−9.493 (−3.605)	0.969	3
	All firms (log)			−0.295 (−0.478)	14.245 (8.779)	0.005	51
Codified knowledge (total spending on machinery and equipment)	All firms (log)		0.514** (5.195)		7.631 (4.437)	0.465	32
	All firms (linear)			79661894.8 (0.546)	1133229245.6 (0.349)	0.009	35
Codified knowledge (total R&D expenditure)	All firms (log)	0.357 (0.131)			14.313 (1.047)	0.002	8
	All firms (log)		0.710** (2.064)		2.434 (0.391)	0.516	5

(continued)

Table 8.9 (continued)

All firms (2005–2008)		Coefficient (t-value)					
Independent variables	Dependent variable	Labour	Capital	Age	Constant	R^2	N
	All firms (linear)			−23,181,080 (−0.687)	884908507.6 (1.220)	0.050	10
	All firms (log)		0.040 [2] (0.433)		−3.519 (−2.069)	0.026	8
Codified knowledge and skill upgrading (general total training expenditure)	All firms (linear)	9493.060 (1.049)	0.001** (2.102)		−557534.396 (−0.169)	0.434	8
	All firms (linear)	2125.4 (0.227)			2708644.021 (0.886)	0.006	10
	All firms (linear)		0.001* (1.855)		1945239.301 (0.846)	0.330	8
Codified knowledge and skill upgrading (ICT training expenditure)	All firms (linear)	87806.5** (5.202)			−7,252,877 (−2.033)	0.643	6
	All firms (linear)	115431.1** (5.145)	0. 04* (1.863)		−14,704,554 (−2.592)	0.727	5
Training employees	All firms (linear)	0.021* (1.711)			1.059 (0.266)	0.246	10

Correlation is significant * at the 0.05 level (one-tailed), ** at the 0.01 level (one-tailed)
[1]Log value for all estimated variables: ICT, labour and capital. [1] The logarithm of the variable is taken.
[2]Correlation between Codified knowledge (share of R&D expenditure) and labour.

codified knowledge to include only the share of public spending on education relative to GDP, the coefficient in the regression equation turns more significant. This result is plausible since the public spending on education relative to GDP has high share when compared to the share of public spending on training, R&D and ICT relative to GDP in Sudan. This result can then be used to argue that an increase in public spending on these components would imply an increase in codified knowledge and therefore, GDP in Sudan. From the perspective of the new growth literature, these results would imply that with the assumption of a potential role for public policies, the government could prevent the decline in economic growth and ensure increasing and dynamic economic growth, mainly through improving tacit knowledge and schooling by stimulating investment in education (basic, secondary and tertiary) and by increasing public spending on education, R&D, training and ICT. Moreover, Table 8.3 shows that tacit knowledge (defined by the number of FTER) and codified knowledge (defined by total spending on R&D) show positive correlations with technology (patents). The correlation between tacit knowledge (defined by the number of FTER) and this variable appears more significant than those with codified knowledge (defined by total spending on R&D). When defining the number of FTER as one form of tacit knowledge, we find a positive correlation between the number of FTER and codified knowledge (defined by total spending on R&D), which can be interpreted as complementary relationship between tacit and codified knowledge (cf. Winter 1987; Brusoni et al. 2002). In addition, we observe from Table 8.1 above that total spending on R&D is associated with an increase in the number of FTER. Moreover, Table 8.3 above shows that total spending on R&D is positively correlated with an increase in the share of public (government), private (business) and higher education institutions in total spending on R&D respectively. These results imply the important supporting role of public (government), private (business) and higher education institutions for enhancing and supporting total R&D spending and activities. Moreover, we observe that the correlation between total R&D spending and public (government) spending on R&D appears more significant than the correlations between total spending on R&D and private (business) spending on R&D and higher education institutions spending on R&D. This result is plausible since the share of public (government) spending on R&D is higher than the share of private (business) spending on R&D and the share of higher education institutions spending on R&D. Therefore, these results verify the first and second stylised facts that at the macro/aggregate level knowledge is positively correlated with GDP (economic growth), schooling years and technology (patents).

Table 8.4 verifies the third stylised fact that at the micro/firms level tacit knowledge shows positive correlations with total profit, total output (defined by total sales value), productivity (defined by total sales value per worker), output diversification (defined by sales diversification), value added, technology (total expenditures on ICT) and skill upgrading (total expenditure on ICT training, total expenditures on general training, training employees staff).[12] Notably, tacit knowledge shows positive significant correlations with technology (total expenditures on ICT) for all, medium and textile firms. From the perspective of the new growth literature, the positive correlation between tacit knowledge and output is important to prevent the diminishing returns to scale and to ensure the increasing returns and dynamic growth in the production function. This would imply that with the assumption of a potential role for public policies, the government could prevent the diminishing returns to scale and ensure increasing returns to scale, mainly through improving tacit knowledge by stimulating investment in education (basic, secondary and tertiary).

Table 8.5 verifies the third stylised fact that at the micro/firms level codified sources of knowledge (as measured by total spending on ICT, R&D and training) shows positive correlations with total output (defined by total sales value), output diversification (defined by sales diversification), total profit and value added. Notably, codified sources of knowledge (as measured by total spending on ICT, R&D and training) shows positive significant correlation with total output (defined

[12] There are also positive correlations between tacit knowledge and output, output diversification, productivity and profit that follow the combined correlations of traditional inputs such as labour and capital not reported in Table 8.4; these results are different from the findings in the literature (cf. Drucker 1998; OECD 1999).

by total sales value) (for all, large, medium, chemical and metal firms), output diversification (for all, medium, chemical and food firms), total profit (for all, medium, small, chemical and metal firms) and value added (for medium and food firms). Moreover, codified sources of knowledge (as measured by total spending on machinery and equipment) shows positive significant correlations with technology (total expenditures on ICT), value added, profit, the use of raw materials, total output (defined by total sales value), and positive correlation with output diversification (defined by sales diversification). From the perspective of the new growth literature, the positive correlation between codified sources of knowledge and output is important to prevent the diminishing returns to scale and to ensure the increasing returns and dynamic growth in the production function. This would imply that with the assumption of a potential role for private industrial policies, the industrial firms could prevent the diminishing returns to scale and ensure increasing returns to scale, mainly by improving codified sources of knowledge by stimulating investment in training, R&D, ICT and technology.

Table 8.6 verifies the third stylised fact that at the micro/firms level codified sources of knowledge (as measured by total spending on R&D) shows positive significant correlations with total profit and total output (defined by total sales value). In addition, codified sources of knowledge (as measured by total spending on R&D) shows positive correlation with other codified sources of knowledge (as measured by total spending on machinery and equipment), value added and the use of raw materials but it shows negative correlation with output diversification (defined by sales diversification). Moreover, codified sources of knowledge (as measured by total spending on R&D) and codified sources of knowledge (as measured by total spending on ICT, R&D and training) show positive significant correlations with dependence on the import of foreign technology (as defined by the percentage of value of imported capital equipment to total capital equipment) and the build of foreign technology (as defined by the percentage of value of capital equipment to total capital equipment built by foreign companies). Furthermore, codified sources of knowledge and skill upgrading (measured by total spending on training) show positive correlation with the number of training employees and with tacit knowledge, in addition the number of training employees shows positive significant correlation with tacit knowledge.

Table 8.7 verifies the third stylised fact that at the micro/firms level concerning the positive and complementary correlation between tacit knowledge and codified sources of knowledge as measured by (total expenditures on ICT, R&D and training), the complementary correlation is particularly significant for medium, small, food and textile firms. In addition to positive and complementary correlation between tacit knowledge and codified sources of knowledge as measured by technology (total expenditures on machinery and equipment). In addition to positive and complementary correlation between codified sources of knowledge as measured by (total expenditures on ICT, R&D and training) and codified sources of knowledge as measured by technology (total expenditures on machinery and equipment), the complementary correlation is particularly significant for all, small, food and metal firms. Moreover, codified sources of knowledge as measured by technology (total expenditures on machinery and equipment) shows positive correlation with codified sources of knowledge as measured by both total and share of expenditures on R&D.

Our results from the firm survey (2010) in Table 8.8 bear out the assumption that the increased use of tacit knowledge, defined by skilled workers, scientists and engineers, shows significant effects across firms. In particular, this contributes towards the increase in firm production, improvement in firm product quality, improvement in the level of competitiveness in the local market, effective utilisation of technology and faster adaptation of foreign technology. Moreover, Table 8.8 indicates that the increased use of scientists and engineers would imply the shortening of development time, as well as additions to acquisition of existing knowledge within the firm and acquisition of new knowledge, the latter regarded as of somewhat less importance.[13,14]

Our findings in Table 8.9 prove part of the fourth stylised fact that at the micro/firm level tacit knowledge is positively correlated with market size: total investment, capital and firm size. Moreover, we find that codified sources of knowledge (as measured by total expenditure on ICT, R&D and training) are positively and significantly correlated with market size: total investment, capital and firm size; the correlation is positive and significant for all, large, small, chemical and food firms. In addition we find that codified sources of knowledge (as measured by total expenditure on R&D, total expenditure on training and total expenditure on machinery and equipment) are positively and significantly correlated with total investment, capital and are positively correlated with market size: firm size. Moreover, we find that codified sources of knowledge and upskilling, as defined by total expenditure on training, are positively and significantly correlated with market size: total capital investment and positively correlated with firm size; and upskilling, as defined by total number of training employees, is significantly and positively correlated with firm size. Therefore, at the micro/firm level an increase in total investment, capital and firm size would coincide with more tacit and codified sources of knowledge and more upskilling. In addition, we find that tacit and codified sources of knowledge are insignificantly and negatively correlated with firm age. This result rejects part of the fourth stylised fact, which implies positive correlation between tacit and codified sources of knowledge and firm age. This result contrasts with our expectation and probably implies that the relative improvement in tacit and codified sources of knowledge has probably been observed more for relatively new and young firms when compared to relatively old firms. This result is somewhat surprising in view of the fact that the accumulation of knowledge and learning often takes or requires more time to develop and improve over time. But we observe positive insignificant correlation between firm age and codified sources of knowledge as measured by spending on technology and machinery and equipment, this implies that total spending on technology and machinery and equipment increases with firm age; improvement in total spending on technology and machinery and equipment has been probably observed more for relatively old firms as compared to young firms.

---

[13] Knowledge includes technical, scientific or marketing knowledge.

[14] These results are consistent with the findings in the UAE as indicated in Nour (2005).

8.5 Conclusions

In this chapter we use the firm survey (2010) data at the micro level and secondary data at the macro level to examine hypothesis 7 in Chap. 1 above concerning the importance/impacts of tacit and codified sources of knowledge at the micro and macro levels respectively in Sudan. Our results prove this hypothesis and show that at the macro level tacit knowledge and codified sources of knowledge are positively and significantly correlated with both schooling years and GDP growth (economic growth rate). Moreover, we find that at the macro level codified knowledge and the number of FTER show positive correlations with technology (patents). Furthermore, at the aggregate level, our results imply a significant positive complementary relationship between the number of FTER and codified knowledge, which we interpret as a complementary relationship between tacit knowledge and codified knowledge. At the micro (firm) level, we illustrate the importance of tacit knowledge, which shows positive significant correlations with technology (expenditures on ICT) and upskilling (expenditures on training), output (defined by total sale value), output diversification, productivity and profit. Finally, we find that at the micro (firm) level, tacit and codified knowledge show positive significant correlations with total investment, capital, and firm size. This can be interpreted that higher levels of total investment, capital and firm size would correspond to more tacit and codified sources of knowledge across firms. Our results at the micro and macro levels verify the four stylised facts presented in the introduction, which are consistent with the general findings in the knowledge literature. The major implication of our findings is that knowledge shows positive significant correlations with many variables at both the micro and macro levels. Therefore, this would imply that public policy should provide further incentives to improve tacit and codified sources of knowledge at both the macro and micro levels. Another implication is that the positive impact of tacit knowledge also underlines the importance of good education, since tacit knowledge is often embodied in educated people and thus human capital. Moreover, from the perspective of the new growth literature, the positive correlation between tacit knowledge and output is important to prevent the diminishing returns to scale and to ensure the increasing returns and dynamic growth in the production function. This would imply that with the assumption of a potential role for public policies, governments could prevent the diminishing returns to scale and ensure increasing returns to scale, mainly through improving tacit knowledge by stimulating investment in education (basic, secondary and tertiary). In addition, at the aggregate/macro level, the positive correlation between GDP and codified knowledge -the share of public spending on education; R&D, training and ICT relative to GDP - would imply a positive role for public policy to support codified knowledge by increasing spending on education, R&D and ICT. These results are consistent with the literature that substantiate the role of public policies to support the creation and accumulation of knowledge, as we explained in Sect. 3.2 of this chapter and Sect. 3.5 in Chap. 3 above. In addition, at the micro level, our findings on the positive correlation between tacit and codified sources of knowledge, output and firms performance would imply that with the assumption of a potential role for private industrial policies, private industrial firms could prevent the diminishing returns to scale and ensure increasing returns to scale, mainly through improving codified sources of knowledge by stimulating investment in ICT, R&D, training, the use of technology and machinery.

References

Abramovitz, M., & David, P. A. (1996). Technological change and the rise of intangible investments: The US economy's growth –path in the twentieth century. In D. Foray & B.-Å. Lundvall (Eds.), *Employment and growth in the knowledge based economy* (OECD documents). Paris: OECD.

Abramovitz, M., & David, P. A. (1998). American macro economic growth in the era of knowledge based progress: The long –run perspective. In S. L. Engerman & R. E. Gallman (Eds.), *An economic history of the United States: The twentieth century* (Vol. 3, pp. 1–92). New York: Cambridge University Press.

Ali, A. A. (2006). *On human capital in post-conflict Sudan: Some exploratory results* (Working Paper Series No. API/WPS 0602). Kuwait: Arab Planning Institute.

Arundel, A., & Geuna, A. (2001). *Does proximity matter for knowledge transfer from public institute and Universities to firms?* (SPRU Electronic Working Paper Series No. 73, pp. 3, 5). University of Sussex: SPRU

Aubert, J-E., & Reiffers, J-L. (Eds.). (2003). *Knowledge economies in the Middle East and North Africa: Toward new development strategies* (World Bank Institute learning resources series, Vol. 1, The World Bank (2003) Report No. 28038). Washington, DC: The World Bank

Barro, R., & Lee, J-W. (2000). *International data on educational attainment: Updates and implications.* Department of Economics, Harvard University. Data set at http://www.cid.harvard.edu/ciddata. Cited in Ali Abdel Gadir Ali (2006) On human capital in post-conflict Sudan: Some exploratory results, API/WPS 0602, p. 14.

Barro, R. J., & Lee, J -W. (2010). *A new data set of educational attainment in the world, 1950–2010* (NBER Working Paper No. 15902).

Barro, R. J., & Sala-i-Martin, X. (1992). Public finance in models of economic growth. *The Review of Economic Studies, 59*(4), 645–661.

Barro, R. J., & Sala-i-Martin, X. (1995). *Economic growth.* New York: McGraw Hill.

Brusoni, S., Marsili, O., & Salter, A. (Eds.). (2002). *The role of codified sources of knowledge in innovation: Empirical evidence from Dutch manufacturing* (SPRU Electronic Working Paper Series No. 80). University of Sussex: SPRU.

Chatterji, M. (1995, September). Training subsidies, technical progress and economic growth. *The Manchester School,* Vol. LXIII(3). Blackwell Publishers Ltd./The Victoria University of Manchester. (pp. 274–282).

Cowan, R., Soete, L., & Tchervonnaya, O. (2001). *Knowledge transfer and the services sector in the context of the new economy* (MERIT-INFONOMIC Research Memorandum No. 2001–021). Maastricht: Maastricht University. 9.

Dasgupta P., & David P. (1994). Toward a new economics of science. *Research Policy, 23*(5): 487–521. MERIT Research Memoranda Series No. 2/94-003. Maastricht, The Netherlands: Maastricht University.

David, P. A., & Foray, D. (1995). *Accessing and expanding the science and technology knowledge base* (STI review). Paris: OECD.

David, P. A., & Foray, D. (2001). *An introduction to the economy of the knowledge society* (MERIT –INFONOMIC Research Memorandum Series No. 2001–041). Maastricht: Maastricht University.

Drucker, P. (1998). From capitalism to knowledge society. In D. Neef (Ed.), *The knowledge economy* (p. 15). Woburn: Butterworth.

Firm Survey. (2010, January–June). *Technological change and skill development: A comparative study of chemical, food, metal and textile small, medium and large scale enterprises in the Sudan.*

Freeman, C., & Soete, L. (1997). The economic of industrial innovation (3rd ed.,), London: Cassell. pp. 2, 24, 404, 405.

Grossman, G. M., & Helpman, E. (1994). Endogenous innovation in the theory of the growth. *Journal of Economic Perspectives, 8*(1), 23–44.

Hagedoorn, J., Link, A. N., & Vonortas, N. (2001). Research partnership. *Research Policy, 29*, 567–586.

Haslinger, F., & Ziesemer, T. (1996). Endogenous growth and distributional conflict. In C. Gore, G. Köhler, U. Peter Reich, & T. Ziesemer (Eds.), *Questioning development* (pp. 223–240). Marburg: Metroplis Verlag.

Klepper, S., & Simons, K. (1997). Technological extinctions of industrial firms: An enquiry into their nature and causes. *Industrial Corporate Change, 6*, 379–460.

Langlois, R. N. (2001). Knowledge, consumption, and endogenous growth. *Journal of Evolutionary Economics, 11*, 77–93.

Levin, R., Cohen, W. M., & Mowery, D. C. (1985). R&D appropriability, opportunity and market structure: new evidence on some Schumpeterian hypotheses. *American Economic Review, 75*, 20–24.

Lööf, H., & Heshmati, A. (2002). Knowledge capital and performance heterogeneity: A firm- level innovation study. *International Journal of Production Economics, 76*, 61–85.

Lucas, R. E. (1988). On the mechanics of economic development. *Journal of Monetary Economics, 22*, 3–42. North Holland.

Lundvall, B.-Å. (Ed.). (1992). *National systems of innovation: Towards a theory of innovation and interacting learning*. London: Printer Publishers.

Mansfield, E. (1991). Academic research and industrial innovation: An update of empirical findings. *Research Policy, 26*, 773–776.

Mansfield, E., & Lee, J.-Y. (1996). The modern university: Contributor to industrial innovation and recipient of industrial R&D support. *Research Policy, 25*, 1047–1058.

Mohammed bin Rashid Al Maktoum Foundation (MBRF) and the United Nations Development Programme/Regional Bureau for Arab States (UNDP/RBAS). (2009). *Towards productive intercommunication for knowledge* (Arab Knowledge Report 2009) Dubai: Al Ghurair Printing and Publishing House L.L.C.

Nelson, R. (Ed.). (1993). *National innovation systems: A comparative study*. New York: Oxford University Press.

Nelson, R. R., & Winter, S. G. (1982). *An evolutionary theory of economic change*. Cambridge, MA: Harvard University Press.

Nour, S. (2005, November). *Technological change and skill development in the Arab Gulf countries*. Doctoral Dissertation, Maastricht University Press, Maastricht, The Netherlands.

Nour, S. (2010, December). The incidence and transfer of knowledge in the Arab countries (UNU-MERIT Working Paper 2010-064). Maastricht. http://www.merit.unu.edu/publications/wppdf/2010/wp2010-064.pdf

OECD (Organisation for Economic Co-operation and Development). (1997). *The second european report on science and technology indicators (1997)* (pp. 37, 90, 91). Paris: OECD

OECD. (1999). *The knowledge based economy: A set of figures and facts*. Paris: OECD.

Romer, P. M. (1990). Endogenous technological change. *Journal of Political Economy, 98*(5, Part 2), 71–102.

Romer, P. M. (1994). The origin of endogenous growth. *Journal of Economic Perspectives, 8*, 3–22.
Smith. K. (2002, June). *What is knowledge economy'? Knowledge intensity and distributed knowledge bases.* (UNU-INTECH Discussion Paper Series No. 2002–6).
Sudan Central Bureau of Statistics Population Census Data (2010): 5th Sudan Population and Housing Census (2008).
Sudan Central Bureau of Statistics. (2009). *Sudan ministry of the Cabinet- central bureau of statistics: Sudan statistical year book: Sudan statistics 1990–2008.* pp. 39–43.
Sudan Ministry of Finance and National Economy- Annual Economic Survey Report (Various Issues: 1997–2007)
Sudan Ministry of Industry. (2005, July). *The Comprehensive Industrial Survey*, Vol. 4. Khartoum, www.industry.gov.sd.
The World Development Indicators database. (2011). Accessed on June 25th, 2011.
UNDP. (2002, July). *Human development report (2002): Deepening democracy in a fragmental World.* New York/Oxford: UNDP/Oxford University Press.
UNDP- AHDR. (2002). *Arab human development report (2002): Creating opportunities for future generations* (Regional bureau for Arab state, UNDP). New York/Amman: Regional Bureau for Arab States (RBAS)/Icons Printing Services.
United Nations Development Programme UNDP- The Arab Human Development Report AHDR. (2003, July). *Arab human development report (2003): Building a knowledge society., UNDP.* New York/Amman: Regional Bureau for Arab States (RBAS)/National Press, p. 56; Table 5.1 p. 102.
UNDP. (2004). *Human development report (2004): Cultural liberty in today's diverse world, UNDP.* New York: Hoechstetter Printing Co.
United Nations Development Programme (UNDP)-Arab Human Development Report (AHDR). (2004). *Towards freedom in the Arab World.* The UNDP. New York/Amman: Regional Bureau for Arab States (RBAS)/National Press.
UNDP United Nations Development Programme UNDP. (2010). *Human development report 2010: The real wealth of nations: Pathways to human development.* Human development report 2010 – 20th Anniversary Edition New York, NY: UNDP, Palgrave Macmillan, Statistical Tables, Statistical Annex, Table 12: Decent Work, p. 191.
UNESCO- UIS. (2000). *World education report (2000): UNESCO's World education indicators: The right to education: towards education for all throughout life*: http://www.unesco.org/education/information/wer/htmlENG/tablesmenu.html
UNESCO-UIS. (2003). *The UNESCO education data and statistics indicators 1998–2001.* www.unesco.org
UNESCO. (2006) *UNESCO R&D statistics (2006)*: www.unesco.org
UNSESCO: *UNESCO Institute of Statistics*: www.uis.unesco.org
UNESCO Institute for Statistics. (2010a, March). *Correspondence on education indicators.* Montreal
UNESCO-UIS. (2010). *Global education digest 2010 Comparing education statistics around the World.* p. 176
UNESCO- UIS. (2011). *Global background information on education statistics: UNESCO- UIS Data Centre: Beyond 20/20 WDS (2011).* Accessed 22 Feb 22 2011: UNESCO- UIS (2011) UIS web site global statistics on education: www.unesco.org: http://stats.uis.unesco.org/unesco/TableViewer/tableView.aspx?ReportId=163
van Zon, A. (2001). *ICT- investment, knowledge accumulation and endogenous growth* (MERIT – INFONOMIC Research Memorandum Series No. 2001–038). Maastricht: Maastricht University.
Verspagen, B., & Schoenmakers, W. (2000, April 19–20). *The spatial dimension of knowledge spillovers in Europe: Evidence from firm patenting data.* Paper prepared for AEA conference on intellectual property econometrics, Alicante.
Von Hippel, E. (1988). *The sources of innovation.* New York: Oxford University Press.

Winter, S. G. (1987). Knowledge and competence as strategic assets. In D. J. Teece (Ed.), *The competitive challenge: Strategies for industrial innovation and renewal* (pp. 159–184). Cambridge, MA: Ballinger Publishing Company.

Ziesemer, T. (1990). *Public Factors and Democracy in Poverty Analysis*. (Oxford economic papers, 1990, Special issue on public economics, Vol. 42, January pp. 268–280). Reprinted in: Sinclair, P. J. N. & Slater M. D. E. (eds.), Taxation, private information and capital, Oxford: Clarendon Press.

Ziesemer, T. (1995). Endogenous growth with public factors and heterogeneous human capital producers. *Finanzarchiv, Neue Folge*, Band 52, 1995, Heft 1, 1–20.

Part IV
Policy Recommendations, Summary and Conclusions

Chapter 9
Education, Training and Skill Development Policies in Sudan: Macro–Micro Overview

Abstract This chapter uses some secondary data and the macro and firm surveys (2010) to analyse the educational, training and skill development policies in Sudan. We show that skill development depends on: reforming the educational system; enhancing the provision of training; planning skill needs and matching educational output with market needs; enhancing the transfer of knowledge/schooling effect; and incentives and collaboration between public and private institutions. We explain that the promotion of local technologies and adoption of appropriate foreign technologies and the interaction between both these to foster economic growth in Sudan depends on skill development. Our results show the low commitment to the standardised international adequacy, equity and efficiency criterion related to the supply and demand sides of educational and training policies. Moreover, we interpret the regional disparity in the demand and enrolment in education due to demographic reason, economic reasons and other reasons across the main regions in Sudan. Notably, our results imply that the incidence of high poverty rate seem to be the most important factor determining or limiting the demand and enrolment, notably, in basic education. We confirm our hypothesis 8 that the effective institutional environment and consistent policies of public and private institutions will enhance upskilling plan and skill development.

9.1 Introduction

Economists of different schools of thought have confirmed the essential role of education and human capital in the creation, acceleration and sustainability of economic growth, and improvement of the quality of life in any society. In particular, endogenous and new growth theories and empirical literature recognise

S. Mohamed Nour, *Technological Change and Skill Development in Sudan*,
DOI 10.1007/978-3-642-32811-4_9, © Springer-Verlag Berlin Heidelberg 2013

the importance of human capital accumulation/formulation for economic growth in both developed and developing countries (cf. Lucas 1988; Romer 1990).[1]

In recent years, the UNDP-AHDR Reports (2002, 2003) and Arab Knowledge Report (2009) highlight the investment in human capital, education and knowledge in the Arab region. Earlier studies in the Arab Gulf and Sudanese literature show the importance of a good education and investigate the causes and consequences of deficient educational and training systems, the lack of knowledge transfer and upskilling in the Arab Gulf countries and Sudan (cf. Muysken and Nour 2006; Nour 2005b, c; Al-Sanousi 1999; Al-Sulayti 2002; Suleiman 2007; Jalal al-Din 2002). From that perspective, therefore, it is convenient in this paper to discuss the educational and training policies and to provide insights to help generate policies to enhance skill upgrading by implementation of sound plans and consistent relevant policies for skills development: enhancing the educational system and provision of training and transfer of knowledge/external schooling effects at the macro–micro levels in Sudan. Thus, our paper is relevant to contribute to the few studies that address some aspects in relation to educational and training systems and policies in Sudan and to go beyond these studies, by providing a more comprehensive analysis to complement the earlier studies in the Sudanese literature (cf. Suleiman 2007; Jalal al-Din 2002).

Different from earlier studies an interesting element in our analysis is that we discuss both the supply and demand sides of educational policies in Sudan. Moreover, a novel element in our paper is that we use new primary macro and micro (firm) surveys data (2010) to discuss and compare the macro and micro views/perspectives concerning plans, policies and mechanisms implemented to improve skill upgrading: education, training and transfer of knowledge. In addition, our new results in this paper are consistent with the results in the literature (cf. Nour 2005c) since we show the lack of effective interaction between educational and training policies and a lack of incentives for provision of training within private firms in Sudan, our analysis presents a new element by showing limited commitment to implementation of training and skill upgrading policies only within the two largest mixed and private firms and a further duality/discrepancy at the micro level/across small-medium and large private firms. Our findings are consistent with the results in the Sudanese and Arab literature concerning the poor quality of education and the Sudanese literature concerning the low commitment to the standardised international adequacy, equity and efficiency criterion related to educational policies. Different from the Sudanese literature (Suleiman 2007; Jalal al-Din 2002), an interesting element in our analysis is that we provide a more elaborate and comprehensive analysis concerning the serious problem of poor quality of education and low commitment to the standardised international adequacy, equity and efficiency

[1] For theoretical and empirical literature on the importance of human capital see for instance, Stokey (1991), Schultz (1961), Mulligan and Sala-i-Martin (1995), Mincer, (1984, 1989), Mankiw et al. (1992), Barro (1991, 1996), Barro and Lee (1993, 1996, 2000, 2010), Becker (1962, 1964), Romer (1990) and Lucas (1988).

criterion related to the supply and demand sides of educational and training policies in Sudan. We explain that the low commitment to the standardised international adequacy, equity and efficiency criterion is obvious not only from the supply side, but also holds from the demand side for education and training policies. Finally, we provide a new contribution and fill important gap in the Sudanese literature by explaining the regional inequality and disparity in the supply and demand sides of education and training. Notably, we explain that the observed regional disparity in the demand for education (defined by the share in total enrolment in education) is most probably interpreted due to economic reasons (defined by per capita income and poverty rate), demographic reasons (defined by the share in total population) and other reasons (defined by the degree of urbanisation) across main regions in Sudan. Notably, we find that the increase in the incidence of high poverty rate and low per capita income seem to be the most important factor limiting enrolment and demand for education, notably, demand for basic (primary) education, mainly for females in Sudan.

Based on the above, this paper aims to give an empirical investigation and policy analysis of skill development at the macro–micro levels. First we discuss the supply–demand sides: the major characteristics and implications of educational and training policies, we highlight the need for prioritising skill development and we provide insights to help generate policies to reform the educational and training systems and upskilling of the labour force to foster economic growth and development in Sudan. Second, we examine the major mechanisms for reforming educational system, upgrading skills, enhancing the provision of training and the external effect of schooling/transfer of knowledge at the macro level in Sudan. Third we substantiate the need for consistent macro–micro/public-private policies to ensure the effective implementation of educational and training policies, skills upgrading and the external effect of schooling. Based on the above objectives, this Chapter aims to examine hypothesis 8 in Chap. 1 above that first Sudan needs to upgrade skill through the relevant policies for enhancing educational system, provision of training and transfer of knowledge/external schooling effect at the macro–micro levels. Second, educational reform will have positive implications on: (a) enhancing training provision; (b) skill upgrading; (c) planning skill needs and matching educational output with the needs in the labour market; (d) enhancing the transfer of knowledge/schooling effect; and (e) collaboration between public and private institutions. Third, effective institutional environment and consistent policies of public and private institutions will enhance upskilling plan and skill development.

To fulfil our objective and test our hypothesis, we follow the new growth theories and literature in viewing and using a more broad definition of human capital and its accumulation, including education, training and external effect of schooling. We use the UNESCO conceptual framework and define education indicators as composed of: (1) Input indicators including both financial or public and private spending on education and human resources allocated in education, and (2) Output (quantitative and qualitative schooling indicators), which is defined by many indicators. We integrate the descriptive and comparative methods of analysis and use a combination of new primary and secondary data and information covering the macro–micro levels to test our earlier hypotheses and draw the major policy implications and conclusions on enhancing the educational and training systems. We use a new primary data based on the firm and macro surveys at the micro and macro levels respectively.[2]

The rest of this paper is organised as follows: Sect. 9.2 discusses the supply–demand sides: the major characteristics and implications of educational policies in Sudan based on data and information obtained from the UNESCO, UNDP, Sudan Ministry of Education and other relevant sources. Section 9.3 explains the major characteristics of training policies and examines the training and skill upgrading policies implemented by the large mixed and private firms in Sudan based on data and information obtained from these firms. Section 9.4 uses the results of the macro and firm surveys held in Sudan (2010) and the follow-up interviews to present the macro–micro views and suggestions for relevant mechanisms and policies for skill development: enhancing the educational system, provision of training and transfer of knowledge/external schooling effect. Section 9.5 provides the conclusions.

9.2 Characteristics of Educational Policies in Sudan: Supply, Demand, Quality and Impacts

This Section discusses the supply–demand sides: the major characteristics and implications of educational policies in Sudan.

9.2.1 Characteristics of Educational Policies in Sudan

Before proceeding to discuss upskilling policies, it may be useful to begin with a brief explanation of the major characteristics of educational policies in Sudan, in particular the structure and pattern of educational policies, the supply side as measured by resources or priority of financial and human investment in education. In addition, we examine the demand for education as indicated by enrolment ratios and access to schooling and the impacts on literacy, school life expectancy, training and quality of education.

[2] The firm survey (2010), 'Technological Change and Skill Development: A Comparative Study of Chemical, Food, Metal and Textile Small, Medium and Large Scale Enterprises in Sudan', covered 100 medium and large size firms active in the chemical, food metal and textile industries in Sudan in 2010. The macro survey, 'Reform of Education, Human Resources Development and Policy Intervention', has been circulated amongst 40 of policy makers and experts in 8 public, university, educational, training and research institutions in Sudan in 2010. The surveys were held in Sudan in 2010, the response rates of the firm and macro surveys are 87 % and 90 % respectively.

The UNESCO-UIS (2011) information on the structure/nature of educational system implies an insufficient duration of compulsory education in Sudan.[3] For instance, the duration of compulsory education in Sudan lasts for 8 years and falls behind the international standard of 12–13 years of compulsory education attendance in the advanced countries such as the US, UK, Belgium, Germany and Netherlands and 9–11 years in Korea, Japan and Canada respectively.[4]

Moreover, in Sudan as in most Arab countries, the structure and pattern of the educational policies is characterised by a centralised bureaucracy, which, as remarked by Al-Sulayti, "implies a high degree of centralization and intervention from the governments/ministries of education to control all the educational institutions".[5] Sometimes, the education and higher education institutions lack independence and initiatives in the area of R&D. They are often subordinate to and/or feel the negative effects of state bureaucracy, routine, institutional rigidity and lack of transparency, dynamism, flexibility, planning, organisational development, monitoring and assessment. They also sometimes, lack a proper articulation of "educational policies, dynamism, flexibility, planning, organizational development, monitoring, assessment, cooperation and problem solving ability".[6] (Cf. Suleiman 2007; Jalal al-Din 2002).

9.2.2 *The Supply Side of Educational Policies: Financial and Human Resources*

We use the UNESCO definition to show the supply side/priority of educational investment as measured by financial resources (public and private educational investment, percentage share of public spending on education in GDP and total government spending) and human resources (teaching staff). Next, we show the demand for education (enrolment ratios) and impact in Sudan.

We discuss the adequacy, equity and efficiency criterion related to the supply and demand sides of educational policies. In particular, we begin with the analysis of the standardised international adequacy criterion which was earlier adopted in the 1960s and focused on the supply side that implies the allocation of either 8 % of GDP on education or 20 % of total government or public spending on education. In addition, we explain the extended international adequacy criterion that later

[3] Suleiman (2007), p. 123, indicates the necessity for making basic education free and compulsory at the same time in Sudan.

[4] See the UNESCO Education Statistics, UNESCO-UIS (2011) website for global statistics on education: http://stats.uis.unesco.org/unesco/TableViewer/tableView.aspx?ReportId=163, 'Background information on education statistics in the UIS Data Centre: Beyond 20/20 WDS', accessed February 22, 2011.

[5] See Al-Sulayti (2002), pp. 29–30.

[6] See Fahmey and Mahmoud (1993), pp. 29–30.

Table 9.1 Public expenditures on education in the Sudan compared to world countries (1990–2001/2002)

	Public expenditure on education as percentage of						
	GDP[a, b, c]			Total government expenditures[a, b]			
Country	1990[a]	1998/1999[b]	1999–2001[a]	1990[a]	1995/1997[d]	2000/2001[b]	2001/2002
Sudan[e]	0.9	0.6 (1991)	1.4 (1996)	2.8	8.4	6.4	6.9
Bahrain	4.2	3.67	3.00	14.6	12	11.41	n/a
Kuwait	4.8	n/a	6.6	3.4	14	n/a	14.8
Oman	3.1	3.87	4.2	11.1	16.4	18.2	21.5
Qatar	3.5	3.58	3.6[c]	11.1	n/a	n/a	n/a
Saudi Arabia	6.5	9.47	8.3[c]	17.8	22.8	22.7	24.8
United Arab Emirates	1.9	1.95	1.9[c]	14.6	20.3	22.2	22.8
United States	5.2	5.01	5.6	12.3	14.4	15.5(1)[a]	17.1
Sweden	7.4	7.98	7.6	13.8	12.2	13.40	12.8
Norway	7.1	7.68	6.8	14.6	16.8	16.18	16.1
Republic of Korea	3.5	4.07	3.6	22.4	17.5	17.38	14.7
United Kingdom	4.9	4.71	4.6	n/a	11.6	11.4	11.4

Note (1) Data refers to the year (1999–2001)
[a]UNDP Human Development Report (2004).
[b]UNESCO–UIS (2003).
[c]UNESCO–UIS (2004c) country profile: data refers to the most recent years between 1998 and 2002.
[d]UNDP Human Development Report (2002).
[e]Sudan Ministry of Finance-Ministry of Education (2003: 12).

adopted by the World Bank in the 1970s and extended to include the demand side that implies the adequacy in intake and enrolment rates in primary and secondary education, gender equity in enrolment in education and literacy rate of population. Furthermore, we then discuss the equity criterion, which implies the equal distribution and allocation of financial resources to achieve the balance between the different education sectors and between different geographical rural and urban areas. Moreover, we then examine the international efficiency criterion which implies that the low efficiency often appears from the low rates of attendance, high rates of dropout, high rates of repetition, weak rates of success in final exams, low rates of trained teachers and overcrowded classrooms as indicated by the rate of students enrolment per education institutions.[7]

We begin with the adequacy of the supply side and priority of public investment in education as measured by the financial resources devoted to education, which is indicated by the share of public spending on education as a percentage of GDP and total government expenditures. For instance, Table 9.1 illustrates that the low adequacy and priority of public spending on education, as measured by public spending on education as a percentage of GDP and of total government spending in

[7] See Suleiman (2007), pp. 121–123.

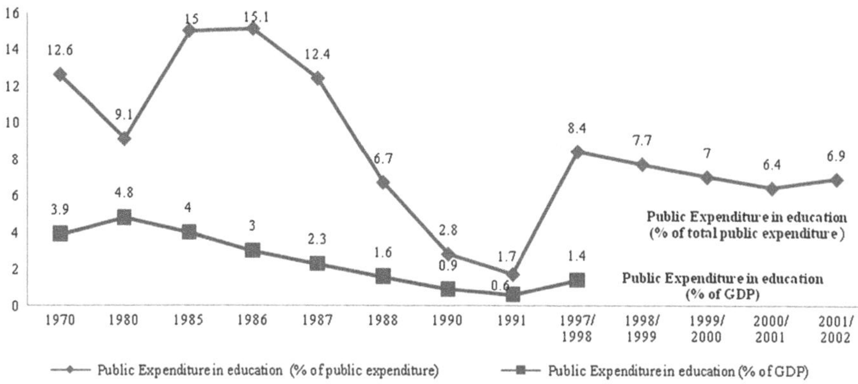

Fig. 9.1 Public expenditures on education as a percentage of total public expenditure and GDP in the Sudan (%) (1970–2001/2002) (Source: Sudan Ministry of Finance cited in Sudan Ministry of Education, Educational Planning Section of Educational Statistics (2003: 12))

Sudan lag far behind the levels prevalent in the Arab Gulf countries and the developed countries. For instance, in the period 1995–2000, the highest public spending on education as percentage of GDP and total government expenditure in Saudi Arabia was close to three and six times those of Sudan respectively. Moreover, public spending on education as a percentage of GDP shows considerable disparity and fluctuation in Sudan over the period 1970–2002. For instance, we observe the great and continuous decline in public spending on education as percentage of GDP from 4.8 % in 1980 to 4 %, 0.9 %, 0.6 % and 1.4 % in 1985, 1990, 1991 and 1996 respectively. In addition, the great and continuous decline in the trend of public spending on education also holds for the trend of public spending on education as a percentage of total government expenditure, for instance, we observe that over the period 1986–2001/02, public spending on education as percentage of total government and public expenditure in Sudan continuously declined from 15.1 % in 1986 to 2.8 %, 8.4 %, 7.7 %, 7 %, 6.4 % and 6.9 % in 1990, 1997/98, 1998/99, 1999/2000, 2000/01 and 2001/02 respectively (see Table 9.1 and Fig. 9.1 below). These findings imply that until recently the low adequacy of the supply side and public investment in education in Sudan remained low and fell below the standardised international adequacy criterion which was earlier adopted in the 1960s and related to the supply side and implies the allocation of either 8 % of GDP on education or 20 % of total government or public spending on education. These results led to increasing debate that the low commitment to standardised international adequacy criterion is somewhat surprising in view of the structural change in Sudan economy that turned into an oil dependent economy in 1999, which implies that the increasing revenues from oil has the potential to enhance increasing spending on social development issues including health and education. Based on this increasing debate, we are aware of the fact that it may be useful to depart from the analysis of general standardise education indicators and to use indepth economic, historical and social evidence to extend our analysis to focus more explicit on whether the production and export of oil (natural resource-based exports) affected the education infrastructure and the growth and development trajectory of Sudan economy. This may be particularly important in view of the fact that the production and export of oil has significant positive impacts on Sudan's economy as it leads to impressive growth in GDP growth rate and change in the structure of Sudan's economy, but unfortunately it is only unsustained growth, mainly because of uncertainty and high fluctuation in oil price in the international market; for instance, the recent global financial and economic crisis led to significant negative impact on Sudan's economy due to high dependence on oil revenues and oil exports. We are aware of the fact that it may be interesting to explain the impact of oil in education and training, but due to practical problems related to availability of adequate and reliable data, unfortunately it will not be possible to discuss this issue in this chapter, so we leave that for a more indepth analysis in our future research. Furthermore, we believe that most probably the impacts of oil in education and training might be still very limited in view of the very recent start of production and exports of oil just before 11 years in 1999. Moreover, although oil leads to increase in public spending and increase in the share of development expenditure as a percentage of total public expenditure from 9 % in 1999 to around 31 % in 2004, its share declined and sustained at 24 % from the total public spending over the period 2006–09. Furthermore, the development expenditures include all public spending in development issues including public spending on education, health, etc. Therefore, this implies that it is not at all clear and it is somewhat problematic to distinguish the share and growth of spending on education that is mainly attributed to production and export of oil, but it is important to realise that at the macro level in the pre- and post-oil periods the share of spending on education as a percentage of GDP most probably remained below the standardised international adequacy criterion of spending 20 % of total government or public spending on education. In addition, also due to practical problems related to availability of adequate and reliable data unfortunately it will not be possible to give an indepth analysis of the impact of the private oil companies spending on education and training at the micro level. So, we hope to cover these issues in our future studies when adequate and reliable data are available; therefore, our analysis focuses on two interpretations of the low commitment to the standardised international adequacy criterion in Sudan. Our first interpretation of the low commitment to the standardised international adequacy criterion in Sudan despite the increasing oil revenues, is that it is probably true that both uncertain public revenues (oil revenues in particular) and increasing competition for these revenues for defence (due to political instability) and infrastructure spending, notably, the increasing public spending on defence and security issues probably put further pressure on public spending on education and make it difficult for the government in Sudan to continue allocating high proportions of public revenues on education.

Moreover, our second explanation for the low commitment to the standardised international adequacy criterion is probably related to the potential limitation of the private spending on education to compensate the declining public spending on education despite the recent increasing expansion and facilities offered for the

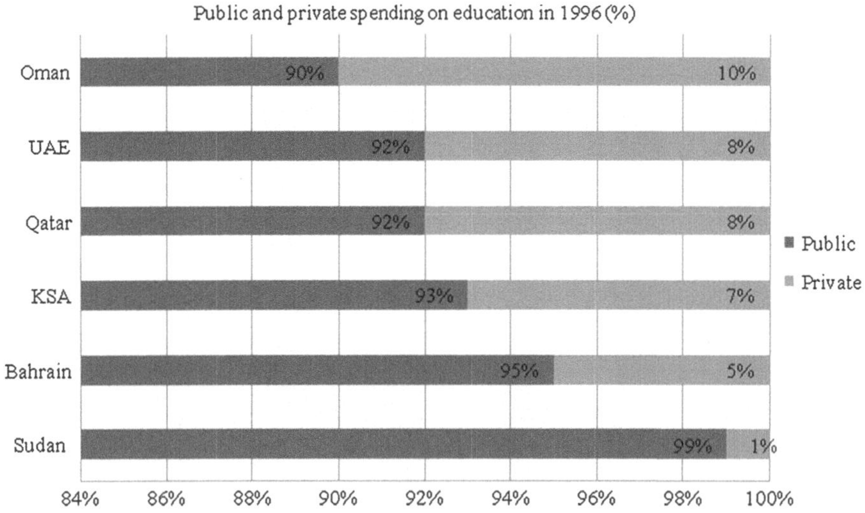

Fig. 9.2 Distribution and share of public and private spending on education in Sudan and Arab Gulf countries (1996) (Source: UNESCO–UIS (2000) World Education Report (2000): UNESCO's World Education Indicators, (b) UNESCO- UIS (2003))

introduction of private education institutions in Sudan. For instance, one important common characteristic of educational policies in Sudan and most Arab countries is the lack of incentives or marginal contributions of the private sector on educational investment (see Fig. 9.2 below). One shared feature of the education policies in the Arab countries is that public education is perceived as being very important for development. However, as indicated by Al-Sulayti, both uncertain public revenues (oil revenues in particular) and increasing competition for these revenues for defence and infrastructure spending make it difficult for oil dependent Arab countries to continue allocating high proportions of public revenues on education.[8] More recently though, following the declining trends of public spending, private spending on education shows an opposite increasing trend, it has increased slightly to fill the funding gap in Sudan as in most of the Arab and Arab Gulf countries; however, educational investment is still almost entirely public. In Sudan the extent of privatisation shows an increasing trend in tertiary education faster than in secondary and basic education respectively (see Figs. 9.2, 9.3, 9.4, and 9.5). In Sudan until 1989/90, tertiary education was publicly provided: there are only 2 private compared to 17 public higher education institutions and universities; the number of private tertiary education institutions increased from 2 in 1989/90 to 26 in 1999/2000 and the share of private tertiary education institutions increased from 11 % in 1989/90 to 50 % in 1999/2000 to be equivalent to the share of public

[8] See Al-Sulyati (2002).

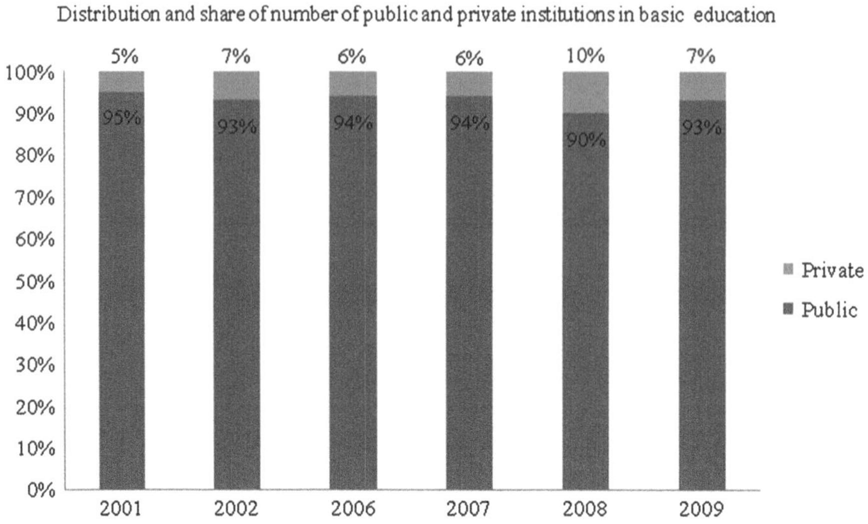

Fig. 9.3 Distribution and share of number of public and private institutions in basic education in Sudan (2001–2009) (%) (Sources: Own calculation based on Table 9.4)

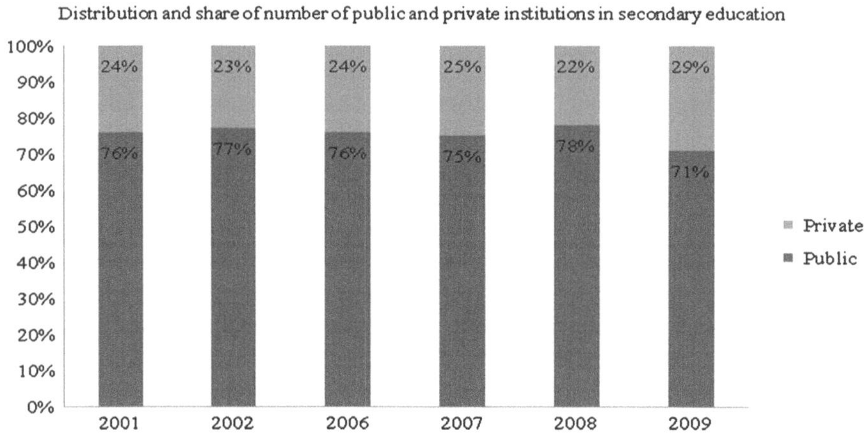

Fig. 9.4 Distribution and share of number of public and private institutions in secondary education in Sudan (2001–2009) (%) (Sources: Own calculation based on Table 9.4)

tertiary education institutions; in 2008/09 the share of private (52 %) is larger than the share of public (48 %) in total higher education institutions in Sudan.[9] However, the increasing private investment on education should not hide the fact that educational investment is almost entirely dependent on the public sector, with a very minimal contribution from private sector in Sudan (see Table 9.4 below).

[9] See Sudan Central Bureau of Statistics (2010), 'Annul Statistical Year Book and Statistical Series (1990–2009)', p. 107.

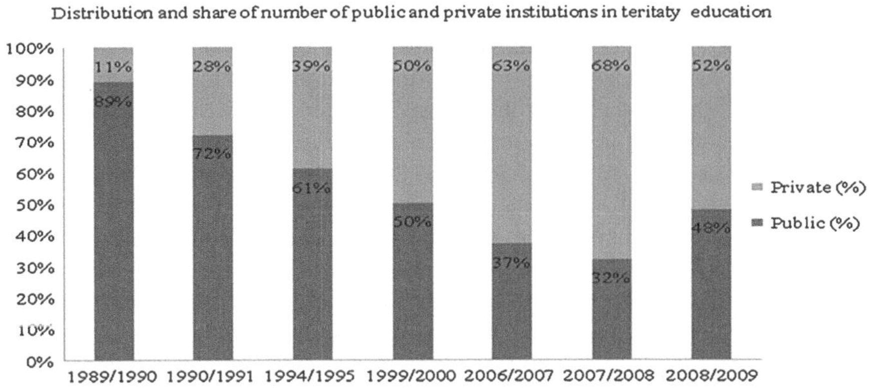

Fig. 9.5 Distribution and share of number of public and private institutions in tertiary education in Sudan (1989/1990–2008/2009) (%) (Sources: Own calculation based on Table 9.4)

Furthermore, the low commitment to the standardised international adequacy and efficiency criterions in the supply side is clear from the inadequacy, poor efficiency and quality of financial resources and physical infrastructure as measured by the overcrowded classrooms as indicated by students enrolment per education institutions in Sudan (see Table 9.2 below). Moreover, we find that the low commitment to the equity criterion in the supply side appears from the priority and trend of distributing the public and private investment in education between the different education levels in Sudan. One common characteristic of educational policies in Sudan and Arab countries is that the distribution/allocation of public investment on various educational levels tends to prioritise either primary or secondary education and seriously neglect tertiary education. Table 9.2 shows that despite the recent gradual increase in total investment in basic, secondary and tertiary educational levels as measured by the number of institutions, number of students and teachers in Sudan over the period 1993/94–2008/09, however, the distribution of increasing investment and physical infrastructure on education as measured by the number of education institutions is still biased towards basic education followed by secondary education, in general the share of investment on tertiary education remains marginal and insufficient and even shows a declining trend in Sudan as in many Arab countries. In our view the distribution of investment by educational levels may be related to both potential share of students in total population and costs of various educational levels as measured by spending per pupils, thus the low investment and spending in tertiary education is probably related to low share of potential student in tertiary education in total population and high costs of spending on tertiary students as compared to secondary and primary pupils and also probably because of high poverty rate in Sudan.[10]

[10] By contrast Jalal al-Din (2002) argues that the tragic reality of public education in some Arab and African countries is due to limited public and private spending that holds for all levels of education. But at least in many cases, the percentage of public spending on basic education is less

In addition we observe the low commitment to equity and the incidence of wide regional disparity between the main geographical regions in Sudan in terms of both supply of and demand for education.[11] As for the commitment to equity criterion in the supply side we observe that the priority and trend of distributing the public and private investment in education varies across the main geographical regions in Sudan. For instance, Tables 9.3 indicates the low commitment to equity and the incidence of regional disparity that appears from the share of the main regions in public, private and total number of schools in basic and secondary education in Sudan over the period 2001–2009. For instance, we observe the large share of the central region followed by Darfur, Kordofan and Khartoum as compared to Eastern, Southern and Northern regions in total numbers of basic schools and the large share of the central region followed by Khartoum, Darfur and Kordofan as compared to Northern, Eastern and Southern regions in total numbers of secondary schools over the period 2001–2009. Furthermore, from Table 9.4 we observe the large share of the central, Darfur, Kordofan and Khartoum regions as compared to Northern, Eastern and Southern regions in total numbers of public basic and secondary schools over the period 2001–2009. Moreover, we observe the low commitment to equity and incidence of wide regional disparity between the main regions in Sudan in terms of the participation of private sector in basic and secondary education as can be seen from the heavy concentration of privatisation in Khartoum region that has the largest share in terms of private basic and secondary schools as compared to other regions. This probably implies the low commitment to equity and incidence of unbalanced or biased distribution of allocation of investment, resources and infrastructure in education as measured by the number of schools for basic and secondary educational levels which is probably related to an unbalanced spending and development planning in education. This low commitment to equity and incidence of unbalanced regional distribution in the supply side includes not only the financial resources, investment and physical infrastructure in education as measured by the share in total and public and private schools, but also includes human resources in education as measured by both the total number of teachers and

than the proportion of spending on higher education, for example, in Sudan, spending on the basic education, which extends to 8 years in Sudan, exclusively accounted only for about 43 % of the total spending on higher education. This is in spite of the fact that basic education includes more than three million students, a number nearly ten times the proportion of students in higher education. However, spending on higher education after the expansion is still very limited unless it is compared to the tragic situation of general education. Hence, there is urgent need to allocate more resources to the entire education sector, notably, by giving high priority to public and general education. See Jalal al-Din (2002), pp. 23–24.

[11] In this research our analysis of the supply and demand for education from regional perspective is based on the classification of Sudan geographical areas according to main seven geographic regions that includes the Northern, Khartoum, Central, Kordofan, Darfur, Eastern and Southern regions. In particular, this classification implies that the Northern region includes Northern and Nahr Alnil states, the central region includes Al Gezira, Sinnar, White Nile and Blue Nile states, the Eastern region includes Red Sea, Algedarif and Kassala states, the Western region includes Kordofan and Darfur regions that include all the states which are located in Kordofan and Darfur respectively and the Southern region includes Bahr Elghazal, Equatoria and Upper Nile states.

Table 9.2 The distribution of current investment in basic, secondary and tertiary educational level in Sudan (1993/1994–2008/2009)

	1993/1994	1995/1996	1996/1997	1997/1998	2000/2001	2003/2004	2004/2005	2005/2006	2006/2007	2007/2008	2008/2009
Number of institutions											
Basic	12,515	10,713	10,969	11,278	12,539	11,541	14,071	15,089	15,907	18,095	18,052
Secondary	641	1206	1422	1648	1835	1642	2268	2877	3224	3478	3455
Tertiary	29	47	45	47	57	68	72	73	78	90	106
Number of students											
Basic	2,823	2,963.7	2,978.8	3,030.7	3,451.6	3,966.9	4,299.7	4,713.4	4,785.9	5,253	5,800
Secondary	292.8	363.7	448.5	485.6	437.5	526.2	611.6	569.7	602.9	648	753
Tertiary	116.2	125.7	171.6	172.3	254	315	447	488	440.2	508.2	519
Number of teachers											
Basic	73,654	94,414	98,747	103,082	125,391	136,401	141,315	143,327	145,999	155,023	161,345
Secondary	7,909	9,902	12,117	13,247	19,783	22,951	32,917	34,222	38,953	39,874	43,028
Tertiary	5,268	3,148	5,394	5,623	7,481	7,804	9,248	10,063	10,251	12,560	12,720
Pupils teacher ratio											
Basic	38	31	30	29	28	29	30	33	33	34	36
Secondary	37	37	37	35	22	23	19	17	15	16	18
Tertiary	22	50	32	31	17			15	15	22	13(1)
Pupils institution ratio											
Basic	226	277	272	269	275	344	306	312	301	290	321
Secondary	457	302	315	295	238	320	270	198	187	186	218
Tertiary	4,007	2,674	3,813	3,666	4,456	4,632	6,208	6,685	5,644	5,647	4,896(1)

Note (1) Data refers to (2009/2010)

Sources: (a) Sudan Ministry of Education, the Annual Educational Statistics Reports Various Issues (2001–2009): (2000–2001: 17), (2001–2002: 11), (2003–2004: 9), (2004–2005: 9), (2005–2006: 9), (2006–2007: 6), (2007–2008: 11), (2008–2009: 28), (b) Sudan Ministry of Higher Education and Scientific Research, the Annual Educational Statistics Reports Various Issues (1993/1994–2009/2010)

Table 9.3 Regional distribution and share of main regions in total number of schools in basic and secondary education in Sudan (%) (2001–2009)

	Number of basic school							Number of secondary school						
Share in total (%)	2001	2002	2004	2006	2007	2008	2009	2001	2002	2004	2006	2007	2008	2009
All Sudan	100 %	100 %	100 %	100 %	100 %	100 %	100 %	100 %	100 %	100 %	100 %	100 %	100 %	100 %
Northern	9 %	9 %	9 %	7 %	7 %	7 %	7 %	10 %	10 %	9 %	8 %	8 %	8 %	8 %
Khartoum	14 %	15 %	12 %	13 %	13 %	12 %	13 %	24 %	28 %	25 %	23 %	24 %	30 %	25 %
Central	27 %	26 %	27 %	24 %	23 %	25 %	22 %	32 %	30 %	35 %	37 %	37 %	35 %	35 %
Kordofan	17 %	17 %	20 %	15 %	15 %	16 %	17 %	8 %	7 %	9 %	8 %	8 %	8 %	9 %
Darfur	20 %	19 %	19 %	21 %	20 %	19 %	21 %	12 %	13 %	11 %	13 %	12 %	9 %	12 %
Eastern	11 %	12 %	10 %	10 %	11 %	12 %	11 %	10 %	10 %	9 %	8 %	8 %	7 %	7 %
Southern	3 %	3 %	3 %	10 %	9 %	9 %	9 %	3 %	2 %	3 %	3 %	3 %	3 %	3 %

Sources: Own calculation based on Sudan Ministry of Education, the Annual Educational Statistics Reports Various Issues (2001–2009): (2000–2001: 39–43, 50–52), (2001–2002: 67–70, 88–89), (2003–2004: 74–77, 97–99), (2004–2005: 77–80, 102–104), (2005–2006: 76–79, 97–98), (2006–2007: 79–82, 114–115), (2007–2008: 79–82, 115–116), (2008–2009: 100–103, 136–137)

Table 9.4 Regional distribution and share of main regions in total, public and private basic and secondary schools and the share in total public and private tertiary education institutions in Sudan (%) (1989/1990–2008/2009)

	Public						Private					
	2001	2002	2006	2007	2008	2009	2001	2002	2006	2007	2008	2009
(a) Basic												
All Sudan	94.9 %	93.2 %	94.1 %	93.8 %	90.2 %	92.6 %	5.1 %	6.8 %	6.0 %	6.2 %	9.8 %	7.4 %
Northern	8.3 %	8.5 %	7.3 %	7.2 %	6.5 %	6.8 %	0.2 %	0.2 %	0.1 %	0.1 %	0.2 %	0.1 %
Khartoum	13.0 %	11.8 %	9.5 %	9.4 %	8.5 %	8.8 %	1.2 %	3.3 %	3.0 %	4.0 %	3.7 %	4.2 %
Central	25.9 %	24.9 %	23.3 %	23.0 %	21.3 %	21.7 %	0.6 %	0.6 %	0.4 %	0.3 %	3.2 %	0.6 %
Kordofan	16.4 %	16.7 %	15.1 %	14.8 %	15.1 %	16.3 %	0.4 %	0.4 %	0.4 %	0.5 %	0.8 %	0.5 %
Darfur	19.0 %	18.3 %	19.2 %	18.9 %	17.8 %	19.9 %	1.0 %	1.2 %	1.3 %	0.7 %	1.4 %	1.5 %
Eastern	9.4 %	10.2 %	9.4 %	10.8 %	11.7 %	9.5 %	1.2 %	0.7 %	0.7 %	0.6 %	0.6 %	0.6 %
Southern	2.9 %	2.8 %	10.3 %	9.7 %	9.4 %	9.8 %	0.5 %	0.5 %	0.0 %	0.0 %	0.0 %	0.0 %
(b) Secondary												
All Sudan	76.5 %	77.2 %	75.8 %	74.6 %	77.8 %	71.2 %	23.5 %	22.8 %	24.2 %	25.4 %	22.2 %	28.8 %
Northern	9.6 %	9.5 %	7.2 %	7.4 %	6.7 %	7.1 %	0.3 %	1.0 %	0.7 %	0.4 %	0.4 %	0.6 %
Khartoum	14.0 %	13.2 %	11.1 %	10.3 %	18.3 %	9.9 %	10.8 %	10.3 %	11.9 %	13.9 %	12.6 %	15.1 %
Central	27.0 %	26.6 %	33.4 %	36.2 %	32.4 %	32.5 %	5.7 %	4.6 %	2.9 %	2.1 %	3.1 %	3.5 %
Kordofan	8.1 %	7.2 %	7.2 %	6.5 %	6.2 %	6.5 %	0.5 %	0.5 %	0.9 %	1.9 %	1.8 %	2.8 %
Darfur	8.7 %	11.7 %	8.1 %	6.5 %	6.9 %	6.5 %	3.0 %	3.2 %	6.2 %	5.4 %	2.3 %	5.2 %
Eastern	7.1 %	6.8 %	6.7 %	5.8 %	4.8 %	5.3 %	2.5 %	2.5 %	1.5 %	1.7 %	2.1 %	1.7 %
Southern	2.1 %	2.2 %	2.1 %	2.0 %	2.6 %	3.4 %	0.7 %	0.7 %	0.0 %	0.0 %	0.0 %	0.0 %

	Public							Private						
(c) Tertiary														
Institutions	1989/ 1990	1990/ 1991	1994/ 1995	1999/ 2000	2006/ 2007	2007/ 2008	2008/ 2009	1989/ 1990	1990/ 1991	1994/ 1995	1999/ 2000	2006/ 2007	2007/ 2008	2008/ 2009
Total	17	18	25	26	29	29	51	2	7	16	26	49	61	55
Share (%)	89 %	72 %	61 %	50 %	37 %	32 %	48 %	11 %	28 %	39 %	50 %	63 %	68 %	52 %

Sources: Own calculation based on Sudan Ministry of Education, the Annual Educational Statistics Reports Various Issues (2001–2009): (2000–2001: 29–38, 44–49), (2001–2002: 52–54, 63–66, 76–77, 82–83), (2003–2004: 59–62, 70–73, 83–85, 90–91), (2004–2005: 58–61, 73–76, 87–89, 94–96), (2005–2006: 61–64, 73–75, 85–86, 91–92), (2006–2007: 59–62, 75–78, 88–89, 94–95), (2007–2008: 62–65, 75–78, 88–89, 95–96), (2008–2009: 84–87, 96–99, 109–110, 116–117)

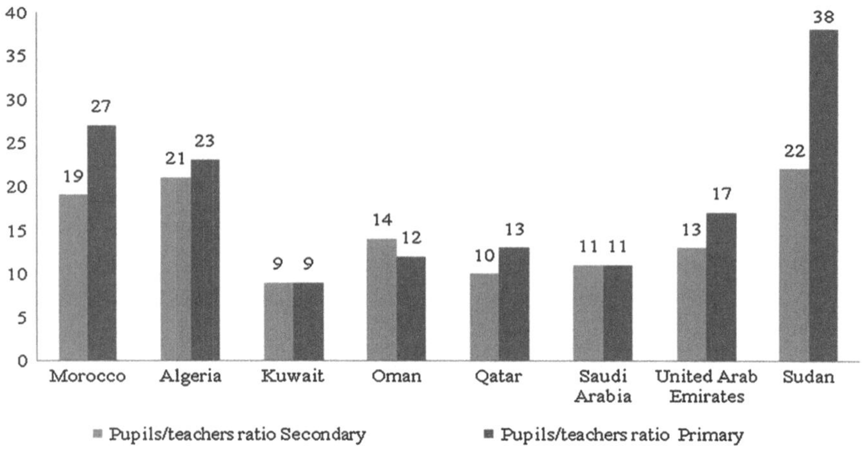

Fig. 9.6 Distribution of pupils teachers ratios in basic and secondary education in Sudan as compared to Arab and Gulf countries (2008/2009) (Source: UNESCO Global Background information on Education Statistics: UNESCO- UIS Data Centre: Beyond 20/20 WDS (2011))

the pupils teachers ratios in basic and secondary education. This low commitment to equity and incidence of unbalanced regional distribution in the supply side probably has further implications for low commitment to equity and incidence of regional distribution in the demand side as measured by the share and distribution of students enrolment in total, public and private basic and secondary education as we will explain below in this section. Suleiman (2007) discusses the equity standard criterion and notes that the imposition of tuition fees does not help to achieve this equity standard criterion and to reduce differences between high income earners and low income earners (see Suleiman 2007: 122).

The standardised international adequacy criterion of human resources in education or teaching staff can be defined by pupil-teacher ratios. Figure 9.6 below shows that the adequacy of human resources and teaching staff varies across the Arab countries including Sudan and the adequacy of human resources in education is generally better for secondary education when compared to primary education and, in sometimes, to tertiary education (see also Table 9.5 below). One serious problem with respect to human resources in education in Sudan is the low commitment to the standardised international adequacy criterion in terms of quantity of teaching staff as measured by the high pupils/teachers ratio in primary and secondary education and in terms of quality and efficiency of teaching staff as measured by the low share of trained teachers in primary, secondary and tertiary education in Sudan as compared to other Arab and Gulf countries (see Figs. 9.6, 9.7, and 9.8 below). As reported in the Sudanese and Gulf literature, "the educational system in Sudan and Gulf countries suffers from serious weak performance/low quality of teachers due to a lack of trained teachers and weak teaching skills and knowledge of recent

Table 9.5 Regional distribution and share of regions in total number of teachers and pupil teacher ratios in basic and secondary education in Sudan (%) (2001–2009)

	Basic number of teachers							Secondary number of teachers					
Total number of teachers	2001	2002	2005	2006	2007	2008	2009	2001	2005	2006	2007	2008	2009
All Sudan	100 %	100 %	100 %	100 %	100 %	100 %	100 %	100 %	100 %	100 %	100 %	100 %	100 %
Northern	13 %	11 %	9 %	9 %	10 %	10 %	9 %	7 %	7 %	8 %	6 %	10 %	10 %
Khartoum	17 %	18 %	19 %	17 %	19 %	18 %	20 %	33 %	45 %	27 %	35 %	28 %	28 %
Central	35 %	29 %	29 %	29 %	29 %	28 %	28 %	31 %	25 %	37 %	29 %	29 %	30 %
Kordofan	10 %	12 %	12 %	12 %	12 %	11 %	12 %	7 %	6 %	7 %	8 %	9 %	8 %
Darfur	11 %	17 %	13 %	13 %	12 %	14 %	12 %	9 %	7 %	9 %	8 %	12 %	9 %
Eastern	12 %	9 %	9 %	10 %	9 %	9 %	11 %	8 %	8 %	10 %	8 %	9 %	10 %
southern	4 %	3 %	9 %	9 %	9 %	9 %	9 %	3 %	2 %	3 %	3 %	5 %	3 %
Pupils teacher ratio	Basic pupils teacher ratio							Secondary pupils teacher ratio					
All Sudan	34	28	30	33	33	34	36	22	19	17	16	16	17
Northern	22	22	23	23	21	20	20	33	25	18	25	13	13
Khartoum	41	31	30	29	27	29	27	20	10	14	12	15	16
Central	28	28	26	27	29	31	30	21	22	15	17	15	16
Kordofan	59	31	34	36	36	36	41	22	30	24	21	21	22
Darfur	51	24	36	47	58	44	54	25	32	30	22	22	29
Eastern	34	31	33	34	34	39	31	27	23	17	18	18	16
Southern	39	39	39	43	46	45	76	16	28	15	15	10	26

Sources: Own calculation based on Sudan Ministry of Education, the Annual Educational Statistics Reports Various Issues (2001–2009): (2000–2001: 39–43, 50–52), (2001–2002: 67–70, 88–89), (2003–2004: 74–77, 97–99), (2004–2005: 77–80, 102–104), (2005–2006: 76–79, 97–98), (2006–2007: 79–82, 114–115), (2007–2008: 79–82, 115–116), (2008–2009: 100–103, 136–137)

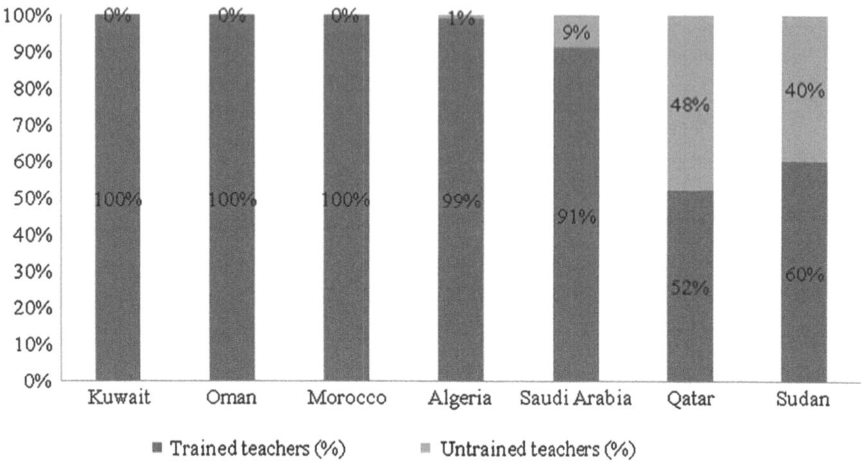

Fig. 9.7 Distribution of trained and untrained teachers ratios in basic education in Sudan as compared to Arab and Gulf countries (2008/2009) (Source: UNESCO Global Background information on Education Statistics: UNESCO- UIS Data Centre: Beyond 20/20 WDS (2011))

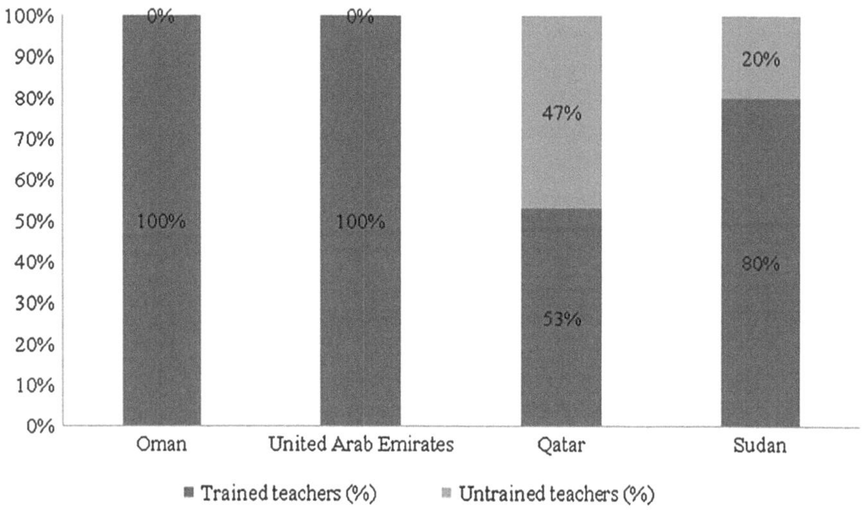

Fig. 9.8 Distribution of trained and untrained teachers ratios in secondary education in Sudan as compared to Arab Gulf countries (2008/2009) (Source: UNESCO Global Background information on Education Statistics: UNESCO- UIS Data Centre: Beyond 20/20 WDS (2011))

teaching and learning techniques".[12] For instance, Table 9.6 and Figs. 9.9, 9.10, 9.11, and 9.12 below investigate the lack of trained teachers and insufficiently trained teaching staff in basic, secondary, technical and tertiary education. For instance, over the period 2002–2008 the trained teachers are respectively 62 %, 60 %, 60 %, 66 % and 61 % of total teachers in basic education in 2002, 2005, 2006, 2007 and 2008 respectively. Moreover, in the period 2002–2009 the trained teachers are respectively 51 %, 66 %, 63 %, 67 % and 62 % of total teachers in secondary education in 2002, 2006, 2007, 2008 and 2009 respectively (see Table 9.6 below). The problem of the lack of trained teachers is more serious in the technical and basic education as compared to secondary and tertiary education, notably, for both basic and secondary education more than one third of teachers are untrained (see Tables 9.5 and 9.6 below). Moreover, we observe the lack of adequately trained teaching staff in tertiary education over the period 1996/97–2007/08. Although the share of total trained teachers in tertiary education improved and increased from 66 % in 1996/97 to 71 %, 75 %, 70 %, 77 % and 81 % in the years 1997/98, 2000/01, 2003/04, 2004/05 and 2007/08 respectively, the share of untrained teaching staff in tertiary education remained high and represents near to one fifth of total teaching staff in tertiary education in 2007/08. We observe the incidence of the sectoral gap in efficiency that appears as the share of trained teaching staff in the public universities is relatively high as compared to private universities and the incidence of the gender gap in efficiency that appears as the share of trained male teaching staff is relatively high as compared to female teaching staff; moreover, the observed gender gap in trained teaching staff in the private universities is higher than in the public universities over the period 2000/01–2007/08 (see Table 9.6 below).

In addition, from Tables 9.5 and 9.6 below we observe the low commitment to equity and incidence of wide regional disparity that appear from the share of the main regions in Sudan in terms of the adequacy and efficiency of human resources allocated to basic and secondary education as measured by the share in total number of teachers, pupils teachers ratios and the share in total number of trained teachers in basic and secondary education in Sudan over the period 2001–2009. For instance, we observe that the quantitative adequacy of teachers as measured by low pupil-teacher ratio is relatively better in the Northern region followed by Khartoum and the Central and Eastern regions as compared to Kordofan, Darfur, and Southern regions. Moreover, we observe the large share of the Central and Khartoum regions as compared to Northern, Kordofan, Darfur, Eastern and Southern regions in terms of quantitative adequacy of human resources, as measured by the share in total number of teachers and in terms of qualitative efficiency of human resources, as measured by the share in total number of trained teachers in basic and secondary

[12] See: Fahmey and Mahmoud (1993), pp. 29–30; Al-Sulayti (2002), pp. 29–30; Jalal al-Din (2002), pp. 12–13, 15–19, 22; and Suleiman (2007), pp. 117, 122–123. These findings are consistent with the observations of Suleiman (2007), who notes that currently a large number of teachers working in various stages of education without any training. See Suleiman (2007), p. 117.

Table 9.6 Regional distribution and share of main regions in trained and untrained teachers in basic and secondary education and the share of trained teachers in tertiary education defined by sector and gender in Sudan (1996/1997–2008) (%)

	Trained					Untrained				
(a) Basic	2002	2005	2006	2007	2008	2002	2005	2006	2007	2008
All Sudan	62 %	60 %	60 %	66 %	61 %	38 %	40 %	40 %	34 %	39 %
Northern	8 %	5 %	5 %	7 %	6 %	2 %	4 %	4 %	3 %	4 %
Khartoum	11 %	11 %	11 %	13 %	11 %	7 %	8 %	7 %	6 %	7 %
Central	17 %	19 %	19 %	22 %	20 %	13 %	10 %	10 %	6 %	9 %
Kordofan	9 %	7 %	7 %	7 %	6 %	4 %	5 %	5 %	5 %	5 %
Darfur	11 %	8 %	8 %	6 %	8 %	6 %	5 %	5 %	5 %	6 %
Eastern	5 %	6 %	6 %	3 %	2 %	2 %	3 %	4 %	4 %	5 %
southern	2 %	0.80 %	2 %			2 %	4 %	5 %	5 %	5 %
	Trained					Untrained				
(b) Secondary	2002	2006	2007	2008	2009	2002	2006	2007	2008	2009
All Sudan	51 %	66 %	63 %	67 %	62 %	49 %	34 %	37 %	33 %	38 %
Northern	6 %	5 %	5 %	7 %	6 %	2 %	3 %	2 %	2 %	4 %
Khartoum	18 %	18 %	25 %	20 %	16 %	10 %	8 %	10 %	9 %	13 %
Central	16 %	25 %	16 %	19 %	20 %	19 %	11 %	14 %	11 %	10 %
Kordofan	3 %	3 %	5 %	6 %	5 %	5 %	3 %	4 %	3 %	3 %
Darfur	5 %	7 %	6 %	8 %	7 %	5 %	2.30 %	3.50 %	4 %	2.30 %
Eastern	3 %	6 %	5 %	6 %	6 %	6 %	4 %	4 %	3 %	3 %
southern	0 %	3 %	3 %	3 %	3 %	3 %	1.80 %	1.60 %	3 %	3 %

(c) Tertiary	Trained 1996/1997	1997/1998	2000/2001	2003/2004	2004/2005	2007/2008	Untrained 1996/1997	1997/1998	2000/2001	2003/2004	2004/2005	2007/2008
All Sudan												
Total	66 %	71 %	75 %	70 %	77 %	81 %	34 %	29 %	25 %	30 %	23 %	19 %
Male	71 %	76 %	78 %	73 %	80 %	84 %	29 %	24 %	22 %	27 %	20 %	16 %
Female	52 %	55 %	64 %	58 %	71 %	75 %	48 %	45 %	36 %	42 %	29 %	25 %
Public												
Total			66 %	55 %	73 %	73 %			34 %	45 %	27 %	27 %
Male			83 %	72 %	81 %	83 %			17 %	28 %	19 %	17 %
Female			77 %	70 %	78 %	80 %			23 %	30 %	22 %	20 %
Private												
Total			65 %	64 %	67 %	64 %			35 %	36 %	33 %	36 %
Male			79 %	75 %	77 %	75 %			21 %	25 %	23 %	25 %
Female			75 %	71 %	73 %	71 %			25 %	29 %	27 %	29 %

Source: Own calculation based on Sudan Ministry of Education, the Annual Educational Statistics Reports Various Issues (2001–2009): (2001–2002: 30–31), (2003–2004: 34–37), (2004–2005: 32–35), (2005–2006: 31–34), (2006–2007: 31–34), (2007–2008: 33–36), (2008–2009: 58)

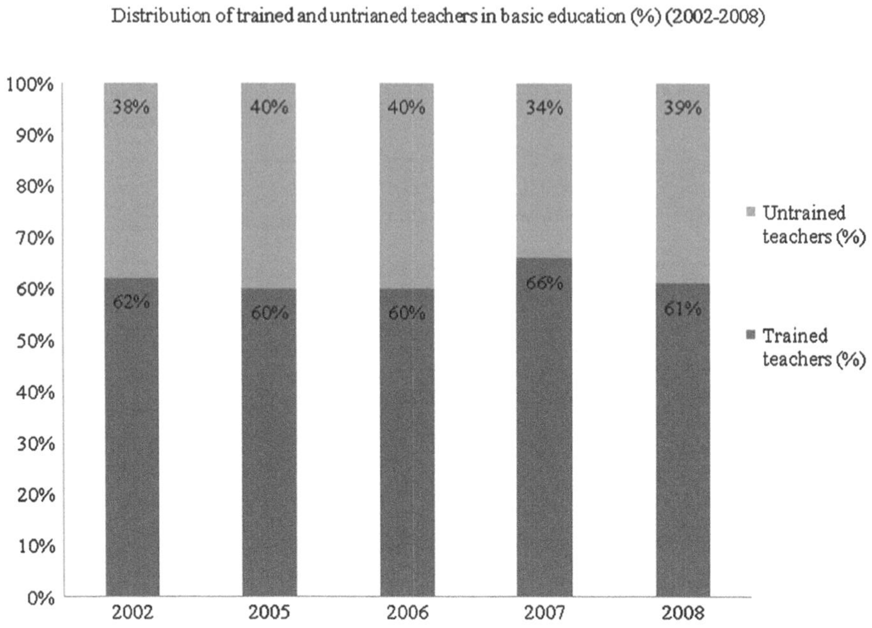

Fig. 9.9 Distribution of trained and untrained teachers in basic education in Sudan (%) (2002–2008) (Source: Own calculation based on Table 9.6)

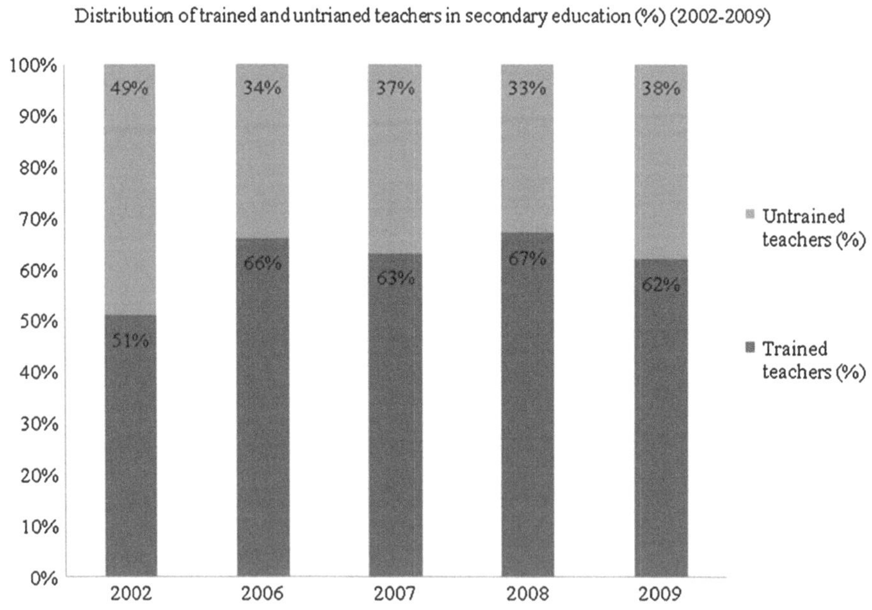

Fig. 9.10 Distribution of trained and untrained teachers in secondary education in Sudan (%) (2002–2009) (Source: Own calculation based on Table 9.6)

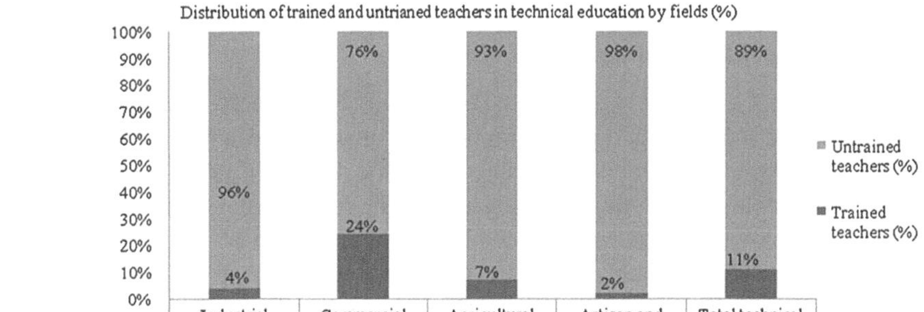

	Industrial science	Commercial science	Agricultural sciences	Artisan and vocational	Total technical education
Untrained teachers (%)	96%	76%	93%	98%	89%
Trained teachers (%)	4%	24%	7%	2%	11%

Fig. 9.11 Distribution of trained and untrained teachers in technical education in Sudan by fields (%) (Source: Own calculation based on Table 9.6. Adapted from Adapted from Table 9.2, p. 11)

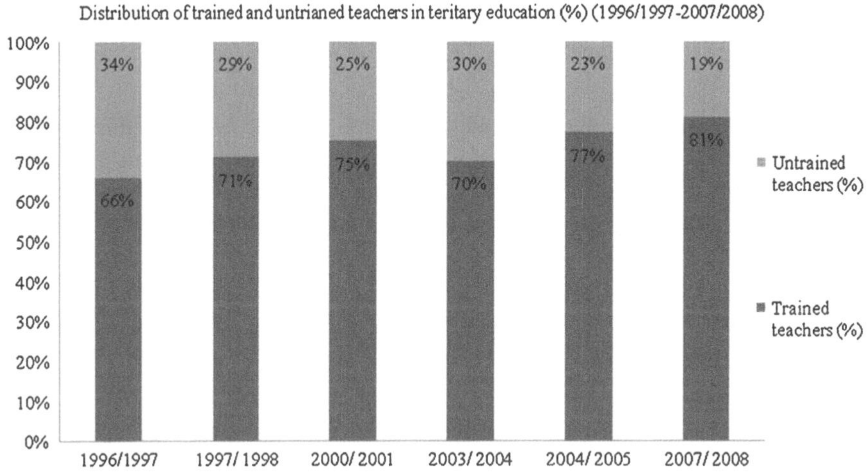

Fig. 9.12 Distribution of trained and untrained teachers in tertiary education in Sudan (%) (1996/1997–2007/2009) (Source: Own calculation based on Table 9.6)

education. This probably implies the low commitment to equity and the incidence of unbalanced regional distribution of trained human resources available for education, as measured by low pupils teachers ratios and large share in the total number of teachers and trained teachers for basic and secondary educational levels. These results may not be surprising in view of the fact that these regions also reported the large share in the financial investment and infrastructure on education as measured by the share in total numbers of basic and secondary schools over the period 2001–2009 as we explained above in this section.

9.2.3 The Demand Side of Educational Policies: The Demand for and Enrolment in Education

Apart from the supply side, it is also important to examine the demand for education as measured by enrolment ratios. Tables 9.7, 9.8, 9.9, 9.10, 9.11, 9.12, and 9.13 below shows that enrolment ratios vary across the main regions in Sudan, decline with the increase of education level and on average, lag behind the levels in Arab, Gulf and developed countries. The large quantitative increase in enrolment in higher education due to higher education revolution in 1990s should not hide the fact that at least until 2000/01 on average gross enrolment ratios in tertiary education in Sudan (6–6.9 %) remain low and fall behind the levels of the Arab (11.4–21 %), Gulf (21.4–24.8 %) and developed countries (58.4–94 %). That also holds for net enrolment ratios in primary and secondary education in Sudan that lag far behind those of the Arab, Gulf and developed countries.[13] A further serious problem is the negligence and declining trend in enrolment in vocational education in Sudan that falls behind Korea, Bahrain and developed countries (see Table 9.7 below). Thus, this implies low commitment to standardised international adequacy criterion in the demand side (or enrolment rate in primary, secondary and tertiary education).

Moreover, another problematic feature on the demand side of education in Sudan and Arab countries is the lack of incentives/minimal enrolment in private education compared to intensive enrolment in public education that is probably related to the high cost of private education and minimal contribution of the private

[13] Probably, the low enrolment at secondary level is attributed to high poverty rate, the high dropout in transition from primary to secondary schooling and the lack of effective actions in educational policy to legitimise compulsory education. For instance, Suleiman (2007) finds that the issue of free education, especially basic education has raised a lot of controversy in Sudan. He argues that in Sudan it is necessary to confirm the commitment for making basic education free and compulsory at the same time and it is the duty of the State to begin soon in the preparation of a plan for a period of 3 years or 5 years at most to achieve free and compulsory basic education, albeit gradually, with implementation to begin in the peripheral and poor regions that suffer from poverty. Other regions can be offered an alternative option of reduced fees that determined by school according to student family income level and the ability to pay for school fees; this can be an option or first step to achieve free education in Sudan in the long run. The plan should ensure that the schools do not have to be compelled to re-impose fees to pay salaries or other new expenses by provision of adequate ongoing or current funds necessary for the establishment and preparation of schools and teachers in light of rapid survey and assessment needs of the most needed states. If there is some shortage of school buildings, female students can be taught in the morning and male students in the afternoon, until a sufficient number of schools are available. Suleiman (2007) indicates that in the Interim Constitution of Sudan for the year 2005 (which is the Constitution that guarantees many rights for all citizens), the article 44 of it indicates that (1) “Education is a right for every citizen and the State to guarantee access to education without discrimination on the basis of religion, race, ethnicity, gender or disability "and (2)“ Primary education is compulsory and the State shall provide it for free.” He suggests cooperation of all people to the enforcement of this Constitution, especially as it is stated the rights of citizens. Suleiman (2007) argues that the education tuition fees and weak potential to get a job is likely paying much of today’s youth to frustration and apathy. See Suleiman (2007), pp. 122–125.

Table 9.7 Enrolment ratios by educational level in the Sudan compared to Arab and world countries (1990–2008/2009) (%)

Educational level	Primary level[a, g]			Secondary level[a, g]			Tertiary level[a, b, c, g, i]			Tertiary students in Science, Math and Engineering[a, b, c, f, h]		Vocational education (1990–1995)[d]		
Year/ country	1990/ 1991[a]	2001/ 2002[a]	2008/ 2009[g]	1990/ 1991[a]	2001/ 2002[a]	2008/ 2009[g]	1998/ 1999[b]	2000/ 2001	2008/ 2009[g, i]	1994/ 1997[a, f]	1996[c(3)] – 2008[f(3), h (4)]	1990/ 1991[d]	2000[g]	2008/ 2009[g]
Sudan	43	46	74	24	18.2	38	6	6.9	5.9[i (1)]	27[f]	30[f] (3)	4.2	2	2
Bahrain	99	91	107	85	81	96	25.2	22	29.9i[(1)]	n/ a	18[h] (4)	13.3	17	16
Kuwait	49	85	95	n/a	77	90	21.1	22	17.6	23	27[c] (3)	0.6(5)	2	2
Oman	69	75	84	n/a	68	91	n/a	8.5	26	31	27[h] (4)	2.8	1.4	n/a
Qatar	89	94	106	70	78	85	27.7	24.6	10	n/a	23[h] (4)	2.9	2	1
KSA	59	59	99	31	53	97	0.71	22.4[(2)]	33	18	24[c] (3)	n/a	2 8	2.3
UAE	100	81	105	58	72	95	12.1	n/a	30	27	33[h] (4)	0.7	1	n/a
Average Gulf	77.5	80.8	99.3	61	71.5	92.3	21.4	18.5	24.8	24.5	25.3	3.9	4.4	5.3
US	97	93	99	85	85	94	75.7	72.6	83	n/a	15[h] (4)	n/a	n/a	n/a
Sweden	100	102	95	85	99	103	62.3	70	71	31	24[h] (4)	n/a	30	30
Korea	104	101	105	86	89	97	65[e]	77.6	98	34	32[h] (4)	18	19	12
UK	100	101	106	81	95	99	58.4	59.5	57	29	22[h] (4)	n/a	20	13
Finland	98	100	97	93	95	110	83[e]	85[e]	94	37	27[h] (4)	n/a	33	29
Arab	81.4	91	96	52.2	61	68	11.4	19	21	12.1	n/a	n/a	15	13

Notes: (1) Data refers to most recent available data between 2001 and 2009, (2) data refers to 1999/2000, (3) data refers to (%) of tertiary students in natural science, engineering, agriculture and medical sciences 1996 (4) data refers to (%) of tertiary graduates in natural science, engineering, agriculture and medical sciences 2008 (5) data refers to 1992/1993

[a]UNDP Human Development Report (2004).

[b]UNDP Human Development Report (2002).

[c]UNESCO–UIS (2000) UNESCO's World Education Report 2000 "The right to education: towards education for all throughout life".

[d]UNESCO (1996) Statistical Yearbook and UNESCO Statistics: www.unesco.org

[e]UNESCO- UIS (2004b) UIS web site global statistics on education: www.unesco.org

[f]Sudan Ministry of Education, the Annual Educational Statistics Reports Various Issues (2001–2009)

[g]UNESCO Global Background information on Education Statistics: UNESCO- UIS Data Centre: Beyond 20/20 WDS (2011)

[h]UNESCO-UIS (2010) "Global education digest 2010 Comparing education statistics around the e world"

[i]UNDP Human Development Report (2010).

Table 9.8 Regional distribution and share of main regions in total number of students enrolled in basic, secondary and tertiary education in Sudan (%) (1996–2009)

	(a) Basic education							(b) Secondary education						
Total	2001	2002	2004	2006	2007	2008	2009	2001	2002	2004	2006	2007	2008	2009
All Sudan	100 %	100 %	100 %	100 %	100 %	100 %	100 %	100 %	100 %	100 %	100 %	100 %	100 %	100 %
Northern	8 %	8 %	7 %	7 %	6 %	5 %	5 %	10 %	6 %	9 %	8 %	8 %	8 %	7 %
Khartoum	21 %	21 %	19 %	15 %	16 %	16 %	15 %	29 %	33 %	24 %	23 %	28 %	27 %	26 %
Central	30 %	29 %	29 %	25 %	26 %	27 %	24 %	31 %	32 %	36 %	31 %	30 %	28 %	29 %
Kordofan	13 %	14 %	14 %	14 %	13 %	13 %	14 %	7 %	8 %	7 %	9 %	10 %	10 %	10 %
Darfur	14 %	14 %	14 %	17 %	19 %	19 %	18 %	10 %	9 %	12 %	17 %	11 %	16 %	14 %
Eastern	12 %	12 %	10 %	10 %	10 %	10 %	9 %	11 %	8 %	10 %	10 %	9 %	9 %	9 %
Southern	3 %	3 %	6 %	11 %	11 %	11 %	16 %	2 %	3 %	3 %	3 %	3 %	3 %	6 %

(c) Higher and tertiary education

Total	1996/1997	1998/1999	1999/2000	2000/2001	2001/2002	2002/2003	2004/2005	2005/2006	2006/2007	2007/2008	1996/2007
Northern	13.1 %	11.7 %	28 %	11.1 %	10 %	11.2 %	10.6 %	10 %	10.3 %	9.5 %	12.5 %
Khartoum	30.8 %	31.8 %	33.8 %	31.1 %	30.4 %	280 %	31.8 %	33.1 %	33.1 %	30.9 %	31.5 %
Central	35 %	35.5 %	32.9 %	33.9 %	32.6 %	29.8 %	30.5 %	30.2 %	29.2 %	30.4 %	32.1 %
Kordofan	5.9 %	4.8 %	4.1 %	4.9 %	6.6 %	6.6 %	5.2 %	6.2 %	6.5 %	7.5 %	5.8 %
Darfur	5.5 %	6.1 %	6.3 %	9.6 %	12 %	11.4 %	9 %	8.2 %	9 %	9.8 %	8.6 %
Eastern	8.4 %	8.9 %	8.4 %	7.4 %	6.7 %	7.8 %	8.5 %	8.1 %	7.4 %	7.7 %	7.9 %
Southern	1.4 %	1.17 %	1.7 %	2.2 %	2.7 %	3.9 %	4.2 %	4.4 %	4.4 %	4 %	3 %
Outside Sudan	0 %	0 %	0 %	0 %	0 %	0.1 %	0.2 %	0.2 %	0.2 %	0.2 %	0.1 %
All Sudan	100 %	100 %	100 %	100 %	100 %	100 %	100 %	100 %	100 %	100 %	100 %

Sources: Own calculation based on (a–b) Sudan Ministry of Education, the Annual Educational Statistics Reports Various Issues (2001–2009): (2000–2001: 39–43, 50–52), (2001–2002: 67–70, 88–89), (2003–2004: 74–77, 97–99), (2004–2005: 77–80, 102–104), (2005–2006: 76–79, 97–98), (2006–2007: 79–82, 114–115), (2007–2008: 79–82, 115–116), (2008–2009: 100–103, 136–137), (c) Sudan Ministry of Higher education statistics on enrolment various issues reports (1996–2007).

Table 9.9 Regional distribution and share of main regions in total number of students enrolment rate in basic, secondary and tertiary education level defined by education levels and gender in Sudan (%) (2001–2009)

	2001			2006			2007			2008			2009		
(a) Basic	M	F	MF	M	F	MF	M	F	MF	M	F	MF	M	F	MF
All Sudan	57.4	49.3	53.4	69.8	59.3	64.5	72	60.3	66.2	73.9	63.4	68.7	67.4	64.6	66.1
Northern	94	90	92	98	91	95	91	82	86	93	84	89	88	86	87
Khartoum	88.4	83.7	86.1	83.2	79	81.1	85.5	81.6	83.6	88.8	88.5	88.6	91.5	95.9	93.7
Central	65	42	60	72	61	66	78	64	71	82	69	75	84	74	79
Kordofan	50	43	46	63	57	60	67	52	60	67	55	61	84	74	79
Darfur	45	33	27	65	46	56	46	51	60	69	56	62	67	60	64
Eastern	48	38	43	67	50	58	70	50	60	67	48	58	51	49	50
Southern	24	17	21	55	45	50	55	52	53	56	52	54	49	48	49
(b) Secondary															
All Sudan	18.1	18.4	18.2	26.2	25.3	25.7	26.9	25.7	26.3	29.6	26.3	28	29.4	29.9	29.7
Northern	39	52	45	45	52	49	42	48	45	46	52	49	46	53	49
Khartoum	35.7	40.8	38.1	44	50.6	47.4	52.9	56.4	54.6	54	55.8	54.9	62.1	68.9	65.4
Central	20	22	21	28	28	28	30	30	30	32	30	31	39	35	37
Kordofan	10	10	10	21	17	19	21	18	20	24	19	21	30	24	27
Darfur	14	9	12	24	18	21	18	12	15	26	15	20	22	18	20
Eastern	20	21	21	24	22	24	23	22	22	23	21	22	19	18	18
Southern	3	2	3	6	4	5	7	5	6	7	5	6	7	6	7

(continued)

Table 9.9 (continued)

	2001			2006			2007			2008			2009		
(a) Basic	M	F	MF	M	F	MF	M	F	MF	M	F	MF	M	F	MF
(c) Tertiary and higher education															
	2001/2002			2004/2005			2005/2006			2006/2007			2007/2008		
	M	F	MF	M	F	MF	M	F	MF	M	F	MF	M	F	MF
All Sudan	49.7	50.3	100	47.8	52.2	100	48	52	100	48.7	51.3	100	47.9	52.1	100
Northern	4.2	5.8	10	4.5	6.2	10.6	4.4	5.6	10	4.4	5.9	10.3	4		9.5
Khartoum	13.3	16.5	30.4	14.4	17.4	31.8	15	18.1	33.1	15.2	17.9	33.1	13.7	5.5	30.9
Central	15.7	17.1	32.6	14.5	16	30.5	14.5	15.8	30.2	14.2	15	29.2	14.4	16	30.4
Kordofan	3.4	3.2	6.6	2.6	2.7	5.2	3	3.2	6.2	3.3	3.3	6.5	3.6	3.9	7.5
Darfur	8.4	3.8	12	4.9	4.1	9	4.4	3.8	8.2	5.2	3.9	9	5.7	4.1	9.8
Eastern	3.5	3.2	6.7	4.1	4.4	8.5	3.8	4.1	8.1	3.5	3.8	7.4	3.6	4	7.7
Southern	2	0.9	2.7	3	1.4	4.2	3.1	1.4	4.4	2.9	1.6	4.4	2.7	1.5	4
outside Sudan	0	0	0	0.1	0.1	0.2	0.1	0.1	0.2	0.1	0.1	0.2	0.1	0.1	0.2

Sources: Own calculation based on (a–b) Sudan Ministry of Education, the Annual Educational Statistics Reports Various Issues (2001–2009), (c) Sudan Ministry of higher education educational survey statistics (2000–2001) cited in ministry of education educational statistics (2001–2002: 49)

Table 9.10 Regional distribution and share of main regions in total number of students enrolled in public and private basic and secondary schools and the share of students enrolled in public and private tertiary education institutions defined by gender in Sudan (1993/1994–2009) (%)

	Public						Private					
(a) Basic	2001	2002	2006	2007	2008	2009	2001	2002	2006	2007	2008	2009
All Sudan	95.6 %	95.2 %	96.4 %	96.0 %	93.8 %	95.7 %	4.4 %	4.8 %	3.6 %	4.1 %	6.2 %	4.3 %
Northern	8.0 %	7.9 %	6.5 %	5.9 %	5.5 %	3.3 %	0.2 %	0.2 %	0.1 %	0.1 %	0.2 %	0.1 %
Khartoum	19.9 %	19.4 %	13.9 %	13.8 %	13.9 %	13.0 %	1.0 %	1.6 %	1.5 %	1.9 %	1.9 %	1.9 %
Central	28.5 %	28.1 %	25.0 %	25.3 %	24.3 %	22.7 %	0.7 %	0.7 %	0.5 %	0.5 %	2.2 %	0.4 %
Kordofan	12.1 %	13.1 %	13.2 %	12.7 %	12.1 %	13.5 %	0.4 %	0.3 %	0.3 %	0.4 %	0.6 %	0.4 %
Darfur	13.8 %	13.1 %	17.3 %	17.4 %	17.8 %	17.2 %	0.6 %	0.9 %	0.7 %	0.6 %	0.8 %	1.0 %
Eastern	10.0 %	10.4 %	9.6 %	9.4 %	9.5 %	8.4 %	1.0 %	0.6 %	0.6 %	0.6 %	0.6 %	0.5 %
Southern	3.4 %	3.3 %	11.0 %	11.5 %	10.7 %	16.0 %	0.6 %	0.6 %	0.0 %	0.0 %	0.0 %	0.0 %
	Public						Private					
(b) Secondary	2001	2002	2006	2007	2008	2009	2001	2005	2006	2007	2008	2009
All Sudan	77.7 %	77.7 %	84.9 %	84.0 %	83.0 %	77.8 %	22.3 %	22.3 %	15.2 %	16.1 %	17.0 %	22.2 %
Northern	8.6 %	8.6 %	7.9 %	7.2 %	7.2 %	6.4 %	2.1 %	2.0 %	0.5 %	0.2 %	0.2 %	0.8 %
Khartoum	19.8 %	19.1 %	16.7 %	19.7 %	18.2 %	14.3 %	10.0 %	9.6 %	6.1 %	8.5 %	8.4 %	12.1 %
Central	25.1 %	24.6 %	28.0 %	28.0 %	26.2 %	26.1 %	6.7 %	6.4 %	1.5 %	2.6 %	2.6 %	3.1 %
Kordofan	6.8 %	7.0 %	8.9 %	9.2 %	9.2 %	8.5 %	0.3 %	0.5 %	0.6 %	1.0 %	1.2 %	0.9 %
Darfur	8.7 %	9.3 %	12.2 %	9.2 %	12.0 %	10.1 %	1.7 %	2.8 %	5.4 %	2.5 %	3.4 %	3.7 %
Eastern	7.3 %	7.7 %	8.2 %	7.7 %	7.1 %	6.5 %	2.8 %	2.5 %	1.2 %	1.2 %	1.1 %	1.6 %
Southern	1.4 %	1.4 %	3.1 %	2.9 %	3.0 %	6.0 %	0.4 %	0.4 %	0.0 %	0.0 %	0.0 %	0.0 %
	Public						Private					
(c) Tertiary	1993/1994	1995/1996	2001/2002	2002/2003	2005/2006	2007/2008	1993/1994	1995/1996	2001/2002	2002/2003	2005/2006	2007/2008
Male	89 %	93 %	83 %	98 %	92 %	94 %	11 %	7 %	17 %	2 %	8 %	6 %
Female	79 %	92 %	79 %	82 %	91 %	93 %	21 %	8 %	21 %	18 %	9 %	7 %
Total	84 %	92 %	81 %	83 %	91 %	93 %	16 %	8 %	19 %	17 %	9 %	7 %

Sources: Own calculation based on (a–b) Sudan Ministry of Education: the Annual Educational Statistics Reports Various Issues (2001–2009): (2000–2001: 29–38, 44–49), (2001–2002: 52–54, 63–66, 76–77, 82–83), (2003–2004: 59–62, 70–73, 83–85, 90–91), (2004–2005: 58–61, 73–76, 87–89, 94–96), (2005–2006: 61–64, 73–75, 85–86, 91–92), (2006–2007: 59–62, 75–78, 88–89, 94–95), (2007–2008: 62–65, 75–78, 88–89, 95–96), (2008–2009: 84–87, 96–99, 109–110, 116–117), (c) Sudan Ministry of higher education and scientific research: the Annual Educational Statistics Reports Various Issues (1993/1994–2007/2008)

Table 9.11 Regional distribution and share of main regions in technical and vocational education and training and ggraduates of the apprenticeship programmes in Sudan (1995–2009)

	Technical education								Vocational education			
School[a]												
	2001	2002	2004	2005	2006	2007	2008	2009	2001	2007	2008	2009
All Sudan	100 %	100 %	100 %	100 %	100 %	100 %	100 %	100 %	100 %	100 %	100 %	100 %
Northern	14 %	16 %	19 %	15 %	9 %	12 %	12 %	14 %	18 %	13 %	12 %	12 %
Khartoum	15 %	17 %	17 %	19 %	14 %	17 %	18 %	16 %	24 %	20 %	21 %	19 %
Central	25 %	23 %	22 %	35 %	42 %	30 %	27 %	35 %	39 %	49 %	46 %	42 %
Kordofan	8 %	8 %	12 %	7 %	8 %	7 %	7 %	6 %	0 %	0 %	0 %	0 %
Darfur	19 %	18 %	15 %	14 %	11 %	13 %	14 %	10 %	6 %	4 %	6 %	8 %
Eastern	15 %	17 %	13 %	11 %	15 %	16 %	18 %	17 %	12 %	13 %	14 %	20 %
Southern	5 %	4 %	0 %	0 %	0 %	3 %	3 %	3 %	0 %	0 %	0 %	0 %
Students[a]												
	Technical education								Vocational education			
	2001	2002	2004	2005	2006	2007	2008	2009	2001	2007	2008	2009
Number of students												
All Sudan	100 %	100 %	100 %	100 %	100 %	100 %	100 %	100 %	100 %	100 %	100 %	100 %
Northern	8 %	8 %	12 %	9 %	6 %	7 %	9 %	11 %	8 %	5 %	6 %	7 %
Khartoum	30 %	30 %	27 %	30 %	23 %	27 %	22 %	22 %	47 %	56 %	37 %	45 %
Central	23 %	24 %	30 %	31 %	45 %	24 %	26 %	25 %	26 %	25 %	41 %	27 %
Kordofan	4 %	4 %	7 %	7 %	6 %	8 %	8 %	8 %	0 %	0 %	0 %	0 %
Darfur	12 %	12 %	9 %	8 %	9 %	14 %	14 %	12 %	5 %	3 %	5 %	7 %
Eastern	19 %	19 %	14 %	12 %	12 %	16 %	17 %	18 %	14 %	9 %	10 %	14 %
Southern	3 %	2 %	0 %	0 %	0 %	3 %	3 %	3 %	0 %	0 %	0 %	0 %
Number of tecachers[a]												

	Technical education								Technical education				Vocational education
									Trained		Untrained		
	2001	2002	2004	2005	2006	2007	2008	2009	2008	2009	2008	2009	2009
All Sudan	100 %	100 %	100 %	100 %	100 %	100 %	100 %	100 %	86 %	59 %	14 %	41 %	100 %
Northern	8 %	8 %	10 %	10 %	10 %	10 %	8 %	3 %	8 %	8 %	0 %	9 %	17 %
Khartoum	28 %	28 %	29 %	39 %	25 %	16 %	15 %	83 %	15 %	18 %	0 %	0 %	18 %
Central	23 %	24 %	30 %	29 %	33 %	41 %	40 %	1 %	35 %	18 %	6 %	6 %	25 %
Kordofan	7 %	6 %	8 %	7 %	6 %	6 %	6 %	2 %	6 %	5 %	0 %	6 %	10 %
Darfur	17 %	17 %	21 %	14 %	12 %	11 %	16 %	0 %	13 %	7 %	2 %	6 %	12 %
Eastern	12 %	12 %	16 %	11 %	16 %	18 %	15 %	9 %	9 %	5 %	6 %	13 %	18 %
Southern	4 %	4 %	0 %	5 %	0 %	0 %	0 %	0 %	0 %	0 %	0 %	0 %	0 %
Training and graduates of the apprenticeship programmes[b]													
	1995	1996	1997	1998	1999	2000	2001	2002	2003	2004	2005	2006	1995–2006
All Sudan	100	100	100	100	100	100	100	100	100	100	100	100	100
Khartoum	63.1	61.7	61.2	60.5	61	57.3	59.5	58.3	57.2	58.4	60.1	80.8	61.2
Central	18.1	19.3	15.5	16	24.9	24.2	22.8	19.8	22.1	21.1	21.4	0	19.1
Darfur	0	0	3.6	2.4	2.3	1.8	0.9	1.3	0.5	0	0	0	1
Kordofan	0	0	0	3.8	1.8	1.8	2.1	2.2	1.7	1.8	5.8	4.5	2.3
Eastern	18.7	19	19.7	17.4	10	11	12	14	13.9	12.8	2	0	11.9
Southern	0	0	0	0	0	0	0	0	0	0	2.1	2.9	0.4
Other	0	0	0	0	0	4	2.8	4.3	4.5	5.9	8.5	11.7	4

Sources: Own calculation based on

[a]Sudan Ministry of Education, the Annual Educational Statistics Reports Various Issues (2001–2009)

[b]Sudan Ministry of Labour (2006) "Statistics on enrolment and graduates for Training of the apprenticeship programmes over the period (1995–2006)," Unpublished Statistics, Department of registration and certificates, Secretariat General of apprenticeship programme, Sudan Ministry of Labour, Khartoum, Sudan.

Table 9.12 Regional distribution, share and value of main regions in urbanization, economic, MDGs, demographic and education indicators in Northern Sudan (2005–2009)

(a) Urbanization, economic and MDGs indicators[a], [b]

	Urbanization [a]	Actual Per capita federal allocation [a]	Poverty[b] Proportion of population below poverty line	Poverty gap ratio	MDG 2.3. Literacy rate of 15–24 years-olds[b] Total	Men	Women
			MDG 1.1	MDG 1.2	(15–24)	(15–24)	(15–24)
	2005	2005	2009	2009	2009	2009	2009
Total			46.5	16.2	77	84	71
Northern	27	9068	33.7	9.4	88	91	86
Khartoum	88	8497	26	6.4	94	96	92
Central	29	4872	45.4	13.8	77	84	70
Kordufan	29	3765	58.7	23.1	69	79	61
Darfur	20	2732	62.7	24.6	74	85	64
Eastern	43	2553	46.3	17.7	63	68	57

(b) Education and demographic indicators[b, c, d]

	Net enrolment rate in primary education[b]	Primary education[c]		Secondary education[c]		Tertiary education[d]		Female enrolment		
		Students (6–13)	Population (6–13)	Students (14–16)	Population (14–16)	Students (16–20)	Population (16–20)	Primary[c] (6–13)	Secondary[c] (14–16)	Tertiary[d] (16–20)
	2009	2008	2008	2008	2008	2008	2008	2008	2008	2008
Total	67	100 %	100 %	100 %	100 %	100 %	100 %			
Northern	83	5 %	4 %	7 %	4 %	9.5	4.5 %	84 %	52 %	9.5 %
Khartoum	85	15 %	10 %	26 %	12 %	30.9	13.6 %	88.5 %	55.8 %	30.9 %
Central	67	38 %	29 %	29 %	19 %	30.4	19.3 %	69 %	30 %	30.4 %
Kordufan	60	14 %	12 %	9 %	10 %	7.5	10.1 %	55 %	19 %	7.5 %
Darfur	62	18 %	22 %	14 %	21 %	9.8	20.0 %	56 %	15 %	9.8 %
Eastern	57	9 %	12 %	8 %	12 %	7.7	12.2 %	48 %	21 %	7.7 %

[a]Elbadawi and Suleiman (2008: 107)
[b]the Sudan Central Bureau of Statistics (2011: 12)
[c]own calculation based on Sudan Ministry of Education, the Annual Educational Statistics Reports (2008/2009)
[d]own calculation based on Sudan Ministry of higher education statistics (2007/2008)

Table 9.13 Correlation between education indicators, urbanization, MDG, demographic and economic indicators in Northern Sudan (2005–2009)

	Coefficient (t-value)						R^2	N
Independent Variable	Proportion of population below poverty line	Poverty gap ratio	Per capita income	Urbanization	Share of population (2008)	Constant		
Dependent variable	MDG 1	MDG12						
Enrolment in basic education (MDG2:)	−0.993**					111.881	0.707	6
	(−3.103)					(7.403)		
		−2.000**				98.419	0.775	6
		(−3.717)				(10.642)		
			0.007**			37.278	0.938	6
			(7.801)			(8.831)		
				0.260*		58.780	0.289	6
				(1.274)		(6.339)		
					1.163**	−0.008	0.836	6
					(4.516)	(−0.172)		
Enrolment in secondary education	−0.002					0.258	0.108	6
	(−0.697)					(1.684)		
		−0.006				0.242	0.172	6
		(−0.910)				(2.341)		
			0.00001			0.116	0.048	6
			(0.448)			(1.209)		
				0.002		0.086	0.204	6
				(1.014)		(1.102)		
					0.844*	0.045	0.294	6
					(1.289)	(0.488)		
Enrolment in tertiary education	−0.445*					36.205	0.301	6
	(−1.313)					(2.261)		
		−0.970*				31.326	0.387	6
		(−1.588)				(2.985)		
			0.002			7.186	0.177	6

(continued)

Table 9.13 (continued)

	Coefficient (t-value)						R^2	N
			(0.926)			(0.678)		
				0.267*		5.460	0.342	6
				(1.442)		(0.648)		
					84.099	4.796	0.185	6
					(0.951)	(0.379)		
Female enrolment in basic education	−0.754*					101.039	0.407	6
	(−1.658)					(4.703)		
		−1.415*				88.798	0.383	6
		(−1.575)				(5.835)		
			0.006**			37.278	0.938	6
			(7.801)			(8.831)		
				0.347*		53.103	0.272	6
				(1.223)		(4.116)		
Female enrolment in secondary education	−0.640					61.261	0.262	6
	(−1.192)					(2.412)		
		−1.213				51.042	0.251	6
		(−1.157)				(2.874)		
			0.006**			0.687	0.952	6
			(8.917)			(0.174)		
				0.434*		15.059	0.380	6
				(1.565)		(1.194)		
Female enrolment in tertiary education	−0.693**					47.517	0.731	6
	(−3.300)					(4.783)		
		−1.268**				35.727	0.652	6
		(−2.736)				(4.551)		
			0.002			7.186	0.177	6
			(0.926)			(0.678)		
				0.267*		5.460	0.342	6

				(1.442)	(0.648)		
Female literacy rate	−0.858**				110.689	0.723	6
	(−3.229)				(8.813)		
		−1.692**			98.449	0.759	6
		(−3.547)			(12.012)		
			0.005**		46.620		
			(11.888)			0.927	6
(7.137)							
Male literacy rate	−0.382*				101.200	0.304	6
	(−1.323)				(7.417)		
		−0.806*			96.591	0.366	6
		(−1.520)			(10.600)		
			0.003**		69.303	0.663	6
			(2.807)		(11.981)		
Total population literacy rate MDG 2	−0.636**				106.409	0.591	6
	(−2.404)				(8.514)		
		−1.277**			97.714	0.644	6
		(−2.689)			(11.978)		
			0.004**		57.778	0.857	6
			(4.885)		(12.808)		

Correlation is significant
*at the 0.05 level (one-tailed)
**at the 0.01 level (one-tailed)

sector in total spending on education compared to the public sector. Enrolment in private education in Sudan is low and falls below the level in some Arab Gulf countries that probably attributed to high poverty rate in Sudan (see Figs. 9.13, 9.14, 9.15, 9.16, and 9.17 below). Similar to public enrolment, private enrolment ratio increases with the increase of educational level, i.e. are higher at secondary level, followed by tertiary level and lower at primary level. It is worth noting that despite the tremendous spread of private education institutions and despite great regional disparity in private enrolment in basic, secondary and tertiary education, private primary, secondary and tertiary enrolment ratios have not shown a large increasing trend over time in Sudan (see Table 9.10 and Figs. 9.18, 9.19, and 9.20 below).[14] Somewhat surprising we observe that different from the large share and increasing trend in the supply side as measured by the share of private sector in total number of higher education institutions, by contrast, the demand side as measured by the share of student enrolment in private higher education shows opposite declining trend and small share in Sudan over the period 1993/94–2007/08 (see Fig. 9.20 below).

Furthermore, we find evidence of the low commitment to the standardised international equity criterion in the demand side that appear in terms of the gender differences in educational attainment as measured by the gross enrolment ratio of female for primary, secondary and tertiary education. For instance, Fig. 9.21 and Tables 9.9 and 9.12 below illustrate that the percentages of female students for all levels of education in Sudan are lower than male students and both are low compared to the average for the Arab countries. For Sudan, female enrolment in secondary education is better than primary education, which implies that the gender

[14] The scarcity of reliable information limited our analysis from discussing two interesting issues related to educational policies: the contribution of private sector in both spending and enrolment in tertiary education; and the enrolment of the citizens from Sudan in overseas educational institutions. Due to the long-term low GDP per capital and high poverty rate a few number of citizens from Sudan have been able to seek higher education abroad; this is apparent from the very limited information documented from a few sources such as the UNESCO and UNDP reports. For instance, according to UNESCO-UIS (2007–2010), 'Global education digest 2007–2010: comparing education statistics around the world', for Sudan, figures on international flows of mobile students (2005–2008) indicate that the gross outbound enrolment ratio is stagnant and estimated at about 0.1, whereas students from Sudan studying abroad (outbound mobile students) was about 3,197, 3,153, 2,793 and 2,792 in 2005, 2006, 2007 and 2008 respectively. The report indicates that the top five destinations (host countries) for outbound mobile students (the number of students from given country studying in the host countries is shown in brackets) are Malaysia (507), Germany (460), UK (339), Saudi Arabia (334) and Australia (313) in 2005; Malaysia (572), USA (320), Saudi Arabia (313), Australia (313) and UK (302) in 2006; Malaysia (479), UK (343), USA (328), Saudi Arabia (313) and Germany (247) in 2007; Malaysia (634), UK (347), USA (224), Germany (209) and Qatar (137) in 2008. See UNESCO-UIS (2007–2010), 'Global education digests 2007–2010'. Moreover, the only available information indicates that during the period 2000/2001–2006/2007, the number of students from Sudan who study in the US declined by 12 % from 366 in 2000–2001 to 321 in 2006–2007. See 'Arab Knowledge Report' (2009), p. 300. This result is consistent with the results from the Arab Gulf countries which indicate that during the period 1999–2002/2003 students from Saudi Arabia, Qatar and Oman who study in the United States declined by 31 %, 26 % and 25 % respectively. See the UNDP-AHDR (2003), Table 9.1, p. 23. This may substantiate the need for improving domestic higher educational institutions to fill the gap and absorb the students who have returned.

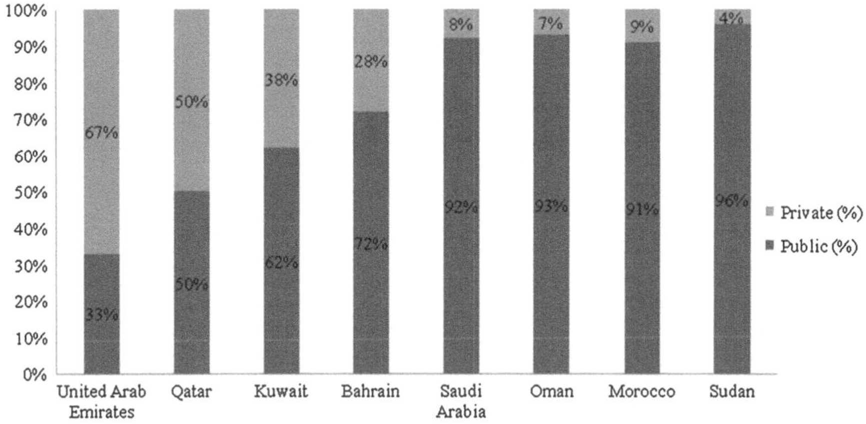

Fig. 9.13 Distribution and share of students' enrolment in public and private basic education in Sudan as compared to Arab and Gulf countries (%) (2008–2009) (Source: UNESCO Global Background information on Education Statistics: UNESCO- UIS Data Centre: Beyond 20/20 WDS (2011))

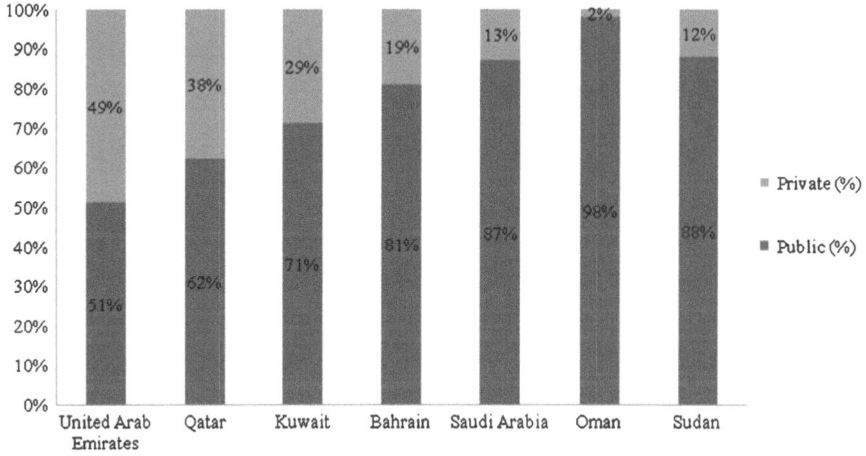

Fig. 9.14 Distribution and share of students' enrolment in public and private secondary education in Sudan and Gulf countries (%) (2008–2009) (Source: UNESCO Global Background information on Education Statistics: UNESCO- UIS Data Centre: Beyond 20/20 WDS (2011))

gap in primary education is higher than in secondary education. We observe the differences in the regional distribution in the incidence of the gender gap across the main regions in Sudan, which probably implies that the presence of gender disparity in primary education, as the gross intake and enrolment rate for females falls below the gross intake and enrolment rate of males in all regions. For secondary and tertiary education, gender disparity exists for the most poor and rural regions as the gross intake and enrolment rate for females falls behind the enrolment rate of males, by contrast for all Sudan and for relatively more urbanised and less poor regions,

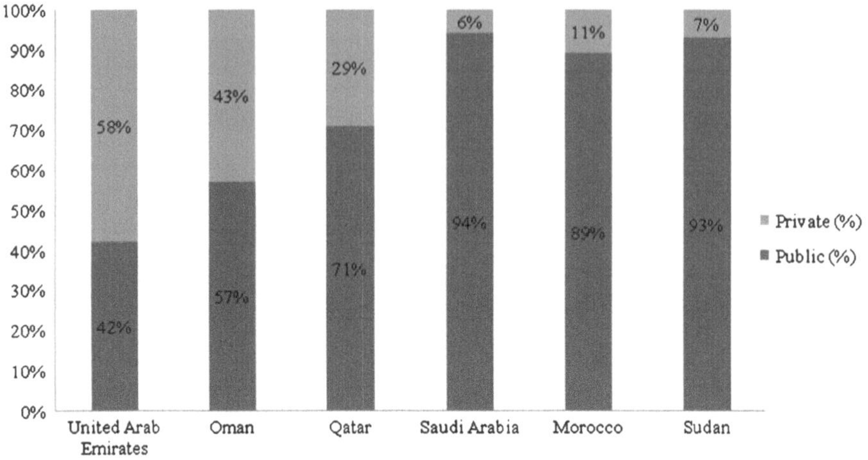

Fig. 9.15 Distribution and share of students' enrolment in public and private tertiary education in Sudan and Arab countries (%) (2007–2009) (Source: UNESCO Global Background information on Education Statistics: UNESCO- UIS Data Centre: Beyond 20/20 WDS (2011))

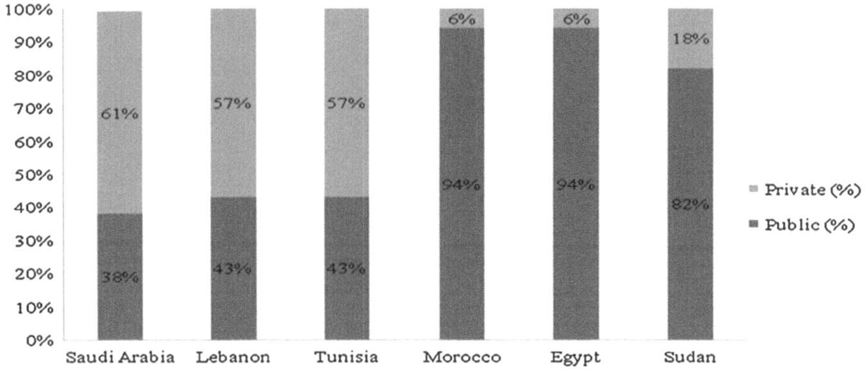

Fig. 9.16 Distribution and share of students' enrolment in public and private technical education in Sudan and Arab countries (%) (2008–2009) (Source: UNESCO Global Background information on Education Statistics: UNESCO- UIS Data Centre: Beyond 20/20 WDS (2011))

the gross intake and the percentage of enrolment rate for females is greater than the enrolment rate of males (see Tables 9.12 and 9.13 below). This implies that gender disparity and gap is more critical for more poor and rural regions and population groups, i.e. poor females and females living in rural areas are facing a serious situation of inequality and are suffering more in terms of net attendance or access to primary and secondary education in Sudan. Somewhat surprisingly, the gender disparity is more serious in primary education compared to secondary education, especially for poor females; this is consistent with the findings based on the data

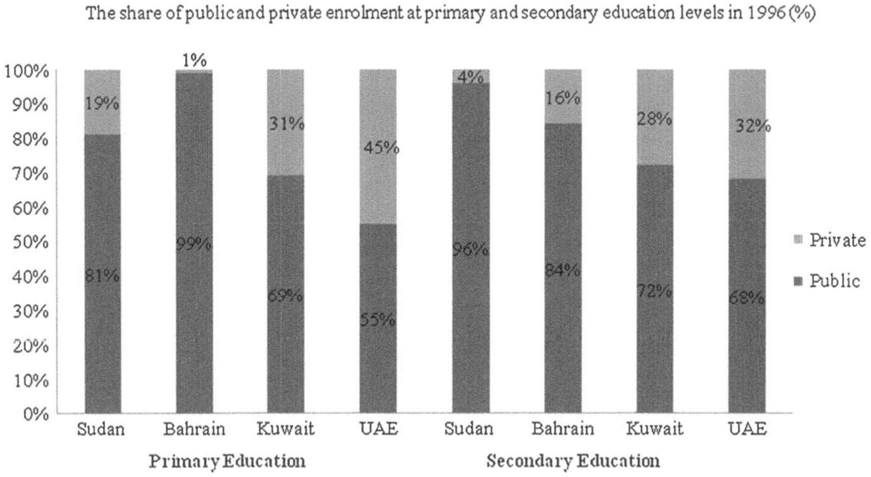

Fig. 9.17 Distribution and share of students' enrolment in public and private basic and secondary education in Sudan and Gulf countries (%) (1996) (Source: UNESCO Global Background information on Education Statistics: UNESCO- UIS Data Centre: Beyond 20/20 WDS (2011))

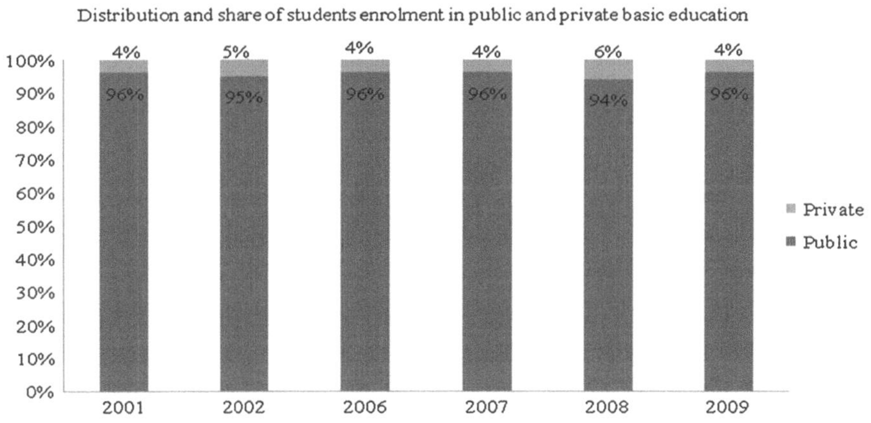

Fig. 9.18 Distribution and share of students' enrolment in public and private basic education in Sudan (2001–2009) (Source: Own calculation based on Table 9.10)

from UNESCO (2006), which we presented in Nour (2011). This implies that, especially amongst the poor, economic reasons were considered to be the most important factor limiting girls' potential to complete their primary (basic) and secondary school education and that the factors preventing males from completing their education differ from those hampering females. It is clear from Tables 9.9, 9.12, and 9.13 below that family economic problems impact more negatively on female than on male education. Likewise, families perceive educating girls to be

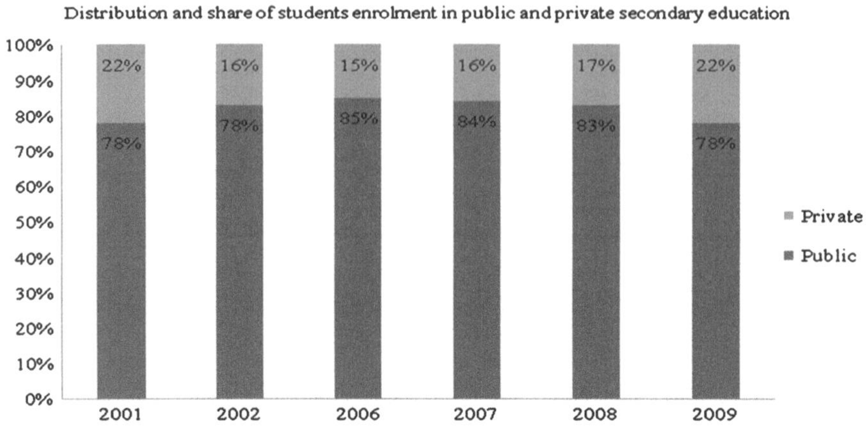

Fig. 9.19 Distribution and share of students' enrolment in public and private secondary education in Sudan (2001–2009) (Source: Own calculation based on Table 9.10)

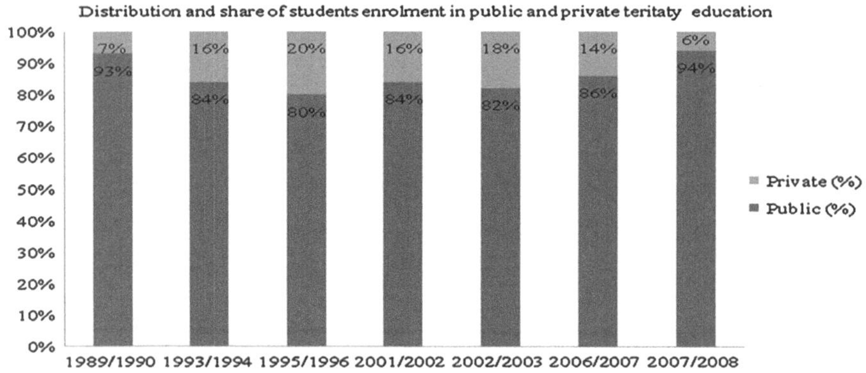

Fig. 9.20 Distribution and share of students' enrolment in public and private tertiary education in Sudan (1989/1990-2007/2008) (Source: Own calculation based on Table 9.10)

less important than schooling boys. It is the need to work that has the largest effect on the withdrawal of boys from school. It is worth noting that despite the tremendous spread of female education in the last five decades, Sudanese women remain poorly prepared to participate effectively and fruitfully in public life by acquiring knowledge through education. This is most clearly manifested in the extent to which girls and women are still deprived of education and knowledge, especially those forms of knowledge that bring high social returns. Sudan has one of the world's lowest rates of female enrolment opportunities at all levels of education, especially higher education. Female access to all levels of education remaining below that of males implies further evidence on the incidence of relatively higher deprivation of girls in terms of educational opportunities at all levels in Sudan. Sudan has one of the highest rates of illiteracy approximately near to one half of

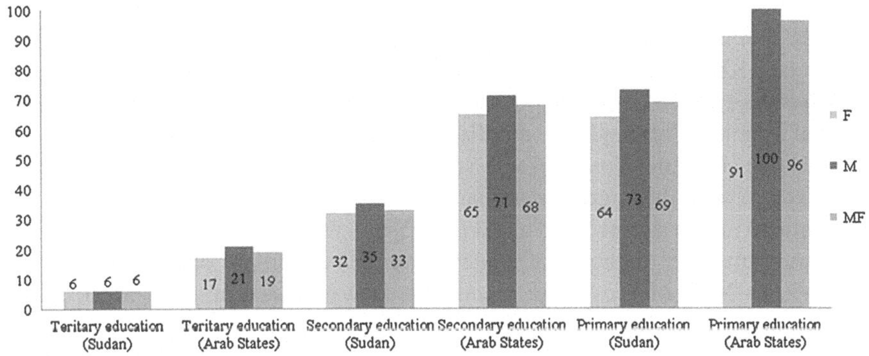

Fig. 9.21 Enrolment ratios in basic, secondary and tertiary education in Sudan compared to the average for the Arab countries defined by educational level and gender (2000) (%) (Source: Arab Knowledge Report (2009: 279))

females are illiterate compared to only one third of males.[15] Moreover, data and information from the Sudan Ministry of Education and Ministry of Higher Education provide further evidence for the incidence of gender gap in education as measured by enrolment of students in primary (basic), secondary and tertiary education (measured by students nominated and admitted for governmental, private and foreign higher education institutes) over the period (2004–07). For instance, the gender gap for basic (primary) education is higher than for secondary and tertiary education. The incidence of gender gap in tertiary education as measured by the share of female students enrolled or admitted in tertiary education in public and governmental higher education institutes is higher than that for private and foreign higher education institutes in 2005. The high share of female students compared to male students in tertiary education, can be interpreted in relation to the observation from the preliminary findings of the Sudan Central Bureau of Statistics on Sudan's fifth population census which indicate that the structure of Sudan's total population according to different age groups implies that for the age group 20–39 the total number of females is slightly higher than the total number of males. Another justification is probably because of the presence of male Sudanese studying abroad (Nour 2011: 7–10). Therefore, these findings imply low commitment to the standardised international adequacy and equity criterions in the demand side as measured by the lack of adequacy and gender equity in enrolment rate in primary, secondary and tertiary education and literacy rate of population.

In addition we observe that the low commitment to equity criterion and the incidence of wide gap and regional disparity between the main regions in Sudan is not only limited to the supply side but also holds for the demand side. For instance, Tables 9.8, 9.9, 9.10, 9.11, 9.12, and 9.13 below indicate the incidence of regional disparity that appear from the share of main regions in total number of students

[15] See for instance, AHDR (2004), pp. 73–74.

enrolment in basic, secondary and tertiary education in Sudan over the period 2001–09. For instance, we observe the large share of the Central and Darfur regions followed by Khartoum in total numbers of students enrolled in basic education, and the large share of the Central and Khartoum regions followed by the Darfur region in total numbers of students enrolled in secondary education as compared to Kordofan, Eastern, Northern and Southern regions over the period 2001–09. Moreover, we note the large share of the Central and Khartoum regions followed by the Northern region as compared to Darfur, Eastern, Kordofan and Southern regions in terms of the total numbers of students enrolled in tertiary education over the period 1996–2007. Moreover, we observe the heavy concentration of privatisation in Khartoum region that has the highest share in terms of total number of student enrolment in private basic and secondary schools as compared to other regions. This low commitment to equity criterion and the incidence of unbalanced distribution in the demand side as measured by the distribution and share of student enrolment in total and in public and private basic, secondary and tertiary education levels can be perceived as an implication that is consistent with the share of these regions in total number and public and private schools in basic and secondary education as we explained above in this section. This probably implies that the low commitment to equity criterion and the incidence of unbalanced development planning and unbalanced distribution in the supply side and investment as measured by the number of schools, number of teachers and the pupil/teacher ratios for basic and secondary educational levels that probably led to further implications in the demand side as measured by the distribution and share in total number of student enrolment in public and private schools and in basic, secondary and tertiary educational levels as we explained above in this section. Moreover, using the ordinary least squares method and E-VIEWS, Tables 9.8, 9.9, 9.10, 9.11, 9.12, and 9.13 below explain that it is probably plausible to interpret the observed regional disparity in the share in demand and enrolment in education due to demographic reasons (measured by the share in total population), economic reasons (measured by per capita income and poverty rate) and other reasons (measured by the degree of urbanisation) across the main regions in Sudan. Starting with the demographic reasons, Table 9.13 indicates significant positive correlation between the share in enrolment in basic and secondary education and the share in total population and positive correlation between enrolment in tertiary education and the share in total population. These results can be used to argue that the share in total population seems to be the first important factor determining the share and regional disparity in enrolment in education. Moreover, as for the economic reasons, Table 9.13 indicates significant positive correlation between per capita income and total, female and male literacy rates, between per capita income and enrolment in secondary and tertiary education and female enrolment in tertiary education, and also significant positive correlation between per capita income and enrolment in basic education and female enrolment in basic and secondary education. In addition, Table 9.13 indicates significant negative correlation between poverty rate and total, female and male literacy rates, significant negative correlation between poverty rate and total and female enrolment in basic and tertiary education and negative correlation between poverty rate and total and female enrolment in secondary education. These results can be used to argue that the economic reasons as measured by per capita income and poverty rate seem to be the second important factor that determining the share and regional disparity in enrolment and demand for education. Notably, our results imply that the incidence of high poverty rate seems to be the most important factor determining or limiting the demand and enrolment, notably, in basic education. These findings imply that especially among the poor regions, economic reasons were considered to be the most important factor limiting poor students and especially, girls' potential to complete their primary (basic), secondary and tertiary education and that region economic problems impact more negatively on female than on male education. These results imply that the increase in the incidence of poverty and the low per capita income limited or led to low demand and enrolment in education, notably, across the poor regions and this probably explains the regional disparity in the demand for education across the main regions in Sudan. In addition, as for the other reasons, Table 9.13 indicates positive correlation between enrolment in secondary education and degree of urbanisation, significant positive correlation between enrolment in basic and tertiary education and degree of urbanisation and between female enrolment in basic, secondary and tertiary education and the degree of urbanisation. These findings can be used to argue that the degree of urbanisation is the third and other factor determining the share and regional disparity in enrolment in education. The major policy implication from our findings is that Sudan has the potential to achieve equity and fulfil the second and third United Nations Millennium Development Goals (UN MDG) on universal access to primary education and gender equality respectively through reduction and elimination of poverty, notably, across the poor regions and population in Sudan, and this implies achievement of equity and international commitment to fulfilment of UN MDG in Sudan.

Moreover, we observe the low commitment to equity criterion and the incidence of regional disparities in terms of technical and vocational education. For instance, the Central and Khartoum regions followed by Eastern region as compared to the Northern, Darfur, Kordofan and Southern regions respectively show the large share in total number of schools, students, total number of teachers and trained teachers in technical and vocational education in Sudan over the period 2001–09. In addition, Khartoum and the Central region followed by the Eastern region as compared to Kordofan, Darfur and Southern regions respectively show the large share in total number of trained and graduated students of the apprenticeship programmes and vocational training in Sudan over the period 1995–2006. This probably implies the low commitment to equity criterion and the incidence of regional disparity and unbalanced distribution of trained human resources available for technical and vocational education. That may not be surprising in view of the fact that these regions also reported the large share in total numbers of basic, secondary and tertiary education over the period 2001–09 as we explained above in this section.

One major problem of the educational system in Sudan and the Gulf countries is the recent serious deterioration in the quality of tertiary education; for instance, after considerable improvement and increase in enrolment in tertiary education in

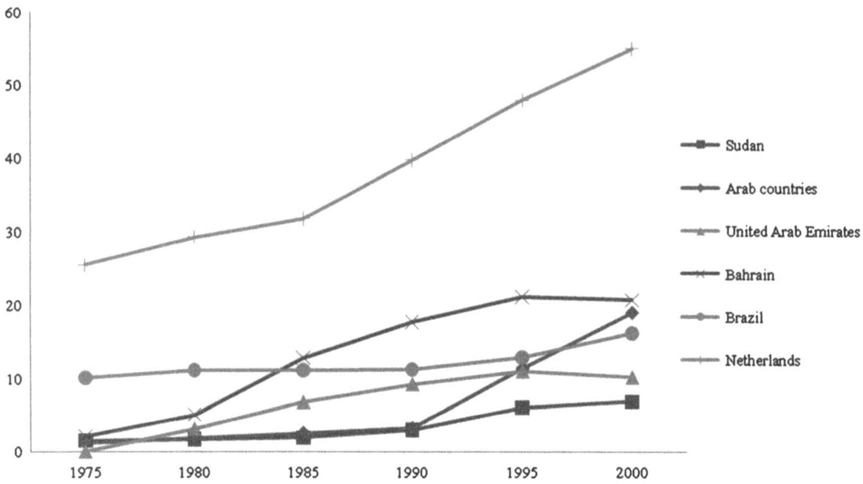

Fig. 9.22 Gross enrolment in tertiary education in the Sudan, UAE, Bahrain, Brazil and the Netherlands (1975–2000) %) (Source: WDI (2004) Database)

Sudan until around 1995 – after the higher education revolution in 1990s – while the gross enrolment figures have increased, the net enrolment ratio of the number of tertiary students of official tertiary education age who are enrolled in tertiary education to the total population of students remains low and below the international standard and also the quality has deteriorated in recent years (see Fig. 9.22 below). Therefore, this implies an ample role for policymaking to improve the quality and enrolment in tertiary education. In addition, as in most other developing countries, one serious problematic feature concerning tertiary education in Sudan and the Gulf is that enrolment and graduation ratios in tertiary education are biased against scientific, technical, engineering, agriculture, medical and natural sciences and are focused on art, humanities, law and social sciences. For instance, in the period 2009–10, enrolment and graduation ratios in medical sciences, natural sciences, engineering and agriculture accounted for only 30 % as compared to 70 % for art, humanities, law and social sciences; these biases remained for enrolment and graduation rates in the period 1994–2009 (see Table 9.7 above).[16]

[16] "The irrelevant wrong policy for admission in higher education leads to focus on humanities and social science and biases against science, technology and engineering studies leads to mismatch, unemployment, shortage of technical skills that leads to dependence on foreign technical skills". See Al-Sulayti (2002), pp. 16–18; Jalal al-Din (2002), pp. 7–10, 15, 23–24; Almagzoub (2007/08); Al-Tuhami (2008), pp. 2–12; Mohammed (2008), pp. 12, 2–20; Ismail (2008), pp. 2–19. "The low share of enrolment in technical education relative to total enrolment in tertiary education is probably attributed to both social/cultural aspect in the society that discourage the involvement in technical work and the weak relationship between educational policies and development planning". See Alfakhery (1999), p. 82; Jalal al-Din (2002), pp. 7, 20; Almagzoub (2007/08); Al-Tuhami (2008), pp. 2–12; Mohammed (2008), pp. 12, 2–20; Ismail (2008), pp. 2–19.

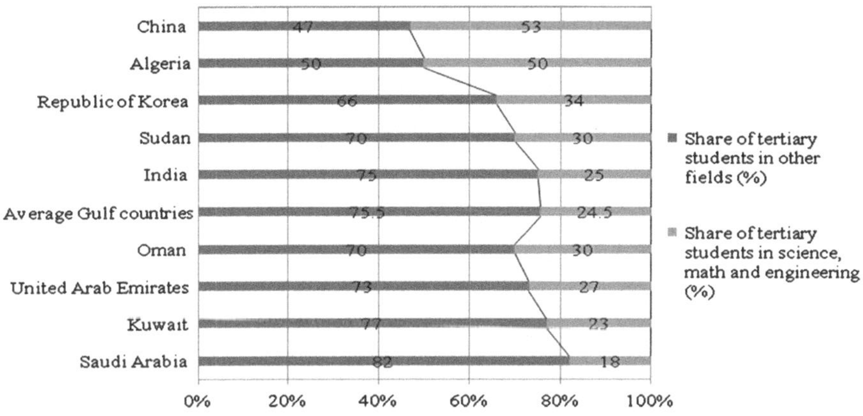

Fig. 9.23 Relative distribution of tertiary education students (%) by fields in the Sudan compared to the Gulf countries, Algeria, India, China and Korea (1994/1997- 2009/2010) (Source: UNDP (2004), UNDP – AHDR (2003) and UNESCO-UIS (2003): UNESCO web site (www.unesco.org))

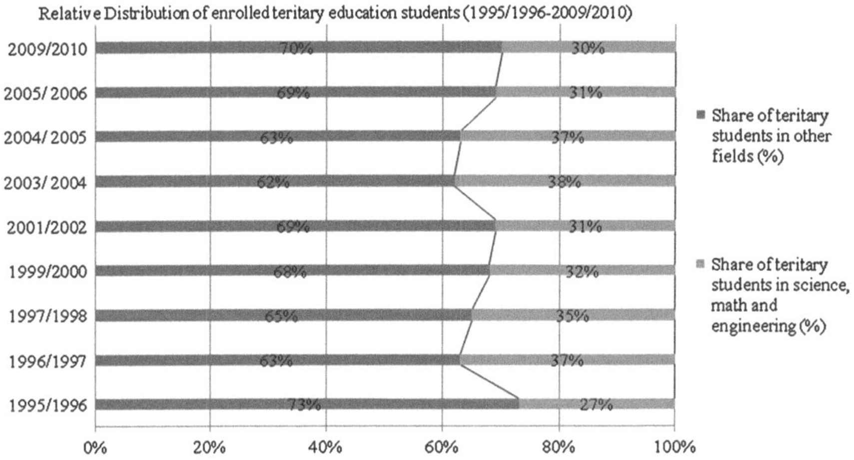

Fig. 9.24 Relative distribution of enrolled tertiary education students (%) by fields in the Sudan (1996/1997- 2009/2010) (Source: Own calculation based on Sudan Ministry of Higher Education and Scientific Research, the Annual Educational Statistics Reports Various Issues (1993/1994–2008/2009))

The share of tertiary students enrolled in sciences, math and engineering in Sudan and the Gulf is low compared to Korea (34 %), Algeria (50 %) and China (53 %) (see for instance, Figs. 9.23, 9.24, and 9.25 below).

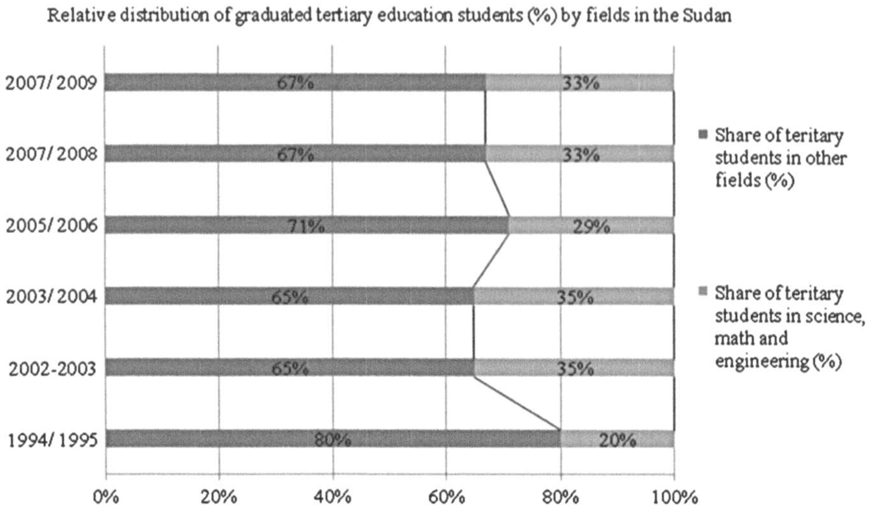

Fig. 9.25 Relative distribution of graduated tertiary education students (%) by fields in the Sudan (1994/1995–2007/2009) (Source: Own calculation based on Sudan Ministry of Higher Education and Scientific Research, the Annual Educational Statistics Reports Various Issues (1993/1994–2008/2009))

9.2.4 Return, Efficiency and Quality of Educational Policies

Another common characteristic of the educational system in Sudan and the Gulf countries is the low commitment to the standardised international efficiency criterion and the weak internal efficiency/quality of primary, secondary and tertiary education; the severity of the problem varies across the main regions in Sudan.[17] For instance, Table 9.14 below illustrates that in the period 2006–09, the percentage of repeaters in primary schooling was high in Sudan and in Saudi Arabia, while the percentages of transition to secondary and tertiary education were low in Sudan and Saudi Arabia.[18]

The UNESCO indicators on the quality of education implies that the average for Sudan and the Gulf countries in terms of quality of education has improved over time; the performance for Sudan is lower than the Gulf as is apparent from the considerable decline in the percentage of repeaters in primary schooling and

[17] No relevant data and information is available to allow an assessment of the quality of tertiary education.

[18] "The poor quality is attributed to: (a) high repetition rates in primary and secondary level, (b) weak absorptive capacity and performance level of students at all levels. (c) failure of educational strategy to motivate innovative skills and problem solving ability/skills, (e) low public and private spending on education, (f) high focus on quantity at the expense of quality; (g) lack of monitoring systems/ institutions to measure and assess the performance of educational and training institutions, (h) lack of trained teachers and (i) lack of materials, infrastructure and facilities". See Suleiman (2007), pp. 121–123; Jalal al-Din (2002), pp. 22–24.

Table 9.14 The quality of education in Sudan and the Gulf countries: and regional distribution and share of main regions in Sudan government basic education defined by the percentage of repetition, transition and dropouts (%) (1995–2009)

(a) Quality of education in the Sudan and Gulf countries: percentage of repetition and transition (1990–2002)

Indicator	Percentage of repeaters primary and secondary education (%):				Percentage of reaching secondary and tertiary education (%)					
	Primary education (%)			Secondary education (%)	Secondary education (%)		Tertiary education (%)		Primary to secondary transition rates (%)	
Country/y ear	1999 [a]	2002[b]	2008/2009[c, d]	2008[d]	1995[a]	2008[d]	1995[a]	2008[d]	1998–2002[b]	2008[d]
Sudan	11	5	4	3	93	94	74–86	93	94	94
Bahrain	4	4	2	5	99		95		98	96
Kuwait	3	3	1	8	97	100	96	100	98	97
Oman	8	4	1	2	99	99	96	100	98	97
Qatar	3		1	3	100	97	98	97	96	99
Saudi Arabia			3[d]–4[c]	4	96	97	89	96	97	94
UAE	3	3	2	5	93	100	83	100	98	98
Average Gulf	4	4	2	5	97	99	93	99	98	97

(b) Repeaters (%)[e]	2006			2007			2008			2009		
	M	F	MF	M	F	MF	M	F	MF	M	F	MF
All Sudan	3.2 %	2.9 %	2.7 %	6.6 %	6.4 %	6.5 %	5.4 %	5.2 %	5.3 %	4.4 %	4.2 %	4.3 %
Northern	7 %	6 %	3 %	10 %	8 %	9 %	8 %	7 %	8 %	12 %	8 %	10 %
Khartoum	1.30 %	1.1 %	1.2 %	0.9 %	0.7 %	0.8 %	0.9 %	0.6 %	0.7 %	0.9 %	0.7 %	0.8 %
Central	3 %	2 %	2 %	4 %	4 %	4 %	4 %	4 %	4 %	4 %	4 %	4 %
Kordofan	9 %	8 %	5 %	10 %	10 %	10 %	10 %	10 %	10 %	6 %	7 %	6 %
Darfur	0 %	0 %	0 %	9 %	10 %	9 %	6 %	6 %	6 %	3 %	3 %	3 %
Eastern	5 %	4 %	2 %	12 %	14 %	13 %	10 %	11 %	10 %	8 %	8 %	8 %

(continued)

Table 9.14 (continued)

(c) Dropouts (%)[e]	2006			2007			2008			2009		
	M	F	MF	M	F	MF	M	F	MF	M	F	MF
All Sudan	1.6 %	1.4 %	1.4 %	3.4 %	3.2 %	3.3 %	2.7 %	2.5 %	2.6 %	1.4 %	1.2 %	1.3 %
Northern	1 %	1 %	1 %	2 %	1 %	1 %	1 %	1 %	1 %	1 %	1 %	1 %
Khartoum	1.2 %	1.2 %	1.2 %	0.3 %	0.5 %	0.4 %	0.3 %	0.4 %	0.4 %	0.3 %	0.4 %	0.4 %
Central	2 %	1 %	1 %	2 %	3 %	3 %	2 %	2 %	2 %	1 %	1 %	1 %
Kordofan	3 %	2 %	2 %	4 %	3 %	4 %	3 %	3 %	3 %	2 %	2 %	2 %
Darfur	0 %	0 %	0 %	6 %	6 %	6 %	4 %	4 %	4 %	2 %	2 %	2 %
Eastern	1 %	1 %	1 %	8 %	8 %	8 %	7 %	7 %	7 %	2 %	2 %	2 %

[a]UNESCO–UIS (2000) UNESCO's World Education Report (2000)
[b]UNESCO –UIS (2004c) country profile: statistics refer to the most recent year between 1998 and 2002
[c]UNESCO Global Background information on Education Statistics: UNESCO- UIS Data Centre: Beyond 20/20 WDS (2011)
[d]UNESCO-UIS (2010) "Global education digest 2010 Comparing education statistics around the world"
[e]Own calculation based on Sudan Ministry of Education, the Annual Educational Statistics Reports Various Issues (2001–2009): (2005–2006: 39–42), (2006–2007: 38–41), (2007–2008: 40–63), (2008–2009: 62–65)

increase in the percentages of transition from primary to secondary education, however, across the main regions in Sudan poor quality is still obvious. For instance, Table 9.14 indicates that throughout the period 2006–09 the percentages of repeaters in primary schooling increased from 2.7 % in 2006 to 6.5 % in 2007 and 5.3 % in 2008 but declined to 4.3 % in 2009. Moreover, Table 9.14 indicates that throughout the period 2006–09 the percentages of dropouts from education in primary schooling increased from 1.4 % in 2006 to 3.3 % in 2007 and 2.6 % in 2008 but declined to 1.3 % in 2009. Therefore, further efforts are needed to enhance the quality of education at all levels, in order to avoid the exacerbation of the problems that will result in the event of a failure to implement some effective policies to improve the quality of education. Moreover, we observe the improvement in the quality as measured by the decline in the percentage of repeaters and dropout despite the decline in the percentage of success in the basic education. We observe the low commitment to the standardised international efficiency and equity criterions and the incidence of considerable regional disparity between the main regions in Sudan. For instance, the efficiency and quality of primary and secondary education, as measured by low percentage of repeaters and dropout in basic and secondary education, are relatively better in Khartoum region followed by the central regions as compared to the Darfur, Kordofan, Eastern and Northern regions respectively. In addition this regional disparity in the quality of education also holds for the percentages of success in the basic education which are reported high in the Northern region followed by the Central, Kordofan, Eastern, Khartoum and Darfur regions respectively. This regional disparity in the quality of education is not surprising in view of the fact that these regions also reported the large share in terms of quantitative supply and demand in basic and secondary education over the period 2001–09 as we explained above in this section. Furthermore, we find evidence of the gender differences that appear in terms of the efficiency and quality of education across the main regions in Sudan as measured by the percentage of repeaters and dropout from basic education. For instance, Table 9.14 indicates that over the period 2006–09 the percentages of repeaters and dropout in primary education for female students are less than those of male students for the majority of the main regions in Sudan. It is the need to work that probably has the largest effect on the withdrawal of boys from school.

In addition, the poor quality of education can be observed from the results of the percentage of success in Sudan's basic education certificate (2000/01–2008/09) and percentage of success in Sudan's secondary school certificate (1996/97–2007/08). For instance, Table 9.15 indicates that throughout the period 2000/01–2008/09 the percentage of success in Sudan's basic education certificate declined from 71.9 % in 2001 to 71.4 % in 2006 and then increased to 73.7 %, and 76.4 % in 2007 and 2008 respectively but declined to 74.9 % in 2009. Moreover, Table 9.15 indicates that over the period 1996/97–2007/08 the percentage of success in Sudan's secondary school certificate increased from 69.4 % in 1996/97 to 73.5 %, 75.4 % and 75.9 % in 2000/01, 2001/02 and 2002/03 respectively but then declined to 72.1 %, 71.4 %, 73.7 %, 74.1 % and 73.7 % in 2003/04, 2004/05, 2005/06, 2006/07 and 2007/08 respectively. Furthermore, the reported annual success in Sudan academic secondary education is higher than Sudan's technical secondary education; the

Table 9.15 The percentage of success in Sudan basic education certificate defined by gender and main regions and Sudan secondary school certificate defined by secondary education type in Sudan (1996/1997–2009)

(a) Sudan basic education certificate basic results percentage of successors (%) by sex and state (2000/2001–2008/2009)

	2001			2006	2007			2008			2009		
	M	F	MF	MF	M	F	MF	M	F	MF	M	F	MF
All Sudan	73.1	72.9	71.9	71.4	72.7	74.5	73.7	72.9	80.4	76.4	73	73.5	74.9
Northern	76	78	66	93	86	92	89	92	96	94	92	97	95
Khartoum	68.5	74.2	71.4	65.3	65.3	59.4	62.8	75.2	82.8	79	63.8	71.1	67.3
Central	76	72	74	76	74	78	76	75	76	75	78	78	78
Kordofan	67	62	64	73	70	70	70	74	73	74	74	74	74
Darfur	75	67	72	60	73	69	72	67	59	64	67	60	64
Eastern	72	64	68	62	66	72	68	63	69	65	68	74	71
Southern	75	80	77										

(b) Sudan secondary school certificate percentage of successors according to education type (1996/1997–2007/2008) (%)

	1996/1997	1997/1998	1998/1999	1999/2000	2000/2001	2001/2002	2002/2003	2003/2004	2004/2005	2005/2006	2006/2007	2007/2008
Government academic secondary education	69.4	70.8	62.1	63.4	73.5	75.4	75.9	72.1	71.4	73.7	74.1	73.7
Non-government academic secondary education	66.0	67.4	73.1	76.3	79.0	78.8	70.9	73.6	74.1	72.0	71.9	73.5
Teacher union	59.6	48.3	60.6	60.0	66.6	66.7	60.6	65.3	65.8	63.5	64.2	65.3
Irregular	75.2	78.9	79.8	78.5	84.1	78.7	66.8	77.5	74.5	74.3	71.3	75.0
All academic education	66.9	66.0	65.1	66.5	74.3	74.1	70.1	71.3	70.8	70.7	70.4	71.3
Technical commercial secondary education	47.9	35.8	52.4	46.0	57.0	52.5	52.4	53.5	52.1	49.4	48.4	49.0
Technical industrial education	55.8	44.1	40.5	37.0	52.0	51.4	44.5	43.8	41.8	43.5	46.0	45.7

Technical agriculture education	37.3	29.2	44.9	26.7	47.0	44.8	46.5	45.8	49.3	52.7	54.2	49.7
Technical home education	35.0	41.3	40.3	23.4	48.7	45.7	42.3	39.8	59.8	62.4	41.0	43.7
All technical education	48.3	37.9	48.9	42.2	55.0	51.8	48.7	48.6	47.3	47.0	47.1	47.2
Vocational training				87.4	92.3	88.5	89.3	85.8			76.6	77.7
Ahlia		67.4	57.8	84.8	85.0	91.6	82.0	79.7	78.3	86.5	84.4	83.2

Source: Own calculation based on Sudan Ministry of Education, the Annual Educational Statistics Reports Various Issues (2001–2009): (a) (2000–2001: 22), (2001–2002: 32), (2003–2004: 26), (2004–2005: 24), (2005–2006: 23), (2006–2007: 23), (2007–2008: 25), (2008–2009: 45), and (b) (2000–2001: 24), (2008–2009: 48).

technical education not only showed a very poor but also a continuous declining annual rate of success over the period 1996/97–2007/08 (see Table 9.15 below). Al-tuhami (2007) finds many reasons for the poor quality problem of technical education in Sudan. For example, the lack of clear vision regarding technical education, lack of central body for organisation, planning and development, the transfer of technical education schools to the States in 1993, weak relationship between technical educational policies and development planning and mismatch between specialisations and content of technical education courses and requirements in the labour market. In addition: the lack of financial resources, facilities and infrastructure (such as buildings, laboratories, workshops, halls, books and references), the critical shortage of qualified and trained Sudanese teachers and trainers specialised in technical education in accordance to the requirements of the labour market and the lack of favourable educational environments and facilities for supporting, hosting and accommodating students in technical education. Further to the social/cultural aspects, due to high preference for enrolment in academic education and low preference for enrolment in technical education, there is a lack of appreciation in the society and lack of attractive working conditions for graduates of technical education that discourage involvement in technical education and work. In addition to: deficiency of the current structure, which implies the sudden move of students from basic school to technical education, lack of a specialised institutional body for preparing technical secondary education courses, lack of linkages between technical education in secondary and tertiary levels, easy transfer from technical to academic education, and from industrial apprenticeships to technological colleges, low rates of enrolment and graduation and low opportunities for postgraduate studies in the fields of technical colleges and tertiary education (Al-tuhami 2007: 2–12).[19]

Therefore, our findings presented above imply that the low commitment to the standardised international efficiency criterion and poor quality and efficiency is obvious not only from the supply side as measured by low rates of trained teachers and over-crowded classrooms as indicated by students enrolment per education institutions as we explained above (see Tables 9.2, 9.6, and 9.11 and Figs. 9.7, 9.8, 9.9, 9.11, and 9.12 above), but also holds from the demand side as measured by low rates of enrolment and attendance, high rates of dropout, high rates of repetition and weak rates of success in the final exams in the basic and secondary, mainly, technical education.[20]

Apart from the problem with regard to enrolment ratios, as with most developing countries, the quality of education in Sudan and the Arab region poses a serious problem. Our findings presented above concerning the poor quality and low

[19] See for instance, Altuhami (2008), pp. 2–12; Mohammed (2008), pp. 2–20; Ismail (2008), pp. 2–19; and Almagzoub (2007/08).

[20] Suleiman (2007) indicates the most important three standard criteria for assessing education in Sudan, notably, the adequacy, equity and efficiency criterion. Suleiman (2007) discusses the equity standard criterion and notes that the imposition of tuition fees does not help to achieve this equity standard criterion and to reduce any significant differences between high earners and low income earners in Sudan. See Suleiman (2007), p. 122.

commitment to the standardised international efficiency and quality criterion are consistent with the results of Sudanese and Arab literature, notably, Suleiman (2007) and Jalal al-Din (2002) are very useful sources to qualify them; we shall here take up part of their description. For instance, Suleiman (2007) argues that there is a common complaint in Sudan about the poor quality of education in all its stages: primary, secondary and university education. Suleiman (2007) indicates the difficulty to identify or measure the poor quality of education, but he observes that the failure to achieve an acceptable standard of educational efficiency criterion takes the form of high rates of absence of pupils from attending school, high rates of repetition and failure in the final examinations, high numbers of unsuccessful educational institutions, poor training for teachers, overcrowded classrooms with students, shortages of teaching equipment and materials needed for the teaching of some courses, absence of textbooks even for undergraduates, as well as the absence of libraries, all of which adversely affect basic, secondary and tertiary education (Suleiman 2007: 122–123). Moreover, Jalal al-Din (2002) notes that all indicators and reports show that the internal efficiency of public education is extremely low and means that the overall system of general education is characterised by waste, high rates of repetition and dropout, and low enrolment ratios (Jalal al-Din 2002: 23–24). Suleiman (2007) finds that the explanation of poor quality of education is related to several reasons, the first reason is the low public spending on education as a proportion of total public spending, and this low ratio has even declined from 8.4 % in 1997/98 to 6.9 % in 2001/02.[21] This ratio does not include public expenditure on above secondary (higher) education and it could drop more and more when adding to public expenditure on defence and security, especially given that public spending in these areas has increased greatly in Sudan because of its circumstances. Also, this proportion may decline significantly if they attribute it to the national income, which is a standard method that is often used for measuring the share of education of a country's resources and to compare between different countries in terms of allocation of appropriate resources for the provision of adequate level of education in terms of both quality and quantity. Suleiman (2007) indicates that another reason for the low quality of education in Sudan is that despite the expansion in public and private education the adequacy standard criterion (as measured by the ratios of spending on education and enrolment in education) is not only still very low, but also still focuses on quantity over quality in Sudan (Suleiman 2007: 122–121). Jalal al-Din (2002) confirms that the commitment to high quality and internal and external efficiency of education so as to respond to the economic and social needs in Sudan, does not imply that it will be rational to stop expansion of enrolment in higher education; on the contrary it implies the need to stop the policy of continued random quantitative expansion in enrolment in higher education that caused a serious negligence of improving qualitative efficiency in higher education in light of the limited material, human and financial resources (Jalal al-Din 2002: 5). Jalal al-Din (2002), argues that the

[21] See 'Educational Statistics for the Academic Year' (2000/02), p. 10.

financial difficulties faced by some Arab governments led to an unacceptable reduction in material resources and financial resources allocated to higher education institutions. He indicates that it seems public and political pressure forced many governments to swap the quality of education by quantitative expansion; while this a swap might seem politically acceptable in the short term, it will have serious repercussions on educational trends and economic and even political directions in the medium- and long-terms. He argues that in Sudan and some Arab countries, the continuous random unregulated expansion, establishment and opening of more universities and colleges and increases in student enrolment rates in the existing universities without allocation and provision of the required material and financial resources, sufficient numbers of qualified faculty members, sufficient facilities for libraries, books, laboratories, materials and equipment for laboratories has led to serious deterioration in higher education institutions. Due to lack of public funding for universities some universities and colleges were even opened in existing old buildings without making the sufficient infrastructural changes, buildings that were not originally suitable for academic higher education institutions. Libraries, which earlier regularly subscribed to important periodicals and journals no longer do so, so they fail to acquire basic references due to lack of funding in both local and foreign currency, which has led to further deterioration in higher education institutions (Jalal al-Din 2002: 22). Jalal al-Din (2002) explains that the deterioration in the quality of higher education in the Arab countries is related to the fact that the majority of plans which have been prepared in many Arab countries in the fields of labour and employment during the past four decades, in particular the last three decades, implies that on the one hand Arab governments were more interested in hiring the holders of diplomas and university degrees more than hiring the holders of lower certificates. This policy has led to wider and greater demand for higher education in various forms and has also led governments to respond to this social demand through the increasingly quantitative expansion of higher education, regardless of the needs of the economy. On the other hand, universities and other higher education institutions seem to have responded to these directives, convictions and plans through exaggeration in the excessive branching and introduction of narrow disciplines (that are increasingly narrowed year after year) in a rigid system in order to respond to the specific needs of professionals and specialists and it seems that this is the most important dilemma in which universities and higher education institutions in the Arab and in other countries are involved.

Moreover, Jalal al-Din (2002) notes that another problem is related to the lack of favourable environments, which exist in Western universities and communities, to fill the gap in higher education curricula, that does not exist in any reasonable amount in any of the Arab countries (Jalal al-Din 2002: 14). Jalal al-Din (2002) attributes the weakness of the general public's education to the seemingly greater focus on lower levels (recognition and understanding) rather than higher levels (analysis and reflection and systematic application); less attention is paid to refinement of imagination, independent thinking and interest in knowledge and research and less emphasis is also placed on modern sciences, including mathematics and basic science (Jalal al-Din 2002: 23–24). Jalal al-Din (2002) discusses the implications of rapid quantitative expansion on education experienced by all Arab countries, he indicates that the first implication is that it has led to qualitative deficit and rapid degradation and deterioration in the quality of higher education because it implies that higher education has become merely an extension of general public education in the Arab countries. He finds that another serious implication is that this situation has led to the graduation of students who are less familiar with and have limited ability and knowledge of their disciplines even closely related to their narrow fields of specialisation; it seems that this situation has been taking root and growing in many Arab countries and it has been producing huge surpluses of graduates in specific areas who are not playing a great role in economic and social development due to using obsolete methods of learning (Jalal al-Din 2002: 7–8). Jalal al-Din (2002) notes that further serious implication is that the trend and direction of universities and other higher education institutions do not only suffer qualitative deficit as explained above, but also do not seem compatible with the local environment and community and practical needs. That is probably because not only have education curricula in universities been taken from the curricula of Western universities, but they are also taught and examined by traditional ways (Jalal al-Din 2002: 23–24).

Suleiman (2007) notes that further reason for the poor quality of education in Sudan is that a large number of teachers are currently working in various stages of education without any training (Suleiman 2007: 117). Suleiman (2007) and Jalal al-Din (2002) explain the reasons of poor quality for tertiary education in Sudan (Suleiman 2007: 122–123). Suleiman (2007) shows the links between the policy of upgrading teaching staff and some issues related to quality of higher education and universities in Sudan. Suleiman (2007) indicates that in the past Sudanese universities implemented a policy with regard to selection, appointment, upgrading by emission of teaching assistants on the basis of academic excellence, similar to the best universities in the UK, USA or elsewhere, as appropriate for their fields of specialisation. This policy allowed the external exposition of excellent international higher education and acquisition of high quality knowledge that also contributed to the promotion of excellence in Sudanese universities. Suleiman (2007) indicates that the problem with regards to the shift from this past policy and recent focus on alternative policy of localisation of knowledge, where the process of "breeding" could be followed by some universities, now means that university graduates are then appointed to teach after studying their Master's or doctorate, yet only know the Arabic language and do not know of any other university except their own. Suleiman (2007) indicates that this policy of "hatching" weakens the universities a lot, because it limits the ability of fulfilling the functions of a faculty member, which is not limited to lectures, but includes regular participation in personal and academic development and revision of the contents of the courses in the area of specialisation in the light of scientific development, selection of new textbooks that are more relevant to the content of new courses, and doing high quality research published in peer-reviewed scientific journals that should assist in economic and social development, in addition to contributing to training and development of

academic rules and regulations to better upgrade the academic level of university over time (Suleiman 2007: 117). Moreover, Jalal al-Din (2002) observes that the lack of funding and deterioration of facilities and deterioration of Sudan's relations with the majority of advanced donor countries, has put some limitations on the regular provision of scholarships and led to interruption of scholarships and limited the opportunities for most professors and graduate students to follow the rapid developments in their area of interests and studies in applied and social sciences. In addition, the low salaries of university professors have limited their ability to acquire scientific books and sometimes forced them to work in other income-generating activities or teaching in several universities to maximise their income; this implies that they do not have any free time to develop their abilities through the follow up of successive developments, even if facilities are available for free. Jalal al-Din (2002) indicates that the deterioration is not only limited to infrastructure or basic standard of living for teachers, but also extends to include teaching staff, as some of universities professors do not originally possess the required qualifications for engagement in teaching or university colleges. Jalal al-Din (2002) argues that part of the continued decline is also due to low capacity of people in charge of higher education, in particular members of the teaching staff as reflected in their weaknesses and limited abilities not only in the field of research and preparation for lectures but also in terms of interaction with students (Jalal al-Din 2002: 12–13, 22).

Jalal al-Din (2002) indicates that the agreed efforts in all Arab countries to achieve the required balance between the tracks of humanitarian and scientific paths, including science, engineering, medicine and agriculture, should not hide the fact that the random expansion in scientific paths, including science, engineering, medicine and agriculture, has led and still leads to serious deterioration in the quality of education, even for disciplines that affect human life such as medicine, pharmacy, architecture, civil engineering and chemical. For instance, the rapid expansion in engineering education in Sudan and some Arab countries has led to rapid decline and continued chronic weaknesses and deterioration in the quality of engineering education that still does not meet the needs of the advanced industrial sector. For example, some of the newly established engineering schools in Sudan completely suffer from the lack of laboratories and workshops, and therefore, in the practical aspects depend mainly on vocational training centres or where available on the laboratory equipment of secondary schools, which are not available in most cases. As a result of this situation, the students of some universities do not have any laboratory experience even after several years of study. For example, the students of the Faculty of Engineering in the Red Sea state had completed seven semesters without conducting any experiments in the laboratory during the period 1993–96, but only three semesters of the ten that are required to achieve a degree in engineering sciences were left until they graduated. This sad situation is accompanied by quantitative and qualitative deficit in the faculty members, where the percentage of students to professors of engineering in some universities has increased to more than 90 students per professor; consequently, these colleges are forced to resort to employing colleagues from abroad, who mostly hold initial university degrees or a Master's degree. Jalal al-Din (2002) observes that the deterioration in the quality is not only limited to the engineering sciences but also extends to the medical sciences. Jalal al-Din (2002) argues that in fact, the expansion in medical education has been slowing not commensurate with the potential facilities of Sudan and its demographic weight. For example, until 1978 there was only one medical college, and it absorbed no more than 50 students per year; until 1990 the number of medical schools in Sudan was around four colleges, all of which were supposed to accommodate less than 200 students annually. Then an expansion in enrolment throughout the 1990s led to the opening of dozens of colleges in a decade, reaching 26 completely by the third millennium. However, the expansion has not only been limited to the further opening of new colleges, but also focused more on further increases in the number of students admitted each year, for example, the number of students admitted to the Faculty of Medicine at the University of Aljazeera during the academic year 1989/90 was only up to 59 students, but this number jumped to about 400 students per year during the next 3 years, 1991/92–1993/94. This expansion of colleges and numbers of enrolment has happened without being accompanied by the provision of the minimum financial, material and human resources to help accommodate the new colleges in their absorption of these numbers and provision of the minimum required acceptable academic standards. Moreover, these new medical colleges not only lack many of the required disciplines but also they do not have close ties with teaching hospitals and other medical and health facilities that provide practical training for medical students and graduates. In general, all teaching hospitals are still lacking adequate and qualified staff, lacking medical practical equipment to an acceptable and satisfactory level, and also lacking all the required museums of pathology and anatomy. For instance, a report prepared by the Sudan Ministry of Higher Education indicates that in Sudan a lot of new universities lack references, books, periodicals, or that these references are too few, the majority of them are obsolete and few of them are relevant to recent developments in medical science and health. In general, the specialised laboratories of pre-clinical medical sciences are either not available or are incomplete in almost all new universities and the laboratory facilities in all new and old universities are very limited and unsatisfactory, lacking in teaching staff members in basic medical sciences and relying either wholly or partly on visiting lecturers, which lead to weak links and interactions between teachers and students (Jalal al-Din 2002: 15–19). In addition, colleges of medicine, pharmacy, nursing and medical laboratories are often face the difficulties to be closely related with hospitals and pharmaceutical industries in order to provide training for students and researchers in these fields. The attainment of medical and pharmaceutical overseas education is not only expensive and attracts only the best students in the Arab countries, but also it has tended more and more to sub-specialities in diseases that affect only a small percentage of citizens, particularly in the poorer Arab countries. This medical and pharmaceutical education is not available in an acceptable level in universities in the Arab countries, but only in universities in Western countries, which makes distinguished Arab students continue their specialties and their training process in these Western universities and often end up settled in Western societies. The migration of Arab medical specialists trained in Western universities with scholarships financed from public spending in education in Arab countries, implies a great loss for the poor people in Arab countries, not only because the poor people in the Arab countries partially and indirectly bear part of the burden of subsidising the high spending on educating the migrant Arab medical specialists; but also because the poor people in Arab countries are still susceptible to highly prevalent diseases such as malaria, tuberculosis, schistosomiasis and other diseases, which are not being sufficiently tackled by most of the medical specialists in the Arab countries. Jalal al-Din (2002) argues that medical science, although significant in the fight against poverty and disease, has not received so far the significant progress it deserves in the Arab countries; it has suffered from poor quality because of qualitative and quantitative deficits in the medical staff in poor regions and rural areas, and qualitative deficit despite quantitative surplus in the medical staff in urban cities. Jalal al-Din (2002) notes that apart from that, the level of medical education on the one hand and the level of wages and incentives for doctors and medical facilities available do not help provide any reasonable degree of public health requirements (Jalal al-Din 2002).

Moreover, the poor quality and efficiency of higher education in Sudan is probably related to the short age of some universities. For instance, Sudan Central Bureau of Statistics (1990–2009) shows that three quarters of Sudanese universities were established in the last period between 1991/92–2008/09, and 58 % are no more than 15 years old having been established between 1996/97–2008/09. These results seem consistent with the results in the Arab countries. For instance, the AHDR indicates that three quarters of Arab universities were established in the last quarter of the twentieth century, and 57 % are no more than 15 years old.[22] This observation is telling since higher education institutions, and universities in particular, require a long time to consolidate their institutional structure, and to foster their role in the dissemination and production of knowledge.[23] Our results presented above concerning poor quality and low commitment to the standardised international efficiency and quality criterion in Sudan are consistent with the results in the Arab literature. For instance, the AHDR authors (2003) argue that: "The quality of higher education institutions in Arab countries is affected by many factors, chief among which is the lack of a clear vision, and, as noted earlier, the absence of well-designed policies regulating the educational process. One of the main features of many universities in the Arab world is their lack of autonomy, i.e., they fall under the direct control of the ruling regime. Nevertheless, universities are often the arenas for political and ideological conflicts, the more so because of restrictions imposed on political participation in general and the promotion of political currents that owe allegiance to the regime more particularly. These contextual features have adverse effects on the degree of freedom allowed for education and research. This lack of autonomy has resulted in a situation where universities run according to the requirements of the governing political rationality and not according to a plan. [Higher education institutions often suffer from a lack of funding that prevents them from implementing sound scientific plans]. Some universities, for example, are

[22] See UNDP-AHDR (2003).

[23] See Nader Fergany (1998), pp. 18–19.

overcrowded on account of the uncalculated increase in enrolment rates, simply because the announcement of enrolment numbers in universities has become a political gesture to appease society. The quality of higher education is also influenced by an ongoing decline in expenditure, reflected in inadequate facilities for students and faculty. Quantitative expansion in higher education came at the expense of quality. University libraries are in a sorry state, laboratories are old and cannot accommodate the increasing numbers of students, and classes are overcrowded, thus creating a wide distance between students and teachers. Moreover, faculty members in many Arab universities earn meager salaries, and therefore cannot devote themselves fully to teaching or research".[24]

The dearth of reliable information precluded an analysis and discussion of interesting policy issues related to the quality of the private sector's contributions to both spending and enrolment in tertiary education in Sudan and the Arab countries. Although in the recent years Sudan and most West Asian Arab countries have been overwhelmingly open to private education, however, in terms of quality, it has not produced yet any visible results. Moreover, in Sudan as in most Arab countries, there is strong debate that the private universities are mainly businesses, the only exceptional cases being those of few older "private" universities, which rely on a history of good teaching and patient growth of research. This implies that the increasing private sector's contribution and the observed shortcomings in the quality and performance of the government or public education institutions, however, should not hide the fact that public sector institutions will remain very important; it would not be rational to absolutely replace them by a massive introduction of private education institutions and establishments in Sudan and all Arab countries, not only because of the uncertainties of a contribution to higher education by the private universities, but also because of the potential failure of private universities when deviating to target mainly business and profit objectives instead of focusing on targeting the conventional intrinsic values of higher education and higher quality. And also it is worth noting that the status of higher education, science and scientists is much better in Sudan and the Arab countries (especially in the Maghreb, see UNESCO) than in other parts of the world (e.g. Africa, see STS)[25] where, neoliberal policies have lead to the withdrawal of governments' support, the collapse of renowned establishments and the ruin of the profession, while such a deinstitutionalisation and the replacement by a global market of scientific skills had no results or disastrous ones in terms of scientific publications. Moreover, the bibliometric data demonstrate that the "newly founded" private establishments in Sudan and Arab countries (with the exception of three or four ancient and proud research universities like AUB or St Joseph in Lebanon and to some extent AUC in Cairo) contribute very little to the research output of the country, and that most of them do not care at all about

[24] See UNDP-AHDR (2003), p. 56.

[25] See Roland Waast (2003), pp. 153–181; see also ESTIME (2006), pp. 1–80.

research (see ESTIME 2006). Jalal al-Din (2002) explains that in light of the increasing economic difficulties, the declining public spending in education and increasing dependence on private spending in education, a serious growing trend appearing in higher education in Sudan is that the ability criterion (ability to pay large expenses by the student's higher income groups) has started to replace the merit and efficiency criterion. This is not only detrimental to society's poorest sections but also impairs the efficiency and quality of higher education itself, when the merit is replaced by the ability to pay and to bear the costs, and this has increased the disruption of the relationship between education and work (Jalal al-Din 2002: 22).

9.2.5 The Impacts of Educational Policies on Literacy and Access to Schooling (School Life Expectancy)

Educational policies in Sudan, Arab and Gulf countries lead to only slight improvements in school life expectancy and enrolment in all educational levels. However, in Sudan the educational policies have insufficient effects on improving school life expectancy, which remains low and lags behind when compared to the Arab and international standard. One important positive implication of educational policies is the increase in literacy rates; however, the educational policies have so far only managed to alleviate rather than fully eliminate the youth illiteracy problem in Sudan compared to the Gulf countries. For instance, Table 9.16 illustrates that in 1990 illiteracy rates amongst the youth population was 40 %, throughout the period 1990–2008 it continuously declined, but in the year 2008 it still remained in excess of 30 % in Sudan (30.7 %). Moreover, the youth illiteracy rate for women was in excess of 40 % (40.4 %) and was higher than for men, which accounted for 21 %. This implies that there is an ample room for policy to increase the literacy rate among the young population, especially for women. Furthermore, when comparing the supply and demand sides, we observe that the supply side or public spending seems to be only one component in the educational policies, because higher public spending per se does not imply higher demand, participation and enrolment ratios at all educational levels, access to schooling/school life expectancy and higher literacy rates.

Table 9.16 Educational outcomes: literacy, youth illiteracy rate and school life expectancy in the Sudan and Gulf countries (1990–2008)

	Adult literacy rate (% ages 15 and above)[a, d]			Youth literacy rate (% ages 15–24)[a, d]			Youth illiteracy rate (% ages 15–24)[a, d]			School life expectancy[b, c]	
Country	1990[a]	2002[a]	2008[d]	1990[a]	2002[a]	2008[d]	1990[a]	2002[a]	2008[d]	1999	2008
Sudan	45.8	59.9	69.3	65	79.1	85.2	35	20.9	14.8	4.4	4.6[(3)]
Bahrain	82.1	88.5	90.8	95.6	98.6	99.7	4.4	1.4	0.3	13.4	
Kuwait	76.7	82.9	94.5	87.5	93.1	98.4	12.5	6.9	1.6	13.6	
Oman	54.7	74.4	86.7	85.6	98.5	97.6	14.4	1.5	3.4	n/a	11.1
Qatar	77.0	84.2	93.1[(4)]	90.3	94.8	99.1[(4)]	9.7	5.2	0.9[(4)]	12.5	12.7
Saudi Arabia	66.2	77.9	85.5	85.4	93.5	97.3	14.6	6.5	2.7	n/a	13.5
UAE	71.0	77.3	90.0[(3)]	84.7	91.4	95.0[(3)]	15.3	8.6	5(3)	10.8	

Note: (1) Data refers to 1991 (2) Data refers to 2000/2001, (1) Data refers to 2005 (2) Data refers to 2007

[a]Calculated from UNDP Human Development Report (2004).

[b]UNESCO–UIS Statistical Yearbook (1999): www.unesco.org.

[c]UIS–UNESCO (2003): www.unesco.org.

[d]UNESCO Global Background information on Education Statistics: UNESCO- UIS Data Centre: Beyond 20/20 WDS (2011). (d) UNESCO-UIS (2010) "Global education digest 2010 Comparing education statistics around the e world".

9.3 Characteristics of Training Policies and The Impact of Educational Policies on Training Policies: Large Mixed and Private Firms and Public Policies of Training and Skills Upgrading in Sudan

Based on the above, this section shows the major characteristics of training policies and the impact of education policies on training policies and skill upgrading in Sudan. Before proceeding to discuss the impact of education policies on training policies and skill upgrading, it is useful to begin with a brief explanation of the major characteristics of training policies in Sudan, in particular the structure and pattern of training policies.

9.3.1 Characteristics of Training Policies in Sudan

Starting with the major characteristics of training policies in Sudan, we observe the biased structure in terms of type, duration and specialisation pattern of training policies and the low commitment to efficiency, adequacy and equity criterion in the provision of training in Sudan over the period 2004–09. In particular, beginning with the biased structure, we find that one major characteristic of training policies in Sudan is the biased structure in terms of type and duration of training, which indicates greater focus on internal type training (99 %, 98 %), particularly training that achieved during the short run (99 %, 96 %) as compared to external type training (2 %, 2 %), particularly training that achieved during the long run (1 %, 4 %). In addition, another characteristic of training policies in Sudan is the biased structure in terms of specialisation pattern of training, which indicates that training seems to be more biased towards practical specialisation pattern of training (88–73 %) as compared to academic specialisation pattern of training (12–27 %) in Sudan over the period 2004–09. This biased structure in terms of type, duration and specialisation pattern of training may imply the low commitment to the efficiency criterion and that has most probably immensely undermined the efficient provision of training. Another major characteristic of training policies in Sudan is the inadequate financial resources allocated for training, for instance, the annual training budget implies serious inadequacy in terms of finance allocated for training, as the real training budget represents only 46.5 % of the total budget approved for training in Sudan in 2009. This implies the low commitment to the adequacy criterion in the provision of financial resources for enhancing training in Sudan. Moreover, another characteristic of training policies in Sudan is significant regional disparity across the main geographical regions as measured by the share of the main regions in terms of supply side as measured by the total number of training centres, the demand side as measured by the capacity of training centres and the impact of training as measured by the share of trained work force as a percentage of total labour force and the share in internal and external short and long run training in Sudan 2004–09. In particular, from the distribution and share of main regions in training we observe the high share and greater concentration of training in the Khartoum region as compared to other regions in terms of the regional share in the supply side as measured by total number of training centres, the demand side as measured by the capacity of training centres and the impact of training as measured by the share of trained work force as a percentage of total labour force and the share in internal and external short and long run training in Sudan over the period 2004–09. This implies the low commitment to the equity criterion in terms of supply, demand sides and impact in the provision of training for different geographical regions in Sudan. Moreover, we observe that the disparity between Khartoum and other regions in terms of capacity of training centres implies the full utilisation of capacity of training in centres located in Khartoum as compared to under-utilisation of capacity of training centres located in others regions that most probably critically undermined the efficient provision of training. Therefore, the low commitment to efficiency, adequacy and equity in the provision of training led to limited effect of training in upgrading skill level for the labour force; this is observable from the low share of trained labour force that accounts for only 16 % and 17 % of the total labour force in Sudan in 2008 and 2009 respectively (see Table 9.17 below and Figs. 9.26, 9.27, 9.28, and 9.29 below).

9.3.2 The Impact of Educational Policies on Training Policies: Large Mixed and Private Firms and Public Policies of Training and Skills Upgrading

Our findings presented above concerning the low commitment to adequacy, equity and efficiency in the provision of education and training imply that the educational policies in Sudan have insufficient effect on training provision and failed to integrate sufficiently with training policies. This is probably because the relationship between the educational and training policies are somewhat separated from each other in Sudan. For instance, technical education and technological education are integrated in secondary and higher educational institutions and are administered by the Ministries of Education and Higher Education respectively, whereas the official general public training (including vocational training and apprenticeships) are separated from educational institutions and integrated within three official public training entities that are administered by independent institutions headed by the Minister of Labour, Public Service and Human Resources Development.[26] This organisational structure implies that educational policies in Sudan are still needed to enhance the fruitful cooperation, coordination and integration with training policies.

Earlier findings in Chap. 5 show that the lack of interaction between educational and training systems hinders the provision of training and upskilling plans within private firms. We will illustrate below that the interaction between educational and training policies appears to be effective only within very large two mixed and private firms that adopt training policies consistent/in line with public policies.

Table 9.18 illustrates the case of two of the largest mixed and private enterprises in Sudan, namely, Kenana Sugar Company (KSC) and DAL Group; both are astonishing Sudanese success stories, which seem committed to implement diversified training and skill upgrading policies that are quite consistent with the line taken by public policies. In particular, they adopt similar strategies that highlight training and upskilling of workers, linkages with universities to absorb graduates, active human resources development units and recruitment policies to set up and implement regular internal and external training plans and wide use of ICT to upskilling workers. Therefore, in contrast to the other private firms, these two large mixed and private firms (KSC and DAL) have successfully contributed to

[26] Suleiman (2007) indicates training institutions and training fields that received some attention in Sudan include for example, the Institute of Public Administration in Khartoum that offers training in public service and public administration for managers and teachers and the Center for Management Development in Khartoum that provides training programmes with limited fees for private and public sectors and the states, for example in 2005 it provided 28 training programmes in the computer field. Suleiman (2007) discusses some problematic issues related to training in Sudan and indicates that since independence and so far training in general has not found the attention it deserves, it has suffered due to lack of resources allocation, it remained weak and almost entirely concentrated in the capital Khartoum – particularly vocational training. Perhaps this reflects the fact that human development is not given the priority it is worthy of. Suleiman (2007) believes that currently (in 2005) the one of the most important difficulties for various kinds and levels of training in Sudan is the critical shortage of well-qualified and experienced trainers. Suleiman (2007) indicates that other problem is that the provision of high quality training with commitment to efficiency is costly and expensive, especially, vocational training, that requires the availability of construction, equipment and materials, and above all trainers who have the ability and knowledge necessary to provide effective training. In Sudan the Supreme Council for vocational training and apprenticeship was established under the law of vocational training and apprenticeships for the year 2001. The council represents a good step, as it has become responsible for the regulation and control of vocational training and apprenticeships, but it seems that it does not obtain the sufficient resources to play its role fully and effectively, as it gets the funding for the first quarter only in addition to training fees, and although the training fees are not high, it does not help to attract people to the training due to their weak resources or income due to the country's high poverty rate. Also, the training fees cannot provide adequate money, which is required for covering the requirements of offering effective training, equipment, materials and salaries high enough to retain the well-qualified and experienced trainers and avoid their migration. Suleiman (2007), pp. 112, 114–115.

serve the public policies for enhancing training and skill upgrading (cf. Table 9.18 below).,[27, 28] However, it is less clear whether these two large mixed and private firms (KSC and DAL) induce positive effects on upskilling workers in private firms. In our view, the interpretation of the serious discrepancy between these two large mixed and private firms and other firms can be attributed to presence of high

[27] Kenana Sugar Company (KSC) is the largest integrated sugar project in the world, it is the world's largest producer of white sugar, and it is located 250 km south of Khartoum, Sudan. Behind KSC success story a combination of factors such as sound plans, advanced western technology, Arab finance, and the rich natural resources of area (climate, fertile soils and the Nile waters). KSC employs up to 16,000 workers (50 % permanent), and provides education and healthcare to the 100,000 people who are estimated to depend on the plant. It also makes Sudan self-sufficient in sugar (the domestic market consumes some 150,000 tonnes) and provides valuable export earnings. Most of the sugar that is exported goes to African and Middle Eastern states, as well as India and Bangladesh, and as far away as Europe. Molasses are sold to Britain and the Netherlands. KSC is also supporting scientific research and vocational training and providing education, health and electricity supply services to the estimated 100,000 population of the area. KSC production activity focuses mainly on growing sugar cane and processing it to produce white refined sugar, molasses and honey, in addition to other products such as milk products. KSC is established in 1975 as a joint venture between the Government of the Republic of Sudan (35.33 %), General Investment Authority of Kuwait (30.64 %), Government of the Kingdom of Saudi Arabia (10.97 %), Arab Company for Investment (6.99 %), Industrial Development Bank (5.69 %), Arab Authority for Agricultural Investment and Development (5.59 %), Group of Sudanese commercial banks (4.47 %), Gulf Fish Company (0.16 %) and Nesho Iwai Corporation (0.16 %). See Arab Authority for Agricultural Investment and Development website: http://www.aaaid.org/arabic/AAAID_COM/KSC.htm, accessed 10 April 2011.

[28] DAL Group started out in 1951 as an engineering dealership company under the name Sayer & Colley founded in 1952 by two British businessmen. The company specialises in trading engineering products, such as bearings and belts, and is later awarded the Caterpillar dealership for Sudan. In 1966, 10 years after Sudan's independence, Caterpillar decides to transfer the dealership to a Sudanese company; the original British owners retained a minority share. The contract was awarded to Mr. Daoud Abdellatif and his newly created Sudanese Tractors Company (SUTRAC). In 1970–71 Sudanese Tractors Company (SUTRAC) and Sayer & Colley were both nationalised. Within a year, the companies had been denationalised with the government retaining a minority share (the British partners were bought out), and in 1978–79 the name of the company had changed from Sayer & Colley to Daoud Abdellatif Engineering, later abbreviated as 'DAL Engineering'. After that DAL Engineering expanded to include other activities and the name has changed to Daoud Abdellatif Group or 'DAL Group', which included SUTRAC (1952), DAL Agriculture (1984), DAL Property Development (1988), DAL Motors (1994), Sayga Flour Mills (1996), the Blue Nile Dairy company (today DAL Dairy) (1997), DAL Medical (1997), Sudanese Liquid Air (SLA) (1998), DAL Food Industries (DFI) (2002) and Khartoum International Community School (KICS) (2004). Due to this expansion since its establishment in 1951, the DAL Group has become one of the largest and most diversified conglomerates in Sudan, operating to international standards, underpinned by strong, clear business principles. The Group businesses play a leading role in their fields and operate ten businesses across six sectors – food, agriculture, engineering, real estate property development, medical services and education (DAL runs a non-profit school). DAL Group has a presence in seven countries, China, Malaysia, UAE, Saudi Arabia, Djibouti, Ethiopia and the UK and represents 38 international brands in Sudan, including some of the world's best known brands. See: http://www.dalgroup.com, Accessed 10 April 2011.

Table 9.17 Regional distribution and share of main regions in total number of training centers, capacity, trained work force and internal and external training in Sudan (2004–2009) (%)

(a) Number of training centers, capacity and trained work force

	Total number of training centers		Share in total number of training centers		Total capacity of training centers		Total number of training centers/total capacity of training centers		Trained work force/total labour force (2008–2009) (%)		
	2008	2009	2008	2009	2008	2009	2008	2009	2008	2009	Change (2008–2009)
All Sudan	64,927	65,929	100 %	100 %	74,984	78,318	86.59 %	84.18 %	16.12 %	17 %	0.88 %
Northern	15	17	0.023 %	0.026 %					64 %	70.00 %	6 %
Khartoum	64,864	65,864	99.903 %	99.901 %		65,864		100 %		13.06 %	
Central	29	29	0.045 %	0.044 %	7000	10,234	0.41 %	0.28 %	2.%	2.90 %	0.9 %
Kordofan	15	15	0.023 %	0.023 %	1050	1050	1.43 %	1.43 %	6.09 %	4.65 %	−1.44 %
Darfur	2	2	0.003 %	0.003 %	720	720	0.28 %	0.28 %	3.22 %	3.37 %	0.15 %
Eastern	2	2	0.003 %	0.003 %	350	350	0.57 %	0.57 %	5.3 %	8.00 %	2.7 %

(b) Internal and external training

	Internal training						External training						Total internal and external training	
	Short t run		Long run		Total		Short run		Long run		Total		Total	
	2008	2009	2008	2009	2008	2009	2008	2009	2008	2009	2008	2009	2008	2009
All Sudan	100 %	100 %	100 %	100 %	100 %	100 %	100 %	100 %	100 %	100 %	100 %	100 %	100 %	100 %
Northern	3 %	6 %	0 %	0 %	3 %	5 %	4 %	13 %	22 %	16 %	10 %	14 %	3 %	5 %
Khartoum	55 %	58 %	0 %	31 %	55 %	57 %	10 %	23 %	54 %	54 %	26 %	40 %	54 %	57 %
Central	23 %	15 %	28 %	10 %	23 %	15 %	17 %	35 %	14 %	10 %	16 %	22 %	23 %	15 %
Kordofan	10 %	7 %	49 %	15 %	11 %	7 %	28 %	10 %	10 %	5 %	22 %	7 %	11 %	7 %
Darfur	1 %	3 %	1 %	7 %	1 %	3 %	9 %	10 %	0 %	0 %	6 %	5 %	1 %	3 %
Eastern	8 %	12 %	22 %	12 %	8 %	12 %	8 %	9 %	0 %	0 %	5 %	4 %	8 %	12 %

Source: (a) Sudan Ministry of Federal Government and the General Secretariat of the National Council for strategic planning (2009) "Performance Digital Reports of the Northern States (2009)". p. 471. (b) Sudan Ministry of Federal Government and the General Secretariat of the National Council for strategic planning (2009) "Performance Digital Reports of the Northern States (2009)". pp. 446–447

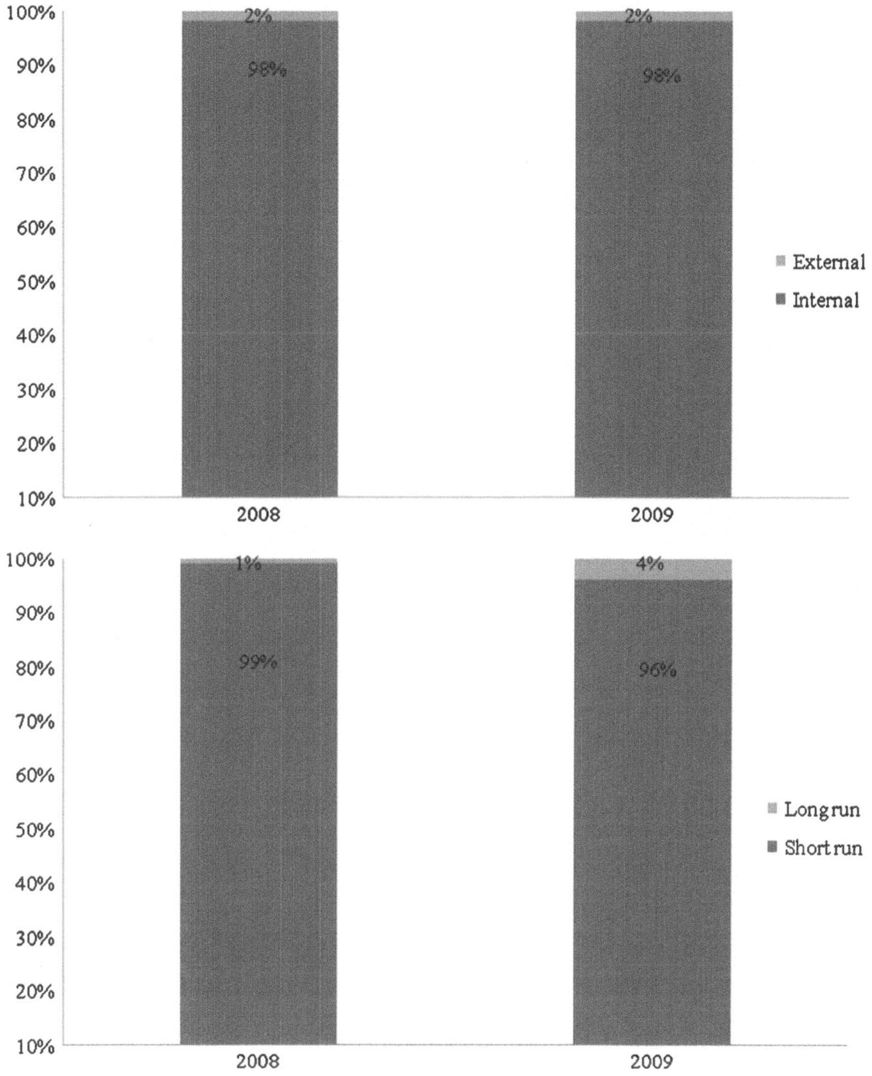

Fig. 9.26 The share of internal and external short and long types of training in total training in Sudan (%) (2008–2009) (Source: The National Center for training -Ministry of Labour, Public Service and Human Resources Development Report (2007–2009) – cited in the General Secretariat of the National Council for strategic planning. p. 439)

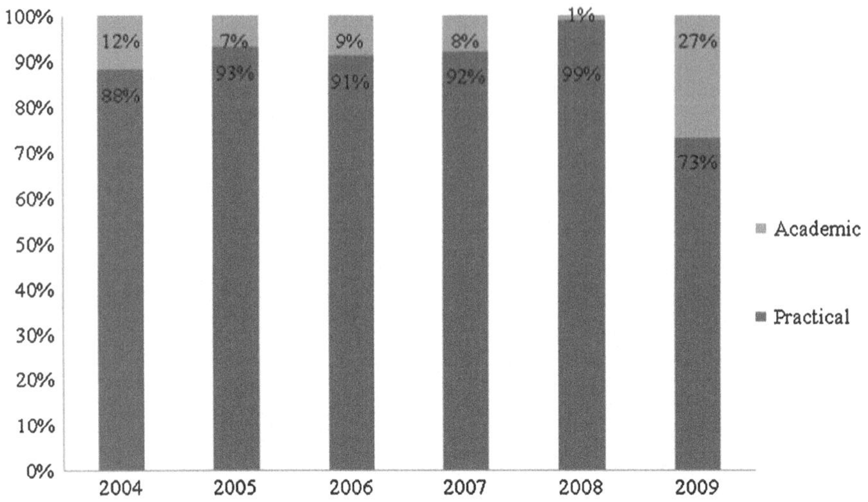

Fig. 9.27 The share of academic and practical types of training in total training in Sudan (%) (2004–2009) (Source: The National Center for training -Ministry of Labour, Public Service and Human Resources Development Report (2007–2009) – cited in the General Secretariat of the National Council for strategic planning. p. 439)

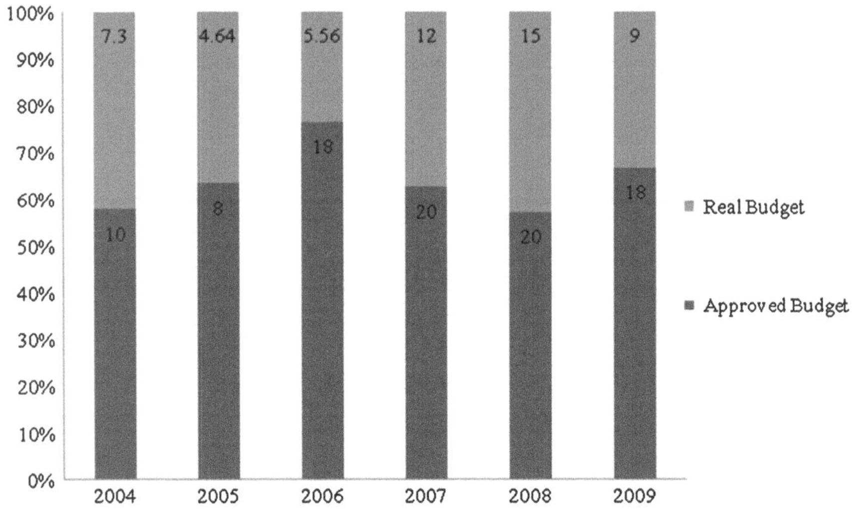

Fig. 9.28 The total approved budget and real budget for training in Sudan (Million Sudanese Pounds) (2004–2009) (Source: The National Center for training -Ministry of Labour, Public Service and Human Resources Development Report (2007–2009) – cited in the General Secretariat of the National Council for strategic planning. p. 439)

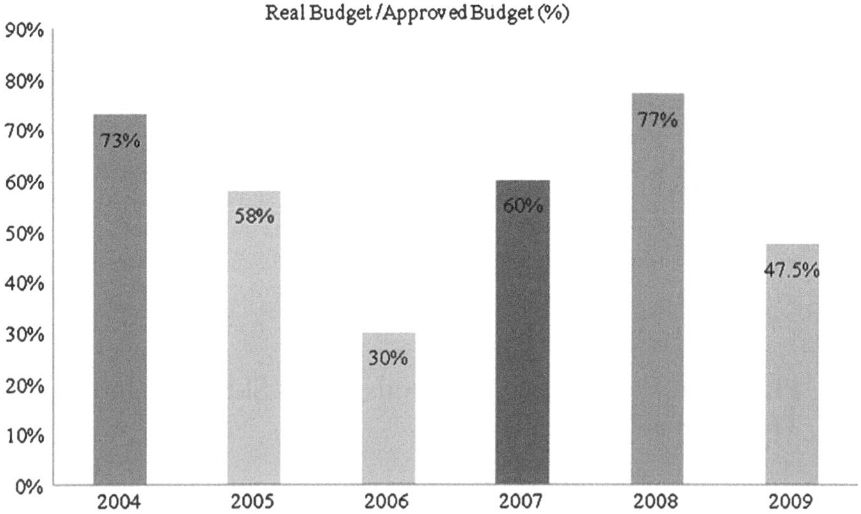

Fig. 9.29 The ratio of total real budget /approved budget for training in Sudan (%) (2004–2009) (Source: The National Center for training -Ministry of Labour, Public Service and Human Resources Development Report (2007–2009) – cited in the General Secretariat of the National Council for strategic planning. p. 439)

resources, support and incentives within these two firms, which are probably lacking within other private firms.,[29, 30]

[29] For instance, the selected two largest mixed and private firms have several common characteristics such as large market size, namely size of employment, capital (local capital), market, products, sales, sales revenues; investment in ICT, use of advanced technologies and active R&D/technology development units.

[30] From the results of the firm survey (2010) and the selected two cases of large mixed and private firms studied in this chapter, it may be true that large firms are systematically more organised than medium and small firms. It is therefore plausible to expect that the large firms to have higher financial capacity to support training and skill upgrading than medium and small firms. However, it is less clear and hard to make generalisations to conclude whether this applies to all other large firms as well, because the available information for Sudan indicates the classification of firms according to size, activities and ownership (nationality of main owner(s)) rather than public-private sectors. Suleiman (2007) notes that in Sudan in the private sector the large firms are relatively willing to train employees from time to time, due to high costs of training programmes and absence of trainees for work during training period. If the number of employees is small, this may adversely affect the production, which can be regarded as indirect or additional costs for training. Despite the fact that workers in small enterprises may be in more need for training of their counterparts in large enterprises, the weakness of the capabilities and the small number of staff may prevent it. Suleiman notes that, in addition the large firms often have the capabilities and attractiveness to attract the well-qualified workers and staff from the outset, who are thus in less need of training. It is thus clear that the training of workers is often determined by the availability of the capabilities and the facilities of the largest firms more than it is determined by the need for it. He point out that the State must help small establishments in the process of training, especially since the small establishment constitutes the largest number and most widespread geographically

The major policy implication from these findings is that the improvement of the educational systems in Sudan is essential and requires improvement of adequacy, equity and efficiency including quality/internal efficiency, supply (investment) and demand (enrolment) sides, particularly with respect to tertiary and technical education. From that perspective, we explain below the views of the policy makers and experts regarding the relevant plans and policies to reform the educational system, which is essential for skill development. After that, we show that policy makers, experts and firms all view the improvement of education as important for enhancing the provision of training and knowledge transfer/external schooling effects.

9.4 Plans and Policies and Mechanisms for Skill Development: The Macro–Micro Views

The major policy implication from these findings is that the improvement of the educational systems in Sudan is essential and requires improvement of adequacy, equity and efficiency including quality/internal efficiency, supply (investment) and demand (enrolment) sides, particularly with respect to tertiary and technical education. From that perspective, we explain below the views of the policy makers and experts regarding the relevant plans and policies to reform the educational system, which is essential for skill development. After that, we show that policy makers, experts and firms all view the improvement of education as important for enhancing the provision of training and knowledge transfer/external schooling effects

It is useful in this section to discuss and compare first the relevant plans and then policies and mechanisms for skill development from both macro and micro perspectives/views.

9.4.1 Plans for Skill Development: The Macro–Micro Views

In light of the above findings, we now use the results of the macro and firm surveys (2010) to provide insights to help generate policies to enhance skill levels by implementation of short- and long-term plans at the macro–micro levels. We then compare and integrate the macro (official: policy makers and experts) and micro (firms) views concerning the important tools and plans for skills development in the current, short run and long run.

The policy makers and experts suggest several important instruments for skill development, some of which have already been implemented and others being

---

in Sudan and also because the private sector in Sudan in general is still emerging and needs the assistance and protection in the face of competition from imports at home and competition abroad in the area of export. See Suleiman (2007), pp. 109–110.

implemented now or in the near future. For instance, Table 9.19 shows that the policy makers and experts highlight investment in training of existing employees, investment in vocational training, improving the quality of teachers, trainers and mentors and encouraging learning on the job. This is to be coupled with investment in formal education, using ICT to upgrade skill levels, enhancing the system or programme of apprenticeship, sending teachers and mentors abroad to acquire knowledge and skills and sending workers abroad to acquire skills.[31] Measures such as supporting long distance learning, enhancing the system of accreditation and licensing, sending students abroad to acquire knowledge and skills and bringing/attracting new foreign skills, scientists and engineers from abroad are viewed as somewhat less important factors. [32] In addition to the official perspective, other less important factors for enhancing skill upgrading in Sudan include: increasing motivations, follow-up, increasing the effectiveness of the national training strategies and policies, improving work environment and condition and encouragement of interest in other languages, specifically English, to facilitate communication with international communities and learning from the experiences of other countries. In addition: increasing motivation by offering scholarships to encourage continuing education, training and learning, improvement of education and training curriculum, facilitating the ownership of modern information technology and modern methods of learning (computer per learner) to promote single education and distance education, reviewing government strategies in all education levels, specification of an annual fixed percentage of national expenditure for education and training, provision of equal education and training opportunities to reducing the regional disparity in the provision of education and training and unbalanced development in human resources and finally, establishment of a specialised ministry responsible for national human resources development. Plans currently implemented include investment in education, investment in training of existing employees, investment in vocational training, improving the quality of teachers, trainers and mentors, enhancing the system or programme of apprenticeships, using ICT to upgrade skill levels, bringing/attracting new foreign skills, scientists and engineers from abroad and sending workers, teachers and students abroad to acquire knowledge and skills and encouraging learning on the job. However, our earlier discussion in this chapter and in Chap. 5 above, illustrates the serious shortcomings of some of these components, particularly with respect to investment in education, training and vocational education. Hence, the officials' view suggests further efforts in the short run to motivate investment in education, investment in training of existing employees, investment in vocational training, sending workers abroad to acquire skills and improving the quality of teachers, trainers and mentors.

[31] For instance, the use of modern information technology to raise skills level has been implemented in particular by oil companies.

[32] For instance, the use of distance learning for skill upgrading has been used mainly in universities, for example, Sudan Open University provide distance education and training for teachers of basic education during their service.

Table 9.18 Human resources development and training policies in the large mixed and private firms in Sudan

Firms	Human resources development strategies and training policies in Kenana sugar company (KSC) and DAL group
1. KSC[a, b, c]	KSC human resources development unit aims to upgrade skill levels via the following 1. Employing about 16,000 workers (half of them are permanent workers) and providing education and healthcare to the 100,000 people who are estimated to depend on the plant 2. Encouraging an active R&D unit and supporting scientific research to be continuously involved in the development of ways and means for utilisation of sugar industry 3. Supporting vocational training through establishment of a centre for vocational training since the founding of KSC 4. Enhancing access to primary and secondary education by adding to the already existing local government schools and non-government religious (khalawis) education in Kenana area. Notably, through the establishment of Kenana Sugar Company primary and secondary schools. This includes KSC 23 basic schools (which includes 11 mixed basic schools, 5 basic schools for boys and 7 basic schools for girls), in addition to KSC 4 secondary schools (which includes 2 secondary schools for boys and 2 secondary schools for girls) in Kenana area 5. Enhancing access to tertiary education and linking education and industry via the following: (a) Establishment of the KSC company branch of Omdurman Islamic University including a number of colleges (b) Contribution of KSC to the establishment of Kenana technical college in Kenana in 2004. Kenana technical college offers technical university education in 8 fields of specialisation that facilitated with the employment of 12 academic teaching staff, 4 technical teachers and 8 technicians. In 2008/09 Kenana technical college offers opportunity for enrolment of 276 students and graduation of 139 students 6. Offering several internal and external (abroad outside Sudan) training programmes for skill upgrading and upskilling of personnel workers. The various internal and external training opportunities are offered for workers to fit with the KSC current and future business and to fit with skills and technical change and needs in the field of industrial and agricultural technology related to sugar manufacture and cultivation 7. Supporting training provision in cooperation with the universities and local institutes 8. Participating in supporting improvement of the quality of teachers by supporting sending some teachers for training in overseas universities 9. Upgrading the level of KSC local skill by bringing foreign experts and integrating them in the company to help the transfer of knowledge and skill to Sudanese through the establishment of company training centre since 1986 to offer various training programmes for upskilling and upgrading the level and capacity of management and technical skills of local staff so as to maintain the overall efficiency of the company 10. Supporting of training via the establishment of the subsidiary Kenana Engineering and Technical Services Company (KETS) since 1989 with the aim to benefit from the experience and knowledge accumulated by Kenana company

(continued)

Table 9.18 (continued)

Firms	Human resources development strategies and training policies in Kenana sugar company (KSC) and DAL group
	to provide engineering and technical services for agricultural and industrial sectors in Sudan
	11. Encouraging and supporting strategic partnerships and cooperation with the public sector and state institutions to promote employment for graduates in Sudan. In February 2011, KSC subsidiary company KETSC has signed a partnership agreement with the National Fund for the employment of graduates and the Ministry of Human Resource Development that aim to build in partnership to upgrade the skill levels of graduates. Kenana Sugar Company and National Fund for employment of graduates are working together to facilitate access to further practical knowledge offered to graduates to facilitate their engagement in the work environment via the integrated training programmes in KSC and through temporary absorption of graduates in different provisional projects operated by KSC in industrial and agricultural fields
	12. Offering contribution towards the British Council and the Chevening Scholarship scheme (2009–2010). The British Council and the Chevening Scholarship scheme is organised in cooperation with the British Embassy in Sudan and it aims at offering Sudanese nationals with strong academic backgrounds and relevant work experience in the fields of politics, public administration, economics, environmental sciences, law, human rights, international relations, conflict resolution, engineering, development, including rural development and agriculture, an excellent opportunity to study for a 1 year Master's degree in the UK through the Chevening Scholarship scheme
2. DAL Group[d,]	Dal's human resources development unit aims to upgrade skill levels via the following:
	1. Attracting, developing, and retaining the talented local and international workers by employing more than 5,500 employees drawn from 26 different nationalities
	2. Providing employees with world-class learning and development opportunities
	3. Pursuing a long-term vision, plan and strategy for investment in people and providing its employees with a working environment that stimulates diversity, innovation, teamwork, learning and development
	4. Focusing on three main core values including commitment for maintaining high personal and business standards
	5. Offering specialised training for more than 60,000 people across the country in baking and bread making as part of Dal corporate responsibility activities
	6. Offering specialised training via the Dal Sudanese Tractors Company (SUTRAC) training centre, which was the first in Africa and the Middle East to be certified by Caterpillar
	7. Enhancing investment and growing human capital through significant investment in learning and development and talent management to achieve the key future initiatives for the company including broadening geographical footprint in the region, deepening product mix and accelerating speed to market through supply chain optimisation
	8. Encouraging cooperation with several government bodies (including the Ministry of Agriculture, Department of Forestry, Higher Council for Technical and Vocational Training, Ministry of Education, Ministry of Health and Sudanese Standards Metrology Organisation) to promote better standards and create opportunities

(continued)

Table 9.18 (continued)

Firms	Human resources development strategies and training policies in Kenana sugar company (KSC) and DAL group
	9. Supporting the Dal group individual businesses further initiatives to enhance learning and capacity building
	10. Encouraging an active R&D unit via supporting involvement in several activities, introducing new products, processes and exploring opportunities, trailing new technologies and adapting ideas, to develop new products and services, which meet the needs for development in Sudan
	11. Supporting education through DAL Group initiative of establishment of Khartoum International Community School (KICS) in 2004. Founded by DAL Group, KICS is the first and only school to offer the highly acclaimed International Baccalaureate (IB) in Sudan. KICS provides a first world-class international education to the local and international communities in Khartoum. The Dal Group KICS offers enrolment for 400 students representing over 30 nationalities. In December 2010 KICS achieved international recognition; KICS has been awarded accreditation by the prestigious Council of International Schools. This was achieved only 5 years after the opening of the school. The award of CIS accreditation is an internationally recognised quality mark showing that KICS is a school that aspires to the highest possible standards. KICS is the first and only school in Sudan to have achieved both CIS accreditation and authorisation by the International Baccalaureate to offer the IB Diploma and Primary Years programmes in Sudan
	12. Improving human resources effectiveness within the DAL group by supporting several strategic initiatives. These include firstly, strategic workforce planning to ensure that the organisation has access to the necessary resources to fulfil the strategic business objectives, also this includes identifying and monitoring key workforce information and modelling future requirements. Secondly, talent management to focus on recruitment and retention of employees, succession planning, and Dal graduate and future leaders programmes. Thirdly, Learning and Development (L&D) to enhance the knowledge, skills and attitudes of employees through a range of activities from specialist courses and on the job training, to work shadowing and mentoring. DAL Group has invested in a world class L&D centre to ensure its employees receive the best possible professional development. Fourth, performance management and rewards to create and maintain a performance management and appraisal system that encourages a performance driven culture, with rewards based on performance
	13. Adopting and supporting the involvement in corporate responsibility (CR) strategy by ensuring that social contribution and responsibility is integrated into Dal everyday business activities and through Dal heavily investment in learning and development and in programmes that seek to identify and promote talent throughout the group
	14. Encouraging investment in Sudanese youth graduate by offering a month long work placement opportunities for graduates, university and vocational education students
	15. Offering pre-employment training for Sudanese graduate through the recent launch of graduate trainee scheme, whereby 14 graduates (selected from more than 900 applications) have joined Dal Group's specially designed 2 year programme exposing them to a variety of businesses and roles across the group
	16. Involving in offering awards for best Sudanese secondary school graduates to encourage competition and positive impact in society. Dal Group subsidiary company Sayga has introduced the initiative of the Sayga Al-Awal Annual

(continued)

Table 9.18 (continued)

Firms	Human resources development strategies and training policies in Kenana sugar company (KSC) and DAL group
	Award, 'Al-Awal honours the First'. The Al Awal initiative is one of several of Sayga's CR projects that have been integrated permanently into the business strategy. The award initiative programme started 2 years ago in 2009 when top achieving Sudanese secondary school graduates were sent to the UK. Sayga promotes and rewards excellence by celebrating and recognising the top achieving student in each of the fields of the Sudanese High Secondary School Certificate (graduates from the academic, commercial, agricultural, and industrial and women specific studies). The initiative aims to instil a sense of competition in high schools, to reward students' effort and achievement, expose them to different cultures, strengthen English language command, increase self confidence and enhance skills of top Sudanese students. The Al-Awal Initiative involves organising and sponsoring a 1 month trip to the UK to attend intensive English courses at the Anglo-Continental School of English in Bournemouth, providing internship opportunities within the organisation, providing a Sayga medal and a laptop prize for top achieving Sudanese high secondary school graduates

Sources: Adapted from
[a]KSC Achievement Report, publications and web site
[b]Akhbar Newspaper (2011)
[c]Fadlallah (2006) and
[d]DAL Group Achievement Report (1970–2011), publications and web site: http://www.dalgroup.com accessed 10 April 2011

Further efforts considered important in the long run are to enhance the system or programme of apprenticeships, support long distance learning, encourage the system of accreditation and licensing, sending teachers/instructors and trainers abroad to acquire knowledge and skills and to improve the quality of teachers, trainers and mentors.

On the other side, at the micro level/across firms, the results of the firm survey suggest differing points of view with different priorities that highlight learning on the job as main priority,[33] especially in the short run.[34] Whereas, using ICT to upgrade skill levels, investment in training of existing employees, supporting long distance learning, sending trainers, mentors and workers abroad to acquire skills and knowledge and bringing/attracting new foreign skills, scientists and engineers are receiving less attention, particularly in the short run. Firms highlight these components in the long run and highlight learning on the job in the short run (see Table 9.20 below).

From Tables 9.19 and 9.20 and from the macro-firm surveys it appears that views of the policy makers and experts and those of the firms are consistent in highlighting the learning on the job as one of the top priority for skill development. However, on the other hand, we observe that there appears to be clear discrepancies

[33] As reported by 84 % of the respondent firms.

[34] As indicated by 72 % of the respondent firms.

Table 9.19 Plans and tools for skill development: macro-policy makers and experts view

Tools for skill development (%)	Importance (%)	Has been already implemented (%)	Short run/ current plan (%)	Long run/ future plan (%)
Investment in formal education	97	56	17	17
Investment in vocational training	100	44	17	28
Investment in training of existing employees	100	50	17	25
Improving the quality of teachers, trainers and mentors	100	42	22	25
Sending teachers/instructors/trainers abroad to acquire knowledge and skills	94	36	19	25
Sending students abroad to acquire knowledge and skills	81	33	14	36
Sending worker abroad to acquire skills	92	39	19	25
Bringing/attracting new foreign skills, scientist and engineers	61	39	22	17
Using ICT to upgrade skill levels	97	42	28	11
Encouraging learning on the job	100	33	28	19
Supporting long distance learning	89	42	25	19
Enhancing the system of accreditation and Licensing	86	17	31	31
Enhancing the system or programme of apprenticeship	97	44	14	19
Total response				

Source: Own calculation based on the macro survey (2010)

between the macro–micro views (and also across firms) concerning the selection of both tools and plans for skill development. For instance, the macro/policy makers and experts' view tends to highlight investment in training of existing employees, investment in vocational training, improving the quality of trainers and mentors, encouraging learning on the job and investment in formal education as top priorities. Their next priorities are: using ICT to upgrade skill levels, enhancing the system or programme of apprenticeships, and sending trainers, mentors and workers abroad to acquire skills. Supporting long distance learning encourage the system of accreditation and licensing, sending students abroad to acquire skills and bringing/attracting new foreign skills, scientists and engineers from abroad are viewed as somewhat less important tools. On the other side, the micro (firm) view highlights learning on the job as main priority. The rank of the firms' other priorities are: using ICT to upgrade skill levels, investment in training of existing employees, supporting long distance learning, sending trainers, mentors and workers abroad to acquire skills and knowledge and bringing/attracting new foreign skills, scientists and engineers.

From the policy makers, experts and firms' perspectives the top priorities in the short run are encouraging the system of accreditation and licensing, learning on the job, using ICT to upgrade skill levels, supporting long distance learning, improving the quality of teachers, trainers and mentors and bringing new foreign skills, scientists and engineers; this is followed by investment in education, training (including vocational training) and sending trainers, mentors and workers abroad to acquire skills. From the policy makers, experts and firms' perspectives the top priorities in the long run highlight sending students abroad to acquire knowledge and skills, encouraging the system of accreditation and licensing, investment in training (including vocational training) of existing employees, improving the quality of teachers, trainers and mentors and sending teachers, mentors and workers abroad to acquire knowledge and skills.

From the macro/policy makers and experts' perspective, the top priorities in the short run are encouraging the system of accreditation and licensing, learning on the job, using ICT to upgrade skill levels, supporting long distance learning, improving the quality of teachers, trainers and mentors and bringing new foreign skills, scientists and engineers. Less emphasis would be placed on sending trainers, mentors and workers abroad to acquire skill, investment in education, training (including vocational training) of existing employees and sending students abroad to acquire skill. On the other side, firms highlight encouraging learning on the job as top short run priority, followed by using ICT to upgrade skill levels, investment in training of existing employees, supporting long distance learning, and bringing new foreign skills, scientists and engineers. Less emphasis would be placed on sending trainers, mentors and workers abroad to acquire skills.

Furthermore, from the macro/policy makers and experts' perspective, the top priorities in the long run would be sending students abroad to acquire knowledge and skills, encouraging the system of accreditation and licensing and investment in vocational training. This is followed by investment in training of existing employees, improving the quality of teachers, trainers and mentors and sending teachers/instructors, trainers and workers abroad to acquire knowledge and skills. Less emphasis would be on supporting learning on the job, supporting long distance learning, enhancing the system or programme of apprenticeships, investment in education, bringing new foreign skills, scientists and engineers and using ICT to upgrade skill levels. On the other hand, firms tend to highlight sending trainers, mentors and workers abroad to acquire skills as main long run priority, followed by investment in training of existing employees, supporting long distance learning and bringing new foreign skills, scientists and engineers. Less emphasis would be placed by firms on using ICT to upgrade skill levels and encourage learning on the job.

In addition to the above observed discrepancies between macro–micro views concerning the selection of plans and tools and arrangement of priorities and policies for enhancing skill, we explain below the visible differences in the macro–micro perspectives in suggesting policies for improving the provision of training and transfer of knowledge. Therefore, this implies that further efforts are needed to enhance the consistency between the macro–micro views and public-private sectors, particularly with respect to the arrangement of priorities and plans to ensure more successful and consistent implementation of policies for skills development and encouraging private sector participation in education and training.

Table 9.20 Plans and tools for skill development: micro-firm view

Tools for skill development	All	Industry				Size		
Response rate	firms	Chemical	Food	Metal	Textile	Large	Medium	Small
Of special importance (%)								
Investing in training to train existing employees	65 %	71 %	63 %	55 %	60 %	78 % %	52 %	67 %
Sending trainers and mentors abroad to acquire skills	49 %	51 %	41 %	64 %	40 %	59 %	44 %	39 %
Sending workers abroad to acquire skills	46 %	49 %	44 %	45 %	40 %	56 %	44 %	33 %
Bringing/attracting new foreign skills, scientists and engineers	33 %	26 %	48 %	27 %	20 %	47 %	26 %	17 %
Using ICT to upgrade skill level	73 %	77 %	70 %	73 %	60 %	72 %	70 %	78 %
Encouraging learning on the job	84 %	92 %	74 %	82 %	80 %	88 %	78 %	84 %
Supporting long distance learning	49 %	60 %	44 %	36 %	20 %	59 %	44 %	39 %
Others [please specify below]								
Number of respondents	79	36	27	11	5	32	27	19
To be pursued now/ in the short run (%)								
Investing in training to train existing employees	32 %	29 %	33 %	27 %	60 %	47 %	19 %	28 %
Sending trainers and mentors abroad to acquire skills	9 %	9 %	11 %	9 %	0 %	3 %	22 %	0 %
Sending workers abroad to acquire skills	15 %	14 %	15 %	18 %	20 %	22 %	15 %	6 %
Bringing/attracting new foreign skills, scientists and engineers	24 %	17 %	30 %	36 %	20 %	34 %	15 %	17 %
Using ICT to upgrade skill level	42 %	40 %	37 %	55 %	60 %	63 %	26 %	28 %
Encouraging learning on the job	72 %	75 %	63 %	82 %	80 %	75 %	67 %	74 %
Supporting long distance learning	26 %	29 %	19 %	36 %	20 %	31 %	22 %	22 %
Others [please specify]								
Number of respondents	78	35	27	11	5	32	27	19
To be pursued in the near future/long run (%)								
Investing in training to train existing employees	44 %	51 %	33 %	55 %	20 %	41 %	52 %	33 %
Sending trainers and mentors abroad to acquire skills	56 %	57 %	48 %	73 %	60 %	72 %	41 %	50 %

(continued)

Table 9.20 (continued)

Tools for skill development	All	Industry				Size		
Response rate	firms	Chemical	Food	Metal	Textile	Large	Medium	Small
Sending workers abroad to acquire skills	51 %	51 %	44 %	64 %	60 %	53 %	52 %	44 %
Bringing/attracting new foreign skills, scientists and engineers	31 %	34 %	33 %	18 %	20 %	41 %	30 %	17 %
Using ICT to upgrade skill level	28 %	37 %	22 %	18 %	20 %	22 %	41 %	22 %
Encouraging learning on the job	16 %	17 %	15 %	18 %	20 %	16 %	19 %	16 %
Supporting long distance learning	37 %	37 %	33 %	45 %	40 %	41 %	37 %	28 %
Others [please specify]								
Number of respondents	78	35	27	11	5	32	27	19

Source: Own calculation based on the firm survey (2010)

9.4.2 *Policies and Mechanisms for Skill Development: The Macro–Micro Views*

The implementation of the above plans for skills development requires an integration of the macro–micro policies; the results of the macro and firm surveys (2010) are useful for discussing and integrating these policy perspectives. From the macro survey we find that the policy makers and experts' view concerning skill development policies highlights the mechanisms/policies for enhancing the efficiency of educational system, enhancing the external schooling effects/transfer of knowledge and planning skill needs. Additional mechanisms/policies identified include promoting of resources allocation, enhancing social partnership and collaboration between educational and training institutions, employers, workers and the state to determine skill needs and the most effective ways of meeting and financing them, enhancing the provision of training, monitoring skill needs on a regular basis and importing skills from abroad.[35]

Our analysis below discusses the mechanisms for enhancing the educational system, transfer of knowledge/external schooling effects and provision of training. Other components include: planning skill needs, monitoring skill needs on a regular basis, enhancing social partnership in skill development, promoting of resources allocation and importing skill from abroad are somewhat integrated in the above components. We begin with the reform of educational system because we want to argue that both training provision and transfer of knowledge can be enhanced by an efficient educational system.

[35] As reported by 100 %, 100 %, 97 %, 97 %, 97 %, 94 %, 92 % and 75 % of the respondent policy makers and experts respectively.

(a) Reform of educational system

Beginning with the reform of the educational system, Table 9.21 summarises the official view concerning the reform of the educational system, which highlights improvement of internal efficiency/quality of tertiary education, encouragement of modernisation and dynamism in the educational system and enhancing planning for educational need. In addition, the official prioritise the improvement of internal efficiency/quality of basic and secondary education, improvement of the quality of teachers and mentors, improvement of infrastructure, increasing the harmony/consistency between educational output and market needs by focusing on particular future skill needs, increasing public spending on education, increasing spending and incentives to encourage enrolment in technical education, encouraging the system of flexibility of educational institutions and encouraging the use of new technologies for improving education and skill. Further reform measures include monitoring educational needs on a regular basis, increasing motivation for improving education and skill, increasing motivation and incentives to change student attitudes, enhancing the linkages (network) between universities, colleges, technical and training institutes and increasing private sector involvement on education. Additional reform measures include encouragement of apprenticeship education, establishment of the Academy of Educational Sciences and establishment of a national council to establish links between the policies of population, education, training and employment. Further reform measures include provision of free education, increasing awareness of the value of e-learning, periodical review of education policies, development of curriculum to ensure consistency with labour market requirements (consistency between education output and labour market needs), enhancing enrolment in technical education by provision of incentives such as monthly support (bursaries) for students and incentives for technicians employed in government jobs. Finally, we suggest an improvement of duration of compulsory education and autonomy of educational institutions and serious intervention from the government to improve education in all its levels needs.

In recent years, there have been several recent initiatives in Sudan aimed at long-term solutions to develop human resources, reform educational and training programmes and the labour market. For instance, Sudan has established the National Center for Training and Administrative Systems, the Selection Committee for the National Civil Service and the Supreme Council for vocational training and apprenticeships – affiliated to the Sudanese Ministry of Labour, Public Service and Human Resources Development to help improve the skills of Sudanese national workers and help with looking for jobs. For instance, the National Center for Training and Administrative Systems is implementing policies linked and consistent with the government strategies that aim to enhance the development and modernisation of training, development of capacity, development of the civil service, address the gaps in the performance of the civil service, activate and develop the workers training programmes using modern methods of training, planning and determining of training needs and preparation of annual plans and budget for the establishment of training programmes and courses aimed at raising the skill level of workers in the civil service intermediate cadres and executive managers. Moreover, the Supreme Council for vocational training and apprenticeships is established in 2001 to be responsible of controlling and monitoring of vocational training and apprenticeships. It is implementing specialised policies aimed to develop vocational training and apprenticeships, preparation of plans and vocational training programmes, examine the need for vocational training in different units, the establishment of specialised committees to assist in implementing its duties, the adoption of training curricula for various categories, development of standards and levels of different professional, adopting approach to training of trainers and observers and any other categories, cooperation to arrange and conduct vocational and professional performance testing, encourage research in all vocational training and apprenticeships fields, certification of national vocational training and apprenticeship certificates or any other certificates.[36] In our view, these recent initiatives would be more effective if the government in Sudan collaborates with the private sector to work actively to influence both the supply and demand sides by implementing more effective policies to increase incentives, for example, through subsidies to improve both education and training. For instance, public policies can influence the demand side for education and change the low enrolment ratios at the tertiary level, especially technical education, by providing more fellowships, scholarships and prizes for engineering and science students, and increasing incentives for students to increase attraction for enrolment into science and engineering at secondary schools levels. The government should continue to upgrade schooling and increase enrolment at all levels, especially in higher education.

In our view Sudan can benefit from the experiences of other advanced countries to improve the coordination and planning to avoid the mismatch between supply and demand and to meet critical skills needs. For instance, Sudan can benefit from the experiences of the European countries, where the government limits itself to pay teachers' salaries and leaves the coordination problem to employers' federations. Sudan should continue to upgrade schooling and increase enrolment in all levels especially in higher education and should also induce firms to organise in a federation, which has the task to organise branch specific education, by using taxes as a stick and the payment of teachers' salaries as a carrot.

(b) Enhancing the transfer of knowledge/external schooling effect

The reform of the educational system is expected to have a direct positive effect on motivating/enhancing the transfer of knowledge/external schooling effects. For instance, Table 9.22 shows that the macro–micro views highlight the potential positive implications of improving the qualifications of skilled and unskilled workers, the quality of education and training in enhancing the transfer of knowledge/external schooling effects. Furthermore, from the official perspective, other important factors

[36] See: http://www.mol.gov.sd/index.php?option=com_content&view=article&id=16&Itemid=10, accessed 21 March, 2011.

Table 9.21 Policies and mechanisms for skill development: (a) macro/official view: Reform of educational system

Macro policies and mechanisms for enhancing the efficiency of education system	(%)
Improve the quality of teachers or mentors	97
Encourage the system of modernization and dynamism	100
Improve the infrastructures	97
Better planning for educational needs	100
Improve the internal efficiency/quality of basic education	97
Improve the internal efficiency/quality of secondary education	97
Improve the internal efficiency/quality of tertiary education	100
Enhance the linkages [networks] between universities, colleges, technical and training institutes	92
Monitoring educational needs on a regular basis	94
Encourage the system of flexibility of educational institutions	97
Increase the harmony/consistency between educational output and market needs by focusing on particular future skill needs	97
Increase public spending on education	97
Increasing the motivation and incentives to change the attitudes of educated economically active population	94
Increase spending and incentives to encourage enrolment in technical education	97
Increase private sector spending and involvement on education	86
Encouragement of the use of new technologies for improving education and skill	97
Increase motivation for improving education and skill	94
Total response	

Source: Own calculation based on the macro survey (2010)

for enhancing the external effect of schooling include enhancing the role of personal internal motivation and leadership in raising the level of skills, the relevant selection of more appropriate trainees for internal and external training programme to acquire training that match with their skills and enable them to transfer a positive external effect in their institutions. Moreover, from the firms' perspective the other important factors for enhancing the external effects of schooling includes provision of incentives, supporting provision of training opportunities at all levels of at the firms and supporting stability and continuity of workers in the firms. The macro–micro views differ with respect to the potential effect of improving firm conditions to encourage external effects and sponsoring educational scholarship. In addition, the macro survey shows that the policy makers and experts' view highlight the importance of improvement of awareness for enhancing the incidence and the transfer of knowledge/external schooling effect. Moreover, the macro survey shows that the policy makers and experts' view indicates that the transfer of knowledge/external schooling effect can be motivated via minimisation of education, learning and training costs in addition to increasing the interaction to market needs and increasing the information about future educational, training and skill needs in the productive sectors. There is also a need to increase awareness about the future value of investments in education and training to minimise the risk aversion: preference of more certain short term returns to available jobs than long-term skill investments.

Both the provision of adequate incentives for trainers and enhancing a system of certification of skills acquired are expected to have somewhat less important potential effects in the transfer of knowledge/external schooling effect.[37]

(c) Enhancing training provision

From the macro-firm surveys it appears that views of the policy makers and experts and those of the firms are consistent in highlighting the reform of educational system as an important mechanism to improve the provision of training, and the reform of educational and training systems for the enhancement of knowledge transfer/external schooling effects. However, on the other hand, there appears to be clear discrepancies between the macro–micro views (and also across firms) concerning the arrangement of priorities of other mechanisms for improving the provision of training. For instance, Table 9.23 presents the policy makers and experts' view to improve the provision of training that highlights enhancing training programmes to fit the changing technical needs, regular/adequate assessment and monitoring of training needs and improving the quality of trainers and mentors. Other measures towards enhancing planning for training needs, include: increasing the appreciation of/information on the benefits of training, enhancing the availability of finance to cover training costs, improve quality, efficiency and comprehensiveness and modernity of training programmes and enhancing the interactions between training institutions and firms, are also highlighted. Further, measures aimed at enhancing the educational qualifications of workers, enhancing the specialised training institutions, enhancing the appropriability of the return from investment in training, increasing the availability of training materials and equipment, increasing availability of trainers and mentors, enhancing training programmes to fit the changing skill needs, enhancing the system of training certification, increasing the participation of private training institutions and decentralisation of decision-making, are also mentioned. Furthermore, from the firms' perspective others enhancing factors for provision of training includes the promotion of awareness, encouragement of modernisation and development amend the restricting laws, reduce government control and censorship, enhancing adequate availability of trainers rather than restricting the movement of trainers and trained persons.

Moreover, from the official perspective, the other important factors for enhancing training include linking training by administrative development for staff, making training a prerequisite for career advancement, increasing the financial rate of return from training for the participants during the training period and after obtaining a certified training, enhancing continued practical training, encouraging training institutions to adopt certificates of competency and quality from international centres of excellence and finally increasing government concern for prioritising training by prioritising financing training and human resources

[37] As indicated by 97 %, 94 %, 94 %, 92 %, 89 % and 89 % of the respondent policy makers and experts respectively.

Table 9.22 Policies and mechanisms for skill development: (b) macro–micro views: factors enhancing the transfer of knowledge/external schooling effect

Factors enhance external schooling effect/knowledge transfer	All firms	Chemical	Food	Metal	Textile	Large	Medium	Small	Official
Improves the qualifications and ability of unskilled workers to learn from skilled workers	87 %	89 %	85 %	80 %	100 %	94 %	88 %	72 %	100 %
Improves the qualifications of skilled workers to permit the positive effects on unskilled workers	88 %	86 %	89 %	90 %	100 %	87 %	96 %	78 %	100 %
Improves the quality of training to coincide with international standard	82 %	83 %	78 %	80 %	100 %	84 %	81 %	78 %	97 %
Improves the quality of education	82 %	86 %	74 %	80 %	100 %	87 %	81 %	72 %	97 %
Improves firm conditions to encourage the external effects	82 %	86 %	81 %	60 %	100 %	84 %	73 %	89 %	97 %
Sponsors educational scholarship	74 %	80 %	67 %	60 %	100 %	74 %	77 %	72 %	100 %
Improves firm selection in both recruitment and termination	74 %	77 %	70 %	60 %	100 %	74 %	73 %	78 %	92 %
Improve awareness of the importance of external effect									100 %
Total response	76	35	27	10	4	31	26	18	

Sources: Own calculation based on the macro survey (2010) and firm survey (2010)

development apart from current government strategy that focus in prioritising financing roads, bridges, dams, etc.

The firms' view indicates that the provision of training could be improved by enhancing educational qualifications of workers, increasing the appreciation of/ information on the benefits of training, enhancing training programmes to fit the changing technical needs, enhancing the availability of finance to cover training costs and enhancing encouraging the specialised training institutions. Mechanisms such as, increasing availability of trainers and mentors, enhancing the availability of training materials and equipment, enhancing the system of training certification and restricting the mobility of trainers are viewed by the firms as being of somewhat less importance. That also holds for regular/adequate assessment and monitoring of training needs, and enhancing planning for training needs, enhancing the interaction between training institutions and firms, enhancing training programmes to fit the changing skill needs, improving the quality of trainers and mentors and enhancing the appropriability of the return from investment in training. Since training is costly, firms prefer policy interventions to finance training; however, it is less clear to what extent firms have a sound policy to contribute to training costs, as only 54 % of all respondent firms have upskilling plan (see for instance Chap. 5 above).

Our findings in this research are consistent with the findings of Suleiman (2007). For instance, Suleiman (2007) argues that the good provision of training required the provision of sufficient resources to meet all the needs of the Sudanese economy by enhancing the provision of training and variety of skills for the labour force in Sudan. In addition to the need to benefit from the technical assistance offered by friendship cooperation agreements with partner countries that in the past have been already long established in the field of vocational training for instance, in the past Germany has offered a good contribution to vocational training. Furthermore, the training in business management in various specialties requires cooperation to be undertaken between the public official bodies in charge of training with the private sector to provide a significant contribution to training in management according to the needs, in addition to investment incentives to encourage investors to train skilled workers, and by linking them with the quantity and quality achieved by the training facility for Sudanese workers. Finally, it is important to point out the importance of broadening and deepening the training of teachers by establishing institutes for education, especially in rural areas as one of the necessary steps to improve the quality of primary and secondary education, as well as considering training of Sudanese women. Suleiman (2007) notes that it is true that there are faculties of education in some universities in Sudan, but these colleges focus mainly on the theoretical, while the training of teachers needs more focus on the practical or applied sides. Therefore Suleiman (2007) indicates that it is necessary to create several institutes to train teachers at different levels of education and curriculum in order to achieve all goals of education, based on the lessons to be learned from the rich experiences of the Bakhtelrdah Institute, which was established in 1964 and enables the connection between teacher training and curriculum reform and scientific research in the development of primary and intermediate education. Suleiman (2007) believes that for a better future for training in Sudan, the country needs to start sooner in the preparation of these trainers both internally and externally through scholarships – each according to his qualifications and the type of training will do in the future (Suleiman 2007: 112, 114–115, 117).

Finally, in view of the complementary relationship between skills, skill upgrading and technological progress (see Chapter 7 above), the development of education, training, transfer of knowledge and skill levels may have further positive

Table 9.23 Policies and mechanisms for skill development: (c) macro–micro views: promotion of training

Policies intervention for enhancing training provision	Official	All firms	Chemical	Food	Metal	Textile	Large	Medium	Small
Enhancing the educational qualifications of workers	94 %	86 %	83 %	91 %	75 %	100 %	96 %	84 %	71 %
Enhancing the availability of training materials and equipment	94 %	71 %	83 %	67 %	50 %	75 %	87 %	63 %	62 %
Enhancing training programmes to fit the changing technical needs	100 %	78 %	78 %	81 %	88 %	33 %	86 %	74 %	69 %
Increasing the appreciation of or information on the benefits of training	97 %	81 %	87 %	82 %	75 %	50 %	87 %	79 %	71 %
Regular/adequate assessment and monitoring of training needs	100 %	69 %	83 %	67 %	50 %	33 %	86 %	53 %	69 %
Improving the quality of trainers and mentors	100 %	64 %	70 %	67 %	50 %	50 %	78 %	53 %	62 %
Enhancing training programmes to fit the changing skill needs	92 %	65 %	70 %	71 %	50 %	33 %	86 %	58 %	46 %
Increasing availability of trainers and mentors	94 %	71 %	74 %	67 %	75 %	75 %	87 %	58 %	69 %
Enhancing adequate planning for training programme/needs	97 %	69 %	78 %	67 %	63 %	33 %	82 %	63 %	62 %
Enhancing the availability of finance to cover training costs	97 %	76 %	83 %	81 %	50 %	67 %	86 %	74 %	69 %
Enhancing/encouraging the specialized training institutions	94 %	76 %	83 %	81 %	63 %	33 %	91 %	63 %	77 %
Enhancing the interactions between training institutions and firm	97 %	67 %	70 %	76 %	50 %	33 %	82 %	58 %	62 %
Enhancing the full appropriability of the return from investment in training	94 %	62 %	70 %	62 %	38 %	67 %	73 %	58 %	54 %
Enhancing the system of training certification of skills acquired	89 %	71 %	78 %	67 %	75 %	33 %	82 %	63 %	62 %
Increasing the participation of private training institutions	89 %								
Decentralization of decision-making	75 %								
Restriction the mobility of trainees	%	71 %	83 %	71 %	50 %	33 %	82 %	63 %	69 %
Improve quality, efficiency and comprehensiveness and modernity of training programmes	97 %								
Total response		57	23	22	8	4	23	19	14

Sources: Own calculation based on the macro survey (2010) and firm survey (2010)

implications on the development of local technologies. Accordingly, the promotion of local technologies depends on skill upgrading, the promotion of R&D activities and enhancement of networks systems, collaboration between universities, firms, public and private sectors and the implementation of an explicit technology policy.

Therefore, our findings in this section support our three hypotheses presented in Sect. 9.1 above. We verify our first hypothesis that Sudan needs to upgrade skill through the relevant policies for enhancing educational system, provision of training and transfer of knowledge/external schooling effect at the macro–micro levels. We prove our second hypothesis that the educational reform will have positive implications on: (a) enhancing training provision; (b) skill upgrading; (c) planning skill needs and matching educational output with the needs in the labour market; (d) enhancing the transfer of knowledge/schooling effect; and (e) collaboration between public and private institutions. Finally, we confirm our third hypothesis that the effective institutional environment: consistent policies of public and private institutions will enhance upskilling plan and skill development.

9.5 Conclusions

In this chapter we use some secondary data and information and the macro and firm surveys (2010) to analyse the educational, training and skill development policies in Sudan. We prove our hypothesis 8 in Chap. 1 above that Sudan's needs to enhance skill upgrading through the reform of the educational and training systems/policies and the transfer of knowledge. In particular, we show that skill development depends on: (a) reforming the educational system; (b) enhancing the provision of training; (c) planning skill needs and matching educational output with market needs; (e) enhancing the transfer of knowledge/schooling effect; and (d) incentives and collaboration between public and private institutions. We explain that the promotion of local technologies and adoption of appropriate foreign technologies and the interaction between both these to foster economic growth in Sudan depends on skill development. Particularly, on an enhancement of: (a) skill upgrading: educational and training systems; (b) R&D activities; (c) the transfer of knowledge/schooling effect; (d) networks system; and (e) incentives to motivate collaboration between universities and firms and between public and private institutions.

Our findings in Sect. 9.1 show that despite the enormous variations with respect to the supply and demand sides of educational policies, the educational policies in Sudan and Arab and Gulf countries share several problematic features such as an insufficient duration of compulsory education, the dominance of public sector and the lack of incentives/marginal contribution of the private sector on educational investment. Additional problems include poor quality, insufficient demand (enrolment ratios), an insufficient supply (spending) and the biased structure of tertiary education. We find that the priority of investment in education, as measured by public expenditures on education as a percentage of total government expenditures

the priority of investment as a percentage of GDP in Sudan lags far behind the level prevalent in the Arab and Gulf countries and the level of developed countries.

Our results show the low commitment to the standardised international adequacy, equity and efficiency criterion related to the supply and demand sides of educational policies. We explain the low commitment to the standardised international adequacy criterion in the supply and demand sides. On the supply side, the low commitment to the standardised international adequacy implies the allocation of less than 8 % of GDP on education and less than 20 % of total government or public spending on education and in the demand side the low commitment to the standardised international adequacy implies the inadequacy in intake and enrolment rates in primary and secondary education, gender equity in enrolment in education and literacy rate of population. Furthermore, we then discuss the equity criterion, which implies the equal distribution and allocation of financial resources to achieve the balance between the different education sectors and between different geographical rural and urban areas. Moreover, we then examine the international efficiency criterion which implies that low efficiency often appears from the low rates of attendance, high rates of dropout, high rates of repetition, weak rates of success in final exams, low rates of trained teachers and over-crowded classrooms as indicated by the rate of students enrolment per education institutions.

Moreover, we explain that it is probably plausible to interpret the observed regional disparity in the share in demand and enrolment in education due to demographic reason (as measured by the share in total population), economic reasons (as measured by per capita income and poverty rate) and other reasons (as measured by the degree of urbanisation) across the main regions in Sudan. Starting with the demographic reason, our results can be used to argue that the share in total population seems to be the first important factor determining the share and regional disparity in enrolment in education. Moreover, our findings can be used to argue that the economic reasons as measured by per capita income and poverty rate seem to be the second important factor that determining the share and regional disparity in enrolment and demand for education. Notably, our results imply that the incidence of high poverty rate seem to be the most important factor determining or limiting the demand and enrolment, notably, in basic education. These findings imply that especially among the poor regions, economic reasons were considered to be the most important factor limiting poor students and especially, girls' potential to complete their primary (basic), secondary and tertiary education and that region economic problems impact more negatively on female than on male education. These results imply that the increase in the incidence of poverty and the low per capita income limited or led to low demand and enrolment in education across the main, notably, poor region and this probably interpret the regional disparity in the demand for education across the main regions in Sudan. In addition, our findings can be used to argue that the degree of urbanisation is the third and other factor determining the share and regional disparity in enrolment in education. The major policy implication from our findings is that Sudan has the potential to achieve equity and fulfil the second and third MDG on universal access to primary education and gender equality respectively through reduction and elimination of poverty, notably, across the poor regions and poor population in Sudan, and this implies achievement of equity and international commitment to fulfilment of UN MDG in Sudan

We observe that while the educational policies in Sudan and the Gulf countries have raised enrolment ratios and literacy rates, they have failed to show satisfactory outcomes with respect to access to schooling/school life expectancy and training. This is due to serious deficiencies concerning the quality of education, coupled with the serious problems of biased structure and inadequate spending and enrolment in tertiary education in these countries. Hence, the major policy implication from our findings is that the improvement of the educational policies in Sudan and Gulf countries is vital and requires an improvement in the quality/internal efficiency, in the supply (investment) and demand (enrolment) sides, particularly in tertiary and technical education, and encouraging private sector investment in education.

Our results in Sect. 9.2 explain the major characteristics of training policies, in particular the structure and pattern of training policies and the impacts of education policies on training policies and skill upgrading in Sudan. First, we observe the biased structure in terms of type, duration and specialisation pattern of training policies and the low commitment to efficiency, adequacy and equity criterion in the provision of training and under utilisation of capacity of training centres all of which most probably critically undermined the efficient provision of training and led to limited effect of training in upgrading skill level for the labour force, this observable from the low share of trained labour force out of total labour force in Sudan. Next, we show that the implication and interaction between educational and training policies seem to be effective but limited to only within the largest two mixed and private firms, which appear more committed to implement skill upgrading policies that are consistent with the line of public policies. These two largest mixed and private firms successfully contribute to serve public policies of training and skill upgrading via establishing active human resources development units, recruitment policies and specialised training centres to implement various regular and special internal and external training programmes, especially for national workers. In addition they encourage the use of ICT to upgrade skill levels, offer scholarships and collaborate with universities to absorb young national graduates. These results support our earlier findings in Chap. 5 above, which indicate a lack of effective interaction between educational and training policies and a lack of incentives for provision of training within private firms. Hence, these findings imply a further duality/discrepancy at the micro level/across small-medium and large private firms.

In Sect. 9.3 we use the results of the macro and firm surveys (2010) to integrate the divergent macro–micro views concerning plans and mechanisms for skill development in the short and long run and propose some policies and recommendations. The short run plans include encouraging the system of accreditation and licensing, learning on the job, using ICT to upgrade skill levels, supporting long distance learning, improving the quality of teachers, trainers and mentors and bringing new foreign skills, scientists and engineers; this is followed by investment in education, training (including vocational training) and sending trainers, mentors and workers abroad to acquire skills. The long run plan highlights sending students abroad to acquire knowledge and skills, encouraging the system of accreditation and licensing, investment in training (including vocational training) of existing employees, improving the quality of teachers, trainers and mentors and sending teachers, mentors and workers abroad to acquire knowledge and skills.

The policy makers and experts' view concerning the reform of the educational system highlights improvement of internal efficiency/quality of tertiary education, encouragement of modernisation and dynamism in the educational system and enhancing planning for educational need. They also prioritise the improvement of internal efficiency/quality of basic and secondary education, improvement of the quality of teachers and mentors, improvement of infrastructure, increase in the harmony/consistency between educational output and market needs by focusing on particular future skill needs, increase in the public spending on education, increase in the spending and incentives to encourage enrolment in technical education, encouragement of the system of flexibility in educational institutions and encouragement of the use of new technologies for improving education and skill. Further reform measures include: monitoring educational needs on a regular basis; increasing motivation for improving education and skill; increasing motivation and incentives to change student attitudes; enhancing linkages (network) between universities, colleges, technical and training institutes; and increasing private sector involvement in education.

Moreover, the macro–micro suggestions with respect to knowledge transfer/ external schooling effects stress the improvement of quality of educational and training systems, qualifications of skilled and unskilled workers. In addition, recommendations include increasing information about future skill needs and the value of investments in education and training, interaction/consistency to market needs and a certification system and improvement of awareness for enhancing the incidence and the transfer of knowledge/external schooling effect.

Moreover, the macro–micro views regarding improvement of the provision of training vary in arranging priorities, but generally emphasise enhancement of training programmes to fit both the changing skill and technical needs, enhancing planning for training needs and quality and availability of trainers and mentors. Other measures include enhancing of training materials and equipment and educational qualifications of workers, assessing and monitoring of training needs regularly/adequately, increasing appreciation of/information on the benefits of training, enhancing availability of finance to cover training costs and enhancing specialised training institutions.

Therefore, our findings in this paper support our hypothesis 8 presented in Chap. 1 above. We verify part of our hypothesis 8.a that Sudan needs to upgrade skill through the relevant policies for enhancing educational system, provision of training and transfer of knowledge/external schooling effect at the macro–micro levels. We prove part of our hypothesis 8.a that the educational reform will have positive implications on: (a) enhancing training provision; (b) skill upgrading; (c) planning skill needs and matching educational output with the needs in the labour market, (d) enhancing the transfer of knowledge/schooling effect; and (e) collaboration between public and private institutions. Finally, we confirm our hypothesis 8 that the effective institutional environment and consistent policies of public and private institutions will enhance upskilling plan and skill development.

In addition, in view of the complementary relationship between skills, skill upgrading and technological progress (see earlier discussion in Chap. 1 above) the development of education, training, transfer of knowledge and skill levels may have further positive implications on the development of local technologies. Accordingly, this proves the second part of our hypothesis 8.b that the promotion of local technologies depends on skill upgrading, promotion of R&D activities and enhancement of networks system, collaborations between universities, firms, public and private sectors and the implementation of an explicit technology policy

Finally, our results show a serious discrepancy between small-medium and large private firms regarding the implementation of public policies of training and skill upgrading and also divergent macro–micro views concerning the arrangement of priorities to implement plans, mechanisms and policies for enhancing skill, provision of training and transfer of knowledge. Therefore, we recommend further efforts to be made to enhance the consistency between the macro–micro views and public-private sectors, particularly with respect to arrangement of priorities, plans and mechanisms to ensure more consistent, effective and successful policies for skill development and encouraging private sector participation in education and training. We find that the percentages of female students in all levels of education in Sudan are low compared to most other Arab countries. So, more efforts are required to improve the percentage of female student enrolment ratio for all levels of education in Sudan and Arab countries.

References

Akhbar Newspaper (2011). *Signing a partnership agreement between the fund of graduates employment and Kenana engineering company*. Akhbar newspaper, archive (805) 27 Feb 2011 . http://alakhbar.sd/details.php?articleid=933.

Alfakhery, J. J. (1999). *Human development indicators in the UAE: A comparative study of UNDP human development reports 1994–1998* (*Labour and Social Research and Studies Series* No. 1) (pp. 81–82). Dubai: The UAE Ministry of Labour and Social Affairs, Department of Studies, Research and Statistics, (in Arabic).

Almagzoub, A. (2007/2008). Study of the need in labour market for technician. In S. A. Mohamed (2008) (Ed), *Economic globalization and requirements of improving the efficiency and outputs of technical and technological education*. Paper prepared for the third conference of managers and senior directors of technical education in States. Khartoum: Ministry of Education, General Directorate of Technical Education, June 2008, Table 4, p. 12.

Al-Sanousi, T. A. (1999). Literacy and adult education in Sudan: Problems and future prospects, *Mahawir, Journal for Sudanese Studies*, (2), 26–35, Feb 1999. Omdurman, Sudan, M. O. Beshir Centre for Sudanese Studies, Omdurman Ahalia University.

Al-Sulayti, H.A. (2002). *Education and human development in the GCC countries: An analytical study*, the Emirates Center for Strategic Studies and Research (ECSSR) Strategic Studies No. 71, Abu Dhabi: ECSSR (in Arabic).

Al-Tuhami, O. A. (2007, November). *The independent approach of technical and technological education*. Report of the scientific committee, the National Council for Technical and Technological Education, Ministry of the Cabinet, Khartoum, pp. 2–12.
Barro, R. (1991). Economic growth in a cross section of countries. *Quarterly Journal of Economics CVI, 106*(2), 407–443.
Barro, R. (1996). *Determinants of economic growth: A cross-country empirical study*. (NBER Working Paper Series No. 5698).
Barro, R.J., & Lee, J.-W. (2010). *A new data set of educational attainment in the World, 1950–2010*. (NBER Working Paper No. 15902).
Barro, R. J., & Lee. W. (2000). International data on educational attainment updates and implications. (NBER Working Paper Series No. 7911).
Barro, R. J., & Lee, W. (1993). International comparison of educational attainment. *Journal of Monetary Economics, 32*(3), 363–394.
Barro, R. J., & Lee, W. (1996). International measures of schooling years and schooling quality. *American Economic Review Paper and Proceedings, 86*(2), 218–223.
Barro, R., & Lee, J.-W. (2000). International data on educational attainment: Updates and implications. Department of Economics, Harvard University. Data set at http://www.cid.harvard.edu/ciddata. Cited in Ali Abdel Gadir Ali. (2006). On human capital in post-conflict Sudan: Some exploratory results, API/WPS 0602, p. 14.
Becker, G. S. (1962). Investment in human capital: A theoretical analysis. *Journal of Political Economy, 70*(1), 9–49.
Becker, G. (1964). *Human capital*. Chicago: The University of Chicago press.
DAL Group Achievement Report (1970–2011). http://www.dalgroup.com/media/115-al-awwal-honours-the-first.html. Accessed Apr 2011.
Elbadawi, I., & Suleiman, K. (2008). Toward an equitable inter-governmental transfer system for the Sudan chapter 3. In S. Abd El Rhman (Ed.), *Generating growth and making fiscal decentralization work in post-conflict Sudan* (p. 107). Khartoum: Ministry of Finance and National Economy. Table 3.
ESTIME (2006). Towards science and technology evaluation in the Mediterranean countries' sixth framework programme project: Evaluation of scientific, technology and innovation capabilities in Mediterranean countries (pp. 1–80). www.estime.ird.fr.
Fadlallah, Y. A. (2006). Economic and social effects of the concentration of the sugar industry in Kenana area. Unpublished partial research for B.Sc. degree in geography, Al-zaim Al-azhari University, Khartoum, Sudan 2006, Research center, Kuwait, 13 Feb 2006. http://www.kuwait25.com/ab7ath/view.php?tales_id=651. Accessed Apr 2011.
Fahmey, M. S., & Mahmoud, H. A. (1993).Problems of educational administration in the gulf cooperation council countries. *The Arab Education office for the Gulf countries* – Riyadh (1993) In H. A. Al-Sulayti (2002) (Ed) Education and Human Development in the GCC countries: An analytical study (pp. 29–30). (ECSSR Strategic Studies No. 71), Abu Dhabi, ECSSR, (in Arabic).
Fergany, N. (1998).*Ro'iah Mustaqbaliyah lil Ta'leem fi-l-Wattan Al-'Arabi: Al-Watheeqa Ar-ai'eesiyah (A Future Vision for Education in the Arab World: The Main Document)* (pp. 18–19). Cairo: Almishkat .
Firm Survey. (2010, June). *Technological change and skill development: A comparative study of chemical, food, metal and textile small, medium and large scale enterprises in the Sudan*. Khartoum.
Ismail, S. A. (2008, August). Technical quality in the preparation and training of technical student and teacher.Wworking paper for the third conference of managers and senior directors of technical education in States, (pp. 2–19). Khartoum: Ministry of Education, General Directorate of Technical Education.
Jalal Al-Din, M. A. (2002). *The relationship between higher education and the world of work and production*. Omdurman: Omdurman Ahlia University, Mohamed Bashier Omer Centre for Sudanese studies and Friedrich Ebert Stiftung, Omdurman Ahlia University, Omdurman, Sudan.

Lucas, R. E. (1988). On the mechanics of economic development. *Journal of Monetary Economics, 22*, 3–42. North Holland.

Macro Survey. (2010, June). *Skill creation, human resources development and policy intervention: Interviews with policy makers and experts in the Sudan*. Khartoum.

Mankiw, N. G., Romer, D., & Weil, D. (1992). A contribution to the empirics of economic growth. *The Quarterly Journal of Economics, 107*(2), 407–437.

Mincer, J. (1984). Human capital and economic growth. *Economics of Education Review, 3*(3).

Mincer, J. (1989), Human Capital Response to Technological Change in the Labor Market, (NBER Working Paper Series No. 3207).

Mohamed, S. A. (2008, June). *Economic globalization and requirements of improving the efficiency and outputs of technical and technological education*. Paper prepared for the third conference of managers and senior directors of technical education in States, Ministry of Education, General Directorate of Technical Education, Khartoum, pp. 2–20.

Mohammed bin Rashid Al Maktoum Foundation (MBRF) and the United Nations Development Programme/Regional Bureau for Arab States (UNDP/RBAS) (2009). Arab knowledge report. 2009 *Towards productive intercommunication for knowledge*, Dubai: Al Ghurair Printing and Publishing House L.L.C.

Mulligan, C. B.,& Sala –i– Martin, X. (1995). *Measuring aggregate human capital*. (NBER Working Paper Series No. 5016), p.2.

Muysken, J., & Nour, S. (2006). Deficiencies in education and poor prospects for economic growth in the Gulf countries: The case of the UAE. *Journal of Development Studies, 42*(6), 957–980. Taylor and Francis Ltd.: UK.

Nour, S. (2005a, August). *Science and technology (S&T) development indicators in the Arab region: A comparative study of Arab Gulf and Mediterranean countries* (Published in UNU-INTECH Discussion Paper Series UNU-INTECHDPS2005-3), Maastricht.

Nour, S. (2005b). Science and technology (S&T) development indicators in the Arab region: A comparative study of Arab Gulf and Mediterranean countries. *The Journal of Science, Technology and Society (SAGE Publications), 10*(2), 249–274. Published in *UNU -INTECH Discussion Paper Series* UNU-INTECHDPS2005-3, Maastricht, the Netherlands, August 2005.

Nour, S. (2005c, November). Technological change and skill development in the Arab Gulf countries. *Doctoral dissertation*, Maastricht: *Maastricht University Press*.

Nour, S. (2011). Assessment of gender gap in Sudan. (UNU-MERIT Working Paper 2011–04). http://www.merit.unu.edu/publications/wppdf/2011/wp2011-004.pdf.

Romer, P. M. (1990). Endogenous technological change. *Journal of Political Economy, 98*(5), 71–102. Part 2.

Schultz, T. W. (1961). Investment in human capital. *American Economic Review, 51*(1), 1–17.

UNESCO -UIS Data and Statistics (2004). Country profile: country profile: statistics refer to the most recent year between 1998 and 2002. www.uis.unesco.org.

Stokey, N. L. (1991). Human capital, product quality and growth. *Quarterly Journal of Economics, 106*(2), 587–616.

Sudan Central Bureau of Statistics (2010). Annul statistical year book and statistical series (1990–2009), p.107.

Sudan Central Bureau of Statistics (2011: 12).

Sudan Central Bureau of Statistics Population Census Data. (2010d). *5th Sudan population and housing census (2008)*. Khartoum: Sudan Central Bureau of Statistics.

Sudan Ministry of Education (2000–2001). The annual educational statistics report. pp. 17, 22, 24, 29–52.

Sudan Ministry of Education (2001–2002). The annual educational statistics report. pp. 10–12, 30–32., 49, 52–54, 63–70, 76–77, 82–83, 88–89.

Sudan Ministry of Education (2003–2004). The annual educational statistics report. pp. 9, 12, 26, 34–37, 59–62, 70–77, 83–85, 90–91, 97–99.

Sudan Ministry of Education (2004–2005). The annual educational statistics report. pp. 9, 24, 32–35, 58–61, 73–80, 87–89, 94–96, 102–104.

Sudan Ministry of Education (2005–2006). The annual educational statistics report. pp. 9, 23, 31–34, 39–42, 61–64, 73–79, 85–86, 91–92, 97–98.

Sudan Ministry of Education (2006–2007). The annual educational statistics report. pp. 6, 23, 31, 34, 38–41, 59–62, 75–82, 88–89, 94–95, 114–115.

Sudan Ministry of Education (2007–2008). The annual educational statistics report. pp. 11, 25, 33–36, 40–65, 75–82, 88–89, 95–96, 115–116.

Sudan Ministry of Education (2008–2009). The annual educational statistics report. pp. 4, 28, 45, 58, 62–65, 84–87, 96–103, 109–110, 116–117, 136–137.

Sudan Ministry of Federal Government and the General Secretariat of the National Council for strategic planning (2009). Performance digital reports of the Northern States (2009) pp. 446–447, 471.

Sudan Ministry of Finance cited in Sudan Ministry of Education, Educational Planning Section of Educational Statistics (2003: p. 12).

Sudan Ministry of Higher Education and Scientific Research, the Annual Higher Educational Statistics Report: Various Issues (1993/1994–2009/2010).

Sudan Ministry of Labour. (2006). *Statistics on enrolment and graduates for Training of the apprenticeship programmes over the period (1995–2006)*. Unpublished statistics, Department of registration and certificates, Secretariat General of apprenticeship programme, Sudan Ministry of Labour, Khartoum.

Sudan National Council for Strategic Planning- General Secretariat (2010: 493, 497): Five-Year Plan (2007–2011): Report of the performance for three years (2007–2009) of the five-year plan, Khartoum, Sudan, Jan 2010.

Suleiman, A. A. (2007). Employment of training and education to serve economic and social development chapter five in *Future of economic and social development in Sudan* (pp. 121–123). Khartoum, Sudan: Sudan Currency Printing Press Company Ltd., (in Arabic).

The World Development Indicators database. (2004). http://data.worldbank.org/data-catalog/world-development-indicators. Accessed 15 Dec 2004.

The World Development Indicators database. (2005). http://data.worldbank.org/data-catalog/world-development-indicators. Accessed 15 Dec 2005.

The World Development Indicators database. (2011). http://data.worldbank.org/data-catalog/world-development-indicators. Accessed 25 June 2011.

UNDP. (2002). Human Development Report (2002): Deepening democracy in a fragmental World. UNDP- New York/Oxford: Oxford University Press, July 2002.

UNDP (2004). *Human Development Report (2004): Cultural liberty in today's diverse World.* UNDP. New York: Hoechstetter Printing July 2004.

UNDP- AHDR (2002). Arab Human Development Report (2002): Creating opportunities for future generations. *Regional Bureau for Arab State,* UNDP. New York/Amman: Regional Bureau for Arab States (RBAS), Icons Printing Services.

UNDP United Nations Development Programme UNDP (2010). Human Development Report 2010: The real wealth of nations: Pathways to human development. Human Development Report 2010 – 20th Anniversary Edition New York: UNDP, Palgrave Macmillan, Statistical Tables, Statistical Annex, Table 12: Decent Work, p.191.

UNESCO- UIS (2000). World Education Report (2000): UNESCO's World education indicators. *The right to education: towards education for all throughout life.* http://www.unesco.org/education/information/wer/htmlENG/tablesmenu.htm.

UNESCO- UIS (2004b). UIS web site global statistics on education: www.unesco.org.

UNESCO- UIS (2011). Global background information on education statistics: *UNESCO- UIS data centre: Beyond 20/20 WDS (2011)*. UNESCO- UIS (2011) UIS web site global statistics on education. www.unesco.org: http://stats.uis.unesco.org/unesco/TableViewer/tableView.aspx?ReportId=163. Accessed 22 Feb 2011

UNESCO Institute of Statistics Data and Statistics (1996). UNESCO statistical yearbook. *The World Education Indicators on school life expectancy 1992*. www.unesco.org.

UNESCO –UIS (2006). Educational statistics (2006)

UNESCO- UIS Data Centre (2011). Global background information on education statistics: Beyond 20/20 WDS (2011). UNESCO- UIS (2011) UIS web site global statistics on education. www.unesco.org: http://stats.uis.unesco.org/unesco/TableViewer/tableView.aspx?ReportId=163. Accessed 22 Feb 2011.

UNESCO-UIS, (2003). The UNESCO Education Data and Statistics Indicators 1998–2001. www.unesco.org.

UNESCO-UIS (2007). Global education digest 2007 comparing education statistics around the world, p. 138

UNESCO-UIS (2010). Global education digest 2010 comparing education statistics around the world, p. 176

United Nations Development Programme UNDP (2010). UNDP Sudan Country profile of Human Development Indicators (2010): http://hdrstats.undp.org/en/countries/profiles/SDN.html. Accessed 22 Dec22 2010.

United Nations Development Programme UNDP- The Arab Human Development Report AHDR (2003). Arab Human Development Report (2003): Building a knowledge society. UNDP. New York, Regional Bureau for Arab States (RBAS), Amman, Jordan: National Press, July 2003, P. 56; Table 5.1 p.102.

Waast, R. (2003). Science in Africa: From institutionalisation to scientific free market – what options for development? *Science, Technology and Society, 8*(2), 153–181.

Chapter 10
Summary and Conclusions

This concluding chapter summarises the major findings and contributions of the research. Sect. 10.1 briefly identifies the research problem and the major findings of the research; in Sect. 10.2 we show the relevance of the research findings to the general literature and contribution to the Sudanese literature; and, finally, Sect. 10.3 presents short outlines on policy recommendations.

10.1 Research Problem, Methodology and Major Findings

This research is composed of four parts and ten chapters: Part One includes both Chaps. 1 and 2 and presents the introduction and motivation of the research; Part Two contains Chap. 3, which presents the conceptual and theoretical frameworks; Part Three comprises Chaps. 4, 5, 6, 7, and 8 and presents the empirical application; and finally, Part Four encompasses Chaps 9 and 10 on policies, recommendations and conclusions. We explain below the main findings of each chapter.

Chapter 1 presents a brief introduction that gives a general overview of the research problem, its importance, relevance, objectives, questions, hypotheses and the structure of the research. We explain that the central themes of discussion in this research are the required skills formation and upskilling of the workers, together with their interaction with technological change in Sudan. In particular, we intend to provide an empirical investigation of the causes and consequences of deficient educational system, their interaction with the excessive share of low educated workers and their implications on skill levels, skills mismatch, transfer of knowledge, provision of training, level of local technologies and dependence on foreign technologies. In addition, we aim to present an indepth macro-micro analysis to assess technology and skill levels using a more comprehensive set of indicators.

Following the identification of the research problem in Chap. 1, we then present the background that motivated the research in Chap. 2. In particular, we show some stylised facts about Sudan, to examine the research problem more extensively, along with other strategic problems confronting economic development in Sudan.

S. Mohamed Nour, *Technological Change and Skill Development in Sudan*,
DOI 10.1007/978-3-642-32811-4_10, © Springer-Verlag Berlin Heidelberg 2013

We explain that oil greatly contributed to economic development in Sudan by satisfying domestic consumption and achievement of self sufficiency, increasing government and public revenues, rapid and impressive economic growth as measured by the growth in the GDP and its composition and structure, increasing foreign direct investment (FDI) and increasing the volume of foreign trade as measured by the volume and structure of exports. We find that however, the recent heavy dependence on oil, may lead to negative impacts and serious challenges for Sudan since oil is an exhaustible resource and because of the instability of oil prices in the international market the revenue from oil is uncertain and volatile and may lead to instability of economic growth. Moreover, the increasing dependence on oil leads to increasing debate for and against the incidence of the Dutch Disease phenomenon in the Sudanese economy, the lack of diversification and the challenges related to potential north-south conflict and division of the country. Hence, economic growth and sustainable development strategy in Sudan depends on economic diversification, which in turn is contingent upon the availability of adequate and appropriate skills and technologies.

We also illustrate the other serious structural problems in Sudan that relate to the Dutch Disease, structural imbalances in the labour market, poverty, unbalanced development strategy, growing unemployment, slowdown in economic growth and declining labour productivity. We illustrate the low skill and technology indicators and the substantial gap prevalent in Sudan when compared to the world's rapidly advanced countries. In our view, Sudan needs to upgrade skill levels and encourage the development of local technologies to narrow the technological gap and achieve economic development in Sudan. In particular, skill upgrading is imperative to facilitating economic diversification, restructuring the labour market, enhancing productivity of labour, lessening dependence on foreign workers and, consequently, poverty and unemployment in the labour market. Skill upgrading through enhancing educational and training systems is essential to facilitate the adoption of appropriate foreign technologies in the short run, and to encourage the development of local technologies through encouraging R&D activities to reduce the technological gap and dependence on foreign technologies in the long run. Therefore, our findings in Chap. 2 confirm our first hypothesis in Chap. 1 above that Sudan needs to promote the local skill and local technologies to face the challenges created by the depletion of oil resources. Sudan needs to implement the three strategies of diversification, building local technological capacity and restructuring the labour market. Our results confirm part of the second hypothesis in Chap. 1 above that in the short- and medium-term, Sudan is unable to rely on local skills and local technologies and remains heavily dependent on both foreign skills and foreign technologies at the macro level.

In light of the findings in Chap. 2, it becomes more plausible to highlight the need for improvement of education or skill upgrading and the development of local technologies or technological progress to facilitate economic diversification and ensure long run economic growth and sustainable development in Sudan. Before commencing with the empirical analysis, Chap. 3 briefly explains the concepts, measures and theoretical and empirical literature in relation to human capital (education), technological change and economic growth. We provide a background for the empirical analysis in the subsequent chapters by surveying the theoretical and empirical literature that emphasise the positive endogenous growth effects of technical change and human capital in increasing and sustaining economic growth. We explain that economic growth theories recognised and provided different perceptions and analytical frameworks for modelling the various effects of technical change, innovation and human capital on economic growth. The major differences arise because exogenous growth theories perceive and model technical progress and human capital as exogenous variables in growth accounting model, while, in contrast, the endogenous growth theory envisages and models technical progress and human capital as endogenous variables determining the rates and differences of economic growth across countries. The endogenous growth theory contributes towards improving the understanding of the interaction between technological change, human capital and economic growth and fills the gap in earlier growth theories by considering the important endogenous effects of human capital, technological progress and innovation. It also predicts that in the long run economic growth at the aggregate level is determined by endogenous sources of technological change: human capital, learning by doing, spillovers of knowledge and external effect of human capital. The presence of increasing returns to scale and externalities prevent diminishing returns to the accumulation of capital and thereby ensure the steady state of growth in the long run. While the feature of spillovers of knowledge supports endogenous growth, it also creates a form of externality and implies that private investments generate a positive external effect and the private returns from investment tend to be lower than the social returns. The outcomes tend not to be Pareto optimal but sub optimal and they require government intervention to correct the distortion using various instruments, such as providing subsidies (which can be financed by taxation) to improve the accumulation of technology and human capital, the incentives and returns from investment for private investors. We illustrate that the inclusion of human capital and technological change in growth accounting models motivate the endogenous growth literature to provide several interesting explanations of the relationship between human capital and technical progress. In particular, it stimulates considerable debate about the complementary relationship between human capital and technical progress, skilled biased technical change, the role of technical progress in skill upgrading and the role of skills and improvement in the accumulation of human capital in skill upgrading. These explanations imply that next to the important endogenous effects of technical progress and human capital in economic growth, the complementary relationships amongst and between them and skill upgrading are also important for enhancing economic growth. Finally, we show the advantages and limitations of various measures of technological change and human capital that have been used in the theoretical and empirical literature, and we then select the most relevant measures for our empirical analysis in the subsequent chapters according to the availability of data.

We consider the endogenous growth framework as a useful background for the empirical investigation in the following chapters. Before we go into the empirical analysis, we define the methods of data collection including both surveys and interviews in Chap. 4. We utilise the surveys data in our analysis and use the descriptive, comparative and statistical (OLS regression) methods of analysis. We explain that the basic objective of performing the macro and firm surveys is to obtain specific information to provide insights into the factors influencing or the causes and consequences of low skill and technology and to help to generate policies to improve skill and technology at the macro-micro levels. The macro survey examines the causes and consequences of the deficiency of the educational system and the firm survey discusses the implications of the excessive use of unskilled workers. The field research to collect our primary data was held in the period from January to June 2010, in Sudan as a case study of the Arab countries. The selection and focus of our analysis on Sudan was related to the easy accessibility to data and information and facilities for the fulfilment of the fieldwork/surveys, which were offered by the Department of Economics, Faculty of Economic and Social Studies at the University of Khartoum. Moreover, the case of Sudan is both important and interesting as in recent years, Sudan shows significant increase in terms of ICT diffusion and shows rapid increase in the net inflow of foreign direct investment, for instance, Sudan is the fourth country in Africa and fifth in the Arab regions in terms of attraction of foreign investment.

In Chap. 4 we explain the selection of the sample, its composition, operation, coverage, advantages and limitations of the survey data, and show the structure and design of the questionnaire. The firm survey (2010) on 'Technological Change and Skill Development' covers 100 of the small, medium and large size firms working in four industries in the manufacturing sector: the chemical, food, metal and textile industries. The selection of these four industries was based on many reasons, the most important of which is that the argument for both upskilling and technological upgrading is promising in these four sectors and can be used to reduce poverty and unemployment problems in Sudan and also due to the important contribution of these sectors in total output, value added, capital investment, employment, exports, imports and number of industrial establishments in the manufacturing sector. The sample in the firm survey was drawn from the small, medium and large size firms active in the chemical, food, metal and textile industries, which are located in Khartoum state. The selection of Khartoum state was based on its significant and highest average share in total employment, capital investment and total number of factories and industrial establishments engaged in the chemical, food, metal and textile industries and also because the manufacturing industries in Khartoum state are characterised by being more diversified compared to other states in Sudan. The macro survey (2010) on 'Skill Creation, Human Resources Development and Policy Intervention' was sent to 40 of policy makers and experts in 8 public, university, educational, training and research institutions in Sudan. The number of respondent firms and policy makers are 87 and 36 respectively. In addition to support the firm and macro surveys we collected primary data by conducting interviews and small survey on R&D based on 25 face-to-face interviews with the official policy makers and experts in the government and the academic staff in the public and private universities. The main purpose of this survey is to collect primary data to investigate the causes and consequences of poor R&D activities, to examine the main factors hindering and those contributing towards the promotion of R&D and to give recommendations to improve R&D and S&T in Sudan.

The data from the firm and macro surveys provide us with the required information, which is particularly useful for presenting a macro-micro comparative analysis to identify the causes and consequences of the skills problem and the policies for skill development from macro-micro perspectives. The results of the macro survey seem quite representative, since the selection covers governments, universities, and educational and training institutions. One advantage of the macro survey is that it examines the problem after integrating the two different perspectives of policy makers and experts and also integrating two different perspectives from both the fields of education and training. The results of the firm survey are also quite representative, since the selection and coverage of firms in the survey includes a broad range of firms working in the chemical, food, metal and textile industries, which provides us with relevant data and information that is of considerable use in our analysis. Such coverage also has the advantage of enabling us to compare between firms according to two criteria: the size of employment and the industrial activity. One major limitation with respect to the firm survey is the low response rate for some questions, especially when the answers or data required quantitative measurement. Such problems arise because some of the respondent firms were unwilling to provide complete and reliable quantitative data, while others offered somewhat selective answers.

Apart from this limitation, the data from the firm and macro surveys remains useful, not only for the empirical investigation in Chaps. 5, 6, 7, 8 and 9, but also for the policy analysis and suggestions in Chap. 9. We begin our analysis in Chap. 5, by using the results from the macro and firm surveys to verify our third hypothesis in Chap. 1 above, about the serious implications of the interaction between the deficient educational system and the high incidence of unskilled workers. In particular, the results from the macro survey show that the deficient educational system is attributed to many causes such as the poor quality of education that leads to many serious consequences including low skill levels, poor provision of training, skills mismatch and low transfer of knowledge at the macro level. In addition, the results from the firm survey illustrate that the excessive use of low educated workers leads to several serious implications such as low skill levels, poor provision of training, skills mismatch, poor technology indicators, weak adaptation of imported technologies and a heavy dependence on foreign technologies. Our findings from the surveys and follow-up interviews indicate that the poor technology indicators/indigenous capability to build the local technology and heavy dependence on foreign technology can be attributed to low skill levels, lack of R&D activities, weak linkages, lack of networks systems, and low transfer of knowledge. These findings at the micro level seem consistent with those at the macro level.

Chapter 6 shows the status of S&T input and output indicators at the macro level in Sudan; it explains that the combination of poor S&T inputs/resources together with an inadequate economic system as a whole results in Sudan producing poor S&T outputs/performances. Moreover, we find that most R&D and S&T activities and FTER employment in Sudan occur within the public and university sectors, while the private sector and industry make only a minor contribution. When comparing the same S&T input and output indicators of Sudan with those of the Arab countries and world's other developed and developing countries, our findings indicate that Sudan lags behind in terms of most S&T input indicators (both financial and human resources). That also holds for the average share of high-technology exports, GDP per capita growth, number of scientific publications, level of share in international publication and number of patent filings. Our findings indicate that despite the important role of R&D in satisfying the needs for economic development, development of local technologies and adaptation to imported foreign technologies. However, the contribution of R&D seems to be constrained mainly by the lack of finance to cover the high costs of R&D as the main problem, moreover, the lack of human resources (researchers and qualified workers in R&D fields) is also mentioned but of somewhat less importance. We find that from the policy makers and experts' perspective the main problems hindering R&D include: the lack of finance from public sector; lack of management and organisational ability; lack of coordination and weak relationships, networks and consistency and cooperation between universities and higher education institutions on the one side and the productive sector (agriculture, industry, services) on the other side; lack of R&D culture; lack of finance from the private sector; lack of favourable conditions and the necessary facilities; lack of awareness and appreciation of the economic values of R&D, and lack of human resources (researchers and qualified workers in R&D fields) respectively.

When distinguishing between firms according to firm size and industry level, we find that skill and technology indicators show considerable variation across firms. Our findings show one surprising contradicting macro-micro view. Notably, the contradicting optimistic-pessimistic micro and macro view regarding the incidence and success of knowledge transfer/external schooling effect implies that the transfer of knowledge/the external effects of schooling is probably successful within firms, but is unsuccessful within society at large. This is probably because the transfer of knowledge is hindered by: the low quality of education; the weak linkages and a lack of networks between universities, colleges, technical and training institutes and the productive sectors; the incidence of high illiteracy rate; and incidence of high mismatch between educational output of population and labour market. Our observation show a consistent optimistic macro and micro view, concerning the self-reliance on local skill and the role of both technological upgrading and upskilling in reinforcing it, implies that the self-reliance strategy is probably not only a preferred government strategy but probably is also one followed by private firms. Though driven by profit-maximising considerations, private firms are likely to continue in hiring cheap readymade skilled workers rather than in hiring, training and upskilling workers with expensive costs. From these observations, our results accept hypothesis 8.c. in Chap. 1 above about the consistency of upskilling and transfer of knowledge at the macro-micro levels. These results corroborate a part of the sixth hypothesis in Chap. 1 above with respect to the failure and the factors hindering the transfer of knowledge/external schooling effects at the macro level and also corroborate part of hypothesis 8.c. in Chap. 1 above about the consistency of upskilling at the macro-micro levels. But, on the other hand, our findings surprisingly reject a part of the sixth hypothesis concerning the failure of the transfer of knowledge/external schooling effect at the micro level, and also reject part of hypothesis 8.c. in Chap. 1 above about the consistency of the transfer of knowledge at the macro-micro levels.

The surprising results from Chap. 5 motivate our research to attempt a more comprehensive analysis of skills problem and the implications of unskilled workers at the micro level/across private firms. Therefore, in Chap. 7, we then use the data from the firm survey (2010) to broaden our earlier analysis in Chap. 5 by providing an indepth analysis of skill indicators, their implications and relationships with wages, upskilling (ICT training) and technology (ICT) indicators at the micro/firm level. Our findings illustrate the low skill levels – due to the excessive share of unskilled workers - and the implications on skills mismatch, and industrial performance indicators and productivity decline across firms. These results are consistent with the micro-macro findings in Chap. 5, which indicate the low share of high skilled workers in total population and employment (measured by both educational and occupational levels) and the serious implications for skills mismatch. Furthermore, we show that difference between required and actual education indicates severe skills mismatch across all firms and all skill levels, especially, across high and low skilled workers respectively and across medium, small, chemical, food and metal firms respectively. These findings, together with those in Chap. 5, verify hypotheses 3 and 4.a in Chap. 1 above regarding the implications of low skill levels and the argument, earlier in Chap. 2, about the pressing need for upskilling, mainly in the private sector.

Our results show positive correlations between skill (actual and required education, experience and its square) and average wages. We show that differences in skill levels/stock of human capital (share of high skilled in total employment) across firms are related to market size (share in total employment, capital, output/sales and profit). We also find that an increase in skill level (share of high skilled in total employment) and firm size lead to improved relationships between actual and required education, experience and its square and wages. Next, our results show positive complementary relationships between technology (ICT), skill and upskilling (ICT training). We illustrate that an increase in skill level (share of high skilled in total employment) and firm size lead to improved in the complementary relationships between skill, upskilling and technology (ICT). The relationships between skill indicators and wages and between skill, technology (ICT) and upskilling (ICT training) substantiate our fourth hypothesis (4.b–4.c) in Chap. 1, and agree with the findings in Chap. 5 concerning the differences in skill and technology indicators across firms according to firm size and industry. These results imply the importance of good education/skill levels for bridging differences between firms and also for improving skills, technology and upskilling complementarity at the micro level.

These results guide us in Chap. 8 to use the firm survey (2010) data at the micro level and secondary data at the macro level to examine and verify hypothesis 7 in

Chap. 1 above concerning the importance/impacts of tacit and codified knowledge at the micro-macro levels respectively. We find that at the macro level, tacit knowledge and codified sources of knowledge are positively and significantly correlated with both schooling years and GDP growth (economic growth rate). Moreover, we find that at the macro level codified knowledge and the number of FTER show positive correlations with technology (patents). Furthermore, our results at the macro level show significant positive complementary relationships between codified knowledge and the number of FTER, which we interpret as a complementary relationship between tacit knowledge and codified knowledge. Moreover, at the micro (firm) level, we illustrate the importance of tacit knowledge, and we illustrate that tacit knowledge is positively and significantly correlated with technology (expenditures on ICT) and upskilling (expenditures on training), output (defined by total sales value), output diversification, productivity and profit. In addition, we find that at the micro (firm) level, tacit and codified knowledge show positive significant correlations with total investment, capital, and firm size. The major implication of our findings is the positive correlation between knowledge and various variables at both the micro and macro levels. Therefore, further incentives should be provided to improve tacit and codified sources of knowledge at the macro and micro levels. Another implication is that the positive impact of tacit knowledge also implies the importance of a good education since tacit knowledge is often embodied in educated people and so in human capital.

In view of the findings in Chaps. 2, 3, and 5, 6, 7, 8, which indicate the importance of a good education, in Chap. 9 we use the data from the firm and macro surveys to conclude our study with policy analysis. We corroborate hypothesis 8 (8.a–8.b) in Chap. 1 above concerning the need for skill and technological upgrading through the reform of the educational and training systems and the transfer of knowledge. Moreover, we use secondary data and information and the macro and firm surveys (2010) to present a policy analysis of the educational (supply-demand sides) and training systems in Sudan. We show that the educational policies in Sudan and Arab and Gulf countries share several problematic features such as an insufficient duration of compulsory education, the dominance of public sector and the lack of incentives/marginal contribution of the private sector on educational investment. Additional problems include poor quality, insufficient demand (enrolment ratios), an insufficient supply (spending) and the biased structure of tertiary education. However, despite these similarities, we also observe enormous variations, particularly with respect to the supply and demand sides of educational policies. Differences in the supply side include financial resources or priority of public expenditures on education relative to GDP and total government expenditures, allocation/distribution of public spending and investment at various educational levels, human resources or availability of teaching staff and the extent of privatisation. Differences regarding the demand side include enrolment ratios and outcomes or implications on literacy rates, access to schooling/school life expectancy and interaction with training. We find that the priority of investment in education, as measured by public expenditures on education as a percentage of total government expenditures and priority of investment as a percentage of GDP in Sudan lags far behind the level prevalent in the Arab and Gulf countries and the level of developed countries. When comparing supply-demand sides, it turns out that the supply side or public spending seems to be only one component in educational policies, because higher private spending in tertiary education per se does not lead to higher demand, participation, access and enrolment ratios in private tertiary education, which shows moderate private spending, but low demand/enrolment ratios, most probably because of high cost of private tertiary education and high poverty rates in Sudan. We observe that while the educational policies in Sudan, Arab and Gulf countries have raised enrolment ratios and literacy rates, they have failed to show satisfactory outcomes with respect to access to schooling/school life expectancy and training. This is due to serious deficiencies concerning the quality of education, coupled with the serious problems of biased structure and inadequate spending and enrolment in tertiary education in these countries. Hence, the major policy implication from our results is that the improvement of the educational policies in Sudan, Arab and Gulf countries is vital and requires an improvement in the quality/internal efficiency, in the supply (investment) and demand (enrolment) sides, particularly in tertiary and technical education, and encouraging private sector investment in education.

Our results show the low commitment to the standardised international adequacy, equity and efficiency criterion related to the supply and demand sides of educational policies. We explain the low commitment to the standardised international adequacy criterion on the supply side that implies the allocation of less than 8 % of GDP on education and less than 20 % of total government or public spending on education and on the demand side that implies the adequacy in intake and enrolment rates in primary and secondary education, gender equity in enrolment in education and literacy rate of population. Furthermore, we then discuss the low commitment to the standardised international equity criterion, which implies the lack of equal distribution and allocation of financial resources to achieve the balance between the different education sectors and between different geographical rural and urban areas. Moreover, we then examine the low commitment to the standardised international efficiency criterion which implies that the low efficiency often appears from the low rates of attendance, high rates of dropout, high rates of repetition, weak rates of success in final exams, low rates of trained teachers and over-crowded classrooms as indicated by the rate of students enrolment per education institutions. Moreover, we explain that it is probably plausible to interpret the observed regional disparity in the share in demand and enrolment in education due to three important determining factors or reasons, notably, the demographic reason (as measured by the share in total population), economic reasons (as measured by per capita income and poverty rate) and other reasons (as measured by the degree of urbanisation) across the main regions in Sudan. Our results imply that the increase in the incidence of high poverty rate and the low per capita income seem to be the most important factors determining, limiting or that led to low demand and enrolment in education across the main, notably, poor regions, particularly in basic education. These results imply that especially among the poor regions, economic reasons are considered to be the most important factor limiting poor students' and especially, girls' potential to complete their primary (basic), secondary and tertiary education and that region economic problems impact more negatively on female than on male education. Our results imply that these factors probably interpret the regional disparity in the demand for education across the main regions in Sudan. The major policy implication from our results is that Sudan has the potential to achieve equity and fulfil the second and third MDG on universal access to primary education and gender equality respectively through reduction and elimination of poverty, notably, across the poor regions and poor population in Sudan, and this implies achievement of equity and international commitment to fulfilment of UN MDG in Sudan.

Our results show that the implication and interaction between educational and training policies seem to be effective but limited to only within the largest two mixed and private firms, which appear more committed to implement skill upgrading policies that are consistent with the line of public policies. These two largest mixed and private firms successfully contribute to serve public policies of training and skill upgrading via establishing active human resources development units, recruitment policies and specialized training centres to implement various regular and special internal and external training programmes, especially for national workers. In addition they encourage the use of ICT to upgrade skill levels, offer scholarships and collaborate with universities to absorb young national graduates. These results support our earlier findings in Chap. 5, which indicate a lack of effective interaction between educational and training policies and a lack of incentives for provision of training within private firms. Hence, these findings imply a further duality/discrepancy at the micro level/across small-medium and large private firms.

We next discuss the macro-micro views concerning plans, mechanisms and policies for skill development through enhancing the educational system, provision of training, transfer of knowledge/external schooling effects, effective collaboration between public and private institutions and increasing incentives for private sector investment in education and training in Sudan. Our results show a serious discrepancy and divergence in arranging priorities to implement plans, mechanisms and policies for enhancing skill levels, provision of training and transfer of knowledge at the macro-micro levels. These results are opposite to earlier observations in Chap. 5 concerning the contradicting optimistic-pessimistic macro and micro views concerning upskilling efforts and/or the self-reliance on local skill and the role of both technological upgrading and upskilling in reinforcing it. Therefore, we recommend further efforts be made to enhance the consistency between the macro-micro views and the public-private sectors, particularly in the arrangement of priorities, plans and mechanisms to ensure more consistent, effective and successful policies for skills development.

10.2 Relevance and Contribution of the Research

Most of our findings in this research are consistent with the new growth literature, Arab literature and the Sudanese literature. Compared to the endogenous growth framework, we provide further evidence in support of the endogenous and new growth literature, in particular with respect to the positive correlation between actual education, experience, its square and wages. Our results in Chap. 7 show positive complementary relationships between technology (ICT), skill and upskilling (ICT training), these findings seem consistent with the theoretical framework in Chap. 3, endogenous growth framework and the stylised facts in the new growth literature concerning the relationships between human capital, technical progress and upskilling. Our findings are broadly consistent with and provide further evidence in support of the findings in the new growth literature concerning the skilled biased technical change theorem. In addition, our results concur both with the general literature that defines both skill and technology in relation to firm characteristics (size and industry), and also the recent literature highlighting the growing effects of new technologies, especially ICT diffusion. Our results in Chap. 8 verify four stylised facts about the importance/impacts of knowledge at the micro and macro levels, and are in line with the recent general findings in the knowledge literature. In particular, the complementary relationship between tacit knowledge and codified sources of knowledge at the macro level and the significant correlations between both tacit knowledge and codified sources of knowledge and output and growth at the micro and macro levels respectively. Tacit knowledge is important not only through its direct effects, but also through its further effects on upskilling and the use of technology (ICT). Our findings about the knowledge components have less significant impacts on output (total sales value) that follow the effects of traditional variables (i.e. labour and capital) are in contrast to the recent results in the knowledge literature, which indicate that knowledge components have more significant impacts on output (total sales value) that exceed the effects of traditional variables (i.e. labour and capital).

On the other hand, we find a positive significant correlation between the use of/total spending on ICT and total profit and value added, but an insignificant correlation between the use of/total spending on ICT and output at the micro/firm level. This result proves our fifth hypothesis in Chap. 1 above and the observations about the insignificant/inconclusive effect of ICT at the macro level in Sudan and the recent literature in the Arab and developing countries. However, our results with respect to ICT should be interpreted more carefully as they probably have two-way causality and may imply a possibility for reversed causality. Mainly because more profit and output would imply more financial capacity that permits more spending on ICT, on the other hand, more spending on ICT implies higher costs and lower profit.

Our findings about the significant correlation between the required education and wages are consistent with the findings in the new growth literature on the importance of job characteristics (skills required) in wages determination.

Compared to the Sudanese literature, we provide new evidence and add to the few recent studies in Sudan that highlight the need for upskilling, and the low skill, low technological level and dependency on foreign technologies. Compared to the Arab and Sudanese literature, our research is important for elaborating and providing a more indepth analysis, not only for assessing Sudanese technology-skill indicators using a more comprehensive set of indicators than often used in the new growth literature, but also for analysing the causes and consequences of low skills and technology, the relation between them at both the macro and micro levels, and for addressing policy aspects aiming to enhance them. Basically, we identify upskilling as an essential element for building adequate human resources needed for the fulfilment of long run economic growth and sustainable development strategies in Sudan: achieving economic stabilisation; balanced development; economic diversification; reducing poverty; reducing unemployment; restructuring and reducing imbalances in the labour market; creation of adequate and appropriate employment opportunities; enhancing self-reliance on domestic capital and workers; and building institutional reform and technological development. We show that the low skill level is basically attributed to the deficient educational system – due to low quality of education – and high incidence of unskilled workers at the macro and micro levels respectively. The importance of our analysis is the identification of the numerous implications of the interaction between a poor educational system and an excessive share of low educated workers that leads to low skill levels, poor provision of training, skills mismatch, low transfer of knowledge, poor technology indicators and high dependence on foreign technologies. We add to the results of Hassan (2009) and Elamin (2009) regarding the lack of technology policy, technical skills, R&D, technology culture in Sudanese society and the mismatch in the labour market due to deficient educational system and cultural reasons. Using new primary data based on the R&D survey (2010) and firm survey (2010) we provide a new contribution and fill the gap in the Sudanese literature by examining the major factors hindering R&D at the macro and micro levels in Sudan respectively. At the macro level our results from the R&D survey (2010) show that despite the importance of R&D in satisfying the needs for economic development, development of local technologies and adaptation to imported foreign technologies; however, the contribution of R&D seems to be constrained mainly by the lack of finance to cover the high costs of R&D as the main problem, moreover, the lack of human resources (researchers and qualified workers in R&D fields) is also mentioned but is of somewhat less importance. We find that from the policy makers and experts' perspective the main problems hindering R&D at the macro level include: lack of finance from the public sector; lack of management and organisational ability; lack of coordination and weak relationships, networks and consistency and cooperation between universities and higher education institutions on the one side and the productive sector (agriculture, industry, services) on the other side; lack of R&D culture, lack of finance from the private sector; lack of favourable conditions and necessary facilities; lack of awareness and appreciation of the economic values of R&D; and lack of human resources (researchers and qualified workers in R&D fields) respectively. At the micro level our results from the firm survey (2010) show that the lack of local efforts for technology development is basically related to low R&D efforts that are attributed to low skill level, lack of networks systems, fruitful cooperation between universities and firms, lack of resources and lack of social awareness and concern.

We provide basic and new contributions and fill the gap in the Sudanese literature by investigating the significance of the incidence and transfer of knowledge/external schooling effects, the factors hindering and those contributing towards enhancing them at the macro and micro levels in Sudan. We show the significance of tacit and codified sources of knowledge at the micro-macro levels. Unlike the few recent studies of knowledge in the Arab countries, one advantage of our analysis is that we provide a more specific analysis that focuses only on Sudan as a new case in the Arab countries. Different from earlier studies, we provide new empirical investigation of both the importance (impacts) of tacit knowledge at the micro level – see our discussion in Chap. 8 – and the discrepancy in the transfer of knowledge/ external schooling effects at the macro-micro levels – see our discussion in Chap. 5. We show the positive correlation between skill indicators (actual and required education and experience) and between these indicators and wages; the distribution of average wages is significantly correlated with the actual education, experience and its square, and also seem to be sensitive to the required education. We add to the literature in indicating the implications of the poor educational system and the high incidence of unskilled workers in the labour market, particularly the skills mismatch problem at the macro level. A novel element of our research, distinguishing it from the few earlier studies, is that we manage to measure the skills mismatch at both the macro and the micro level/across firms. Our new findings show that at the micro level although occupations are improving with education, schooling requirements are seldom significantly match with the actual/attained schooling, especially within both the high and low skilled worker groups. The gap that appears between the required and actual/attained schooling indicates a mismatch at the micro level, which is notably higher within both high and low skilled workers respectively.

To complement the micro analysis, at the macro level an important contribution in our analysis is that we examine the relationships between the low skill level, skill mismatch, structure of labour market and unemployment in Sudan. One advantage of our analysis is that we explain several stylised facts on labour market using new secondary data on population, employment and unemployment based on the Sudan Central Bureau of Statistics (2010) 'Fifth Sudan Population and Housing Census (2008)'. An interesting element in our analysis is that we explain several stylised facts on the relation between structure of labour market and demographic structure, labour force, participation rates, economic activities, low skill level and high unemployment rate, defined by gender and mode of living in Sudan. Different from the results in the empirical literature in support of Phillips curve on the negative correlation between inflation and unemployment rates, we find positive and significant correlation between unemployment and inflation rates in Sudan during the period 2000–2008. Moreover, different from the analysis in the Sudanese literature we present a more comprehensive analysis of four stylised facts on unemployment problem in Sudan, these include: distinction of several types of unemployment; interpretation of the unemployment problem from two different endogenous and exogenous perspectives due to endogenous and exogenous causes; analysis of high incidence of unemployment among youth population; and high mismatch between educational qualifications – supply – and labour market requirements – demand. In addition, our analysis is useful from a policy perspective to address the relevant mechanisms and policy issues to reduce unemployment, highlight the role of both public and private labour market institutions and educational policies and the need for incentives, agreement and collaboration between public and private institutions in upgrading skill and reducing unemployment problem in Sudan. The major policy implication from our findings indicate that since the unemployment problem is related to these endogenous and exogenous causes, therefore, policy interventions for reducing unemployment should deal with these endogenous and exogenous causes, notably, improvement of job creation and quality of educational policies and consistency between educational qualifications (output) and labour market requirements. Another major policy implication from our result on the significant positive correlation between increase in unemployment and inflation rates (2000–2008), implies that macroeconomic policies aimed at targeting reducing inflation rates would also contribute to reduce unemployment rates in Sudan.

Compared to the Sudanese literature, a new element in our analysis is that we show that the low skill levels may contribute to productivity decline across firms; we illustrate considerable variation in the value and trend of labour productivity (total output/labour ratio) in physical terms, in particular, considerable decline in labour productivity (output/labour ratio) for numerous firms over the period 2005–2008. In addition we show that the low skill levels may contribute to the trend of decline of industrial performance indicators across firms over the period 2005–2008, which we measure by three different sets of economic and productivity indicators, activity indicators and profitability indicators. Compared to the Sudanese literature, a new and novel element in our analysis is that we illustrate and assess the value and trend of industrial performance indicators across firms over the period 2005–2008 that we measure by three different sets of economic and productivity indicators, activity indicators and profitability indicators. We assess the industrial performance by the first set of economic indicators including the degree of industrialisation indicator, capital intensity level indicators and a set of productivity indicators such as labour productivity, capital productivity, fixed capital productivity, wage productivity and raw materials productivity indicators. In addition, we assess the industrial performance by the second set of activity indicators including both capital and fixed capital turnover ratios and the third set of profitability indicators including the rate of return on labour or profit/labour ratio, the rate of return on capital or profit/capital ratio and profit margin or profit/sales ratio. Our results imply that in most cases an increase in skill level (share of high skill in total employment), firm size and industry most probably leads to an improvement in most of industrial performance indicators.

Compared to the Sudanese literature, we provide a new contribution and we improve the understanding by explaining the important potential contribution of the industrial sector in enhancing economic development in Sudan from the perspective of the industrial firms based on our results from the firm survey (2010). Our results from the firm survey (2010) are consistent with the findings in the developing country and Sudanese literature that indicate several problems of industrialisation in Sudan and similar to those in the typically developing countries. Different from the studies in the Sudanese literature which provide somewhat general overview concerning the problems of industrialisation in Sudan, an interesting and novel element in our analysis is that our findings is based on recent micro primary data based on the firm survey (2010) and the follow-up interviews with firms managers and we present new and a more elaborate interpretation of the main problems of industrialisation in Sudan from the perspective of the different industrial firms considering the opinions of a more diversified sample of industrial firms, defined by industry and size. In addition we provide a new contribution since our findings from the firm survey (2010) support our argument that the low skill levels may contribute to the declining of labour productivity and other industrial performance indicators including economic, productivity, activity and profitability indicators across firms. We provide a new contribution since our results recognise that improving skill level is an important factor for facilitating improvement of labour productivity and other industrial performance indicators, so upskilling or improving skill level and adequate availability of skill and trained labour force are amongst the important factors facilitating improvement of industrial firms performance and contribution towards economic development in Sudan.

Our findings concerning the channels of technology transfer and the wide variation between the level of technology transfer in the different industrial scales and activities/sectors are not only consistent with some findings in the Arab literature (e.g. Nour 2005) but also go beyond the other findings in the Arab literature (e.g. Elsabaa 1997). This is because we identify wide variations in the preferred channels of technology transfer that include not only joint ventures and foreign industrial projects, but also strategic alliances, hiring foreign skill/ technologically advanced workers and consultants, technology licensing and FDI. We show that across firms not only the level of technology use and channels of transfer are determined by firm size and industry, but also skill and technology indicators (the use of ICT, R&D, patent, product and process innovations) and most of the industrial performance indicators are significantly defined by firm characteristics, i.e. size and industry. These results are consistent with the general literature, which illustrates that both large size firms and high intensive/active industry (e.g. chemical) are more intensive in terms of the use of technology and skills. However, one should also expect that these results might imply a possibility for reversed causality, mainly because R&D is a fixed cost that requires high financial capacity, which is most likely to be stronger amongst large size firms.

We add to the very few studies in the Arab literature concerning the positive impacts the technology transfer brings to output/production (Nour 2005; Elsabaa 1997) and the negative impact the use of technology induces in the demand for unskilled workers/labour saving effect (Nour 2005; Haan 1999). We provide a more elaborate analysis, relying not only on the qualitative effects of the use of technologies, particularly ICT diffusion on the demand for labour and the effects of increasing use of technologies on product and process innovations, but also on the quantitative effects of ICT on output and profit. These finding together with our results concerning the surprising contradicting views relating to the transfer of knowledge and impact of ICT are all seem to be consistent with the results in the Arab literature (e.g. Nour 2005).

Our new results from Chap. 9 is consistent with the results in the Arab literature since we show that the implication and interaction between educational and training policies seem to be effective but limited to only within the largest two mixed and private firms, which appear more committed to implement skill upgrading policies that are consistent with the line of public policies. These two largest mixed and private firms successfully contribute to serve public policies of training and skill upgrading via establishing active human resources development units, recruitment policies and specialized training centres to implement various regular and special internal and external training programmes, especially for national workers. In addition they encourage the use of ICT to upgrade skill levels, offer scholarships and collaborate with universities to absorb young national graduates. These results support our earlier findings in Chap. 5, which indicate a lack of effective interaction between educational and training policies and a lack of incentives for provision of training within private firms. Hence, these findings imply a further duality/discrepancy at the micro level/across small-medium and large private firms. Therefore, we observe that, in contrast to medium and small size private firms, large mixed and private firms (Kenana Sugar Company (KSC) and DAL Group) in Sudan have successfully contributed to serve the public policies for enhancing training and skill upgrading, especially amongst national workers. However, it is less clear whether these two large mixed and private firms (KSC and DAL) induce positive effects on upskilling workers in private firms. In our view, the interpretation of the serious discrepancy between these two large mixed and private firms and other firms can be attributed to presence of high resources, support and incentives within these two firms, which are probably lacking within other private firms.

Our findings are consistent with the results of the Sudanese and Arab literature concerning the poor quality of education and the Sudanese literature concerning the low commitment to the standardised international adequacy, equity and efficiency criterion related to educational policies. Different from the Sudanese and Arab literature, we provide a more elaborate analysis and new evidences concerning the serious problem of the poor quality of education, and an interesting element in our analysis is that our results are more comprehensive and show the low commitment to the standardised international adequacy and poor equity and efficiency criterion related to the supply and demand sides of educational policies in Sudan. We explain that the low commitment to the standardised international adequacy, equity and efficiency criterion is obvious not only from the supply side as measured by low rates of trained teachers and overcrowded classrooms as indicated by students enrolment per education institutions, but also holds from the demand side as measured by low rates of enrolment and attendance, high rates of dropout, high rates of repetition, weak rates of success in the final exams in the basic and secondary, mainly, technical education. We provide new a contribution to the Sudanese literature by explaining the regional inequality and disparity in the supply and demand sides of education, notably we explain that it is probably plausible to interpret the observed regional disparity in the share in demand and enrolment in education due to demographic reason (as measured by the share in total population), economic reasons (as measured by per capita income and poverty rate) and other reasons (as measured by the degree of urbanisation) across the main regions in Sudan. Notably, our results imply that the increase in the incidence of high poverty rate and low per capita income seem to be the most important factors determining, limiting or leading to low demand and enrolment, notably, in basic education, especially for females across the main regions in Sudan. Our results imply that these factors probably interpret the regional disparity in the demand for education across the main regions in Sudan. The major policy implication from our findings is that Sudan has the potential to achieve equity and fulfil the second and third MDG on universal access to primary education and gender equality respectively through reduction and elimination of poverty, notably, across the poor regions and poor population in Sudan, and this implies achievement of equity and international commitment to the fulfilment of the UN MDG in Sudan.

10.3 Policy Recommendations

The major policy recommendation from this research is that skill development policies can be enhanced by making improvements to the educational and training systems and enhancing the transfer of knowledge/external schooling effects. As for improving of education, the major policy recommendations include the improvement in the quality of teachers or mentors, infrastructure, planning for educational needs, the internal efficiency/quality of basic, secondary, technical and tertiary education, encouraging the systems of modernisation and dynamism. In addition, measures should be undertaken to enhance the linkages (networks) between universities, colleges, technical and training institutes, monitor educational needs on a regular basis, encourage the system of flexibility of educational institutions and increase the harmony/consistency between educational output and market needs. In addition, further measures include: increasing incentives for enrolment and spending on education by both public and private sectors particularly in tertiary and technical education; changing the attitudes of educated economically active population; improvement of regulations/laws to legitimise sufficient duration of compulsory education; and decentralisation of decision-making.

Concerning the improvement of training, the major policy recommendations include enhancing the educational qualifications of workers, availability of training materials and equipment, planning and regular/adequate assessment and monitoring of training programmes to fit the changing technical and skill needs. Further recommendations include increasing the appreciation of or information on the benefits of training, improving the quality and availability of trainers and mentors, enhancing availability of finance to cover training costs. In addition, there should be measures aimed at encouraging specialised training institutions, the interactions between training institutions and firms, enhancing the full appropriability of return from investment in training, enhancing the system of training certification of skill acquired, increasing the participation of private training institutions and decentralisation of training provision. With respect to improvement of the transfer of knowledge, the major policy recommendations include an improvement of the quality of education and training and the qualifications of both skilled and unskilled workers to permit the positive effects of skilled workers on unskilled workers. In addition, there are recommendations that aim at the improvement of firms conditions to encourage the external effects, sponsoring educational scholarships, increasing the interaction to market needs through increasing the information about future educational, training and skill needs, especially in the productive sectors and their demand for graduate students. The policy recommendations also aim at increasing awareness about future value of investments in education and training, enhancing a system of certification of skill acquired, providing adequate incentives for trainers and minimisation of education, learning and training costs.

Since the skills problem is partially attributed to high presence of unskilled workers, skill upgrading requires both a reduction in numbers and an upgrading of unskilled workers. There is much to be learned from the successful stories in the rapidly advanced countries, in particular, the experiences of skill upgrading in Singapore and Korea (see our results in Chap. 3).

In view of the complementary relationship between skills, skill upgrading and technological progress (see our discussion in Chaps. 3 and 7) the development of education, training, transfer of knowledge and skill levels may have further positive implications on the development of local technologies. Accordingly, the promotion of local technologies depends on skill upgrading, promotion of R&D activities and the enhancement of networks systems, collaboration between universities, firms, public and private sectors and implementation of an explicit technology policy.

Finally, our results show that the main suggestions to improve R&D includes: availability of sufficient finance from public sector; availability of sufficient finance from private sector; offering incentives and motivation and facilitate availability of sufficient human resources (researchers and qualified workers in R&D fields); improvement of management and organisational ability and coordination; improvement and strengthening the relationships, networks, consistency and cooperation between universities and higher education institutions on the one side and the productive sector (agriculture, industry, services) on the other side; and improvement of awareness and appreciation of the economic values of R&D. This is followed by creation of more favourable conditions and offering all facilities and improvement of R&D culture. Hence, our analysis indicates that in order to improve S&T performance, Sudan needs to invest heavily in both financial and human resources and to learn from the lessons of the advanced and developing S&T nations. Furthermore, investment in science and technology can be more effective if it is made according to targeted and well-defined comprehensive national plans for improvement of economic performance covering all productive sectors (agriculture, industry, and services) and adopting new policies for partnership with the private sector. Sudan needs to form a body to formulate a policy on manpower resources for S&T, suggest measures to minimise brain-drain impacts, to continue building well-developed S&T infrastructure, mainly, sufficient number of highly qualified university and R&D personnel to put Sudan in a good position in terms of globally competing in S&T. So far Sudan does not possess all the human and financial resources necessary to promote S&T; however, Sudan could have a wider range of capabilities to promote S&T if it pooled and integrated its resources, restructuring the economic system, encouraging the private sector and implementing effective S&T cooperation with regional, global and international S&T institutions, which will most likely enhance S&T development and hence long-term harmonious development in Sudan.

Our findings from the firm survey imply several recommendations for the development of the industrial sectors in Sudan. We recommend the government to seriously address the problems hindering accelerating industrial development. In addition, others recommendations include facilitating the use of new and advanced technologies, supporting stimulating environment to encourage investment in industrial sector, improving managerial and organisational capability and improving regulations and laws related to improving adequate supportive work conditions for stability of workers in employment and reducing potential movement of workers to other work after short period by increasing sufficient wages for workers and their families. In addition, we recommend the government support domestic industry by implementing policies for protecting domestic industry, preventing dumping of the local market and preventing presence of intermediary and middlemen in industry. Moreover, we recommend the government to encourage industrial sector by reducing prices of energy, electricity and water, providing raw material locally, facilitating clearance and reduced customs duties for imported raw materials and removing taxation, levies, customs and fees imposed on industrial firms. Furthermore, we recommend the government to support R&D activities and cooperation between industrial firms and universities in Sudan. In addition we recommend the government support provision of training to Sudanese trainers and trainees by reducing training costs, rehabilitation and increasing specialised training centres, focusing on practical rather than theoretical training, increasing vocational and management training, introducing modern training methods and linking human resources development strategies with economic development strategies. In addition we recommend the industrial firm owners and managers to support a major change from adopting the current short-sighted commercial perspective that aims only to make high profits to recover expenses on investment, to adopting a new alternative far-sighted entrepreneurial perspective to focus on qualitative development of the industrial sector by enhancing R&D, training and skill upgrading in industrial firms. Furthermore, we recommend the government encourage national capital to invest in human resource development, and use of modern methods of production and development of technology by creating stimulating environments that beg for industrial development in Sudan. Moreover, we recommend the government encourage industry by provision of infrastructures, modern machinery, qualified administrative staff, and financial capital by offering sufficient loans, funding through banks and encouraging foreign investment, in addition to improvement of internal and external marketing opportunities and prevention of bureaucratic and slow procedures. In addition, we recommend the government enhance relationships between industrial and agricultural sectors and to increase support for manufacturing industries such as textiles, leather and animal products because of their comparative advantages and direct effect on the economy and employment. In addition, we recommend the government improve the culture of cooperation and coordination between industrial firms, the Ministry of Industry and other ministries, to encourage implementation of economic policies to prevent duality of decisions or resolutions, prevent discrimination and confirm unification and equal treatment for workers in public and private sectors. Finally, we recommend the government adopts a far-sighted vision, implements a more consistent industrial development strategy and encourages learning from the successful stories in the East Asian industrial renaissance.

Our findings from the macro survey imply several recommendations for skill development in Sudan. We recommend the government seriously review policies and planning for development and modernisation of education and training systems at all levels, to increase spending on education and training, encourage participation of private institutions in enhancing education and training, improve appropriate curricula for each education level to fit with capacity of students and their families and encourage free education. We recommend the government to enhance the culture and awareness of the importance of skills development among employees and encouraging commitment of public and private employers to training and skill development strategies according to specialisation of workers. Furthermore, we recommend the government improve provision of training for trainers by facilitating internal and external training in internal and external training institutions and to improve coordination for training by supporting and providing adequate resources and facilities to the National Council for Training to coordinate and organise public and private training activities and institution. In addition, we recommend the government support expansion of vocational and technical training and continuing specialised training internally and externally, taking into account labour market requirements. Moreover, we recommend the government adequately incorporate skill development strategies in the current national strategic plan and increase commitment of training units of central ministries to skill upgrading strategies by supporting specialised investment in education and training in all related institutions. Furthermore, we recommend identification of training needs on three levels; first at the enterprise-level, for example, training for all employees on a new system of organisation, second at the job level, for example, specification of the requirements of the job level and necessary skills of the job and third, at the individual level, for example, determining the level of the individual needs of the training compared with the requirements of the job undertaken by the individual. Moreover, we recommend encouraging the transfer of knowledge by supporting and directing the culture of the community towards the interest in education and training and learning from the experience of other countries that promoted skill, education, training and human resources.

10.4 Direction for Future Research

We plan to utilise and extend the major findings of this study for future empirical research to improve understanding of the causes and consequences of low skill and technological level in other Arab countries with similar circumstances. It is hoped that our future research makes reference/would be relevant to more than one country and the results could be generalised and extended to be of value, use and benefit to other developing countries. It is also hoped that the results will generate some useful insights for international comparisons across developing countries and contribute to enhancing the accumulation of human capital, external schooling effects/transfer of knowledge, upskilling, technological capacity, social welfare and economic development in the developing countries.

References

Elamin, H. B. (2009). Science, technology and innovation indicators. UNESCO sub-regional training workshop. Cairo: Sudan Ministry of Science and Technology, 28–30 Sept 2009.

El Sabaa, S. (1997, November). Linking technology to industrial development in the United Arab Emirates. *Trade and Industry Magazine – Dubai*, 22(263), 16–29, 651.

Firm Survey. (2010). Technological change and skill development: A comparative study of chemical, food, metal and textile small, medium and large scale enterprises in the Sudan. Khartoum. June 2010.

Haan, H. (1999, January–March). The UAE: Assessment of technology use in some selected sectors – Towards introduction and explaining capital and skill intensive technologies (Specialist report no. 6). Dubai/Amsterdam: National Human Resources Development and Employment Strategy, Project.

Hassan, A. O. (2009). Sudan national innovation system. Paper presented at the Presidential Initiative on the Reform of Science and Technology System in the Sudan. Friendship Hall, 5–6 Jan 2009.

Hassan, M. H. A. (2009). Promoting excellence in STI for sustainable development in Africa. Paper presented at the International Workshop on Science and Technology Systems in Sudan, Khartoum, 5–6 Jan 2009.

Macro Survey. (2010). Skill creation, human resources development and policy intervention: Interviews with policy makers and experts in the Sudan. Khartoum, June 2010.

Nour, S. (2005). Technological change and skill development in the Arab Gulf countries. Doctoral dissertation, Maastricht University Press, Maastricht.

Sudan Central Bureau of Statistics Population Census Data. (2010). 5th Sudan population and housing census (2008). Khartoum: Sudan Central Bureau of Statistics.

Appendices

Appendix A: Questionnaire of the Firm Survey (2010)

Technological Change and Skill Development: A Comparative Study of Chemical, Food, Metal and Textile Small, Medium and Large Scale Enterprises in the Sudan

1. General Background Information

Code (File No.): .... (For Coding Only: Please do not write in this item)

1. Name of firm: ....
2. Years in operation (since the establishment): ....
3. If your firm has started in the Sudan but has changed names, changed management, changed ownership, merged with other firms, or experienced any other changes, or if it is started in another country and then moved to the Sudan, please specify the years in operation?.........................
4. Location: Name of Emirate where the firm is located (please tick one box)

Khartoum	Khartoum north	Omdurman	Other (please specify)
1	2	3	

5. If your firm has branches in many parts of the country, please name the location?.............................
6. Sector (please tick one box)

Public	Private	Mixed
1	2	3

DOI 10.1007/978-3-642-32811-4,

7. Main Product(s): (please tick up to four boxes)

Chemical	Petroleum	Petrochemical	Plastic	Rubber	Metal/ iron–steel	Metal/ aluminium	Others (please specify below)
1	2	3	4	5	6	7	8

8. Name up to four main products and the approximate proportion that each contributes to your total sales and total employment?

Product name
% Of total sales
% Of employment

9. Is your firm a subsidiary or an affiliate of a multinational enterprise? (Please tick one box)

Yes	No
1	2

10. If your firm is an affiliate of a multinational enterprise, please give the origin of your parent company? (Please tick one box)

Arab	Non–Arab Asian	European	North American	Australian or New Zealand	Others (please specify below)
1	2	3	4	5	6

11. Ownership of the firm. (Please give the percentage)

Local	Foreign	Total
. . . . . . %	. . . %	100 %

12. Nationality of the main owner(s). (Please tick the relevant box/boxes)

Local	Arab	Non–Arab Asian	European	North American	Australian or New Zealand	Others (please specify below)
1	2	3	4	5	6	7

13. Firm size (Employment size and capital size). (Please tick one box)

Employment–labour size			Capital size		
50–99	100–499	500 +	Less than 10 millions	10–100 million	More than 100 million +
1	2	3	1	2	3

14. Firm Production and Firm Performance: (please give an approximate estimate of the performance of your firm in the following items over the period (2005–2008)) (Please give estimation in Sudanese pounds)

	2005	2006	2007	2008
Total employment[a]				
Net Worth[b]				
Total Profit (approximately)				
Total Sale (approximately)				
Total Output (approximately)				
Total fixed capital for machinery and equipment				
Total value of exports				
Total value added				
Total spending on raw materials				
Total spending on wages and salaries				

[a]This term includes total number of workers including subcontractors, support services and other workers employed by the firm
[b]Net worth equals total assets minus total liabilities; this item refers to Capital which reflects the current market value of Fixed Assets, paid up capital in operation and investment/inflow

2. The Use of Technology and Firm Product

According to the definition that technology is more than just machinery and equipment. One can distinguish the following elements which in one form or combination constitutes a technology:

- Hard ware refers to materials, production tools and end-product.
- Soft ware concerns the documented – processes, techniques, methods, blue prints, etc.
- Human ware is the know- how [i.e. knowledge, skills and experiences] held by group of people.
- Institution ware includes institutions and wider networks/systems of management structures, business services [e.g. consulting and design firms], research and development [e.g. laboratories], policies and regulations [i.e. the political and legal framework], etc., that activate the interaction of the production system with the physical and social environment.

Based on the above definition, please answer the following questions regarding the use of technology and firm product and technology indicators.

2.1 General Information: The Use of Technology, Spending on ICT, Dependence on Foreign Technology

15. Firm's own appreciation of its level of technology: (Please tick one box)

Very advanced production technology.	1
Advanced production technology.	2
Basic/simple production technology.	3
Mixed: simple and advanced production technology.	4

16. Compared to the international standard of technology used in the activity in which you operate, how do you rate the level of technology used in firm production in the last 3 years? (Please tick one box)

Higher than the International Standard.	1
Similar to the International Standard.	2
Lower than the International Standard.	3

17. In the last 3 years, approximately how much did your firm spend on ICT: Computer and Telecommunication hardware and software technologies and other ICT related Services (Please give estimation in old Sudanese pounds)?

Expenditure on:	1999	2000	2001 (approximately)
Computer.			
Telecommunication			
Training and Software Development			
Maintenance services			
Hosting and other relevant ICT services			

18. Does your firm have an adequate capacity/ability to produce and develop local technologies[1] and to adapt with foreign imported technologies at this moment? (Please tick one box)

Develop local technologies		Adapt with foreign imported technologies	
Yes	No	Yes	No
1	2	1	2

[1] Local technology refers to ability/capability to build, develop, master, enhance and utilize the production and promotion of technology locally/for local purposes.

19. Does your firm purchase equipment, machines and techniques from abroad? (Please tick one box)

Yes	No
1	2 ⟶ skip to question (22)

20. If any, what are the main reasons (s) of purchasing from abroad? (Multiple Answers Possible [MAP])

Not available from local supplier	1
Better quality	2
Better price	3
Others.....(please specify below)	4

………………………………………………………………………………

21. What was the percentage value of imported capital equipment to total capital equipment in the year 2001?

22. What was the percentage value of capital equipment to total capital equipment that has been built by foreign companies?

23. What kind(s) of technology transfer[2] did your firm use over the last 3 years? (Multiple Answers Possible [MAP])

Transfer that based on direct investment made by foreign firm(s) to facilitate the inflow of technologies	1
Transfer that based on an earlier contractual agreement to jointly share the losses (risks/ dangers) and profits	2
Transfer that required an issuance of a formal licensing	3
Transfer that based on an allied between foreign firm and local firm to transfer strategic technologies to local firm	4

(continued)

[2] The transfer of technology from the developed countries to the developing countries takes different forms, such as:

- Foreign Direct Investment [FDI]: refers to a direct investment made by foreign firm to facilitate the inflow/transfer of technology to the hosting/local firm.
- Joint ventures is a contractual agreement to facilitate the transfer/inflow of technology to local firm based on an earlier agreement to share the [risk/ danger] losses and profits between local firm and foreign firm.
- Technology Licensing is a form of technology transfer that required a permission or consent to permit or authorize the transfer of technology by issuance of a formal license.
- Strategic Alliances is a state or condition under which the local firm and the foreign firm formed an allied to transfer strategic technologies to the hosting/local firm.
- Import and hire foreign skills and foreign technologically advanced workers/consultants to build the local technology.

Transfer that made by hiring foreign skills/technologically advanced workers/consultants to build the local technology	5
Others. . ...(please specify below)	6

. . . .

24. How important is/are the effect(s) of technology transferred in enhancing firm production and enhancing the capacity to develop the local technologies in your firm over the last 3 years? (Please tick one box for each statement)

	Very important	Important	Important only in restricted filed	Unimportant
1. Enhancing firm production	1	2	3	4
2. Enhancing the capacity to develop the local technologies	1	2	3	4

25. Did your firm hire technologically advanced workers/consultants over the last 3 years? (Please tick one box)

Yes	No
1	2
	→ skip to question (28)

26. If any, how important is/are the effect(s) of technologically advanced workers/consultants in enhancing firm production and enhancing the capacity to develop the local technologies in your firm over the last 3 years? (Please tick one box for each statement)

	Very important	Important	Important only in restricted filed	Unimportant
1. Enhancing firm production	1	2	3	4
2. Enhancing the capacity to develop the local technologies	1	2	3	4

27. Does your firm have a plan to introduce, develop or improve the local technologies benefiting from the imported technologies[3]? (There are three

[3] Short run refers to next 3 years, medium term refers to the next 3 to 5 years and long run refers to the next 10 years.

statements below, Please respond to each one by ticking yes or no box corresponding to each one).

	Yes	No
In the short run your firm needs to depend on imported technology.		
In the medium term the use of imported technologies and accumulation of knowledge and learning will allow your firm to begin produce its own technology or to use other sources of local technologies.		
In the very long run to some extent your firm will manage to partially develop its own technologies or to use other sources of local technologies.		

28. Does your firm have a plan to pursue the following strategies to accelerate the development/promotion of local technologies? (There are three statements below, Please respond to each one by ticking yes or no box corresponding to each one).

	Yes	No
Autonomous strategies (national – led strategy) to guide learning in domestic firms.		
Foreign Direct Investment- dependent strategies that rely on foreign companies to guide learning/promotion of local technologies subject to government policies/regulations.		
Foreign Direct Investment- dependent strategies that rely on foreign companies to guide learning/promotion of local technologies subject to market conditions.		

29. How important are the effects of technology upgrading in fulfilling the following long run strategies? Please, tick the relevant answer(s) in the respective columns (Multiple Answers Possible [MAP])

	Importance			Not relevant
	Extremely	Moderately	Slightly	
	3	2	1	0
Enhancing firm production				
Raising skill level				
Reinforcing firm ability to promote its own technologies				
Upskilling national workers in the firm				
Hiring more skilled national workers				
Others (please specify below)				

…………………………………………………………………………………

30. In your opinion how important are the effects of the following factor in constraining the development of local technologies and adaptation with foreign

imported technologies? Please, tick the relevant answer(s) in the respective columns (Multiple Answers Possible [MAP])

	Importance			
	Extremely	Moderately	Slightly	Not relevant
	3 ⟵	2	1 ⟶	0
a. Constraints for development of local technologies				
Inadequate availability of human resources (skill and high skill worker)				
Inadequate availability of financial resources (lack of adequate fiancé for covering high costs for building technologies)				
b. Constraints for adaptation with imported technologies				
Inadequate availability of human resources (skill and high skill worker)				
Inadequate availability of financial resources (lack of adequate fiancé for covering high costs for building technologies)				

2.2 Technology Indicators

31. How many patents[4] did your firm apply for in the last 3 years? (Please tick one box)

Zero.	1	2	3	4–10	11 and above
1	2	3	4	5	6

32. Does your firm perform research (including outsourcing research activities) for the purpose of development of production, development and adaptation of technology (R & D[5])? (Please tick one box)

Yes, Continuously	Yes, Occasionally as needed	Not at all
1	2	3
		⟶ skip to section (c), question (40)

[4] Patent is a protection of property rights especially for inventions.

[5] R & D is a combination of activities/efforts or practical researches carried out with the aim to (1) develop and enhance the productive capacity; (2) introduce new technologies, or improve the old technologies and (3) allow optimal adaptations of imported technologies to industrial/local needs. This item includes the outsourcing research activities.

If the answer is (1) or (2) for question (32) continues in answering questions (33)–(39).

33. If any, approximately how much did your firm spend on these research activities (including outsourcing research activities) during the year 2001(Please give estimation in old Sudanese pounds)?

……………………………………………………………………………

34. Approximately what was the percentage of research expenditure (including outsourcing research activities) to total output expenditure in the year 2001?

……………………………………………………………………………

35. In the year 2001 approximately what percentage of your research efforts and percentage of research expenditures (including outsourcing research activities) is devoted to the following activities? Please approximate the percentage.

	New product	Improved product	New process	Improved process	Total
1. % Of Research Efforts	……… %	……… %	……… %	……… %	100 %
2. % Of Research Expenditures	……… %	……… %	……… %	……… %	100 %

36. How many employees are engaged in research (including outsourcing research activities) in your firm? (Please tick one box)

Full – Time				Part – Time			
Zero	1–5	6–10	11+	Zero	1–5	6–10	11+
1	2	3	4	1	2	3	4

37. Does the performance of the research unit (including outsourcing research activities) in your firm contribute to adaptation of imported technologies to fit the industrial need in your firm? (Please tick one box)

Yes, to large degree	Yes, to acceptable degree	Yes, to minimum degree	Not at all
1	2	3	4

38. If the research unit in your firm did not contribute to adaptation of imported technologies, how important is the shortage of finance and shortage of skilled & qualified workers in the performance of the research activities in your firm? (Please tick one box)

(a) Shortage of finance			(b) Shortage of human resources skilled & qualified workers		
Very important	Important	Unimportant	Very important	Important	Unimportant
1	2	3	1	2	3

39. How important is the effect of research output in your firm in affecting research output in other firms working in your sector? (Please tick one box)

Very important	Important	Important only in restricted filed(s)	Unimportant.
1	2	3	4

2.3 Technology and Firm Product/Process Innovation

In the literature Innovation is defined as firm ability to involve in one or all of the following activities: introduction/production of a new product/process, a new combination of old output, a new method of production, a new organizational method, a significant improvement in the quality of old product/process, ability to produce at low costs and to open a new market. Based on the above definition please answer the following questions regarding the use of technology and firm product/process innovation.

40. In the last 3 years has the use of both technology and the ICT in firm production . . . (please tick one box)

(a) The use of technology			(b) The use of ICT technology		
Increased	Decreased	Remained stable	Increased	Decreased	Remained stable
1	2	3	1	2	3

If the answer is (1) for question (40) continue answering questions (41)–(43), if not proceed to section 3.a, question (44)

41. If the use of both technology and the ICT increased, did they require more workers with higher skills? (Please tick one box for each)

(a) The use of technology required more workers with higher skills		(b) The use of ICT required more workers with higher skills	
Yes	No	Yes	No

42. In the last 3 years did the increasing use of technology enable your firm to introduce any new product/process[6] or any product/process changes? If any, what were they? Please tick the relevant answer(s) in the respective columns· (Multiple Answers Possible [MAP])

	Yes	No
To produce a new product		
To produce a new process		
To produce a new combination of old output		
To produce a new service		
To produce a new method of production		

(continued)

[6] New product/process even just for your firm or for local market and not necessarily new for the international market.

	Yes	No
To produce a new organizational method		
To produce more output with low cost		
To produce the same output with low cost		
To open a new market		
To improve the quality of firm product		
To improve the process of personal selection		
To improve training within the firm		
To improve communication within the firm		

43. If any, approximately, how much did these products/processes decrease/increase firm's costs, sales and profits? Please tick the relevant answer in the respective columns.

	Decrease by			Increase by			No change
	1%–10%	11%–25%	Over 25%	1%–10%	11%–25%	Over 25%	No change
Per unit energy costs							
Per unit material costs							
Total costs							
Total sales							
Total profits.							

3. *Human Capital (Skill) and Firm Production*

Based on the definition in the literature that skilled workers are educated workers with college/university degree (i.e. sixteen or more years of schooling), please answer the following questions regarding human capital (skill) indicators and firm production.

3.1 Human Capital (Skill) Indicators

44. Please give the approximate percentage of workers according to educational level in your firm in the year 2001?

Level of Education	National	Foreign
With Post Secondary education		
With Secondary education		
With Less than Secondary education		
Total Percentage	100 %	100 %

45. Please give the approximate percentage of workers according to occupation classification in your firm in the year 2001?

Group of Occupation	National	Foreign
Managers		
Professional/management executive/scientific/technical and engineers		
Clerical/administrative		
Skilled craftsmen		
Plant machinery operators, assemblers and elementary occupation		
Other workers		
Total Percentage	100 %	100 %

46. Please approximate the average years of experience and average wages according to educational level in your firm in the year 2001?

Level of Education	Average years of experiences		Average wages	
	National	Foreign	National	Foreign
With post secondary education				
With secondary education				
With less than secondary education				

47. Please approximate the qualifications (level of education), average years of experience and average wages according to occupation classification in your firm in the year 2001?

Group of occupation	Qualifications (level of education)	Average years of experience	Average wages
Managers			
Professional/management executive/scientific/technical and engineers			
Clerical/administrative			
Skilled craftsmen			
Plant machinery operators, assemblers and elementary occupation			
Other workers			

3.2 Skilled Workers and Firm Product

48. Did scientists & engineers hired by your firm contribute to firm activities in any of the following ways?

	Yes	No	Partially
Shorten the development time			
Add technical, scientific or marketing knowledge to areas where your firm already had expertise			
Add new technical, scientific or marketing knowledge in area where your firm lacked expertise			
Other types of knowledge (please specify below)			

……………………………………………………………………………

49. Over the last 3 years has your firm experienced any shortage of skilled workers? (Please tick one box)

Yes	No
1	2
	→ skip to question (51)

50. If any, over the last 3 years has at least one project been seriously affected by the shortage of skilled workers, qualified personnel and expertise? (Please tick one box) (Multiple Answers Possible [MAP]).

	Yes	No
Seriously delayed		
Abolished		
Not even started		

51. In the last 3 years has the proportion of skilled workers in your firm ……… (Please tick one box)

Increased	Decreased	Remained Stable
1	2	3
	→ skip to question (55)	→ skip to question (55)

If the answer is (1) for question (51), continue answering questions (52), (53) and (54).

52. If the use of skilled workers in your firm has increased, in the last 3 years, has this increase lead to any of the following effects? Please tick the relevant answer(s) in the respective columns. (Multiple Answers Possible [MAP])

	Yes	No
Increase firm production		
Effective utilization of technologies		
Easier and faster adaptation of technologies		
Improve product quality		
Improve the level of competitiveness in the local market		
Improve the level of competitiveness in the international market		
Others (please specify below)		

………………………………………………………………………………

53. Does the increasing use of skilled workers induce any negative effect(s) in your firm? (Please tick one box)

Yes	No
1	2
	→ skip to question (55)

54. If any, please specify the effect(s)?

………………………………………………………………………………

3.3 The Incidence of External Effect of Schooling/Transfer of Knowledge the Hindering and Other Promoting Factors

55. In your opinion, does the use of skilled workers induce a significant positive effect[7] on unskilled workers in your firm? (Please tick one box).

Yes	No
1	2
→ skip to question (57)	

[7] This is a non-intentional external effect made by skilled worker and lead to an enhancement of the productivity of unskilled worker.

56. If no, in your opinion, how important are the following factors in reducing the effect in your firm? Please tick the relevant factors in the respective columns. (Multiple Answers Possible [MAP])

	Importance: Extremely 3	Moderately 2	Slightly 1	Not relevant 0
Skilled workers failed to deliver their knowledge and experiences to benefit unskilled workers				
Unskilled workers were unable to acquire the knowledge and experiences from skilled workers				
Low return from/quality of education				
Low return from/quality of training compared to international standard				
Firm conditions do not encourage the external effects				
Others (please specify below)				

……………………………………………………………………………………

57. In your opinion, how important are the following policies/solutions in inducing the effect in your firm? Please tick the relevant answer(s) in the respective columns. (Multiple Answers Possible [MAP])

	Importance: Extremely 3	Moderately 2	Slightly 1	Not relevant 0
Improves the qualifications of skilled workers to permit the positive effects on unskilled workers				
Improves the qualifications and ability of unskilled workers to learn from skilled workers				
Improves the quality of education				
Improves the quality of training to coincide with international standard				
Improves firm conditions to encourage the external effects				
Improves firm selection in both recruitment and termination				
Sponsors education scholarship				
Others (please specify below)				

……………………………………………………………………………………

3.4 Plan and Strategy for Skill Upgrading and Potential Effects

58. Does your firm have a plan for the next 3 years to raise the general level of skill? (Please tick one box).

Yes	No
1	2 → skip to chapter (4) question (61)

59. If any, how important is the effect of upskilling (raising skill level) in fulfilling the following strategies in your firm? Please tick the relevant answer in the respective columns.

	Importance			Not relevant
	Extremely	Moderately	Slightly	
	3 ←	2	→ 1	0
Enhance/Increase firm production				
Upskilling national Sudanese workers in the firm				
Hiring more skilled national workers.				
Reduce future demand for foreign skilled workers				
Others (please specify below)				

……………………………………………………………………………………

60. If the strategies (2) and (3) in question (59) are important, how important are their effects in fulfilling the following strategies in your firm? Please tick the relevant answer in the respective columns.

	Importance			Not relevant
	Extremely	Moderately	Slightly	
	3 ←	2	→ 1	0
a. Upskilling Sudanese national workers in the firm:				
Increase the employment of national workers.				
Reduce future demand for foreign skilled workers.				
b. Hiring more skilled Sudanese national workers:				
Increase the employment of national workers.				
Reduce future demand for foreign skilled workers				
Others (lease specify below)				

……………………………………………………………………………………

4. The Use of Technology and the Demand for Skilled Workers

61. Firm demand for skilled workers: (please tick one box for each)

(a) In the last 3 years has the demand for skilled workers hired by your firm			(b) In the long run do you expect the demand for skilled workers needed in your firm to		
Increased	Decreased	Remained stable	Increase	Decrease	Remain stable
1	2	3	1	2	3

62. If the long run demand for skilled workers needed in your firm is expected to increase, decrease or remain stable, please give the reason(s) for the selected choice: increase, decrease or stable?

....

63. In the last 3 years, if your firm has used several new technologies, please name them and specify their various effects on the demand for skilled workers?

....

64. In the last 3 years, has the use of new technologies induced a significant effect/change on skill level and the demand for skilled workers? (Please choose the more appropriate description of the effect on the use of new technology)

(a) Change in skill level			(b) Change in the demand for skilled workers		
Increased	Decreased	Has no effect on the skill level of workers	Increased	Decreased	Has no effect on the demand for skilled workers
1	2	3	1	2	3

65. If any, in your opinion, how important is the use of new technologies in changing both the skill level and the demand for workers hired by your firm? Please tick the relevant answer(s) in the respective columns. (Multiple Answers Possible [MAP])

	Importance			Not relevant
	Extremely	Moderately	Slightly	
	3 ←	2	1 →	0
Increasing the general skill level				
Increasing the skill level mainly for unskilled workers				
Increasing the demand for more educated, trained/experienced workers				
Increasing the demand for more professional workers				
Decreasing the demand for less educated, trained/experienced workers.				
Decreasing the demand for production workers				
Substituting the demand for unskilled (less educated and production) workers				
Others (please specify below)				

....

66. If the use of new technologies reduces the demand for unskilled workers, in your opinion, how important are the following factors? (Please tick the relevant answer(s) in the respective columns). (Multiple Answers Possible [MAP]).

	Importance			Not relevant
	Extremely	Moderately	Slightly	
	3 ←	2	1 →	0
Unskilled workers lack the skills required for filling the jobs				
Unskilled workers are less productive				
Unskilled workers failed to use technology effectively				
Increasing demand for more professional workers				
The use of technology leads to reduction in some unskilled jobs				
The use of technology leads to substitution/elimination of some unskilled jobs				
Others (please specify below)				

....

67. Does your firm have adequate availability of Sudanese national skilled workers at this moment? (Please tick one box).

Yes	Yes, in restricted field(s)	No
1	2	3
→ skip to question (70)	→ skip to question (70)	

68. If the answer is (3) for question (67), does your firm have a plan to rely on Sudanese national skilled workers over the coming period? Please tick the relevant answer in the respective columns.

Yes	Yes, in restricted field(s)	No
1	2	3
		→ skip to question (70)

69. If the answer is (1) or (2) for question (68), what would be the percentage of Sudanese national skilled workers in your firm in the short, medium and long term? Please approximate the percentage for each time period.

Short term (Up to 3 years)	Medium term (3–5 years)	Long term (5–10 years)
........%	........%	........%

70. How important are the effects of Upskilling (raising skill level) in fulfilling the following long run strategies? Please, tick the relevant answer(s) in the respective columns. (Multiple Answers Possible [MAP])

	Importance			Not relevant
	Extremely	Moderately	Slightly	
	3 ←	2	1 →	0
Facilitating an effective utilization & upgrading of technologies				
Reinforcing the employment of Sudanese national skills				
Others (please specify below)				

....

5. Training and Short and Long Run Plans for Skill Development

71. How important are the following types of skill development activities in your firm? Please tick the relevant answer(s) in the respective columns. (Multiple Answers Possible [MAP])

	Importance: Extremely 3	Importance: Moderately 2	Importance: Slightly 1	Not relevant 0
Investing in training to train existing employees.				
Sending trainers and mentors abroad to acquire skills.				
Sending workers abroad to acquire skills.				
Bringing/attracting new foreign skills, scientists and engineers.				
Using ICT to upgrade skill level.				
Encouraging learning on the job.				
Supporting long distance learning.				
Others (please specify below)				

……………………………………………………………………………

72. If any of the above types are important, please specify if these activities are to be pursued now or in the near future?

	To be pursued now	To be pursued in the near future
Investing in training to train existing employees		
Sending trainers and mentors abroad to acquire skills		
Sending workers abroad to acquire skills		
Bringing/attracting new foreign skills, scientists and engineers		
Using ICT to upgrade skill level		
Encouraging learning on the job		
Supporting long distance learning		
Others (please specify)		

73. Does your firm have an in-house training unit? (Please tick one box).

Yes	No
1	2
	→ skip to question (76)

74. If any, how much did your firm spend on it in the year 2001(Please give estimation in old Sudanese pounds)?

……………………………………………………………………

75. How many staffs are there in the unit?

………………………………………………………………………………

76. In the year 2001, what was the percentage of government subsidies offered to support training in your firm? (Please tick one box).

Zero	1%–10%	11%–20%	21%–25%.	Over 25%
1	2	3	4	5

77. How does your firm find information about training opportunities? (Multiple Answers Possible [MAP])

Government departments or semi government institutions	1
Sudanese Chamber of Commerce	2
Public education and training institutions (Universities and specialized colleges)	3
Information from other firms in your sector	4
Foreign universities	5
UNIDO	6
Private trainers (local and foreign companies)	7
Others (please specify below)	8

……………………………………………………………………………

78. How important are the following functions for the training unit in your firm? Please tick the relevant answer(s) in the respective columns. (Multiple Answers Possible [MAP])

	Importance			Not relevant
	Extremely	Moderately	Slightly	
	3	2	1	0
Encouraging technological/organizational improvement				
Upgrading skill level				
Encouraging the acquisition of new knowledge				
Monitoring skills and conducting a regular skill needs assessment				
Others (please specify below)				

……………………………………………………………………………

79. In the year 2001 did your firm provide a regular training to the following types of employees? (Please tick one box)

	Yes	No
Production workers	1	2
Service workers	1	2
Production engineering staffs	1	2
Management staffs	1	2
Other staffs not included above	1	2

80. If your firm provide a regular training, how important are the following categories of training modes in maximizing the return/efficiency of training in your firm?

	Importance			Not relevant
	Extremely	Moderately	Slightly	
	3 ←	2	1 →	0
1. On – the job				
2. On – the job and Off – the job combined				
3. Off – the job within the firm (training centre)				
4. Off – the job outside the firm (specialist training centre inside the Sudan)				
5. Off – the job outside the firm outside Sudan/abroad				
6. Others (please specify below)				

……………………………………………………………………………………

81. Please rank the most important type of training against each of the occupation category listed below?

	Managerial	Professional/scientists/technical and engineers	Clerical/administrative and skilled workers	Others
1. On – the job				
2. On – the job and Off – the job combined				
3. Off – the job within the firm (training centre)				
4. Off – the job outside the firm (specialist training centre inside the Sudan country)				
5. Off – the job outside the firm outside the Sudan/abroad				

82. If any of that training offered exclusively for the national workers, please rank the most important type of training offered exclusively for the national workers against each of the occupation category listed below?

	Managerial	Professional/ scientists/technical and engineers	Clerical/ administrative and skilled workers	Others
1. On – the job				
2. On – the job and Off – the job combined				
3. Off – the job within the firm (training centre)				
4. Off – the job outside the firm (specialist training centre inside the Sudan)				
5. Off – the job outside the firm outside the Sudan/abroad				

83. In the last 3 years, has the performance of training unit contributed significantly in raising skill level of workers hired at your firm? (Please tick one box).

Increased	Decreased	No notable change
1	2	3
→ skip to question (86)		

84. If the training provision did not contribute significantly in raising skill level of workers, how important are the following factors in making the training provision unsuccessful in your firm? Please tick the relevant answer(s) in the respective columns. (Multiple Answers Possible [MAP])

	Importance			Not relevant
	Extremely	Moderately	Slightly	
	3 ←	2	→ 1	0
Low educational qualifications of workers				
Lack of appreciation of or information on the benefits of training				
Lack of training materials and equipment				
Lack of trainers and mentors				
Low quality of trainers and mentors				
In adequate/in accurate assessment of training needs				
In adequate planning for training programmes				
Mismatch between training programmes & changing skill needs				

(continued)

	Importance			Not relevant
	Extremely	Moderately	Slightly	
	3	2	1	0
Mismatch between training programmes and changing technical needs				
Lack of specialized training institutions				
Lack of interactions between training institutions and firm				
Lack of finance to cover the costs of training				
Lack of full appropriability of the return from training investment				
High rate of mobility of trainees to leave for better – paid jobs after training				
Lack of a system of training certification of skills acquired				
Others (please specify below)				

…………………………………………………………………………………………

85. If any of the above problem(s) is/are important, how important are the following factors in making the training provision successful in your firm? Please tick the relevant answer(s) in the respective columns. (Multiple Answers Possible [MAP])

	Importance			Not relevant
	Extremely	Moderately	Slightly	
	3	2	1	0
Enhancing the educational qualifications of workers.				
Increasing the appreciation of/information on the benefits of training.				
Enhancing the availability of training materials and equipment.				
Enhancing the availability of trainers and mentors				
Enhancing the quality of trainers and mentors.				
Enhancing better/accurate assessment of training needs.				
Enhancing adequate planning for training programmes.				
Enhancing training programmes to fit the changing skill needs.				
Enhancing training programmes to fit the changing technical needs.				

(continued)

	Importance: Extremely 3	Moderately 2	Slightly 1	Not relevant 0
Enhancing availability of specialized training institutions				
Enhancing the interactions between training institutions and firm.				
Enhancing the availability of finance to cover training costs.				
Enhancing the full appropriability of the return from training investment.				
Restriction the mobility of trainees.				
Enhancing a system of training certification of skills acquired.				
Others (please specify below)				

……………………………………………………………………………………

6. The Importance of Industrial Firms, Constraints and Solutions for Problems Facing Industrial Firms in Sudan

86. In your opinion how rate important are the contribution of your firm in fulfilling the following economic development goals in Sudan? Please tick the relevant answer(s) in the respective columns. (Multiple Answers Possible [MAP])

	Importance: Extremely 3	Moderately 2	Slightly 1	Not relevant 0
Achieving profit without reducing output or raising price				
Increasing output and income				
Increasing employment opportunities for present and future labour force (population increase) as economic goal				
Utilisation of natural resources and local available raw materials that are unutilised				
Optimal and full utilisation of local raw materials by good way				
Increase exports				
Decrease imports				
Local industrialisation of local raw materials that exported in form of raw materials				

(continued)

	Importance: Extremely 3	Moderately 2	Slightly 1	Not relevant 0
Contribution to making available the basis for industrial and economic development through enhancing industrial linkages				
Making available the needs for economic development in other sectors especially agriculture				
Contribution to reform the structural imbalances in Sudanese economy				
Contribution development and improvement of other regions in Sudan and increasing the degree of urbanisation in them.				
Making available the basic and necessary goods for Sudanese				
Contribution to improve equitable distribution of wealth and income through creation of productive employment opportunities				
Creation of improved production relationships between workers based on team and common work spirits				
Creating and enhancing political independence through its contents of economic independence				
Enhancing capability and raise the political status of the country at the national and international levels				
Perceiving work from political value as well as economic and social values				
Contribution to technological and scientific development in Sudan and reducing its technological dependence on advanced countries through				
Adaptation of imported technologies to fit with local needs				
Development of local and imported technologies				
Development of local technologies that fit local development needs				
Others (please specify below)				

……………………………………………………………………………………

87. In your opinion if the performance of your firm did not contribute significantly in fulfilling the above economic development goals in Sudan, how rate important are the contribution of the following factors in constraining the performance of your firm to contribute significantly in fulfilling the above economic development goals in Sudan? Please tick the relevant answer(s) in the respective columns. (Multiple Answers Possible [MAP])

	Importance			Not relevant
	Extremely	Moderately	Slightly	
	3 ←→	2	1	0
Inadequate finance and inappropriate conditions for industrial development				
Inadequate skill and lack of trained labour force				
Lack of raw materials.				
Weak and narrow marketing opportunities				
In adequate infrastructure				
Weak and in adequate economic visibility studies				
Interruption and inadequate availability and high costs of electricity and water				
Inadequate transportation equipments				
Inadequate management and organizational facilities				
Weak maintenance capability and lack of spares parts				
Weak industrial awareness				
Spread of routine and bureaucracy and slow procedures related to industrial needs				
Others (please specify below)				

....

88. In your opinion if the performance of your firm did not contribute significantly in fulfilling the above economic development goals in Sudan and if the above factors are important in constraining the performance of your firm how important are the contribution of the following factors in overcoming the above constraints and in enhancing the performance of your firm to contribute significantly in fulfilling the above economic development goals in Sudan? Please tick the relevant answer(s) in the respective columns. (Multiple Answers Possible [MAP])

	Importance: Extremely 3	Moderately 2	Slightly 1	Not relevant 0
Improving and enhancing adequate availability of finance appropriate conditions for industrial development				
Improving and enhancing adequate availability of skill and trained labour force				
Improving and enhancing adequate availability of raw materials				
Improving and enhancing adequate availability of marketing opportunities				
Improving and enhancing adequate availability of infrastructure				
Improving and enhancing adequate availability of economic visibility studies				
Improving and enhancing adequate availability of availability with cheap and subsidised price for electricity and water				
Improving and enhancing adequate availability of transportation equipments				
Improving and enhancing adequate availability of management and organizational facilities				
Improving and enhancing adequate availability of maintenance capability and spares parts				
Improving and enhancing adequate availability of industrial awareness				
Avoiding of routine and bureaucracy and speed up the procedures related to industrial needs				
Others (please specify below)				

……………………………………………………………………………

7. Conclusions and Further Recommendations

89. Do you want to add any other general comments or suggestions for enhancing the effort for the development of local technologies, skill upgrading and human resources development in your firm and in the industrial sector in the Sudan?

……………………………………………………………………………

……………………………………………………………………………

……………………………………………………………………………

90. Would you be willing to participate in a follow – up interview as part of this research project?

Yes	No
1	2

We would like to extend to you sincere thanks for your kind cooperation and for finding the time in completing this questionnaire. Please return it by 21/February/2002.

Name of the person completing the questionnaire:

Telephone number:

Fax number:

E-mail:

Position in the company:

Date:

Appendix B: Questionnaire of the Macro Survey (2010)

Interview with Policy Makers in the Government and Experts: Skills Creation, Human Resources Development and Policy Intervention in the Sudan

Please answer the following questions:

1. General Assessment of Upskilling Efforts

1. In your opinion, which of the following statements give the more appropriate description/judgment of skills creation efforts in the Sudan? (Please choose one answer)

Skills creation efforts have been fully/absolutely successful in all fields/sectors	1
Skills creation efforts have been largely successful in all fields/sectors	2
Skills creation efforts have been moderately successful in some fields/sectors	3
Skills creation efforts have been slightly successful in some fields/sectors	4
Skills creation efforts have been unsuccessful	5

2. Please specify in what field(s)/sector(s) the Sudan has managed to create appropriate/relevant skills and in what sectors/fields it has made a relative progress to the development of local skills?

	Sudan has made a relative progress to develop the local skills
Education and Higher Education sector	
Technical and Engineering field	
Information and Communication sector	
Basic science sector	
Health and Medical Care sector	
Engineering and technological fields	
Industrial and Manufacturing sector	
Agriculture and Live Stock sector	
Services sector	
Construction sector	
Petroleum and Petrochemical field	
Others (please specify below)	

……………………………………………………………………………………

2. *Policies for Enhancing Skill Creation and Upgrading*

3. How important are the following factors/components in promoting and enhancing skills creation efforts in the Sudan? Please tick the relevant answer(s) in the respective columns. (Multiple Answers Possible [MAP])

	Importance			Not relevant
	Extremely	Moderately	Slightly	
	3	2	1	0
Factors related to the trainee/educated workers				
Education system				
Training system				
Resources allocation				
Social partnership and collaboration between educational and training institutions, employers, workers and the state to determine skill needs and the most effective ways of meeting and financing them				

(continued)

	Importance: Extremely 3	Moderately 2	Slightly 1	Not relevant 0
Planning skills needs				
Monitoring skill needs on a regular basis				
Import skill from abroad				
Others (please specify below)				

……………………………………………………………………………………

4. If skills creation is linked/related to the side of trainee/educated workers, how important are the following factors in encouraging trainee/educated workers to acquire skills/enhancing skills creation? Please tick the relevant answer(s) in the respective columns. (Multiple Answers Possible [MAP])

	Importance: Extremely 3	Moderately 2	Slightly 1	Not relevant 0
Enhancing the positive external effects of both education and training.				
Increasing the information about future skill needs/ trends Increasing the information on educational and training needs in the productive sectors and their demand for graduate students.				
Increasing the information about the future value of investments in education and training to minimize the risk aversion: preference of more certain short term returns to available jobs than long term skill investments				
Enhancing a system of certification of skills acquired.				
Minimization of education, learning and training costs.				
Providing adequate incentives for trainers.				
Increasing the interaction to market needs.				
Others (please specify below)				

……………………………………………………………………………………

5. If skills creation is linked/related to enhancement of education system, how important/effective are the following factors/intervention policies in enhancing education system and so skills creation? Please tick the relevant answer(s) in the respective columns. (Multiple Answers Possible [MAP])

Factors/possible policies intervention	Importance: Extremely 3	Moderately 2	Slightly 1	Not relevant 0
Improve the internal efficiency/quality of basic education				
Improve the internal efficiency/quality of high secondary education				
Improve the internal efficiency/quality of tertiary education				
Encourage the system of flexibility of educational institutions.				
Encourage the system of modernization and dynamism				
Improve the quality of teachers or mentors				
Improve the infrastructures				
Monitoring educational needs on a regular basis				
Better Planning for educational needs				
Increase public spending on education				
Increase private sector spending and involvement on education				
Increase spending and incentives to encourage enrolment in technical education				
Enhance the linkages (network) between universities, colleges, technical and training institutes and the productive sectors				
Increase the harmony/consistency between education output and market needs by focusing on particular future skills needs				
Increasing the motivation and incentives to attract/change the attitudes of educated economically active population				
Increase motivation for improving education and skill				
Encouragement of the use of new technologies for improving education and skill				
Others (please specify below)				

…………………………………………………………………………………………

6. If skill creation is linked/related to enhancement of training system, how important/effective are the following factors/policies intervention in making training provision more successful and so enhancing skills in the Sudan? Please tick the relevant answer(s) in the respective columns. (Multiple Answers Possible [MAP])

	Importance			Not relevant
	Extremely	Moderately	Slightly	
	3 ←	2	1 →	0
Enhancing the educational qualifications of workers				
Increasing the appreciation of or information on the benefits of training.				
Enhancing the availability of training materials and equipment				
Increasing the availability of trainers and mentors				
Improving the quality of trainers and mentors.				
Regular/adequate assessment and monitoring of training needs				
Enhancing adequate planning for training programmes/needs				
Improve quality, efficiency and comprehensiveness and modernity of training programmes				
Enhancing training programmes to fit the changing skill needs				
Enhancing training programmes to fit the changing technical needs				
Enhancing/encouraging the specialized training institutions				
Enhancing the interactions between training institutions & firm				
Enhancing the availability of finance to cover training costs				
Enhancing the full appropriability of the return from investment in training				
Decentralization of decision-making				
Increasing the participation of private training institutions				
Enhancing the system of training certification of skills acquired				
Others (please specify below)				

↙ ……………………………………………………………………………………

3. Causes of Low Skill Levels and Failure of Skill Creation/Upgrading Efforts

7. If there is some failure in skills creation, how important are the following factors/components in causing/explaining this failure? Please tick the relevant answer(s) in the respective columns. (Multiple Answers Possible [MAP])

	Importance: Extremely 3	Moderately 2	Slightly 1	Not relevant 0
Factors related to trainee/educated workers				
Deficient education system				
Deficient training system				
Factors related to resources allocation				
Others (please specify below)				

……………………………………………………………………………

8. If there is some failure in skills creation linked/related to the side of trainee/educated workers, how important are the following factors? Please tick the relevant answer(s) in the respective columns. (Multiple Answers Possible [MAP])

	Importance: Extremely 3	Moderately 2	Slightly 1	Not relevant 0
Externalities and failure to recoup all the benefits of educational investments				
Lack of information on educational and training needs in the productive sectors and their demand for graduate students				
Uncertainty about the future value of investments in education and training				
Uncertainties about future skill needs/trends				
Risk aversion: preference of more certain short – term returns to available jobs than long - term skill investments				
Lack of a system of certification of skills acquired				
High costs to finance education, learning and training				
Inadequate incentives for trainers				
Lack of interaction to market needs				
10. Others (please specify below)				

……………………………………………………………………………

9. If there is some failure in skills creation linked/related to education system, how important are the following reasons in lowering the returns from education? Please tick the relevant answer(s) in the respective columns. (Multiple Answers Possible [MAP])

	Importance			Not relevant
	Extremely	Moderately	Slightly	
	3 ←	2	1 →	0
A. Deficiency of basic education due to:				
Low quality/internal efficiency				
Lack of flexibility of educational institutions				
Lack of modernization and dynamism				
Lack of teachers or mentors				
Lack of infrastructures due to inadequate investment (public spending on education)				
Inadequate assessments and monitoring of educational needs				
Inadequate planning for educational needs				
Low involvement and spending by private sector				
Others				
B. Deficiency of technical education due to				
Low quality/internal efficiency				
Lack of flexibility of educational institutions				
Lack of modernization and dynamism				
Lack of teachers or mentors				
Lack of infrastructures due to inadequate investment (public spending on education)				
Inadequate assessments and monitoring of educational needs.				
Inadequate planning for educational needs				
Low involvement and spending by private sector.				
Weak incentives for enrolment in technical education.				
C. Deficiency of Secondary education				
Low quality/internal efficiency.				
Lack of flexibility of educational institutions.				
Lack of modernization and dynamism.				
Lack of teachers or mentors.				
Lack of infrastructures due to Inadequate investment (public spending on education)				
Inadequate assessments and monitoring of educational needs.				
Inadequate planning for educational needs				
Low involvement and spending by private sector.				
Others				
D. Deficiency of tertiary education				
Low quality/internal efficiency.				

(continued)

	Importance: Extremely 3	Moderately 2	Slightly 1	Not relevant 0
Lack of flexibility of educational institutions.				
Lack of modernization and dynamism.				
Lack of teachers or mentors.				
Lack of infrastructures due to inadequate investment (public spending on education)				
Inadequate assessments and monitoring of educational needs.				
Inadequate planning for educational needs				
Low involvement and spending by private sector.				
Weak linkages (network) between universities, colleges, technical and training institutes and the productive sectors.				
Others (please specify below)				

....

10. If there is some deficiency in education system mainly attributed to low quality (internal efficiency) of education, how important are the following factors in reducing the internal efficiency/quality of education? Please tick the relevant answer(s) in the respective columns. (Multiple Answers Possible [MAP])

	Importance: Extremely 3	Moderately 2	Slightly 1	Not relevant 0
Low quality at basic education level relative to international standard.				
Low quality at technical education level relative to international standard.				
Low quality at secondary education level relative to international standard.				
Low quality at tertiary education level relative to international standard.				
Low rates of accomplishments and motivation at basic education level relative to international standard.				
Low rates of accomplishments and motivation at technical education level relative to international standard.				
Low rates of accomplishments and motivation at higher secondary education levels relative to international standard.				

(continued)

	Importance Extremely 3	Moderately 2	Slightly 1	Not relevant 0
Low rates of accomplishments and motivation at tertiary education level relative to international standard.				
Low survival rates and high drop –out.				
High repetition rates.				
High pupil/teacher ratios.				
Low public current expenditures per pupil.				
Low quality of teachers or mentors.				
Others (please specify below)				

……………………………………………………………………………

11. If there is some deficiency in education system, did both deficiency of basic education system and deficiency of tertiary education system lead to mismatch between the output of education and market needs in the Sudan? [Please tick one box for (a) and (b)].

(a) Deficiency of basic education lead to mismatch between the output of education and market needs.		(b) Deficiency of tertiary education lead to mismatch between the output of education and market needs.	
Yes 1	No 2	Yes 1	No 2

12. If there is some failure in skills creation linked/related to training system, how important are the following factors in making the training provision unsuccessful in the Sudan? Please tick the relevant answer(s) in the respective columns. (Multiple Answers Possible [MAP])

	Importance Extremely 3	Moderately 2	Slightly 1	Not relevant 0
Low educational qualifications of workers.				
Lack of appreciation of or information on the benefits of training.				
Lack of training materials and equipment				
Lack of trainers and mentors.				
Low quality of trainers and mentors.				
In adequate assessment of training needs.				
In adequate planning for training programmes.				
Mismatch between training programmes and changing skill needs.				

(continued)

	Importance: Extremely 3	Moderately 2	Slightly 1	Not relevant 0
Mismatch between training programmes and changing technical needs.				
Lack of specialized training institutions.				
Lack of interactions between training institutions and firm.				
Lack of finance to cover the costs of training.				
Lack of full appropriability of investment in training.				
High rate of mobility of trainees to leave for better–paid jobs after training.				
Lack of a system of training certification of skills acquired.				
Others (please specify below)				

…………………………………………………………………………………

4. Short and Long Terms Mechanisms for Skill Upgrading and Effects on Technological Upgrading and Fulfilling Socio-Economic Objectives

13. How important are the following types of skill formation in fulfilling/accelerating the long run economic growth strategies in the Sudan? Please tick the relevant answer(s) in the respective columns. (Multiple Answers Possible [MAP])

	Importance: Extremely 3	Moderately 2	Slightly 1	Not relevant 0
Investing in formal education.				
Investing in vocational training.				
Investing in- firm training to train existing employees.				
Improving the quality of teachers, trainers and mentors.				
Sending teachers/instructors/trainers abroad to acquire knowledge and skills.				
Sending students abroad to acquire knowledge and skills.				

(continued)

	Importance: Extremely 3	Importance: Moderately 2	Importance: Slightly 1	Not relevant 0
Sending workers abroad to acquire knowledge and skills.				
Attracting new foreign skills, scientists & engineers.				
Using ICT to upgrade skill level.				
Encouraging learning on the job.				
Supporting long distance learning.				
Enhancing the system of accreditation and Licensing.				
Enhancing the system or programme of apprenticeship.				
Others (please specify below)				

....

14. If any of the above types are important, please specify if these activities have been already implemented, are to be pursued now or in the near future in the Sudan? (Please tick one box for each statement below)

	Has been already implemented	To be Pursued now	To be Pursued in the near future
Investing in formal education			
Investing in vocational training			
Investing in training to train existing employees			
Improving the quality of teachers, trainers and mentors			
Sending teachers/instructors/trainers abroad to acquire knowledge and skills			
Sending students abroad to acquire knowledge and skills			
Sending worker abroad to acquire skills			
Bringing/attracting new foreign skills, scientist and engineers			
Using ICT to upgrade skill level			
Encouraging Learning on the job			
Supporting long distance learning			
Enhancing the system of accreditation and Licensing			
Enhancing the system or programme of apprenticeship			
Others (please specify)			

15. How important are the effects of upskilling in fulfilling the following socio-economic goals? Please, tick the relevant answer(s) in the respective columns. (Multiple Answers Possible [MAP])

	Importance: Extremely 3	Moderately 2	Slightly 1	Not relevant 0
Increasing/reinforcing economic growth				
Enhancing/accelerating technological upgrading				
Increasing the match between educational and training output to match with the needs in labour market				
Reducing future demand for foreign labour				
Reducing unemployment rate				
Reinforcing self-reliance strategy				
Others (please specify below)				

……………………………………………………………………………

16. How important are the effects of upgrading of technologies in accelerating upskilling and fulfilling the following socio-economic goals? Please, tick the relevant answer(s) in the respective columns. (Multiple Answers Possible [MAP])

	Importance: Extremely 3	Moderately 2	Slightly 1	Not relevant 0
Increasing/reinforcing economic growth.				
Enhancing/accelerating upskilling.				
Increasing the match between educational and training output to match with the needs in labour market.				
Reducing future demand for foreign labour.				
Reducing unemployment rate.				
Reinforcing self-reliance strategy.				
Others (please specify below)				

……………………………………………………………………………

5. The Incidence of External Effect of Schooling/Transfer of Knowledge the Hindering and Other Promoting Factors

17. Does the external effect of schooling/education[8] occur and contribute to upskilling in the Sudan? (Please tick one box).

Yes	No
1	2
→ Skip to question (20)	

If the answer is (2), please continue in answering questions (18) and (19).

18. If the external effect of schooling/education does not yet occur, how important are the following factors in reducing the external effect of schooling/education and in lowering its effects on upskilling? Please tick the relevant factors in the respective columns. (Multiple Answers Possible [MAP])

	Importance			Not relevant
	Extremely	Moderately	Slightly	
	3	2	1	0
Skilled workers failed to deliver their knowledge and experiences to benefit unskilled workers.				
Unskilled workers were unable to acquire the knowledge and experiences from skilled workers.				
Low return from/quality of education.				
Low return from/quality of training compared to international standard.				
Firm conditions do not encourage the external effects.				
Lack of awareness on the importance of the external effect				
Others (please specify below)				

……………………………………………………………………………………………

[8] This is a non-intentional external effect made by skilled worker and lead to an enhancement of the productivity of unskilled worker.

19. If any of the above factor(s) is/are important, how important are the following policies/solutions in inducing the external effect of schooling and in enhancing the upskilling? Please tick the relevant answer(s) in the respective columns. (Multiple Answers Possible [MAP])

	Importance			Not relevant
	Extremely	Moderately	Slightly	
	3	2	1	0
Improves the qualifications of skilled workers to permit the positive effects on unskilled workers.				
Improves the qualifications and ability of unskilled workers to learn from skilled workers.				
Improves the quality of education.				
Improves the quality of training to coincide with international standard.				
Improves firm conditions to encourage the external effects.				
Improves firm selection in both recruitment and termination.				
Sponsors education scholarship.				
Improve awareness of the importance of external effect				
Others (please specify below)				

……………………………………………………………………………………

6. *Conclusions and Further Recommendations*

20. Do you want to add any other general comments or suggestions for skill formation and human resources development in the Sudan?

……………………………………………………………………………

……………………………………………………………………………

………………………………………………………………………………

We would like to extend to you sincere thanks for your kind cooperation and for finding the time in completing this questionnaire.

Name of the person completing the questionnaire:

Position in the institution:

Telephone number:

E-mail:

Date:

Appendix C: Questionnaire of the R&D Survey (2010)

Interview with Policy Makers in the Government and Academic Staff in Public and Private Universities: Efforts for Enhancing R&D in the Sudan

Please answer the following questions:

1. In your opinion, did the R&D activities contribute to economic development in Sudan? (Please choose one answer)

	Yes, to large degree	Yes, to acceptable degree	Yes, to minimum degree	Not at all
	1	2	3	4
Satisfying the needs for economic development				
Development of local technologies				
Adaptation to imported foreign technologies				

2. In your opinion, how important is the contribution of R&D activities to fit with economic development needs in Sudan? (Please choose one answer)

	Very important	Important	Important only in restricted filed(s)	Unimportant.
	1	2	3	4
Satisfying the needs for economic development				
Development of local technologies				
Adaptation to imported foreign technologies				

3. In your opinion how important are the effects of the following factor in constraining the contribution of R&D activities to fit with economic development needs, the development of local technologies and adaptation with foreign imported technologies in Sudan? Please, tick the relevant answer(s) in the respective columns (Multiple Answers Possible [MAP])

	Importance			Not relevant
	Extremely	Moderately	Slightly	
	3 ←	2	→ 1	0
1. Constraints for fitting with the economic development needs				
a. Inadequate availability of human resources (R&D researchers, R&D support staff and high skilled qualified workers)				

(continued)

	Importance: Extremely 3	Moderately 2	Slightly 1	Not relevant 0
b. Inadequate availability of financial resources (lack of adequate finance for covering high costs for economic development)				
2. Constraints for development of local technologies				
a. Inadequate availability of human resources (R&D researchers, R&D support staff and high skilled qualified workers)				
b. Inadequate availability of financial resources (lack of adequate finance for covering high costs for building technologies)				
3. Constraints for adaptation with imported technologies				
a. Inadequate availability of human resources (R&D researchers, R&D support staff and high skilled qualified workers)				
b. Inadequate availability of financial resources (lack of adequate finance for covering high costs for adaptation with imported technologies)				

4. In your opinion if R&D activities did not contribute to fit with economic development needs how important are the effects of the following factor and problems in constraining and hindering the contribution of R&D activities to satisfy the economic development in Sudan? Please, tick the relevant answer(s) in the respective columns (Multiple Answers Possible [MAP])

	Importance: Extremely 3	Moderately 2	Slightly 1	Not relevant 0
Lack of finance from public sector				
Lack of finance from private sector				
Lack of human resources (researchers and qualified workers in R&D fields)				
Lack of management and organization ability and lack of coordination				
Weak relationship, network and consistency and cooperation between universities and higher education institutions on the one side and the productive sector (agriculture, industry, services) on the other side				
Lack of favorable conditions and the necessary facilities				
Lack of R&D culture				

(continued)

	Importance: Extremely 3	Moderately 2	Slightly 1	Not relevant 0
Lack of awareness and appreciation of the economic values of R&D				
Others (please specify below)				

……………………………………………………………………………………

5. In your opinion how important are the effects of the following factors and suggestions for overcoming the above constraints and in enhancing the development performance of R&D activities to fit with economic development needs in Sudan? Please, tick the relevant answer(s) in the respective columns (Multiple Answers Possible [MAP])

	Importance: Extremely 3	Moderately 2	Slightly 1	Not relevant 0
Availability of sufficient finance from public sector				
Availability of sufficient of finance from private sector				
Offering incentives and motivation and making availability of sufficient human resources (researchers and qualified workers in R&D fields)				
Improvement of management and organization ability and coordination				
Improvement and strengthen the relationship, network and consistency and cooperation between universities and higher education institutions on the one side and the productive sector (agriculture, industry, services) on the other side				
Creation of more favorable conditions and offering all the necessary facilities				
Improvement of R&D culture				
Improvement of awareness and appreciation of the economic values of R&D				
Others (please specify below)				

……………………………………………………………………………………

6. Do you want to add any other general comments or suggestions for enhancing the contribution of R&D activities to fit with economic development needs in the Sudan?

………………………………………………………………

………………………………………………………………………………

………………………………………………………………………………

We would like to extend to you sincere thanks for your kind cooperation and for finding the time in completing this questionnaire.

Name of the person completing the questionnaire:

Position in the institution:

Telephone number:

E-mail:

Date:

About the Author

Dr. Samia Satti Osman Mohamed Nour obtained her first degree (BSc Hons.) and second degree (MSc) in Economics from the University of Khartoum (Sudan) in 1994 and 1999 respectively, and her doctorate (PhD) in Economics from the University of Maastricht (the Netherlands) in 2005. Currently, she is an affiliated researcher (and former Visiting Research Fellow and former Ph.D. Fellow) at UNU-MERIT, School of Business and Economics, Maastricht University, the Netherlands. She is also an affiliated research fellow at ASC, Leiden University, Leiden, the Netherlands. She is an Assistant Professor of Economics at the Department of Economics, Faculty of Economic and Social Studies, Khartoum University, Sudan (currently on sabbatical leave). At the Department of Economics, Khartoum University, she teaches Macroeconomics, Labour Economics, Development Economics and Industrial Economics for the B.SC. (undergraduate class) and Microeconomics for the M.Sc. (postgraduate class). She was employed as the coordinator of the Ph.D. programme in Economics (2006–2009) and as a teaching assistant (1995–2005) at the Department of Economics, Faculty of Economic and Social Studies, Khartoum University. She received five different best student faculty prizes during her undergraduate studies and several research awards and grants during her postgraduate studies. She completed several research projects supported by several regional and international institutions. She has worked as Economic Consultant for the United Nations Economic Commission for Africa (UNECA, Ethiopia), Research Project "Assessment of MDGs in Africa-Sudan" from May to June 2010. Dr. Samia has worked as Economic Consultant for the European Investment Bank, (EIB, Luxemburg) Research Project "Overview of Knowledge Transfer in MENA Countries-Egypt" from November 2011 to April 2012. Dr. Samia stayed as a Ph.D. Fellow at Maastricht University and UNU-INTECH, now UNU-MERIT, Maastricht, the Netherlands from September 1999 to November 2005; as a Ph.D. Intern at UNU-WIDER, Helsinki, Finland from May to July 2002; as a visiting research fellow at ASC, Leiden University, Leiden, the Netherlands from May to July 2009 and as a visiting research fellow at UNU-MERIT, Maastricht, the Netherlands from October 2010 to December 2011. The research presented in this book is conducted during her stay as a visiting research

S. Mohamed Nour, *Technological Change and Skill Development in Sudan*,
DOI 10.1007/978-3-642-32811-4, © Springer-Verlag Berlin Heidelberg 2013

fellow at UNU-MERIT Maastricht, the Netherlands from October 2010 to December 2011 and is supported and fully sponsored by a grant offered by the Arab Fund for Economic and Social Development Distinguished Scholar Awards and Post-Doctoral Fellowships Program, the Arab Fund for Economic and Social Development (AFESD), Kuwait. Dr. Samia's main research interests are in the fields of Economics, Economics of Technical Change, Macroeconomics, Microeconomics, Labour Economics, Development Economics, Human Capital Development, Endogenous Growth and Knowledge Economy. Dr. Samia has published several discussion and working papers and has contributed to six articles published in international refereed journals.

The manufacturer's authorised representative in the EU is Springer Nature Customer Service Centre GmbH, Europaplatz 3, 69115 Heidelberg, Germany. If you have any concerns regarding our products, please contact ProductSafety@springernature.com

Printed and bound by CPI Group (UK) Ltd, Croydon, CR0 4YY
01/10/2026
02225968-0017